Peptides in Neurobiology

Edited by

Harold Gainer

*National Institute of Child Health
and Human Development
Bethesda, Maryland*

PLENUM PRESS • NEW YORK AND LONDON

Library of Congress Cataloging in Publication Data

Main entry under title:

Peptides in neurobiology.

(Current topics in neurobiology)
Includes bibliographies and index.
1. Neuroendocrinology. 2. Peptides. 3. Hypothalamic hormones. 4. Central nervous
system. I. Gainer, Harold. [DNLM: 1. Peptides. 2. Neurochemistry. W1 CU82P v. 3
/WL104 P424]
QP356.4.P36   591.1'88   76-54766
ISBN 0-306-30978-5

© 1977 Plenum Press, New York
A Division of Plenum Publishing Corporation
227 West 17th Street, New York, N.Y. 10011

Printed in the United States of America

Contributors

Jeffery L. Barker, Behavioral Biology Branch, National Institute of Child Health and Human Development, National Institutes of Health, Bethesda, Maryland 20014.

Michael J. Brownstein, Section on Pharmacology, Laboratory of Clinical Science, National Institute of Mental Health, Bethesda, Maryland 20014.

Robert E. Carraway, Laboratory of Human Reproduction and Reproductive Biology and Department of Physiology, Harvard Medical School, Boston, Massachusetts.

Ian M. Cooke, Laboratory of Sensory Sciences and Department of Zoology, University of Hawaii, Honolulu, Hawaii 96822.

D. de Wied, Rudolf Magnus Institute for Pharmacology, Medical Faculty, University of Utrecht, Vondellaan 6, Utrecht, The Netherlands.

P.D. Edminson, Department of Neurochemistry, The University Psychiatric Clinic, Vinderen, Oslo 3, Norway.

Nora Frontali, Istituto Superiore di Sanitá, Viale Regina Elena, 299, Rome, Italy.

Harold Gainer, Section on Functional Neurochemistry, Behavioral Biology Branch, National Institute of Child Health and Human Development, National Institutes of Health, Bethesda, Maryland 20014.

W.H. Gispen, Rudolf Magnus Institute for Pharmacology, Medical Faculty, University of Utrecht, Vondellaan 6, Utrecht, The Netherlands.

Werner A. Klee, Laboratory of General and Comparative Biochemistry, National Institute of Mental Health, Bethesda, Maryland 20014.

Susan E. Leeman, Laboratory of Human Reproduction and Reproductive Biology and Department of Physiology, Harvard Medical School, Boston, Massachusetts.

Y. Peng Loh, Section on Functional Neurochemistry, Behavioral Biology Branch, National Institute of Child Health and Human Development, National Institutes of Health, Bethesda, Maryland 20014.

Neville Marks, New York State Research Institute for Neurochemistry and Drug Addiction, Ward's Island, New York 10035.

Edmund A. Mroz, Laboratory of Human Reproduction and Reproductive Biology and Department of Physiology, Harvard Medical School, Boston, Massachusetts.

K.L. Reichelt, Department of Neurochemistry, The University Psychiatric Clinic, Vinderen, Oslo 3, Norway.

Yosef Sarne, Department of Behavioral Biology, Technion Medical School, Haifa, Israel.

Berta Scharrer, Department of Anatomy, Albert Einstein College of Medicine, 1300 Morris Park Avenue, Bronx, New York 10461.

Stanley Stein, Roche Institute of Molecular Biology, Nutley, New Jersey 07110.

Ludwig A. Sternberger, Immunology Branch, Edgewood Arsenal, Maryland 21010.

Eugene Straus, Solomon A. Berson Research Laboratory, Veterans Administration Hospital, Bronx, New York 10468.

Rosalyn S. Yalow, Mt. Sinai School of Medicine, The City University of New York, New York.

Preface

Elucidation of the important roles played by peptides as hypothalamic–adenohypophyseal releasing factors, or regulatory hormones, has in recent years led to the recognition that peptides may also be of significance as intercellular messengers in other regions of the nervous system. In this regard, it is interesting that Substance P, which has been proposed as a putative neurotransmitter in the spinal cord, was rediscovered by Leeman and her co-workers during their search for the corticotropin-releasing factor in the hypothalamus. Indeed, with the widespread availability and use of radioimmunoassay techniques, it has become apparent that various "hypothalamic releasing factors" are localized in extrahypothalamic areas of the central nervous system as well. This book represents an expression of the belief that the impact on neurobiology of research into neuropeptides will be comparable to, if not greater than, the recent achievements obtained with the biogenic amines. As already appears to be the case, future investigations on brain peptides will undoubtedly uncover a host of new transmitter candidates, with obvious implications for neuropharmacology. Perhaps the most dramatic developments in this field have been the discoveries of the endogenous opiate peptides (enkephalin and endorphin), and the profound physiological and behavioral effects of specific peptides. The functions and mechanisms of action of these peptides are not entirely clear at present, but expectations are high that the answers to these questions will inaugurate a conceptual revolution in neurobiology. Thus, what had once been considered the exclusive province of neuroendocrinology has now emerged as a major issue in neurobiology as a whole.

The principal aim of this book is to provide the neurobiological community with a baseline of information and perspectives in the rapidly moving field of neuropeptides. Some of the central questions that have been addressed are: What evidence exists to suggest that peptides are involved in the control of the nervous system? Are peptides credible candidates as neurotransmitters and neuromodulators? If so, how do they differ from the conventional neurotransmitter agents? Do peptidergic neurons exhibit distinctive cell biological properties vis-à-vis "conventional" neurons? What are the roles of proteolytic enzymes in peptide biosynthesis and inactivation? Finally, how do peptides influence or mediate behavior? Detailed considerations of the neuroendocrine functions of the releasing factors and of peptides such as nerve growth factor have been omitted, primarily because they have already been dealt with extensively in the literature. Instead, the principal focus has been directed at the strategies and tactics that have been found useful in the characterization of neuropeptides and their functions. In addition to providing a systematic exposition of peptides in neurobiology, the authors have proposed various models and hypotheses, with the expectation that they will provoke new ideas and experimental challenges.

This book is organized into four major areas: (1) biochemical methodology, (2) the anatomical localization and biochemical analysis of various biologically active peptides in the central nervous system, (3) the metabolism of peptides, and (4) the physiological, pharmacological, and behavioral effects of peptides. Overlap of material between the chapters has been kept to a minimum by having each of the authors refer to relevant sections in other chapters. In her historical introduction, Scharrer discusses the conceptual movement towards peptides in neurobiology. Since the biochemical analysis of peptides is to some extent unique and particularly problematic, several chapters have been devoted almost exclusively to methodology. The chapter by Stein deals with the application of fluorescent techniques to the separation and detection of peptides, while Straus and Yalow discuss in detail the potential pitfalls and benefits in the use of the powerful technique of radioimmunoassay. In addition to describing immunocytochemical methodology, Sternberger also reviews the current state of knowledge with regard to the cellular localization of peptides in the central nervous system. Leeman, Mroz, and Carraway discuss their extensive work on Substance P and neurotensin in the context of a general paradigm for the analysis of peptides in nervous tissue. The chapter by Brownstein considers the regional localization of biologically active peptides in the central nervous system, and discusses the difficulties in interpretation of much of the data in this area. Three chapters are concerned with the metabolism of neuropeptides, their biosynthesis by "synthetase" (Reichelt and Edminson) and by ribosomal mechanisms (Gainer, Loh, and Sarne), and their inactivation in nervous tissue (Marks). The study of biologically active peptides in invertebrate nervous systems has been an active field for some time. Since neurobiologists are generally unfamiliar with this work, a chapter devoted entirely to invertebrate neuropeptides was written by Frontali and Gainer. In addition, three other chapters describe work in which invertebrate nervous systems have provided valuable models for the examination of the general properties of peptidergic neurons. Among these are the chapters by Barker and by Cooke, who discuss experiments in which intracellular electrical recordings were made from invertebrate peptidergic neurons and nerve terminals, respectively. The relationship of these data to vertebrate studies is also treated in these chapters. Klee's chapter on the endogenous opiate peptides attempts to place this exciting field in historical context. Finally, the effects of various peptide hormones and their analogues on behavior is presented in a chapter by de Wied and Gispen.

The study of peptides in neurobiology is still only in its infancy. However, sufficient evidence exists at present to indicate that within this class of molecules, there resides a cornucopia of surprises, joys, and clues to brain function for neurobiologists with inclinations from the molecular to the behavioral level. It is hoped that this volume will be the first of many to document the progress of peptides in neurobiology.

Harold Gainer

Bethesda, Maryland

Contents

Chapter 3

Specific Problems in the Identification and Quantitation of Neuropeptides by Radioimmunoassay

Eugene Straus and Rosalyn S. Yalow

Chapter 4

Immunocytochemistry of Neuropeptides and Their Receptors

Ludwig A. Sternberger

Chapter 5

Substance P and Neurotensin
Susan E. Leeman, Edmund A. Mroz, and Robert E. Carraway

Chapter 6

Biologically Active Peptides in the Mammalian Central Nervous System

Michael J. Brownstein

Chapter 7

Peptides Containing Probable Transmitter Candidates in the Central Nervous System

K.L. Reichelt and P.D. Edminson

Chapter 8

Biosynthesis of Neuronal Peptides

Harold Gainer, Y. Peng Loh, and Yosef Sarne

Chapter 9

Conversion and Inactivation of Neuropeptides

Neville Marks

Chapter 10

Peptides in Invertebrate Nervous Systems
Nora Frontali and Harold Gainer

Chapter 11

Physiological Roles of Peptides in the Nervous System
Jeffery L. Barker

Chapter 12

Electrical Activity of Neurosecretory Terminals and Control of Peptide Hormone Release
Ian M. Cooke

Chapter 13

Endogenous Opiate Peptides
Werner A. Klee

Chapter 14

Behavioral Effects of Peptides
D. de Wied and W.H. Gispen

Peptides in Neurobiology: Historical Introduction

Berta Scharrer

Considerable current interest in neurobiological research centers on modes of information transfer that digress from standard synaptic transmission and involve nonconventional chemical mediators, particularly peptides. This first chapter, a retrospective overview of the discovery and interpretation of "peptidergic" neurons, is primarily concerned with the phenomenon of neurosecretion, which has the distinction of constituting the most unorthodox sector in neuroscience.

1. The Neurosecretory Neuron and the Concept of Neurosecretion

Our present view of this distinctive class of neuron, and of its place in the spectrum of neurochemical mediation, has evolved by gradual modulation of a concept that originally appeared more restrictive and more revolutionary than it does today. The story of this evolvement need not be documented here in great detail, since it has been previously presented in various contexts (see, for example, B. Scharrer, 1969, 1970, 1972, 1974a,b, 1975a,b, 1976), and also since many of its facets will be treated individually in subsequent chapters.

A simple and currently operative definition is that neurosecretory neurons are nerve cells that specialize in the manufacture of chemical mediators to a degree greatly surpassing that of the more conventional neurons, and that have secretory activity comparable to that of gland cells. A further characteristic of classic neurosecretory neurons is the proteinaceous nature of their secretory products. In fact, this feature was responsible for the discovery of these cells, since the secretory

Berta Scharrer · Department of Anatomy, Albert Einstein College of Medicine, Bronx, New York.

material stains selectively with certain dye mixtures, and thus stands out in microscopic preparations.

The prototypes of peptidergic neurosecretory neurons are those of the vertebrate hypothalamus. Their discovery, in the preoptic nucleus of the diencephalon of a teleost fish, *Phoxinus laevis,* was made by Ernst Scharrer (1928), who, on the basis of cytological characteristics, proposed that they are of endocrine nature, and that their activity might be related to hypophyseal function. Subsequent studies established the presence of the same kind of "secretory" centers in homologous hypothalamic nuclei throughout the vertebrate series. Among these nuclei, the nuclei supraopticus and paraventricularis of mammals received most of the attention, which eventually established their close functional relationship with the neurohypophysis and resulted in the formulation of the important concept of neurosecretory systems.

Such systems are composed of groups of neurosecretory somata within the nervous system proper, their axons, and neurohemal organs in which the secretory products are stored for release into the circulation. For example, the hypothalamic nuclei already referred to give rise to the hypothalamo-neurohypophyseal tract, which terminates in the posterior lobe. Later, a similar but separate system (the tubero-infundibular tract) was found to connect certain hypothalamic nuclei, such as the arcuate, with the median eminence, the gateway to the adenohypophysis (see below). Furthermore, neurosecretory complexes analogous to the hypothalamic-hypophyseal systems were also demonstrated among invertebrates, especially arthropods (see Hanström, 1941; B. Scharrer and E. Scharrer, 1944).

The salient feature of these and comparable systems is that the terminals are not in synaptic contact with other neurons or nonneural effector cells, but instead end on the walls of vascular channels. Since this arrangement enables the neurochemical messengers released at such sites to reach effector cells at some distance, i.e., via the general circulation, these messengers are justifiably referred to as *neurohormones.* The functional implications of these unique spatial relationships are self-evident. It is the entire organ complex, rather than any of its parts *per se,* that is responsible for the manufacture, intracellular transport, storage, and release of the respective neurochemical messengers.

It should be pointed out here, however, that not all neurosecretory neurons are part of such an organ complex. Groups of peptidergic cells occur in areas of the nervous system that are spatially remote from, and unrelated to, the pituitary, or to its analogues among invertebrates. While the earliest such report deals with "glandlike" nerve cells in the spinal cord of skates (Speidel, 1919), the search among invertebrates has yielded a particularly rich harvest, starting with neurosecretory centers in opisthobranch snails (B. Scharrer, 1935), and encompassing representatives of all other phyla as far down as the coelenterates (Lentz, 1968).

The accumulation of this impressive comparative material has been of considerable heuristic value, in that it revealed a virtually universal phenomenon and a remarkable degree of similarity throughout the metazoan series (see, for example, Hanström, 1941; B. Scharrer and E. Scharrer, 1944; E. Scharrer, 1956). What all these striking neurons have in common is a high degree of versatility, based on the combination of neuronal and glandular properties.

This dual nature found expression in terms such as "glandlike nerve cells," "neuroglandular cells," "diencephalic gland," and "neurosecretory cells." The latter designation had come into general use before the elucidation of chemical synaptic transmission (Bacq, 1975) called attention to the fact that the capacity for "secretory activity" exists in the great majority of conventional neurons, and is not unique for peptidergic cells. Nevertheless, the term *neurosecretory*, as applied to the class of distinctive neuron defined above, remains useful, not only because its meaning is by now firmly established, but also because no satisfactory term to replace it has materialized. One suggestion, i.e., "neuroendocrine cells" (Cross *et al.*, 1972), has the disadvantage of not being inclusive enough, since not all neurosecretory products operate in the manner of (neuro)hormones.

Why was there such long-lasting and nearly unanimous rejection of the principle that neurons may also function as glands of internal secretion? One reason is that the substantiation of the initially cytological evidence by unequivocal physiological data was slow in forthcoming. To many investigators, it made little sense to search, in a class of cells as well defined as neurons, for a functional capacity that seemed so "foreign" to them. Moreover, it was reasoned logically that there would be no need for the kind of activity envisioned, since it is germane to another integrative system, the endocrine apparatus proper, the structural and functional attributes of which have long been as clearly defined as have those of neurons. One exception, the posterior pituitary "gland" with its nonglandular-appearing cells, remained an engima until the inherent capacity of neuronal elements for hormone production was recognized.

2. The Hypothalamic Origin of the Posterior Lobe Hormones

The first breakthrough in the functional interpretation of neurosecretory phenomena was the realization that the "posterior lobe hormones," vasopressin, oxytocin, and their analogues, are synthesized by hypothalamic neurons the cell bodies of which, in higher vertebrates, are located in the nuclei supraopticus and paraventricularis (see Bargmann and Scharrer, 1951). The formulation and subsequent experimental documentation of this concept became possible after the demonstration by Bargmann (1949) and his collaborators that classic neurosecretory neurons can be traced throughout their entire course by the selective staining of their products with dye mixtures originally introduced by Gomori for other purposes.

Evidence for the intraneuronal transport of the active peptides and their "carrier substances" (neurophysins) to the posterior pituitary was obtained by severance of the hypothalamo-neurohypophyseal tract, which resulted in accumulation of secretory material in the proximal and depletion in the distal part of the axon (Hild, 1951). Parallel transection experiments in insects had the same results (B. Scharrer, 1952). The degree of pharmacologically determined hormone activity in various tissue extracts was found to be in good agreement with the amount of stainable neurosecretory material present in these tissues.

In short, a satisfactory explanation had been found for a number of formerly paradoxical data, both structural and experimental. The posterior lobe makes sense when viewed as a neurohemal organ rather than as an endocrine gland in its own right. Yet, the clarification of the site of origin of posterior lobe hormones failed to provide a rationale for the operation of neuron-derived endocrine principles with direct effects on "terminal target cells" such as those of the kidney or mammary gland. As indicated earlier, one-step control operations of this kind could be carried out equally well by the endocrine apparatus proper, not only in the vertebrates, but also in higher invertebrates, such as the insects. The most logical explanation for the existence of such first-order neurosecretory activities in higher animals can be gained from their evolutionary history (see B. Scharrer, 1976).

Since cytological evidence of both neurosecretory activity and hormone-controlled mechanisms has been found in all metazoans, including those that lack "regular" glands of internal secretion, it follows that neurohormones are the phylogenetically oldest and also the only bloodborne messengers operating in these lower forms. Therefore, the conceptual model of the ancestral neuron has the attributes of a functionally versatile structure that is equally endowed for the dispatch of long-distance and strictly localized signals. Viewed in this context, the kind of control operation exemplified by vasopressin makes sense as a carryover from a time in the distant past when this mode of control was the only one available.

Neurohemal organs that are structurally and functionally analogous to the posterior pituitary occur in various invertebrates. Best known among them are the corpus cardiacum of insects and the sinus gland of crustaceans (see B. Scharrer and E. Scharrer, 1944).

3. The Hypothalamic Control of the Adenohypophysis

A new and far-reaching phase in the elucidation of neurosecretory phenomena originated with the realization that the existence of an endocrine apparatus proper in higher organisms requires mechanisms for interaction between the two systems of integration. Owing to its highly specialized dual properties, the neurosecretory neuron was recognized as being singularly endowed for bridging the gap between neural and endocrine regulatory centers, each of which operates in its own way (E. Scharrer, 1952, 1965, 1966). Closer examination of the mode of operation of the neuroendocrine axis now became a major concern of the new discipline of neuroendocrinology (see E. Scharrer and B. Scharrer, 1963; Weitzman, 1964–1977).

Once more, a combination of diverse experimental approaches was required for the clarification of mechanisms that turned out to be more complex than anticipated. A large share in the success of this effort must be accorded to the development and imaginative use of new techniques, especially in the areas of electron microscopy, biochemistry, immunoenzyme histochemistry, and electrophysiology.

An important initial step forward was the demonstration of the role of the hypophyseal portal system as a "directed" pathway for the class of neurohormones controlling adenohypophyseal function (Harris, 1955). This demonstration led to the recognition of the median eminence as another neurohemal organ. Progress in the electron-microscopic identification of various types of neurosecretory products, and in the elucidation of their biosynthesis (see Chapter 8) and "synaptoid" release, provided a powerful tool in the search for the localization of these and other neurohormonal activities.

The wealth of information obtained subsequently from immunocytochemical methods, adapted to both the light- and electron-microscopic levels, is impressive (see Chapter 4). More specifically, the analysis of the differential distribution of posterior lobe hormones as well as hypophysiotropic "regulating" or "releasing factors" became possible once their chemical identities had been established (see Chapter 3). At the same time, the ready availability of synthetically produced principles and of a variety of their analogues opened the way for differential tests relating chemical configuration to function. With this development, the exploration of potential diagnostic and therapeutic uses of such compounds in clinical neuroendocrinology was also initiated (Besser and Mortimer, 1976).

In higher invertebrates that possess nonneural sources of hormones, such as the corpus allatum of insects, a neurosecretory pathway remarkably similar to that of vertebrates links the brain with the respective endocrine gland. In fact, as early as 1917, Kopeć had shown that the brain of lepidopteran larvae contains a neurohormone that brings about pupation. But it took many years before neurosecretory cells were recognized as sources of this and other endocrine factors, some of which function as second-order systems.

4. Nonhormonal Neurosecretory Signals to Endocrine and Nonendocrine Effector Cells

The "semiprivate" portal route, highly advantageous as it might seem, is not universally used by all vertebrates for the neurochemical control of the pituitary. Detailed comparative studies of this neuroendocrine link have revealed a remarkable capacity of the nervous system to adapt itself to special requirements of certain hormone-producing effector cells, thereby apparently enhancing its efficiency.

Lower forms, especially teleosts, show several variants in which hypophysiotropic factors take alternate routes to reach their goals (Vollrath, 1967). In some of these cases, a considerably shortened extracellular pathway, consisting of noncellular stromal material, may separate neurosecretory release sites from adenohypophyseal cells. In other instances, including the mammalian pars intermedia (Bargmann et al., 1967), even this narrow zone of separation is virtually absent. Here, peptidergic "synaptoid" contacts with the effector cells afford the same degree of intimacy as do standard types of synaptic intervention. It is of interest that in insects the same range of spatial relationships exists in the neurosecretory control systems over the endocrine glands.

Furthermore, mechanisms for such "directed delivery" of peptidergic neuronal messengers are not confined to the neuroendocrine axis. They also seem to occur in nonendocrine somatic cells of vertebrates and invertebrates (see B. Scharrer and Weitzman, 1970; B. Scharrer, 1975*b*). The electron-microscopic demonstration of comparable "neurosecretomotor junctions," as well as synaptoid release sites in close vicinity to some muscle and exocrine gland cells, has opened a new area that invites exploration.

For the time being, it must remain an open question whether junctional complexes resembling regular neuroeffector sites, except for the peptidergic nature of their putative messenger, should be considered as a new class of chemical synapse. The answer will have to await the elucidation of the mode of operation of these peptides and the mechanisms available for their inactivation.

5. Peptidergic Interneuronal Communication

The most recent and perhaps also the most fascinating development in the realm of nonconventional neuronal activities is the finding that information transfer occurs between neurosecretory neurons and other neurons that may be either conventional or peptidergic. In other words, there is evidence, both structural and functional, indicating that neurons may not only dispatch but also receive nonconventional neurochemical signals. The sites of origin and modes of conveyance of the neurosecretory messengers have not been clearly ascertained in all these cases. Cytochemically identified peptides in extrahypothalamic areas of the brain (see Chapters 6 and 13) may have originated there, or they may have been transported to these sites by ascending collaterals of hypothalamic neurosecretory neurons, as indicated by the presence of synaptoid contacts (see, for example, Sterba, 1974). In addition, transport by vascular channels or the cerebrospinal fluid is a possibility.

There is a growing body of experimental information on neuronal responses to peptidergic messengers, among them changes in behavior, learning capacity, and metabolic functions (see Chapter 13). What is so intriguing about the mode of operation of such peptidergic interneuronal messengers is that they seem to serve as modulators of neural activities, rather than as neurotransmitters. For example, by applying peptides such as vasopressin to certain molluscan neurons, Barker and Gainer (1974) elicited responses that suggest a membrane regulatory role and thus are "indicative of a new form of information transfer in the nervous system."

6. Conclusion

With the current upsurge of interest in the neurobiological roles of various peptides, interest in the phenomenon of neurosecretion has intensified. Recent conceptual as well as methodological advances in this area have added new dimensions to the range of possibilities for neurochemical communication. Viewed in an evolutionary perspective, the present diversity seems to result from progressive specialization by the descendants of an originally pluripotential neuron, whereby an ancestral protein molecule may have gradually given rise to several more or less related entities that have acquired multiple and varied capacities. The developmental and functional relationships among proteinaceous neurosecretory products, Substance P, endogenous analgesics, and other neural peptides that may operate in as yet undisclosed neural activities remain to be ascertained.

ACKNOWLEDGMENTS

Part of the work referred to in this review was aided by USPHS Grants NB-00840 and 5 PO1-NS-07512, and by NSF Grant BMS 74-12456.

7. References

Aside from some of the classic contributions cited for historical reasons, this list by necessity contains only a small selection of references intended to provide the reader with sufficient guidelines for further orientation in this field.

Bacq, Z.M., 1975, *Chemical Transmission of Nerve Impulses. An Historical Sketch,* 106 pp., Pergamon Press, Oxford.

Bargmann, W., 1949, Über die neurosekretorische Verknüpfung von Hypothalamus and Neurohypophyse, *Z. Zellforsch.* **34:**610–634.

Bargmann, W., and Scharrer, E., 1951, The site of origin of the hormones of the posterior pituitary, *Amer. Sci.* **39:**255–259.

Bargmann, W., Lindner, E., and Andres, K.H., 1967, Über Synapsen an endokrinen Epithelzellen und die Definition sekretorischer Neurone. Untersuchungen am Zwischenlappen der Katzenhypophyse, *Z. Zellforsch.* **77:**282–298.

Barker, J.L., and Gainer, H., 1974, Peptide regulation of bursting pacemaker activity in a molluscan neurosecretory cell. *Science* **184:**1371–1373.

Besser, G.M., and Mortimer, C.H., 1976, Clinical neuroendocrinology, in: *Frontiers in Neuroendocrinology* (L. Martini and W.F. Ganong, eds.), Vol. 4, pp. 227–254, Raven Press, New York.

Cross, B.A., Dyball, R.E.J., and Moss, R.L., 1972, Stimulation of paraventricular neurosecretory cells by oxytocin applied iontophoretically, *J. Physiol.* **222:**22P–23P.

Hanström, B., 1941, Einige Parallelen im Bau und in der Herkunft der inkretorischen Organe der Arthropoden und der Vertebraten, *Lunds Univ. Årsskr. N.F., Avd.* 2:**37**(4):1–19.

Harris, G.W., 1955, *Neural Control of Pituitary Gland,* Arnold, London.

Hild, W., 1951, Experimentell-morphologische Untersuchungen über das Verhalten der "Neurosekretorischen Bahn" nach Hypophysenstieldurchtrennungen, Eingriffen in den Wasserhaushalt und Belastung der Osmoregulation, *Virchows Arch. Pathol. Anat. Physiol.* **319:**526–546.

Kopeć, S., 1917, Experiments on metamorphosis of insects, *Bull. Acad. Sci., Cracovie, Cl. Sci. Math. Nat. Sér.* **B:**57–60.

Lentz, T.L., 1968, *Primitive Nervous Systems,* Yale University Press, New Haven, Connecticut.

Scharrer, B., 1935, Über das Hanströmsche Organ X bei Opisthobranchiern, *Pubbl. Staz. Zool. Napoli* **15:**132–142.

Scharrer, B., 1952, Neurosecretion. XI. The effects of nerve section on the intercerebralis–cardiacum–allatum system of the insect *Leucophaea maderae, Biol. Bull.* **102:**261–272.

Scharrer, B., 1969, Neurohumors and neurohormones: Definitions and terminology, *J. Neuro-Visc. Rel., Suppl. IX,* 1–20.

Scharrer, B., 1970, General principles of neuroendocrine communication, in: *The Neurosciences: Second Study Program* (F.O. Schmitt, ed.), pp. 519–529, The Rockefeller University Press, New York.

Scharrer, B., 1972, Neuroendocrine communication (neurohormonal, neurohumoral, and intermediate), *Prog. Brain Res.* (J. Ariëns Kappers and J.P. Schadé, eds.), **38:**7–18, Elsevier Co., Amsterdam—London—New York.

Scharrer, B., 1974*a,* The concept of neurosecretion, past and present, in: *Recent Studies of Hypothalamic Function,* International Symposium, Calgary, Canada, 1973, pp. 1–7, Karger, Basel.

Scharrer, B., 1974*b,* The spectrum of neuroendocrine communication, in: *Recent Studies of Hypothalamic Function,* International Symposium, Calgary, Canada, 1973, pp. 8–16, Karger, Basel.

Scharrer, B., 1975*a,* The concept of neurosecretion and its place in neurobiology, in: *The Neurosciences: Paths of Discovery* (F.G. Worden, J.P. Swazey, and G. Adelman, eds.), pp. 231–243, The MIT Press, Cambridge, Massachusetts, and London.

Scharrer, B., 1975*b,* The role of neurons in endocrine regulation: A comparative overview, *Amer. Zool.* **15**(Suppl. 1):7–11.

Scharrer, B., 1976, Neurosecretion—comparative and evolutionary aspects, in: *Perspectives in Brain Research* (M.A. Corner and D.F. Swaab, eds.), *Prog. Brain Res.* 45:125–137, Elsevier/North Holland Biomedical Press, Amsterdam.

Scharrer, B., and Scharrer, E., 1944, Neurosecretion. VI. A comparison between the intercerebralis–cardiacum-allatum system of the insects and the hypothalamo-hypophyseal system of the vertebrates, *Biol. Bull.* **87:**242–251.

Scharrer, B., and Weitzman, M., 1970, Current problems in invertebrate neurosecretion, in: *Aspects of Neuroendocrinology* (W. Bargmann and B. Scharrer, eds.), pp. 1–23, Springer-Verlag, Berlin—Heidelberg—New York.

Scharrer, E., 1928, Die Lichtempfindlichkeit blinder Elritzen (Untersuchungen über das Zwischenhirn der Fische), *Z. Vgl. Physiol.* **7:**1–38.

Scharrer, E., 1952, The general significance of the neurosecretory cell, *Scientia* **46:**177–183.

Scharrer, E., 1956, The concept of analogy, *Pubbl. Staz. Zool. Napoli* **28:**204–213.

Scharrer, E., 1965, The final common path in neuroendocrine integration, *Arch. Anat. Microsc. Morphol. Exp.* **54:**359–370.

Scharrer, E., 1966, Principles of neuroendocrine integration, in: *Endocrines and the Central Nervous System, Res. Publ. Assoc. Nerv. Ment. Dis.* **43:**1–35.

Scharrer, E., and Scharrer, B., 1963, *Neuroendocrinology,* Columbia University Press, New York.

Speidel, C.C., 1919, Gland-cells of internal secretion in the spinal cord of the skates, *Carnegie Inst. Wash. Publ. No. 13, 1*–31.

Sterba, G., 1974, Ascending neurosecretory pathways of the peptidergic type, in: *Neurosecretion—The Final Neuro-Endrocrine Pathway* (F. Knowles and L. Vollrath, eds.), VIth International Symposium on Neurosecretion, London 1973, pp. 38–47, Springer-Verlag, Berlin—Heidelberg—New York.

Vollrath, L., 1967, Über die neurosekretorische Innervation der Adenohypophyse von Teleostiern, insbesondere von *Hippocampus cuda* und *Tinca tinca, Z. Zellforsch.* **78:**234–260.

Weitzman, M. (ed.), 1964–1977, *Bibliographia Neuroendocrinologica,* Vols. 1–13, Albert Einstein College of Medicine, New York.

2

Application of Fluorescent Techniques to the Study of Peptides

Stanley Stein

1. Introduction

The detection and measurement of naturally occurring peptides requires sensitive and specific analytical procedures such as bioassay and radioimmunoassay. This chapter discusses the use of fluorescent techniques, which also offer high sensitivity and specificity, for the analysis of peptides. The introduction of two fluorogenic reagents,* 4-phenylspiro[furan-2(3H),1'-phthalan]-3,3'dione (fluorescamine) and 2-methoxy-2,4-diphenyl-3(2H)furanone (MDPF), has been followed by the development of novel methodologies and instrumentation for the assay of amino acids, peptides, proteins, and other primary amines at the picomole level.

The development of these reagents was originally based upon a fluorometric assay for serum phenylalanine (McCaman and Robins, 1962), using ninhydrin and peptides. Udenfriend (1969) proposed that ninhydrin and phenylalanine could be employed for the measurement of peptides. During the investigation of this reaction, Samejima et al. (1971a) found that at elevated temperatures, phenylalanine was oxidized by ninhydrin to phenylacetaldehyde, which combined with additional ninhydrin and peptide or other primary amine, to yield highly fluorescent ternary products. Samejima et al. (1971b) demonstrated the applicability of ninhydrin–phenylacetaldehyde to the assay of a variety of amino compounds. Weigele et al. (1972a) elucidated the structures of the reaction products and subsequently

---

* Fluorescamine is available from Hoffmann-La Roche Diagnostics Division, Nutley, New Jersey 07110, as well as from several scientific supply houses. Samples of MDPF may be obtained from Research Technical Services, Hoffmann-La Roche, Inc., Nutley, New Jersey 07110.

Stanley Stein · Roche Institute of Molecular Biology, Nutley, New Jersey.

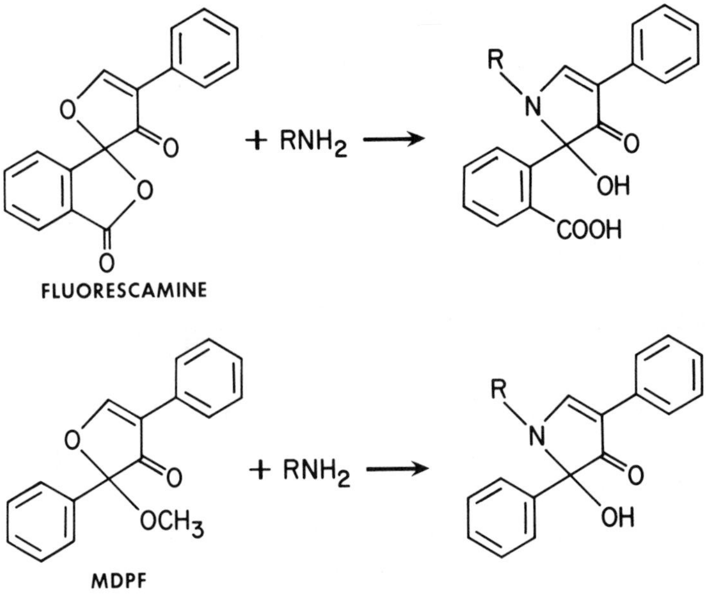

FIG. 1. Fluorogenic reaction of fluorescamine and MDPF with primary amines.

synthesized a series of possible fluorogenic reagents. Fluorescamine (Weigele *et al.*, 1972*b*) was shown to give the same major fluorophor with each amine as the ninhydrin–phenylacetaldehyde reagent, whereas MDPF (Weigele *et al.*, 1973*b*) yielded a similar fluorophor (Fig. 1). Fluorophors from both reagents have excitation maxima at 390 nm and emission maxima at about 475 nm.

Fluorescamine has several unique properties (Udenfriend *et al.*, 1972). A stable fluorophor is produced within seconds at room temperature on mixing fluorescamine, dissolved in a nonhydroxylic solvent, with an aqueous solution of a primary amine. Excess reagent is concomitantly hydrolyzed to nonreactive, water-soluble products within minutes (Stein *et al.*, 1974*a*). Both the reagent and its hydrolysis products are nonfluorescent. Studies with peptides (Weigele *et al.*, 1972*b*; Gruber *et al.*, 1976) have shown that the reaction goes to near completion (about 90% of theoretical yield), producing a monosubstituted derivative with each primary amino group. The influence of various parameters on the fluorogenic reaction, as well as the properties of the fluorophors, have been studied by DeBernardo *et al.* (1974). Below pH 4 and above pH 10, the fluorophors obtained with fluorescamine rearrange to nonfluorescent derivatives. In addition to having properties similar to those mentioned above, MDPF yields fluorophors that are stable at these extremes of pH.

Although other fluorogenic reagents are mentioned, this chapter deals primarily with the use of MDPF and fluorescamine for the detection and quantification of peptides. Procedures are presented in which the fluorogenic reaction is performed either before or after the peptides are resolved. Both column-chromatographic and noncolumn methods are included. Several applications, representative of the research being conducted in this laboratory, are described. There are sections dealing with the preparation of materials and with the extraction of peptides from tissue, since suitable precautions must be taken in order to obtain reproducible results at the subnanomole level. There is also a section on chemical characterization, which includes recent advances in performing amino acid analysis at the picomole level. The intent of the author is to present approaches for the detection, isolation, characterization, and study of peptides in biological material.

2. Preparation of Materials and Equipment

Contamination of buffers and solvents, rather than the instrumentation itself, is often the limiting factor in sensitivity. The presence of trace quantities of amines in water, ammonia, and hydrochloric acid was demonstrated by Hamilton and Myoda (1974). Additional contamination, introduced by the researcher, can arise from fingerprints (Hamilton, 1965) or from growth of microorganisms. Furthermore, detergents may contain appreciable quantities of amines and fluorescent compounds, which are difficult to rinse completely from the glassware. Fortunately, the MDPF and fluorescamine derivatives of ammonia are poorly fluorescent (about $^1/_{1000}$ that of peptide fluorophors), and micromolar levels are generally tolerable.

The degree of purity required depends on the particular application as well as on the peptide levels being assayed. To check for contaminants, amino acid analyses are periodically run on hydrolysates of the residues obtained from evaporated solvents and buffers. Several simple precautions are suitable for analysis at the picomole level. Different grades and sources of chemicals should be checked for purity. For example, in the original fluorescamine–protein assay (Böhlen, *et al.* 1973), histological-grade dioxane produced a lower blank fluorescence than reagent-grade dioxane. Constant boiling hydrochloric acid may be freed of amine contaminants by distilling over sodium dichromate (Schwabe and Catlin, 1974). Pyridine should be distilled and then redistilled in freshly cleaned apparatus over ninhydrin. Acetone may be distilled over fluorescamine. A sophisticated purification system (Hydro Supplies and Service, Durham, North Carolina) yields water with a resistance greater than $10^6\ \Omega$. In this system, house-distilled water is first passed through a 0.2-μm filter to remove microorganisms, then through one activated charcoal cartridge for organics, and then through two deionizer cartridges. Growth of microorganisms in buffers that must be kept at room temperature can be minimized by the addition of capryllic acid (0.02% vol/vol) or sodium azide (0.02% wt/vol).

Glassware is washed with dilute hydrochloric acid and rinsed thoroughly with purified water. Samples are collected, centrifuged, and stored in disposable polypropylene tubes (Brinkman, Westbury, New York). Disposable surgical gloves are worn during sample-handling in order to prevent contamination due to fingerprints. In some instances, it may be necessary to filter (on nitrocellulose or steel) or centrifuge samples that are to be injected into chromatographic columns to remove particulate matter, which can lead to pressure buildup. An in-line filter for column buffers (Altex, Palo Alto, California) should be utilized. Proper care of equipment contributes to accuracy and reproducibility at high sensitivities.

3. Isolation of Peptides from Tissues

Although a full discussion of extraction and purification procedures is beyond the scope of this chapter, general factors, as well as specific illustrations, are presented. The first consideration for an isolation procedure is recovery. In a quantitative assay, it may be necessary to compare several extraction methods, as Saifer (1971) has done for the determination of amino acids in human brain. The best extraction procedure may vary with the particular tissues, as well as with the peptide or peptides being isolated. Completeness of extraction should be checked by reextracting the tissue residue with more of the same solvent, and then with a different solvent.

A factor that may prevent good recovery is that peptides are often found to be closely associated with binding proteins. The peptide nonsuppressible insulinlike activity (NSILA), which is isolated from blood, is separated from binding proteins by passage through Sephadex G-75 in 5 N acetic acid containing 0.15 M KCl (Jakob *et al.*, 1968). In the procedure of Sachs *et al.* (1971), the neurophysin–oxytocin and neurophysin–vasopressin complexes are solubilized and dissociated by homogenizing the pituitary in a solution containing 0.2 N acetic acid and 0.02 N hydrochloric acid; protein is then precipitated with trichloroacetic acid. The quantitative assay of oxytocin and vasopressin in pituitary (Gruber *et al.*, 1976) includes a preliminary peptide purification by copper–Sephadex chromatography (see Section 4.2). In this procedure, the trichloroacetic acid and copper–Sephadex steps, as well as an intermediate ether extraction, allowed a peptide recovery of over 90%. A peptide with properties similar to those being measured, vasotocin, was added to the pituitary homogenate. This internal standard corrected for errors due to losses at each step of the procedure. An internal standard does not compensate, however, for incomplete extraction from the tissue.

Whenever possible, drying of samples should be avoided, since peptides may be lost on surfaces. Binding to glass may be minimized by treatment with Siliclad (Clay Adams, Parsippany, New Jersey). Polypropylene disposable pipette tips and tubes have been found to be suitable for sample-handling and storage.

Whereas complete recovery is essential in an analytical method, losses are acceptable for preparative work. In the isolation of somatostatin from hypothalami, a recovery of less than 1% was achieved (Vale *et al.*, 1975). The 8.5 mg of purified peptide was sufficient for elucidation of the primary structure.

A second consideration is the elimination of artifacts. An artifact is defined here as the appearance of a peptide in an extract that does not accurately reflect its presence in the tissue. An example is the suggestion that β-MSH does not exist as such in the anterior pituitary, but is produced by proteolysis of β-lipotropin during peptide isolation (Scott and Lowry, 1974). Proteases have been demonstrated to be active in extracts of various regions of the brain, including the pituitary, even at pH 3.5 (Marks *et al.*, 1974). Proteolysis can interfere in two ways: First, a substantial amount of the native peptide can be degraded. Even rapid removal of brain tissue followed by freezing in liquid nitrogen may not be fast enough to prevent a fall in the levels of biologically labile substances (Swaab, 1971). In the extraction of proteins from blood, unless the serine-protease inhibitor trasylol is added to the collection bottle, substantial amounts of immunoassayable protein are lost (Spiegelman, personal communication). Second, the products of proteolysis may interfere with the purification of the physiological peptides.

Another source of artifacts is oxidation of sensitive amino acid residues. Inclusion of an antioxidant such as thiodiglycol (0.01–0.1%) in all solvents has been found to protect cystine, methionine, histidine, and tyrosine. Peroxides may be removed from such solvents as ethyl ether by treatment with aqueous ferrous salts. Other artifacts may conceivably arise from esterification of acidic amino acid residues when extraction is performed with an alcohol–acid solvent system or from deamidation of acid–amide groups.

4. Separation and Detection of Peptides

4.1. Noncolumn Methods for Free Peptides

A sensitive and reproducible procedure for the detection of amino acids and peptides on thin-layer chromatograms has been developed by Felix and Jimenez (1974). After chromatography, the plates are sprayed with triethylamine (10% vol/vol in methylene chloride), then with fluorescamine (0.05% in acetone), and then again with triethylamine. By washing paper chromatograms with acetone, both before and after reacting with fluorescamine, Mendez and Lai (1975a) have achieved a sensitivity for peptides below 100 pmol. To obtain a uniform background, their procedure requires that the paper be dipped into a solution of fluorescamine, rather than being sprayed with the reagent. Mendez and Lai have further shown that when nanomole quantities of peptides are used, only 5–10% of the amino compound in each spot forms a fluorophor. In this way, the major portion of the free peptides may be recovered by elution of each spot. Furlan and Beck (1974) utilized fluorescamine to detect peptides on thin-layer cellulose plates. Nakamura and Pisano (1976a) have achieved a sensitivity of 10 pmol by dipping thin-layer chromatograms into a solution of fluorescamine in acetone–hexane (1:4). Somewhat poorer sensitivity for peptides was found with o-phthalaldehyde, and these peptide derivatives lost their fluorescence after several hours, in contrast to the relatively stable fluorescamine derivatives (Mendez and Gavilanes, 1976). MDPF has also been successfully employed on paper and thin-layer chromatograms (Mendez, personal communication).

Under the reaction conditions given above, only peptides having primary amino groups yield fluorophors. For detecting N-terminal prolyl-peptides, the colorimetric reagent chlorine–tolidine has been utilized by Felix and Jimenez (1974) after staining the plates with fluorescamine. Mendez and Lai (1975a) were able to detect prolyl-peptides fluorometrically by heating the fluorescamine-treated paper chromatograms at 110° for 2–3 hr.

The following two techniques are included even though fluorometric detection procedures are not yet available. Polyacrylamide gel electrophoresis of peptides as small as 1000 daltons was demonstrated by Swank and Munkres (1971). Electrophoresis was performed in 12.5% polyacrylamide gels in the presence of 8 M urea and 0.1% SDS, and the gels were fixed in a solution consisting of 10% acetic acid in 50% methanol. This procedure would not be applicable for small peptides that are soluble in this fixative. Rosenblatt et al. (1975) studied the influence of acrylamide concentration and pH on the gel-electrophoretic separation of small peptides.

Isoelectric focusing is feasible only if the peptides of interest are substantially larger than the carrier ampholytes, which may range to 1000 daltons. Denckla (1975) has successfully applied high-voltage isoelectric focusing (up to 10^5 V) in specially constructed compartmentalized units to concentrate and purify peptides. In one application (Denckla, personal communication), a sample of partially purified thymosin was distributed among 10 20-ml chambers and focused. The contents of the chamber containing the active material were placed into another unit having 10 2-ml chambers, and subjected to a second focusing step. Extreme resolving power is achieved, since the pH gradient may be a fraction of a pH unit.

4.2. Column Methods for Free Peptides

Three methods are routinely used for monitoring chromatography columns. The simplest is a manual procedure. Nakai et al. (1974) have taken aliquots from fractions collected during chromatography, hydrolyzed them in alkali, and then assayed the resulting hydrolysates with fluorescamine. Alkaline hydrolysis of peptides to their constituent amino acids improves sensitivity by generating additional fluorescamine-reactive primary amino groups. In addition, peptides that lack both a lysine residue and a primary amino N-terminal group (e.g., proline or an N-acetyl residue) are then detectable. Mendez and Gavilanes (1976) have demonstrated that assay with o-phthalaldehyde after alkaline hydrolysis has approximately the same sensitivity as fluorescamine assay. An advantage of o-phthalaldehyde is that the fluorogenic reaction may be performed in an all-aqueous medium. Unlike fluorescamine, however, o-phthalaldehyde is sensitive to ammonia in the water and the buffers, thereby inviting high blanks unless suitable precautions are taken.

The properties of fluorescamine are ideal for flow systems. Reaction is complete within seconds, while both the reagent and its hydrolysis products are nonfluorescent. Both MDPF and o-phthalaldehyde also have the properties of being nonfluorescent reagents that react rapidly under alkaline conditions. In the second procedure, the total column effluent is directed into a fluorometric detection system. In this configuration, which is similar to the fluorescamine amino acid analyzer (Stein *et al.*, 1973), borate buffer and fluorescamine are added to the column eluate continuously. As little as 5 pmol carnosine has been measured on such an instrument (Margolis and Ferriero, personal communication). Neidle and Kandera (1974) and Benson and Hare (1975) have employed *o*-phthalaldehyde for peptide detection in similar types of instrumentation. Resins that cannot withstand moderate back-pressures (30 psi), such as Sephadex and Biogel, are difficult, but not impossible, to use in such flow systems (Böhlen *et al.*, 1973).

The third procedure for monitoring columns is also based on a flow-detection system. By employing an automatic discontinuous stream-sampling valve, however, this procedure assays only a portion of the column elute, while the remainder is directed to a fraction collector (Böhlen *et al.*, 1975). This monitoring system is shown in Fig. 2. It contains two independent flow systems, one for column chromatography, the other for fluorescamine detection, with crossover at the sampling valve. Operation of this instrumentation is as follows: On its way to the fraction collector, column effluent passes through loop A of the sampling valve (Fig. 3). Simultaneously, transport solution, after passing through loop B, mixes with reaction buffer and fluorescamine, and the resulting mixture enters the flow-cell of the fluorometer. At a signal provided by a timer or by the event-marker output of a fraction collector, a slider in the sampling valve is activated. The net effect of this action is that column effluent now flows through loop B, and transport solution flows through loop A. Consequently, the sample of column eluate trapped in loop A is delivered to the detection system. At the next signal, the slider in the sampling valve returns to its original position, column effluent is directed through loop A, and the sample trapped in loop B is assayed. Automatic sampling is achieved by repeating this process at regular intervals.

Figure 4 compares the chart-recorder patterns that would be obtained for the same column when monitored by the second and third procedures. The lower panel demonstrates that (under ideal conditions) sensitivity is not sacrificed, even though only a fraction of the column effluent is actually assayed. For the sake of explanation, assume that a column is being pumped at the rate of 100 μl/min, and that the sampling valve delivers a 10-μl aliquot to the detection system at intervals of 1.0 min for a splitting ratio of 10%. In other words, for a span of 6 sec out of every 60 sec, all the column eluate is assayed, while for the remaining 54 sec, none is assayed. In order to define the chromatographic pattern with reasonable accuracy, at least 4 sampling events should occur during elution of the sharpest peak.

With continuous stream-splitters, which employ a peristaltic pump or a back-pressure valve, sensitivity is diminished. Assaying 10% of the column effluent yields peaks with 1/10 the peak height. In addition, stream-splitters are prone to drift in the ratio of sample assayed to that collected.

With the discontinuous-sampling valve, the fraction of column effluent used for fluorometric assay to that collected is adjustable from 0.1 to 100% by changing either the loop size or the interval between samples or both. Both large- and small-diameter columns can be accommodated. Under optimal conditions, less than 100

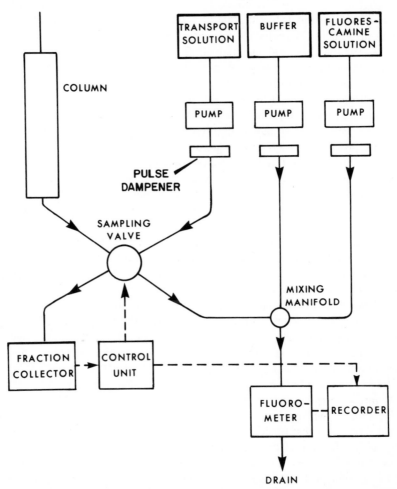

FIG. 2. Diagram of automatic column-monitoring system. (———) Paths of liquid flow (— — —) electrical connections. When the tube-changing mechanism of the fraction collector is activated, a signal (provided by the event-marker of the fraction collector) is transmitted to the control unit. This control unit delivers a 24-V pulse to one of two solenoids. Successive tube changes operate the solenoids in an alternating manner.

pmol peptide (or about 10 pmol in an individual sampling event) is readily detectable. A major drawback thus far is the absence of an in-line alkaline hydrolyzer. Several approaches, including ''stopped-flow'' valving (Stein, 1975) to minimize diffusion, are being considered.

Accompanying the development of detection methodology has been the introduction of high-efficiency resins. Particles of small and uniform size (about 10 μm) can be purchased in prepacked columns. Sulfonated polystyrene (cation exchange) resins have been used with peptides as large as 30 residues (Benson *et*

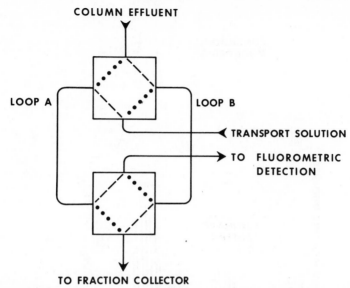

COLUMN EFFLUENT

LOOP A

LOOP B

TRANSPORT SOLUTION

TO FLUOROMETRIC
DETECTION

TO FRACTION COLLECTOR

FIG. 3. Operation of the discontinuous sampling valve. Simultaneous switching of both four-way valves from position I (· · · ·) to position II (———) results in the transfer of column eluate present in loop A into the detection stream. Transport liquid (usually distilled water) present in loop B is concomitantly transferred into the column-effluent stream. Switching from position II back to position I results in the transfer of column eluate now present in loop B into the detection stream. Thus, at each successive switching, 1 vol column eluate is exchanged for 1 vol transport liquid.

al., 1966). Both 7% cross-linked Durrum (Palo Alto, California) resin and 4% cross-linked Hamilton (Reno, Nevada) resin provided identical patterns with small peptides. Peptides larger than a few amino acid residues, especially basic ones, interact strongly with the polystyrene matrix. Such peptides, however, are readily resolved on Partisil SCX (Whatman, Clifton, New Jersey), a resin having sulfonic acid groups on a silica backbone (Radhakrishnan *et al.*, 1976). Several representative physiological peptides, including the basic tetradecapeptide Substance P, have been shown to elute as sharp peaks on Partisil SCX at room temperature with buffers of moderate ionic strength (Fig. 5). Partisil SCX may also be a suitable high-efficiency column for enzyme purification, since chromatography may be carried out at a low temperature. The Partisil SCX column is also useful for resolving polyamines.

Other high-efficiency resins exist, including both polystyrene and silica-based anion-exchangers. There are a variety of polar-bonded phase resins, which include the functional groups $-C\equiv N$, $-NH_2$, and $-N(CH_3)_3^+$. A linear hydrocarbon chain (C_{18}) is the functional group for reverse-phase (based on hydrophobic interactions) resin. The peptide antibiotic bacitracin has been resolved on reverse-phase columns by Tsuji *et al.* (1974). Pellanidon (Whatman) resin effects separations in nonaqueous media by hydrogen bonding. Of these resins, only the cation-exchangers have thus far been used for free peptides in this laboratory.

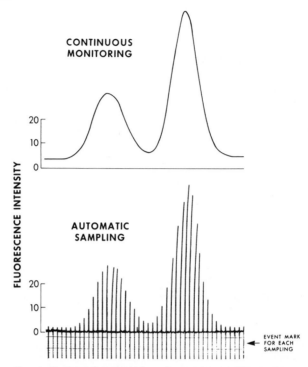

FIG. 4. Automatic monitoring of amines in column eluates. The upper panel shows the pattern obtained when the column eluate is introduced into the fluorescamine-detection system continuously. The lower panel shows the pattern obtained for the same column run under identical conditions when the discontinuous-sampling valve is utilized. Note that although only a fraction of the column eluate is being assayed, this method of sampling preserves both sensitivity and peak resolution.

Separations based on molecular weight have traditionally been conducted on Sephadex (Pharmacia, Piscataway, New Jersey) and on Biogel (Bio-Rad, Richmond, California). There is a great variety of porosites available with either resin. For example, Sephadex G-10, G-15, and G-25 exclude compounds of approximately 700, 1500, and 5000 daltons, respectively. Peptides tend to interact with these resins, and can be eluted only with buffers of high ionic strength or strong acid. Peptide chromatography on controlled-pore glass (Electro-nucleonics, Fairfield, New Jersey) in SDS–urea has been demonstrated by Frenkel and Blagrove (1975). A variety of pore sizes is also available with controlled-pore glass.

When studying small peptides, removal of amino acids and other nonpeptide amines, which are often present in great excess, cannot be accomplished by gel filtration. A procedure for eliminating α-amino acids and polyamines is by passage through a column of copper–Sephadex (Fazakerly and Best, 1965; Gallo-Torres et al., 1975). In the procedure of Gruber et al. (1976), an extract from about 1 mg posterior pituitary tissue in 300 μl, adjusted to pH 11.0, is allowed to drain into a 2.8×0.7 cm column of copper–Sephadex. A polyethylene disc is placed on top of the bed to prevent the resin from drying out (Böhlen et al.,

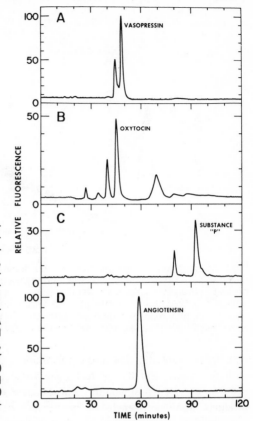

FIG. 5. Chromatography of commercial peptides on Partisil SCX. Vasopressin (Chemical Dynamics), oxytocin (Chemical Dynamics), Substance P (Beckman), and angiotensin (Beckman) were each dissolved at a concentration of 1 nmol/300 μl in 0.01 N HCl and injected onto a 25 × 0.46 cm column. A 60-min linear gradient from 0.005 M pyridine, pH 3.0, to 0.05 M pyridine, pH 4.0, followed by a 60-min linear gradient to 0.5 M pyridine, pH 5.0 (each adjusted to the appropriate pH with acetic acid), was pumped at 8.0 ml/hr at room temperature. From Radhakrishnan et al. (1976).

1973). Lithium borate buffer (0.01 M), pH 11.0, containing 0.01% thiodiglycol (to prevent oxidation) is used for preequilibration of the column, as well as elution of the peptides. Optimal resolution requires a low flow rate provided either by a peristaltic pump (5 ml/hr) or by elution with several small volumes of buffer. As Fig. 6 illustrates, there is a complete separation of the α-amino acids from the peptide; biogenic amines and ω-amino acids elute with the peptides. EDTA is then added to the eluted peptides, since trace quantities of unchelated copper ions can quench the fluorescence of fluorescamine-fluorophors. Larger quantities of tissue extract require more resin. Fresh resin is utilized for each sample.

4.3. Noncolumn Methods for Prelabeled Peptides

Dansyl-tagged peptides have been resolved on polyamide thin-layer plates (Tamura et al., 1973). Use of radiolabeled dansyl reagent accompanied by autoradiography has been reported to achieve sensitivities in the femtomole range (Burzynski, 1975). Seiler and Knödgen (1974) utilized the dibutyl form of this reagent, rather than the dimethyl form, because of the former's higher fluorescence

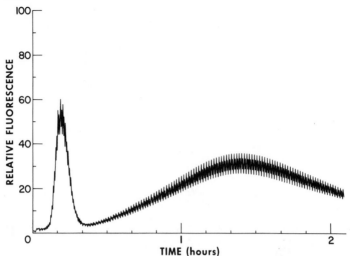

FIG. 6. Separation of peptides from amino acids on copper–Sephadex. A mixture, adjusted to pH 11.0, containing 10 nmol glutamylvaline and 5 nmol of each amino acid in a protein hydrolysate calibration standard was eluted at 5 ml/hr from a 2.8 × 0.7 cm column with 0.01 M lithium borate, pH 11.0, containing 0.01% thiodiglycol. The peptide, but none of the amino acids, eluted within the first 20 min. A 10-μl aliquot of column effluent was assayed every 30 sec with the discontinuous fluorescamine-monitoring system.

and better resolving power on polyamide plates. The disadvantages associated with the fluorescent dansyl technique for peptides include long reaction times, even at elevated temperature, fluorescence of excess reagent and reagent by-products, and heterogeneity of the derivatives. This last problem is due to lack of specificity of the dansyl reagent. Not only are the amino-terminals and the lysine ϵ-amino groups tagged, but fluorescent derivatives are also produced on the histidine imidazole ring and on the tyrosine phenol ring.

Imai *et al.* (1974) demonstrated the feasibility of resolving fluorescamine-labeled peptides by one-dimensional chromatography on silica-gel plates. Unfortunately, fluorescamine-fluorophors rearrange to nonfluorescent derivatives below pH 4 (Weigele *et al.*, 1973*a*). The use of acidic conditions, popular in thin-layer chromatography, is possible with the acid-stable MDPF-fluorophors. Preliminary results indicate that MDPF-labeled amino acids and dipeptides may be resolved on polyamide sheets with a benzene–acetic acid solvent system (Margolis, personal communication). Nakamura and Pisano (1976*b*) have carried out the fluorescamine reaction on samples after spotting on thin-layer plates, but prior to development of the plates. Derivatives obtained from the fluorogenic amine reagent 7-chloro-4-nitrobenzo-2-oxa-1,3-diazole (NBD) have also been chromatographed on thin-layer plates (Fager *et al.*, 1973; Ghosh and Whitehouse, 1968).

Polyacrylamide-gel electrophoresis of fluorescamine- and MDPF-labeled proteins has been carried out by Pace *et al.* (1974). To produce sharp bands with peptides smaller than 10,000 daltons, however, electrophoresis must be conducted in SDS in tight gels (e.g., 12% acrylamide and 6 M urea). Kato *et al.* (1975) have

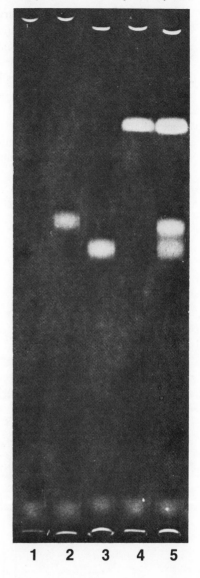

FIG. 7. Isoelectric focusing of MDPF derivatives in polyacrylamide gels. Focusing was performed in 10% gels on a pH 3–6 gradient for 4 hr at 200 V. (1) Blank; (2) MDPF–GABA (1 nmol); (3) MDPF–β-alanine (1 nmol); (4) MDPF–carnosine (1 nmol); (5) mixture of 1 nmol of each of derivatives in 2–4.

performed electrophoresis in gels containing urea and SDS with dansyl-labeled peptides. Gradient-pore gels (Isolabs, Akron, Ohio), cast with increasing acrylamide/bisacrylamide concentration along the length of the gel, seem most promising for resolving components with widely varying molecular weights. For quantitation, the gels can be scanned for fluorescence (Ragland et al., 1974). Alternatively, it may be possible to elute the fluorescent bands from gel slices and determine the concentration of eluted fluorophor by measuring fluorescence in a cuvette.

Isoelectric focusing in polyacrylamide gels has been utilized in preliminary experiments (Brink and Stein, unpublished results) for resolving MDPF-labeled amino compounds (Fig. 7). Salt must be removed in order to achieve focusing. This removal is accomplished by drying the samples after the fluorogenic reaction, and then dissolving the fluorophors in absolute methanol. The methanol extract is dried, and the fluorophors are dissolved in an aqueous solution of sucrose for layering on the gels. A variety of ampholyte solutions, giving both narrow and broad pH gradients, are available (LKB, Hicksville, New York). An advantage of gel electrofocusing over column chromatography is the larger number of samples that can be run simultaneously either on a slab gel or in a tank holding a dozen cylindrical gels.

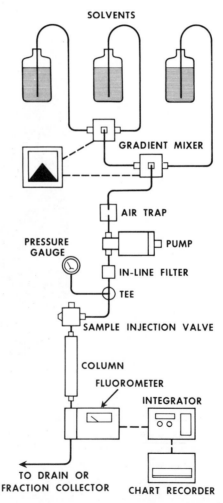

FIG. 8. Diagram of the instrument employed to resolve and detect fluorescamine- and MDPF-labeled peptides. A gradient of increasing acetone content in water is pumped through a high-efficiency reverse-phase column. Fluorescence is measured in a filter fluorometer equipped with a flow-cell, and is plotted on a strip-chart recorder.

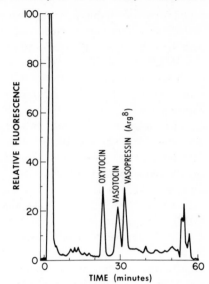

FIG. 9. Chromatography of an extract that had been reacted with fluorescamine, equivalent to 25% of one rat posterior pituitary, on the system described in Fig. 8. An internal standard, vasotocin (1 nmol), was added to the tissue homogenate.

4.4. Column Methods for Prelabeled Peptides

Monitoring of fluorescent derivatives in column eluates requires a filter flow-through fluorometer and a linear recorder, in addition to the column and the column pumping system (Fig. 8). An automated system that provides for 6 buffer changes (including column re-equilibration) and serial injection of multiple samples has been constructed (Techno-Design, Palisades Park, New Jersey). Various filter fluorometers, including ones obtained from American Instrument Company (Silver Spring, Maryland) and prototypes from Gilson Medical Electronics (Middleton, Wisconsin) and Hoffmann-La Roche, Inc. (Nutley, New Jersey), have allowed detection of peptide-fluorophors at the low picomole level.

Reverse-phase chromatography of fluorescamine derivatives was first performed by Samejima (1974) to measure polyamines. This methodology has since been applied (Gruber et al., 1976) to the determination of the nonapeptide hormones, oxytocin and vasopressin, in extracts of individual rat posterior pituitary lobes (Fig. 9). Separations are based on affinity for hydrophobic functional groups. MDPF yields fluorophors that are more hydrophobic than those formed with fluorescamine, since the latter derivatives have an additional carboxylic acid group (see Fig. 1). In addition, there is a stronger affinity for the resin with increasing peptide size. Peptides possessing more than one primary amino group produce fluorophors with a greater affinity for the resin, since each fluorescamine or MDPF moiety contributes significant hydrophobic character (e.g., lysyl–vasopressin-fluorophor requires a more nonpolar solvent than arginyl–vasopressin-fluorophor for elution). Whereas fluorescamine derivatization is useful with peptides greater than 500 daltons, smaller peptide-fluorophors are generally not resolvable from one another or from ω-amino acids, due to their weak binding. On the other hand, MDPF derivatization is suitable for resolving both ω-amino acids and peptides by reverse-phase chromatography. The intensity of fluorescence can vary considerably with either reagent—depending on the particular amine—as well as with the environment (DeBernardo *et al.*, 1974).

Based on the above, general operating conditions have been formulated. Gradients of increasing organic solvent strength, such as acetone mixed with water, have been utilized. Addition to the buffers of a salt, such as 0.03% ammonium formate, is essential. MDPF-carnosine was found not to be retained on Lichrosorb reverse-phase (Altex), even in absolute water, unless this salt was added. Thiodiglycol (0.01% vol/vol) is included in all buffers to prevent artifacts due to oxidation. Since the resin is sensitive to dilute alkali, samples must be adjusted to pH 7.0 or lower before application to the column. The ammonium formate makes the column buffers slightly acidic.

Amino acids, as well as dipeptides and tripeptides, are best resolved as their MDPF derivatives in a 5–25% acetone gradient. MDPF–decapeptides require about 50% acetone for elution. The fluorescamine derivative of a peptide requires about 10–20% less acetone for elution than does the MDPF-fluorophor. Other organic solvents may provide better resolution in certain instances. Fluorescamine and MDPF derivatization should be compared for best sensitivity and resolution. A note of caution is that the fluorogenic reaction of an optically active amino acid yields two diastereomers (Toome *et al.*, 1975). The diastereomeric pair resulting from the reaction of MDPF with L-carnosine has been observed on the reverse-phase column in a methanol/water gradient.

With a 25×0.46 cm column of Partisil ODS (Whatman), flow rates of 15–30 ml/hr and flow-cell volumes of 20–40 μl have been utilized for quantitation of peptides in the 10–500 pmol range. When the sample solution has a lower content of organic solvent than the equilibration buffer, 1.0 ml or more of sample may be applied to the column. Pressures below 500 psi have been obtained, provided the buffers, samples, and column frits are free of particulate matter.

Although resolution of fluorescamine- and MDPF-fluorophors on the other high-efficiency columns mentioned earlier has received little attention, their potential should not be dismissed. Separations based on molecular weight have been achieved for MDPF–peptides on Biogel columns (Fig. 10). To avoid the interaction of peptide-fluorophor with the Biogel (found even in 1 M acetic acid), a strong pyridine–acetate buffer is required. With the great variety of pore sizes available, not only may molecular weights be determined accurately (Fig. 11), but peptide-fluorophor purification may also be achieved.

5. Applications

5.1. Isolation of Pure Peptides

Homogeneous peptides are needed to serve as standards of bioassay, radioimmunoassay, or chemical assay. Most standard peptides are of chemically synthetic origin, but are often impure (see Fig. 5). For example, one highly purified sample of vasopressin (Chemical Dynamics, South Plainfield, New Jersey) yielded two

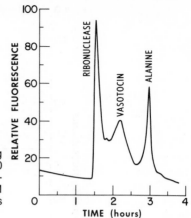

FIG. 10. Chromatography of a mixture containing MDPF-labeled ribonuclease (10 μg), vasotocin (10 nmol), and alanine (10 nmol) on a 48 × 0.9 cm column of Biogel P-2, 37–44 μm. A solution of 2 M pyridine adjusted to pH 5.0 with acetic acid was pumped at 6 ml/hr.

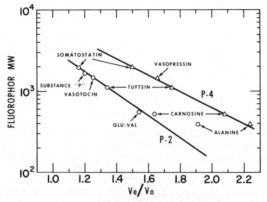

FIG. 11. Relationship of fluorophor molecular weight to mobility on Biogel. Samples containing 200 pmol of each peptide were run on a 25 × 0.46 cm column of Biogel P-4, 200–400 mesh, in the same manner as described in Fig. 10. Data for the P-2 column were obtained under the conditions given for Fig. 10. Elution time of each peptide-fluorophor divided by the elution time of MDPF–ribonuclease is plotted on the abscissa. Note that tuftsin, with a molecular weight of 500 daltons, forms a disubstituted derivative with a total weight of 1100 daltons.

major peaks by ion-exchange chromatography (Fig. 12). Utilizing the discontinuous-sampling valve for monitoring, the latter eluting peak from this column was collected, and several criteria were employed to establish its identity and homogeneity. Rechromatography on the same column showed that the other components had been removed (Fig. 12). On the reverse-phase column, the fluorescamine derivative produced a single peak, which cochromatographed with a pituitary-derived vasopressin-fluorophor. The specific activity in the rat pressor assay was in agreement with published values (Dekanski, 1952). Finally, the purified peptide had the correct amino acid composition (Table 1). The ability to achieve and check purity at the picomole level is especially important in the preparation of radiolabeled peptides, because of the high unit cost of the material.

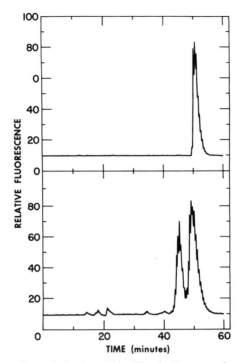

FIG. 12. Purification of a standard peptide. The bottom panel presents the chromatography of a commercial vasopressin standard on a 25 × 0.46 cm column of Partisil SCX. A linear gradient from 5 × 10⁻³ to 5 × 10⁻² M pyridine, each adjusted with acetic acid to pH 3.0 and 4.0, respectively, was pumped at 8.0 ml/hr. Material eluting between 48 and 53 min was collected, dried *in vacuo,* redissolved in 0.01 N HCl, and rechromatographed (top panel). The discontinuous fluorescamine-monitoring system was utilized for detection. From Radhakrishnan *et al.* (1976).

In the isolation of new peptides from tissue extracts, there is often a biological activity that is used for following the purification. In many instances, chemical assay may be simpler to perform or require less sample than bioassay. The bioactivity of the purified material must be checked. Furthermore, there must be many peptides having no known or readily measurable bioactivity. These fluorometric methods also aid in the isolation of other compounds with primary amino groups, which may or may not have known biological activity. Novel compounds may include biogenic amines, polyamines, and amino acids.

5.2. Quantitative Analysis

Subnanomole quantities of peptides are typically determined by bioassay or radioimmunoassay. Fluorometry now brings the requisite sensitivity to the chemical assay of peptides. The foremost advantage of the chemical assay, aside from providing specificity, is that specificity can be proved with the most rigid criteria. In the assay of oxytocin and vasopressin in extracts of individual rat pituitaries

TABLE 1. Amino Acid Analyses [a]

Amino acid	Oxytocin (pmol)		Vasopressin (pmol)	
Aspartic acid	65	54	195	228
Glutamic acid	73	73	211	219
Proline	69	71	229	239
Glycine	68	67	205	197
Isoleucine	63	69	0	0
Leucine	67	69	0	0
Tyrosine	65	64	216	215
Phenylalanine	0	0	207	209
Arginine	0	0	225	226
AVERAGE	67 (3%)	67 (6%)	213 (4%)	219 (5%)

[a] Analyses were performed on the fluorescamine-amino acid analyzer. Two 10-μl aliquots of a sample of oxytocin purified on Partisil SCX were hydrolyzed *in vacuo* in constant boiling hydrochloric acid made 0.1% with respect to thioglycollic acid. Two aliquots of a sample of purified vasopressin were hydrolyzed as above; this is the analysis of the chromatogram shown in Fig. 15 Values reported are the actual quantities present in the column effluent, corrected for the blank, and represent ¼ of each hydrolysate. Amino acids reported as zero, as well as those not listed, were less than 5 pmol for each (less than 10 pmol for serine), when corrected for blank. Relative average deviations are given in parentheses. Cysteine was not measured for the reasons given in the text (Section 5.4).

(Gruber, *et al.* 1976), specificity was ascertained by the following four criteria: (1) The oxytocin- and vasopressin-fluorophor peaks in a column eluate of a pituitary extract (see Fig. 9) were identified by their retention times. (2) Addition of synthetic oxytocin or vasopressin to a pituitary extract resulted in an increase in the peak height of the presumptive peptide hormone with no noticeable peak-broadening. (3) After purification by reverse-phase chromatography, the nonapeptide-fluorophors were collected and hydrolyzed for amino acid analysis; the compositions were in agreement with the known primary structures. (4) Aliquots taken from a pituitary extract gave equivalent vasopressin assay by both the chemical and rat pressor methods.

Proof of specificity by amino acid analysis should be performed whenever any of the experimental parameters is changed. An example of this would be the measurement of the same peptide in a different tissue, such as vasopressin or oxytocin in the hypothalamus, rather than in the pituitary. Another example is the employment of a stress condition designed to modify the level of a peptide. It is conceivable that this stress condition may stimulate the formation, by either degradative or biosynthetic mechanisms, of a previously undetectable peptide coeluting with the peptide being studied.

There are many choices in designing a peptide assay. Preparing fluorescent derivatives before the final resolving step is the preferred approach in this laboratory, as opposed to the utilization of free peptides when the aim is purification of the native material. An advantage of MDPF over fluorescamine is the greater stability of its fluorophors. Another factor is that the fluorescence intensities of the MDPF and fluorescamine derivatives of the same peptide may differ by several-fold.

When a large number of samples are to be analyzed for only one or two components, noncolumn methods may be of greatest utility. For example, thin-layer chromatography of native material, followed by spraying with fluorescamine, has been employed by Tarver *et al.* (1975) to measure GABA in extracts of brain. For determining several components in each sample, however, reverse-phase chromatography of the fluorescent derivatives may be most suitable. Since several components may be determined in an individual sample in a single analysis (on the reverse-phase column), a peptide with properties similar to those being analyzed may be added during tissue extraction as an internal standard to compensate for small losses that occur at each step of the procedure. As with any quantitative assay, optimal reaction conditions and linearity of response with concentration must be checked.

A final point is that compounds that do not produce fluorophors directly may still be measured. Blocked peptides may be hydrolyzed prior to reaction, as mentioned earlier. Proline and *N*-methyl amino acids, as well as the corresponding peptides with these amino-terminals, react with fluorescamine to yield chromophores (Felix *et al.*, 1975; Toome and Manhart, 1975) having high absorbancies in the ultraviolet region (about 320 nm). Because fluorescamine is added continuously, detection is performed at 360 nm due to the end absorption from excess reagent (Felix *et al.*, 1975). Pyroglutamyl and *N*-acetyl peptides do not react at all (Felix, personal communication).

5.3. Physiological Studies

The capability to measure a peptide in an extract obtained from the tissue of a single laboratory-size animal, or even in an aliquot of such an extract, offers many advantages. The use of a chemical assay to elucidate a physiological role is exemplified in the work of Margolis (1974). Using a conventional amino acid analyzer, levels of eluted amines were measured in extracts of olfactory bulb of control mice, as well as in mice with their nasal epithelium selectively destroyed by zinc sulfate irrigation. All amino acid levels were equivalent for the two sets of animals. The level of the dipeptide carnosine, however, was observed to fall drastically in the zinc sulfate–treated animals. Based on these and other studies, Margolis and Ferriero (1975) suggested that carnosine may be a neurotransmitter in the olfactory bulb. In the control animals (Margolis, 1974), carnosine was measured colorimetrically, because it is present at nanomole per milligram levels in olfactory bulbs. With fluorometric detection, however, minor components have been visualized. Injection onto the reverse-phase column of an MDPF-labeled peptide extract, equivalent to ⅛ of an olfactory bulb, is sufficient to drive the carnosine peak beyond the upper detection limit of the fluorometer (Fig. 13). An equivalent sample from a zinc sulfate–treated mouse, as expected, displayed a markedly reduced carnosine peak. The levels of a number of minor components, presumably peptides of higher molecular weight than carnosine, were not appreciably affected by this radical treatment (Fig. 13).

It should also be possible to determine turnover rates of physiologically active peptides such as oxytocin and vasopressin. Although it has been possible to

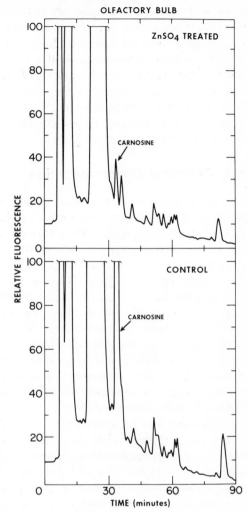

FIG. 13. Chromatography on the system described in Fig. 8 of an extract, which had been reacted with MDPF, equivalent to 12% of the olfactory bulb tissue from one mouse. The bottom panel is the chromatogram of a control mouse; the upper panel illustrates the pattern 4 weeks after the destruction of the nasal epithelium by zinc sulfate irrigation.

label such peptides *in vivo* by intracerebral injection of a radioactive precursor amino acid (Sachs and Takabatake, 1964), the isolation and bioassay procedures for vasopressin did not provide the precision needed for determining specific activity. For this type of study, the eluate from the fluorometer may be collected for measurement of radioactivity (see Fig. 8).

Chemical assay also allows elucidation of biodegradative mechanisms, without resorting to the preparation of a variety of radiolabeled analogues of a given peptide. Degradation of Substance P by extracts of rat brain was studied by resolving the reaction products on an amino acid analyzer, using ninhydrin for detection (Benuck and Marks, 1975). Difficulties associated with the analysis of the products of proteolysis included poor sensitivity, long elution time (18 hr for methioninamide), and inability to chromatograph the large peptide fragments. The importance of identifying the cleavage products is exemplified by the evidence that the degradation products of chemotactic peptides may inhibit the attraction of leukocytes (Goetzl, personal communication). An assay based on the initial cleavage step of a physiologically active peptide may be employed to follow the purification of a specific peptidase. Nardacci *et al.* (1975), assuming that the cleavage of glycinamide from vasopressin is the initial step in the inactivation of this peptide hormone, were able to partially purify a peptidase from renal plasma membrane. They employed thin-layer chromatography of the trinitrobenzene sulfonic acid derivatives for detecting glycinamide.

5.4. Chemical Characterization

The introduction of fluorescamine and its use in a semiautomated analyzer has been shown to give reproducible results with protein hydrolysates at the picomole level (Stein *et al.,* 1973). With the development of a method for proline and hydroxyproline assay (Weigele *et al.,* 1973*a*) and its incorporation into an analyzer (Felix and Terkelsen, 1973*a,b*), complete analyses of protein hydrolysates can be performed. Figure 14 shows the component structure of the amino acid analyzer currently used in this laboratory. A single-column methodology employing three buffers is used for amino acid resolution. Column effluent is adjusted to an alkaline pH with lithium borate buffer, and then fluorescamine, dissolved in acetone, is introduced. After a delay of several seconds (to provide enough time for mixing, reaction, and destruction of excess reagent), fluorescence is measured in a filter fluorometer equipped with a flow-cell. A chromatogram is recorded (Fig. 15), but data-reduction is provided by a computing integrator (Spectra Physics, Parsippany, New Jersey).

For imino acid detection, a solution of *N*-chlorosuccinimide (NCS) is added to the column effluent before the addition of borate buffer (Fig. 14). This oxidizing agent converts proline to *n*-aminobutyraldehyde, but it deaminates α-amino acids. For this reason, the NCS is programmed in only during elution of proline, while a blank solution is pumped at all other times.

The lower detection limit for amino acid analysis with fluorescamine has been about 5–10 pmol, but considerable improvement should be possible. Amino acid contaminants present at these low levels are easily discernible in Fig. 14. Picomole sensitivity has been claimed on the Durrum D-500, using the reagent *o*-phthalaldehyde (Benson and Hare, 1975). Measurement of cystine and proline, however, is problematic with this reagent. Cystine gives a poor response, and proline does not give a fluorescent product. The oxidative conversion of proline to a primary amino compound has not yet been incorporated into the D-500. Colorimetric analysis on the 121 M (Beckman, Palo Alto, California) and on the D-500 is claimed to be applicable at the 100 pmol level. This is undoubtedly at the limits of sensitivity. Maeda *et al.* (1973) employed the fluorogenic reagent pyridoxal-zinc (II) in an automated amino acid analyzer at the nanomole level.

Amino acid analysis is used for determining the absolute quantity of a peptide. From the analyses of an oxytocin and a vasopressin standard (see Table 1), it

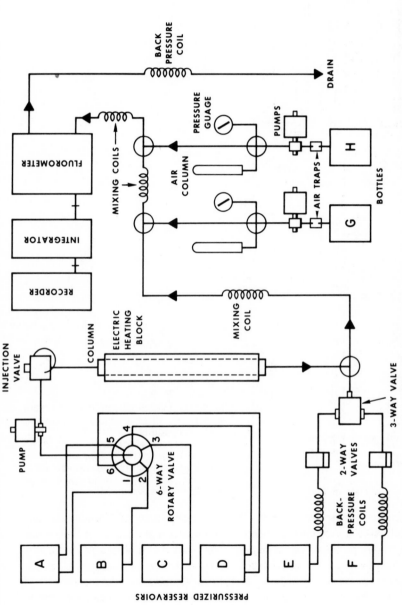

FIG. 14. Diagram of the fluorescamine-amino acid analyzer. Metering pumps are used for all solutions, except for the proline detection system in which gas pressure pumping is utilized. (A) 0.2 N citrate, pH 3.20; (B) 0.2 N citrate, pH 4.20; (C) 0.35 N citrate, pH 8.0, containing 0.65 N NaCl; (D) 0.2 N NaOH; (E) 0.05 N HCl; (F) 0.05 N HCl containing 5×10^{-4} M NCS; (G) 0.18 N lithium borate, pH 9.5; (H) fluorescamine in acetone (20 mg/100 ml).

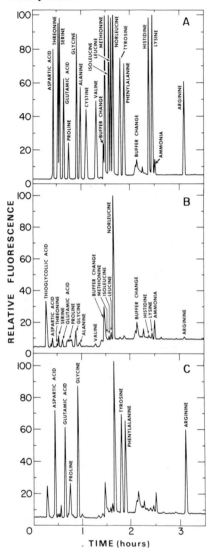

FIG. 15. Chromatography on the instrument shown in Fig. 14 of a calibration standard (A), a hydrolysate of a buffer blank (B), and a hydrolysate of vasopressin purified on Partisil SCX (C). For the calibration mixture, 250 pmol of each amino acid was applied to the column; for the peptide hydrolysate, about 200 pmol was applied.

can be seen that the deviation among the individual amino acid residues is relatively small (3–6%). Accuracy in determining the absolute quantity of the peptide is limited by the problems of incomplete hydrolysis and amino acid destruction during hydrolysis. Addition of a reducing agent, such as thioglycollic acid, alleviates this latter problem. Hydrolysis of replicate samples for varying periods of time, followed by extrapolation of the data to zero time, has been reported to improve accuracy (Robel and Crane, 1972).

The inclusion of thioglycollic acid during hydrolysis results in the reduction of cystine. Under the present chromatographic conditions, cysteine elutes close to proline, and duplicate analyses, with and without the addition of NCS, are therefore required to measure both amino acids. Furthermore, cysteine assay may not be precise (Stein *et al.*, 1974*b*). Simple solutions to this problem, such as amino-ethylation or oxidation to cysteic acid, should be possible if thioglycollic acid is included during hydrolysis.

As discussed before, the amino acid composition is used to prove the identity of a peptide-fluorophor peak. It furthermore provides information about the homogeneity of the peak. After subtraction of the blank (a hydrolysate of the column buffer), unaccountable quantities of amino acid residues may indicate the presence of other peptides.

Except for amino-terminal analysis, fluorometric procedures have not been generally applied to the determination of the primary structure of a polypeptide. With picomole sensitivity for both peptides and amino acids, microgram rather than milligram quantities of purified peptides may be sufficient for structural elucidation. The molecular weight of a newly isolated peptide is first required in order to express the amino acid composition in terms of residues per molecule. Chromatography of an MDPF derivative on a column of Biogel provides this information, using as little as 200 pmol of the peptide. Alternatively, it should be possible to ascertain the molecular weight by gel filtration of the free peptide using a fluorescent detection system.

Dansyl chloride (Seiler and Wiechmann, 1964) and NBD chloride (Fager *et al.*, 1973) have allowed amino-terminal determination at the picomole level. Based on preliminary observations, the amino terminal may be deduced by comparing the amino acid analyses of the free peptide and the fluorescamine derivative of the peptide, since the amino-terminal residue of the derivative is not released on acid hydrolysis (Mendez, personal communication).

For sequencing, Edman degradation is popular, and the use of ^{14}C Edman reagent may be applicable at the picomole level. Recently, Mendez and Lai (1975*b*) have shown that the thiazolinone derivatives are readily convertible to the free amino acids in high yield. Subsequent fluorometric analysis permits sequencing at the picomole level. Another interesting approach for sequencing involves the use of mass spectroscopy of MDPF or fluorescamine derivatives (Pritchard *et al.*, 1975). In this procedure, which does not utilize the fluorescent properties of the fluorophors, mass spectral analysis shows a molecular ion for the dipeptide derivative. Since partial peptide bond cleavage occurs during vaporization, a molecular ion also appears for the *N*-terminal amino acid derivative. The composition and sequence of the dipeptide are thereby determined. For sequencing, after treating the unknown polypeptide with a dipeptidyl aminopeptidase, the resulting dipeptides would be reacted with MDPF. The dipeptide fluorophors could then be separated by reverse-phase column chromatography. To obtain overlapping peptides, this process would be repeated after performing a one-step Edman degradation.

6. Conclusion

Chemical assay using fluorescence is inherently more sensitive than absorptiometric procedures. The introduction of fluorescamine and MDPF has led to the development of a number of applications in peptide chemistry, with improvements in both sensitivity and ease of analysis. Although bioassay and radioimmunoassay are invaluable tools, it is hoped that the researcher will, in addition, take advantage of fluorescent chemical assays.

ACKNOWLEDGMENTS

The guidance and insight of Dr. Sidney Udenfriend, who has advanced the utilization of fluorescence in the life sciences, are greatly appreciated. The excellent technical assistance of Mr. Larry Brink is acknowledged. Dr. A.N. Radhakrishnan, Dr. K.A. Gruber, and Dr. J. Wideman have contributed to the latest research conducted in this laboratory, which includes the chromatography and assay of both free and fluorescent-labeled peptides.

7. References

Benson, J.R., and Hare, P.E., 1975, o-Phthalaldehyde: fluorogenic detection of primary amines in the picomole range. Comparison with fluorescamine and ninhydrin, *Proc. Nat. Acad. Sci. U.S.A.* **72:**619–622.

Benson, J.V., Jones, R.T., Cormick, J., and Patterson, J.A., 1966, Accelerate automatic chromatographic analysis of peptides on a spherical resin, *Anal. Biochem.* **16:**91–106.

Benuck, M., and Marks, N., 1975, Enzymatic inactivation of Substance P by a partially purified enzyme from rat brain, *Biochem. Biophys. Res. Commun.* **65:**153–160.

Böhlen, P., Stein, S., Dairman, W., and Udenfriend, S., 1973, Fluorimetric assay of proteins in the nanogram range, *Arch. Biochem. Biophys.* **155:**213–220.

Böhlen, P., Stein, S., Stone, J., and Udenfriend, S., 1975, Automatic monitoring of primary amines in preparative column effluents with fluorescamine, *Anal. Biochem.* **67:**438–445.

Burzynski, S.R., 1975, Quantitative analysis of amino acids and peptides in the femtomolar range, *Anal. Biochem.* **65:**93–99.

DeBernardo, S., Weigele, M., Toome, V., Manhart, K., Leimgruber, W., Böhlen, P., Stein, S., and Udenfriend, S., 1974, Studies on the reaction of fluorescamine with primary amines, *Arch. Biochem. Biophys.* **163:**390–399.

Dekanski, J., 1952, The quantitative assay of vasopressin, *Brit. J. Pharmacol.* **7:**567–572.

Denckla, W.D., 1975, Isoelectric Focusing Techniques and Devices, United States Patent 3,901,780.

Fager, R.S., Kutina, C.B., and Abrahamson, E.W., 1973, The use of NBD chloride (7-chloro-4-nitrobenzo-2-oxa-1,3-diazole) in detecting amino acids and as an N-terminal reagent, *Anal. Biochem.* **53:**290–294.

Fazakerly, S., and Best, D.R., 1965, Separation of amino acids, as copper chelates, from amino acid, protein and peptide mixtures, *Anal. Biochem.* **12:**290–295.

Felix, A.M., and Jimenez, M.H., 1974, Usage of fluorescamine as a spray reagent for thin-layer chromatography, *J. Chromatogr.* **89:**361–364.

Felix, A.M., and Terkelsen, G., 1973a, Total fluorometric amino acid analysis using fluorescamine, *Arch. Biochem. Biophys.* **157:**177–182.

Felix, A.M., and Terkelsen, G., 1973b, Determination of hydroxyproline in fluorometric amino acid analysis with fluorescamine, *Anal. Biochem.* **56:**610–615.

Felix, A.M., Toome, V., DeBernardo, S., and Weigele, M., 1975, Colorimetric amino acid analysis using fluorescamine, *Arch. Biochem. Biophys.* **168:**601–608.

Frenkel, M.J., and Blagrove, R.J., 1975, Controlled pore glass chromatography of protein–sodium dodecyl sulfate complexes, *J. Chromatogr.* **111:**397–402.

Furlan, M., and Beck, E.A., 1974, Use of fluorescamine in peptide mapping on thin-layer cellulose plates, *J. Chromatogr.* **101:**244–246.

Gallo-Torres, H.E., Ludorf, J., and Miller, O.N., 1975, An ultramicrotechnique for the detection and separation of small molecular weight peptides from amino acids, *Anal. Biochem.* **64**:260–267.

Ghosh, P.B., and Whitehouse, M.W., 1968, 7-Chloro-4-nitrobenzo-2-oxa-1,3-diazole: A new fluorogenic reagent for amino acids and other amines, *Biochem. J.* **108**:155–156.

Gruber, K.A., Stein, S., Radhakrishnan, A.N., Brink, L., and Udenfriend, S., 1976, Fluorometric assay of vasopressin and oxytocin, a general approach to the assay of peptides in tissues, *Proc. Nat. Acad. Sci. U.S.A.* **73**:1314–1318.

Hamilton, P.B., 1965, Amino acids on hands, *Nature London* **205**:284,285.

Hamilton, P.B., and Myoda, T.T., 1974, Contamination of distilled water, HCl and NH₄OH with amino acids, proteins and bacteria, *Clin. Chem.* **20**:687–691.

Imai, K., Böhlen, P., Stein, S., and Udenfriend, S., 1974, Detection of fluorescamine-labeled amino acids, peptides and other primary amines on thin-layer chromatograms, *Arch. Biochem. Biophys.* **161**:161–163.

Jakob, A., Hauri, C.H., and Froesch, E.R., 1968, Nonsuppressible insulin-like activity of human serum. III. Differentiation of two distinct molecules with nonsuppressible ILA, *J. Clin. Invest.* **47**:2678–2688.

Kato, T., Sasaki, M., and Kimura, S., 1975, Applications of the dansylation reaction to the characterization of low molecular weight peptides by dodecyl sulfate–polyacrylamide gel electrophoresis, *Anal. Biochem.* **66**:515–522.

Maeda, M., Tsuji, A., Ganno, S., and Onishi, Y., 1973, Fluorophotometric assay of amino acids by using automated ligand-exchange chromatography and pyridoxal–zinc (II) reagent, *J. Chromatogr.* **77**:434–438.

Margolis, F.L., 1974, Carnosine in the primary olfactory pathway, *Science* **184**:909–911.

Margolis, F.L., and Ferriero, D., 1975, Denervation in the primary olfactory pathway of mice. II. Effects on carnosine and other amine compounds, *Brain Res.* **94**:75–86.

Marks, N., Galoyan, A., Grynbaum, A., and Lajtha, A., 1974, Protein and peptide hydrolases of the rat hypothalamus and pituitary, *J. Neurochem.* **22**:735–739.

McCaman, M.W., and Robins, E., 1962, Fluorometric method for the determination of phenylalanine in serum, *J. Lab. Clin. Med.* **59**:885–890.

Mendez, E., and Gavilanes, J.G., 1976, Fluorometric detection of peptides after column chromatography or on paper: *o*-Phthalaldehyde and fluorescamine, *Anal. Biochem.* **72**:473–479.

Mendez, E., and Lai, C.Y., 1975*a*, Reaction of peptides with fluorescamine on paper after chromatography or electrophoresis, *Anal. Biochem.* **65**:281–292.

Mendez, E., and Lai, C.Y., 1975*b*, Regeneration of amino acids from thiazolinones formed in the Edman degradation, *Anal. Biochem.* **68**:47–53.

Nakai, N., Lai, C.Y., and Horecker, B.L, 1974, Use of fluorescamine in the chromatographic analysis of peptides from proteins, *Anal. Biochem.* **58**:563–570.

Nakamura, H., and Pisano, J.J., 1976*a*, Detection of compounds with primary amino groups on thin-layer plates by dipping in a florescamine solution, *J. Chromatogr.* **121**:79–81.

Nakamura, H., and Pisano, J.J., 1976*b*, Labeling of compounds at the origin of thin-layer plates with fluorescamine, *J. Chromatogr.* **121**:33–40.

Nardacci, N., Mukhopadhyay, N., and Campbell, B.J., 1975, Partial purification and characterization of the antidiuretic hormone-inactivating enzyme from renal plasma membranes, *Biochem. Biophys. Acta* **377**:146–157.

Neidle, A., and Kandera, J., 1974, Carnosine—an olfactory bulb peptide, *Brain Res.* **80**:359–364.

Pace, J.L., Kemper, D.L., and Ragland, W.L., 1974, The relationship of molecular weight to electrophoretic mobility of fluorescamine-labeled proteins in polyacrylamide gels, *Biochem. Biophys. Res. Commun.* **57**:482–487.

Pritchard, D.G., Schute, W.C., and Todd, C.W., 1975, Mass spectrometry of fluorescamine derivatives of amino acids and peptides, *Biochem. Biophys. Res. Commun.* **65**:312–316.

Radhakrishnan, A.N., Stein, S., Licht, A., Gruber, K.A., and Udenfriend, S., 1976, High efficiency cation exchange chromatography of polypeptides and polyamines in the nanomole range, *J. Chromatogr.* in press.

Ragland, W.L., Pace, J.L., and Kemper, D.L., 1974, Fluorometric scanning of fluorescamine-labeled proteins in polyacrylamide gels, *Anal. Biochem.* **59**:24–33.

Robel, E.J., and Crane, A.B., 1972, An accurate method for correcting unknown amino acid losses from protein hydrolyzates, *Anal. Biochem.* **48**:233–246.

Rosenblatt, M.S., Margolies, M.N., Cannon, L.E., and Haber, E., 1975, Peptides: An analytical method for their resolution by polyacrylamide gel electrophoresis applicable to a wide range of sizes and solubilities, *Anal. Biochem.* **65**:321–330.

Sachs, H., and Takabatake, Y., 1964, Evidence for a precursor in vasopressin biosynthesis, *Endocrinology* **75**:943–948.

Sachs, H., Goodman, R., Osinchak, J., and McKelvy, J.M., 1971, Supraoptic neurosecretory neurons of the guinea pig in organ culture. Biosynthesis of vasopressin and neurophysin, *Proc. Nat. Acad. Sci. U.S.A.* **68**:2782–2786.

Saifer, A., 1971, Comparative study of various extraction methods for quantitative determination of free amino acids from brain tissue, *Anal. Biochem.* **40**:412–423.

Samejima, K., 1974, Separation of fluorescamine derivatives of aliphatic diamines and polyamines by high-speed liquid chromatography, *J. Chromatogr.* **96**:250–254.

Samejima, K., Dairman, W., and Udenfriend, S., 1971a, Condensation of ninhydrin with aldehydes and primary amines to yield highly fluorescent ternary products. I. Studies on the mechanism of the reaction and some characteristics of the condensation product, *Anal. Biochem.* **42**:222–236.

Samejima, K., Dairman, W., Stone, J., and Udenfriend, S., 1971b, Condensation of ninhydrin with aldehydes and primary amines to yield highly fluorescent ternary products. II. Application to the detection of peptides, amino acids, amines and amino sugars, *Anal. Biochem.* **42**:237–247.

Schwabe, C., and Catlin, J.C., 1974, Removal of a fluoram-positive impurity from hydrochloric acid, *Anal. Biochem.* **61**:302–304.

Scott, A.P. and Lowry, P.J., 1974, Adrenocortiotrophic and melanocyte-stimulating peptides in the human pituitary, *Biochem. J.* **139**:593–602.

Seiler, N., and Knödgen, B., 1974, Identification of amino acids in picomole amounts as their 5-dibutylaminonaphthalene-l-sulphonyl derivatives, *J. Chromatogr.* **97**:286–288.

Seiler, N., and Wiechmann, J., 1964, Zum Nachweis von Aminosäuren im 10^{-10}-Molmasstab. Trennung von 1-Dimethylamino-naphthalin-5-sulfonyl-aminosäuren auf dünnschicht Chromatogrammen, *Experientia* **20**:559–560.

Stein, S., 1975, Fluorescence Protein and Peptide Analyzer, United States Patent 3,915,648.

Stein, S., Böhlen, P., Stone, J., Dairman, W., and Udenfriend, S., 1973, Amino acid analysis with fluorescamine at the picomole level, *Arch. Biochem. Biophys.* **155**:203–212.

Stein, S., Böhlen, P., and Udenfriend, S., 1974a, Studies on the kinetics of reaction and hydrolysis of fluorescamine, *Arch. Biochem. Biophys.* **163**:400–403.

Stein, S., Chang, C.H., Böhlen, P., Imai, K., and Udenfriend, S., 1974b, Amino acid analysis with fluorescamine of stained protein bands from polyacrylamide gels, *Anal. Biochem.* **60**:272–277.

Swaab, D.F., 1971, Pitfalls in the use of rapid freezing for stopping brain and spinal cord metabolism in rat and mouse, *J. Neurochem.* **18**:2085–2092.

Swank, R.T., and Munkres, K.D., 1971, Molecular weight analysis of oligopeptides by electrophoresis in polyacrylamide gel with sodium dodecyl sulfate, *Anal. Biochem.* **39**:462–477.

Tamura, Z., Nakajima, T., Nakayama, T., Pisano, J.J., and Udenfriend, S., 1973, Identification of peptides with 5-dimethylaminonaphthalen-sulfonyl chloride, *Anal. Biochem.* **52**:595–606.

Tarver, J., Bautz, G., and Horst, W.D., 1975, A new method for the determination of ρ-aminobutyric acid in brain, *Fed. Proc. Fed. Amer. Soc. Exp. Biol.* **34**:283.

Toome, V., and Manhart, K., 1975, A simple simultaneous colorimetric determination of primary and secondary amines with fluorescamine, *Anal. Lett.* **8**:441–448.

Toome, V., DeBernardo, S., and Weigele, M., 1975, A simple method for determining the absolute configuration of α-amino acids, *Tetrahedron* **31**:2625–2627.

Tsuji, K., Robertson, J.H., and Bach, J.A., 1974, Quantitative high-pressure liquid chromatographic analysis of bacitracin, a polypeptide antibiotic, *J. Chromatogr.* **99**:597–608.

Udenfriend, S., 1969, *Fluorescence Assay in Biology and Medicine,* Vol. II. (B. Horecker, N.O. Kaplan, J. Marmur, and H.A. Scheraga, eds.), Academic Press, New York.

Udenfriend, S., Stein, S., Böhlen, P., Dairman, W., Leimgruber, W., and Weigele, M., 1972, Fluorescamine: a reagent for assay of amino acids, peptides, proteins and primary amines in the picomole range, *Science* **178**:871–872.

Vale, W., Brazeau, P., Rivier, C., Brown, M., Boss, B., Rivier, J., Burgus, R., Ling, N., and Guillemin, R., 1975, Somatostatin, *Rec. Prog. Horm. Res.* (R.O. Greep, ed.), **31**:365–396.

Weigele, M., Blount, J.F., Tengi, J.P., Czajkowski, R.C., and Leimgruber, W., 1972*a,* The fluorogenic ninhydrin reaction. Structure of the fluorescent principle, *J. Amer. Chem. Soc.* **94**:4052–4054.

Weigele, M., DeBernardo, S., Tengi, J.P., and Leimgruber, W., 1972*b,* A novel reagent for the fluorometric assay of primary amines, *J. Amer. Chem. Soc.* **94**: 5927–5928.

Weigele, M., DeBernardo, S., and Leimgruber, W., 1973*a,* Fluorometric assay of secondary amino acids, *Biochem. Biophys. Res. Commun.* **50**:352–356.

Weigele, M., DeBernardo, S., Leimgruber, W., Cleeland, R., and Grunberg, E., 1973*b,* Fluorescent labeling of proteins. A new methodology, *Biochem. Biophys. Res. Commun.* **54**:899–906.

3

Specific Problems in the Identification and Quantitation of Neuropeptides by Radioimmunoassay

Eugene Straus and Rosalyn S. Yalow

1. Introduction

The current growth of interest in peptides arising within, or having biological effects on, the nervous system has resulted in a broader application in neurobiology of radioimmunoassay (RIA), a technique that combines great simplicity and sensitivity with the inherent specificity of immunological reactions. Two general strategies in which RIA is a central tool for the study of neuropeptides can be developed, each having its own special considerations and problems. In the first, RIA procedures already developed for well-characterized peptides from other anatomical sites in the same or different species can be applied to the detection of identical or immunologically related peptides within the nervous system. The second approach involves the identification of newly described neuropeptides, development of RIA systems to assist in further characterization, and/or purification of the peptides and then application to the study of their distribution and physiological role.

This chapter will deal with specific problems involved in the application of each of the general strategies outlined above. No attempt will be made here to

Eugene Straus and Rosalyn S. Yalow · Solomon A. Berson Research Laboratory, Veterans Administration Hospital, Bronx, New York. Mt. Sinai School of Medicine, The City University of New York, New York.

discuss general considerations involved in the construction and validation of RIA systems, since this information is readily available elsewhere (Berson and Yalow, 1973).

2. Radioimmunoassay for Detection of Well-Characterized Peptides in Nervous Tissues

In recent years, data have been presented to suggest that peptide hormones that previously were believed to be present only within specific anatomical regions of the CNS have a much wider CNS distribution, and that other peptide hormones that had never been reported to occur in CNS tissues have now been found to reside there. RIA has been used to find thyrotropin-releasing factor (TRF) distributed not only in the hypothalamus, but also within all areas of the rat brain, with the exception of the cerebellum (Winokur and Utiger, 1974), and to detect it even in the brain of a gastropod mollusk (Grimm-Jørgensen et al., 1975). All the components of the angiotensin system have been reported to be widely distributed within mammalian CNS tissues (Ganten et al., 1971a,b; Fischer-Ferraro et al., 1971; Yong and Neff, 1972). The gastrointestinal peptide hormone gastrin, which is the most potent stimulant of gastric hydrochloric acid secretion known, has been reported in aqueous extracts of cerebral cortex from a variety of mammalian species, including man (Vanderhaegen et al., 1975). The role of RIA in the study of the distribution within the CNS of well-known, well-characterized peptides is clear, and it can be anticipated that there will be continued expansion of its usage. It is therefore important to consider some of the problems, pitfalls, and limitations of this approach, including (1) the species-specificity of the immune reaction, (2) techniques of peptide extraction, (3) the possible artifacts that result in the false impression that immunoreactive material is present, and (4) the methods for characterization of immunoreactive material in a variety of physical–chemical systems so as to assure its specific identification.

2.1. Species-Specificity

Currently available RIA systems for peptide hormones frequently employ antisera raised against human, porcine, or other mammalian peptides. The use of these systems for study of peptide hormones in species other than that from which the immunogen was obtained requires consideration of the question of species-specificity of the immune reaction.

The amino acid sequences of peptides from different species have diverged during the course of evolution. This divergence has frequently occurred through single-base mutations, and has generally been conservative, resulting in no significant alteration in the biological function of the molecule. Species differences in a hormone are therefore likely to occur in regions of the molecule that are not directly involved in its biological action. Thus, although all the known physiological actions of heptadecapeptide gastrin are exhibited by its carboxy-terminal tetrapeptide, sequence variations among man, pig, dog, cat, cattle, and sheep gastrins occur only in positions 5, 8, and 10. Immunochemical identity among peptides may require more than identity of the primary structure of a molecule. Thus, while pork, dog, and sperm whale insulins have been reported to have identical amino acid sequences (Smith, 1966), they are distinguishable immunochemically with some, but not with all, antisera (Berson and Yalow, 1961a; Berson and Yalow, 1966).

When applying immunoassay systems across species lines, one must consider that while phylogenetic conservation of the molecular structure of various peptides is required for preservation of biological activity, it may still permit major configurational alterations that greatly affect immunochemical recognition. One cannot predict the hormones and species among which these alterations have been exhibited. A comparison of porcine, bovine, ovine, salmon, and human calcitonins reveals that each consists of 32 amino acids with a 1–7 disulfide bridge at the amino-terminus and prolinamide at the carboxyl-terminus. The sequences are identical, however, only in 9 of the 32 positions (Potts, 1973) (Fig. 1). The differences in structure are reflected in marked differences in immunological activity, so that homologous radioimmunoassay systems are generally required for the assay of calcitonin in plasma. It is of interest that salmon calcitonin has far greater biological potency in every mammalian species than any mammalian calcitonin, presumably because the species-specificity of the *degrading* system permits prolongation of its *in vivo* action.

Guinea pig insulin has almost full biological potency but less than 1/1000 the immunochemical potency of other mammalian insulins in immunoassay systems employing antibodies developed in man or guinea pigs to beef and pork insulins (Berson and Yalow, 1966). Insulins from fish and lower species are even more unreactive in these systems (Yalow and Berson, 1964). Yet antibodies raised

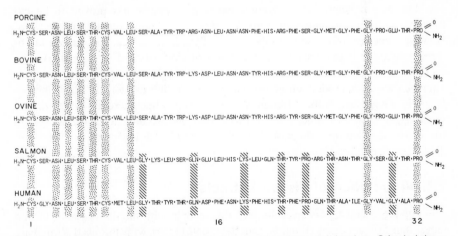

FIG. 1. Comparison of amino acid sequences of various calcitonins. Stippled bars indicate homologies among all five species; cross-hatched bars indicate homologies between human and salmon calcitonin.

against pork and beef glucagons, the other major pancreatic hormone, react strongly with avian and piscine glucagons, as well as with mammalian glucagons, and glucagonlike immunoreactivity can be detected in gastric extracts of various mammals, birds, cyclostomes, and prochordates, using these antisera (see Assan, 1973, for a review).

Comparison of two anterior pituitary peptides, growth hormone (GH) and ACTH, again indicates restrictions imposed by species-specificity that one might not have anticipated. Primate and bovine GHs differ significantly chemically, and bovine and other animal GHs are not active in promoting growth in primates (Knobil and Hotchkiss, 1964). It is therefore not surprising that a high degree of immunological species-specificity is exhibited as well; i.e., only monkey GH, not other animal GHs, has been found to cross-react immunologically with antisera against human growth hormone (HGH) (Glick *et al.*, 1963). On the other hand, HGH does have significant physiological effects in the dog (Raben and Hollenberg, 1960), and the rat tibia responds to GHs from man, ox, sheep, whale, pig, monkey, and certain nonmammalian species (Li, 1960). Yet primate GH does not react in immunoassay systems for the animal GH (Tashjian *et al.*, 1968). Excellent cross-reactivity of human, ovine, and porcine ACTHs has been observed in assay systems employing guinea pig antiporcine ACTH (Yalow *et al.*, 1964), and this observation may be correlated with the fact that the amino acid sequences of the first 24 and the last 7 positions of these 1–39 peptides are identical, differences appearing only in the region of amino acids 25–32 (Lee, 1973). It is more surprising that a larger form of ACTH, called *intermediate ACTH,* with a molecular weight 2–3 times that of the usual 1–39 peptide, which is found to predominate in the pituitaries of rodents such as the rat and mouse (Coslovsky and Yalow, 1974), (Fig. 2), appears to be fully active in the porcine ACTH RIA system. In fact, the immunopotency of their ACTHs is so high that mouse and rat plasma ACTHs are readily measurable in this system (Coslovsky *et al.*, 1975).

The use of heterologous radioimmunoassay systems has proved to be extremely useful in the identification and localization of known peptides in the CNS. However, because of the species-specificity of the immune reaction, failure to observe sought-for immunoreactivity in extracts of CNS tissue does not rule out the presence of peptides of similar structure and biological function when the immunogen used for production of the antibodies and the extract for which the potency is sought are from different species. Even when there is sufficient immunological similarity among species to permit the use of a heterologous system for assay, accurate quantification depends on the availability of appropriate standards (see Section 3.3).

2.2. Preparation of Tissue Extracts

Peptide chemistry received its greatest impetus with the discovery of insulin more than 50 years ago (Banting and Best, 1921) and with the need to prepare large amounts of this peptide in a relatively pure state for therapeutic purposes.

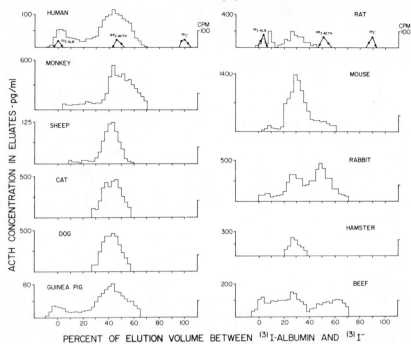

FIG. 2. Sephadex G-50 gel filtration of immunoreactive ACTH in extracts of pituitaries of several mammalian species. Reproduced from Coslovsky and Yalow (1974).

The state of the art of protein chemistry was greatly refined by the time of the final purification and sequencing of insulin by Sanger in the mid 50's (Sanger, 1959), and investigators who are preparing extracts in a search for new hormones or in the detection of well-characterized hormones in the CNS should be guided in the initial steps by methodology already well established for many peptide hormones. (See *Methods in Investigative and Diagnostic Endocrinology, 1973,* Vol. 2, *Peptide Hormones,* for a review.)

The ideal extraction technique would be one that results in the exclusive solubilization of the desired peptide in a protective solvent. This ideal, of course, is never the case, and so care must be taken to determine that the extract does not contain substances that damage the labeled or unlabeled peptide or that interfere nonspecifically with the immune reaction. Although known solubility and stability characteristics of the peptide and considerations relating to the presence of other peptides and/or enzymes contained in the tissues serve as a guide in choosing appropriate extraction methods, the most favorable procedures are frequently arrived at empirically.

Tissues to be used for extraction should be removed from the animal soon after death, and especially immediately after opening the body. They should be extracted immediately after dissection, or, if storage cannot be avoided, they should be kept at $-20°C$ or below. Since most peptides are highly water-soluble, aqueous solutions of low salt concentration are generally effective solvents. Conditions must be modified, however, usually because of the presence in the tissues of proteolytic enzymes that can be coextracted and result in destruction of the desired peptide and/or damage to the necessary constituents of the RIA system (see Section 2.3). This precaution is of special relevance because the CNS contains a variety of proteolytic enzymes (see Chapter 9). If the desired peptide is able to withstand boiling or extremes of pH, extraction procedures can be developed that in general will simultaneously inactivate proteolytic enzymes. Insulin, for example, can be extracted at neutral pH, but is likely to be damaged by the high content of proteolytic enzymes in the pancreas. Boiling in alkaline solution is not useful, since insulin itself will not tolerate boiling at this pH. Insulin is stable at an acid pH, however, even when boiled for a short period, so that insulin can be extracted in boiling dilute acid (pH 3) with simultaneous inactivation of enzymatic activity. The gastrointestinal tract, like the pancreas, is rich in a variety of proteolytic enzymes. Secretin, like insulin, will not be extracted at neutral pH, probably because of proteolytic degradation, and will not tolerate boiling under these conditions. Again like insulin, however, it can be extracted from these same tissues with boiling acid or cold acid–alcohol. Gastrin can readily be extracted simply with boiling water. Extraction is generally completed by maceration or grinding of the tissue with a blender or grinder, and particulate material is removed by centrifugation.

Most peptide hormones have now been shown to exist in more than one molecular form in plasma and tissue extracts (see Yalow, 1974, for a review; see also Section 2.4). The various forms of a peptide may differ with regard to their stability and solubility characteristics, and so it is desirable to determine the total immunoreactivity extracted from tissue by a variety of techniques. In addition, the physical–chemical characteristics of the extracted immunoreactive forms should be characterized, since increases in immunoreactivity may be due to altered molecular forms that are more immunopotent with one antiserum than with another, as has been the case with parathyroid hormone (Silverman and Yallow, 1973) (see Section 2.4).

Purification of the extract is generally not required, because the RIA system usually provides adequate specificity for quantifying the specific substance even in the presence of million-fold or more excess of unrelated peptides. There is always the possibility, however, that the tissue contains an immunochemically cross-reacting material that, because of its high concentration, will be detected in the assay despite its lesser potency in the system. Thus, gastrin, cholecystokinin (CCK), and cerulein share the same C-terminal pentapeptide. Extracts of duodenal mucosa or frog skin, which may contain very high concentrations of CCK or cerulein, respectively, react in the gastrin radioimmunoassay system even though these related peptides in equimolar concentrations may be some hundreds or thousand fold less immunopotent. For this reason also, it is necessary to characterize further material with apparent immunochemical reactivity in a variety of physical–chemical systems (see Section 2.4).

2.3. Sources of Artifact

In the RIA system, the concentration of a peptide being studied in a tissue extract is determined by comparing the inhibitory effect of a known quantity of the extract on the binding of the radiolabeled peptide to a limited amount of antibody with the inhibitory effect of known standards. The tissue extract is thus placed in a milieu containing fixed amounts of specific antibody and of radiolabeled peptide in some standard diluent. The introduction of the tissue extract can result in artifacts due to (1) interference of chemicals in the extract with the antigen–antibody reaction and/or (2) degradation of the antigen and/or antibody.

2.3.1. Chemical Interference with the Antigen–Antibody Reaction

The pH, the ionic environment, and the introduction of chemicals such as anticoagulants or protective substances may influence the antigen–antibody reaction. If standards and unknowns are in the same milieu, then the principal effect might be simply to reduce the sensitivity of the assay. However, if standards and unknowns are not in the same milieu, nonspecific interference in the immune reaction by chemicals in the unknown might be interpretable as due to the specific substance that actually may not be present.

Although antigen–antibody complexes are dissociated by extremes of pH, they are usually independent of pH in the range 7–8.5, and standard diluents used for RIA are generally buffered in this range. If extraction is performed in a highly acid medium, the extract should be neutralized prior to addition to the assay tube, or the volume to be assayed must be kept small enough so that it does not exceed the buffering capacity of the diluent used in the assay. Failure to observe this *caveat* will result in a lowering of the pH, dissociation of the antigen–antibody complex, and falsely attributing this dissociation to high concentrations of antigen in the extract.

High concentrations of salts, even NaCl and barbital buffer at high ionic strengths, may inhibit some peptide antigen–antibody reactions (Berson and Yalow, 1973). This possibility should be evaluated for each antigen and for each antiserum used in such reactions. If an increase in binding of the labeled peptide to antibody occurs simply by reduction of the ionic strength of the buffer in the standard diluent, then the lowest possible ionic strength buffer consistent with adequate buffering capacity should be employed to ensure maximal sensitivity for the assay. If the tissue extract or other biological fluid is to be assayed at high dilution, then it is likely that the salts and other nonspecific proteins would have no effect on the immune reaction. However, if because of low antigen content the extract must be assayed virtually without dilution, then it may be necessary to prepare the standards in similar tissue extracts devoid of antigen. Under all circumstances, it is desirable to evaluate the effect of nonspecific interference in the immune reaction by substances other than the antigen in the extracts and, when necessary, mimic these substances in the diluent used for preparation of standards.

Anticoagulants, bacteriostatic agents, and enzyme inhibitors have in some in-

GP 689

1:25 X 10³ DILUTION OF ANTISERUM

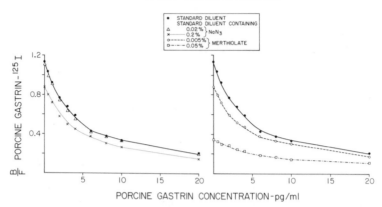

FIG. 3. Effect of sodium azide (left) and merthiolate (right) in incubation mixtures on standard curve for assay. Reproduced from Yalow (1973).

stances been shown to interfere with the immune reaction. We have demonstrated (Yalow and Berson, 1971a) that at concentrations of 40 units heparin per milliliter of incubation mixture, there was a significant nonspecific interference with the immune reaction in the gastrin RIA. This interference was not observed in the same system with heparin concentrations 4-fold lower. Since concentrations of 10–20 units heparin per milliliter of blood are sufficient to prevent clotting, and since we generally assay plasma at a dilution of 1:10 or greater, this interference has not been a problem in our assay, but the potential inhibitory effect of heparin must be considered. In the same system, merthiolate but not sodium azide at concentrations required to be bacteriostatic was inhibitory (Yalow, 1973) (Fig. 3). Trasylol, a commonly used enzyme inhibitor, was shown to be inhibitory in the ACTH RIA (Berson and Yalow, 1973). In each assay system, it is therefore necessary to evaluate the effects of each chemical agent employed. If standards and unknown contain identical concentrations of each of these components, the undesirable effects may be limited simply to reducing the sensitivity of the assay. However, if the unknowns and not the standards contain this agent, its effects might be falsely interpreted as due to the specific substance, the concentration of which is to be determined.

2.3.2. Degradation of the Antigen or Antibody or Both During Incubation

It has long been appreciated that damage to the labeled antigen during incubation will decrease the reliability of the RIA procedure and, depending on the extent of damage, may invalidate the results completely. It has therefore been considered advisable at the very least to use a control mixture containing the labeled antigen and either the unknown sample or the diluent used for the standards, but without antiserum, to evaluate differential damage occurring during the incubation period (Berson and Yalow, 1973). This method of evaluation, while of some value when a specific absorbent method is employed for separation of free labeled antigen from that bound in labeled antigen–antibody complexes, is not suitable in the double-antibody method, in which the antigen–antibody complexes are precipitated by a second antibody. In this system, damaged labeled antigen will not be precipitated by the second antibody, and will therefore appear along with free labeled antigen in the supernatant. Monitoring of the immunological integrity of free labeled antigen can be performed only by adding excess specific antibody to the supernatant at the end of the incubation period, reincubating, and again precipitating with the second antibody. Since this control is wasteful of specific antibody, it is generally not employed. If during incubation either labeled antigen or antibody is destroyed, immune precipitation is reduced, and this reduction is likely to be interpretable erroneously as due to high hormonal content. We have recently demonstrated (Straus and Yalow, 1976) (Fig. 4) that the presence of pancreatic enzymes in unboiled duodenal fluid damaged both labeled antigen and antibody even when diluted 1:25 in the incubation mixture, so that if a double-antibody method had been used for separation, the subsequent failure to achieve immune precipitation of the labeled antigen would have been equated with high hormone concentration in the duodenal fluid. In fact, the latter fluid was devoid of gastrin, the hormone sought, since even if gastrin had been secreted into the duodenal fluid, it would have undergone proteolytic degradation before addition of the fluid to the assay tubes.

When separation is effected by adsorption of free antigen to solid-phase material, incubation damage is usually evaluated by the failure of the damaged labeled antigen to be adsorbed. Charcoal is widely used as a specific adsorbent. However, it does have some disadvantages. In the event of extensive proteolytic destruction of the labeled peptide, small radio-labeled peptides and iodotyrosines are produced (Straus and Yalow, 1976). These fragments adsorb to charcoal, and the "control" appears to be satisfactory. These fragments do not bind to antibody, however, and the lowered fraction of radioactivity found in the labeled antigen–antibody complexes is often falsely interpreted, as in the double-antibody system, to be due to high hormone concentration. The use of charcoal does not permit detection of extensive fragmentation of labeled peptides. However, it is useful for detection of alterations that result in nonspecific binding to serum proteins, such as may occur with less extensive damage to the peptide hormones.

Other adsorbents such as talc, QUSO, and ion-exchange resins that are more discriminating than charcoal may not adsorb the small peptides and iodotyrosines resulting from proteolysis, and their use may therefore be more revealing of alteration of the labeled antigen than is possible with charcoal. Damage to the labeled peptide can also be demonstrated by other techniques such as gel filtration. However, the only certain method for assuring integrity of the labeled antigen is demonstration of its ability to bind to excess antibody after incubation in the tissue extract. The assumption that degradation of antibody during incubation is unlikely to occur unless there is simultaneous degradation of the antigen is usually, though

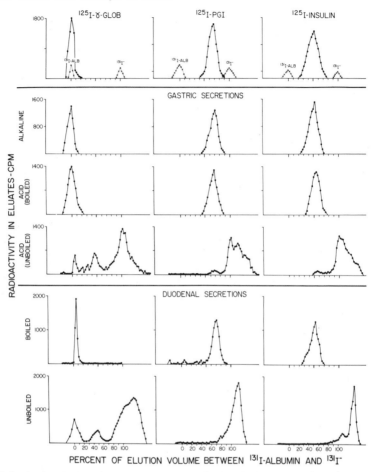

FIG. 4. Sephadex G-50 gel chromatography of ^{125}I-human γ-globulin, ^{125}I-porcine gastrin-I, and ^{125}I-human insulin after incubation in standard diluent and in gastric and duodenal secretions. Incubation in unboiled acid gastric juice and unboiled duodenal secretions resulted in altered gel filtration patterns interpretable as due to extensive alterations in each of the labeled substances. Reproduced from Straus and Yalow (1976).

not always, valid. For instance, trypsin in the incubation mixture would destroy antibody, but not radiolabeled heptadecapeptide gastrin, since the latter does not contain lysine or arginine residues. But this type of exception is quite rare.

2.4. Physical–Chemical Characterization

A necessary but not sufficient condition for establishing the validity of a radioimmunoassay procedure is the demonstration that a dilution curve of an unknown extract is superposable on a curve of known standards, generally over a large range of concentrations, perhaps 100-fold or more. Even if this condition is met, however, the unknown may still not be identical with the standard biologically, chemically, or even immunochemically. In recent years, evidence, derived largely with the use of RIA procedures, has been gathered demonstrating that many if not most peptide hormones are found in more than one form in plasma and tissue extracts (see Yalow, 1974, for a review). Since the different hormonal forms generally differ in biological activity, secretion, and degradation rates, and may even respond differently to diverse secretogogues, measurement of immunological activity *per se* may not suffice for adequate characterization of the unknown peptide.

The presence of heterogeneous forms may be suspected when there are discrepant results for hormonal activity among different bioassay systems, or between bioassay and radioimmunoassay, or between radioimmunoassays employing different antisera. Immunochemical heterogeneity was first shown (Berson and Yalow, 1968) for human parathyroid hormone (hPTH) when it was demonstrated that a constant factor could be used to superpose dilution curves of plasma and standards with several antisera, but not with all antisera (Fig. 5), and that the curves of disappearance of immunoreactivity following parathyroidectomy were dependent on the antiserum employed in the assay. Fractionation in one or more physical–chemical systems identified the reasons for the immunochemical heterogeneity of hPTH. The system most commonly employed for fractionating peptides is Sephadex gel filtration. Labeled or unlabeled marker molecules may be added to samples before fractionation and the position of the marked molecules determined in the eluates by measurement of radioactivity or specific radioimmunoassay. The size of the various hormonal components can then be compared to that of the known marker molecules. It should be appreciated that on Sephadex gel filtration,

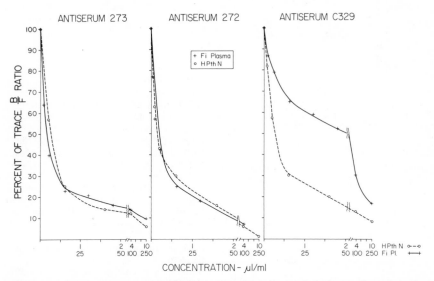

FIG. 5. Inhibition of binding of ^{125}I-hPTH in 3 antisera by pooled plasma from a patient with 2° hyperparathyroidism (+) and by extract of a normal parathyroid gland (o). Reproduced from Berson and Yalow (1968).

elongated molecules elute before globular peptides of the same molecular weight. Some investigators recommend the use of denaturing agents to unfold peptides so as to permit better identification of molecular weight with Sephadex filtration. However, the altered peptides may no longer react in the RIA system, and such agents may not be useful for the determination of molecular forms of picogram to nanogram amounts of peptides in the presence of much higher concentrations of other proteins.

The nature of the hormonal forms that account for the observed immunochemical heterogeneity of hPTH has been elucidated (Silverman and Yalow, 1973). The molecular forms of glandular hormone were studied following extraction of a single adenoma with three different solvents. The immunochemical potencies of the extracts themselves and the eluates following the Sephadex gel filtration of the extracts were determined with two different antisera. It was found that the total immunoreactivity extracted as measured with both antisera was the same if acetone:acetic acid (20%:1%) or 8 M urea were the extracting agents, but that the apparent total immunoreactivity in a NaCl extract depended on the antiserum used for assay. The Sephadex gel filtration pattern of these eluates (Fig. 6) made clear the reason for the difference. The elution patterns of the acetone:acetic acid and 8 M urea extracts were similar and contained a major fraction (Fraction II) that eluted in the same region as intact PTH, i.e., before labeled GH, and that

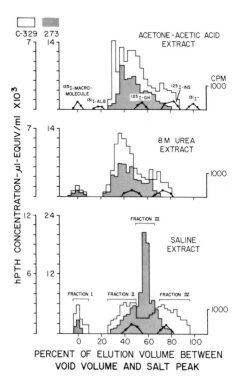

FIG. 6. Sephadex G-100 gel fractionation of acetone–acetic acid (top), 8 M urea (middle), and saline (bottom) extracts of equal portions of a single human parathyroid adenoma. Note that in each fractionation, the ordinate scale is twice as great for antiserum 273 as for antiserum C329, so that a column fraction containing equal concentrations of human parathyroid hormone (hPTH) in both antisera is represented by a hormonal peak in antiserum C329 that is twice the vertical height of the same peak in antiserum 273. The positions of the marker molecules are shown in the top frame. Reproduced from Silverman and Yalow (1973).

showed similar immunoreactivity with both antisera. The most prominent peak of immunoreactivity in the saline extract (Fraction III) had an elution volume between labeled insulin and labeled growth hormone, and reacted with only one of the antisera (273). Thus, the relative proportion of immunoreactive hormonal forms depended on the extraction solvent and the antiserum used for assay. The most likely interpretation of these findings is that there is enzymatic conversion from intact hPTH to smaller forms in the NaCl glandular extract. The increased immunochemical potency of that extract is then due to an unusually high sensitivity of antiserum 273 for a fragment, now known to correspond to a C-terminal fraction about ⅔ the weight of the intact molecule. The sensitivity for detection of this fragment with antiserum 273 is greater than is observed when that fraction is in the intact molecule.

Although the behavior of hPTH is particularly complicated, similar problems may obtain with other peptides, and consideration must be given to the need to use several extraction methods followed by fractionation in different systems for identification of immunoreactive components extracted from CNS tissue.

Fractionation using Sephadex or Bio-rad gel colums is the method most commonly used for demonstrating heterogeneity. However, it is usually advisable to use other techniques as well. These techniques may include ion-exchange resin chromatography, electrophoresis on paper or on starch or polyacrylamide gels, electrofocusing, ultracentrifugation, and others. The observation of the existence of multiple hormonal forms and their identification is made more certain by an increase in the number of independent systems in which the forms can be demonstrated and in the number of tests for their stability.

It is generally advisable when apparent heterogeneity is found to exclude artifacts that may arise from aggregation of peptide molecules or nonspecific binding of peptide to tissue or plasma proteins. The noncovalent bonds in these complexes can usually be dissociated with the use of high concentrations of salt such as 6–8 M urea, 5 M guanidine hydrochloride, or sodium dodecylsulphate (SDS). Consideration must also be given to whether dimers or polymers of the peptide may be formed by disulfide interchange. If the substance can tolerate addition of mercaptoethanol or other SH agents without loss of immunoreactivity, buffers using such agents may also be employed.

Where multiple hormonal forms exist, it is of interest to ascertain the interrelation among them. Where the hormonal form is smaller than the intact, well-known and -characterized form, it is generally presumed to be a metabolic fragment. Larger molecular forms are generally presumed to be precursor molecules. The precursor nature of the larger form has been proved by use of biosynthetic studies for proinsulin (Steiner *et al.,* 1967) and proparathyroid hormone (Wong and Lindall, 1971; Cohn *et. al.,* 1972), and inferred for big ACTH (Yalow and Berson, 1973*a*) and big gastrin (Yalow and Berson, 1971*b*) because of suggestive evidence based on enzymatic conversion of the larger to the small form of known characteristics. The relationship between the usual 1–39 ACTH and intermediate ACTH described earlier (Coslovsky *et al.,* 1975), or between β-MSH and β-lipotropin within which its entire amino acid sequence is contained (Chretien, 1973), has yet to be fully elucidated. Positive demonstration of precursor–product relationships can come only from biosynthetic studies.

3. Development of Radioimmunoassay Systems for Newly Described Peptides

In Section 2, some of the problems and pitfalls of applying already established RIAs for known peptides to neurobiology were discussed. In this section, we shall consider problems in the development of RIAs for newly discovered peptides associated with distinct regions of the CNS or with specific neurons with possible hormonal, neurotransmitter, or other biological activity. The assumption made is that there are limited supplies of the material of interest in pure or even impure form, so that the RIA would be useful in the further purification and chemical characterization of the peptide, as well as in quantitating its presence in tissue extracts or body fluids, or both. Once a properly validated RIA is made available, it is now generally agreed that this method has many advantages in simplicity, sensitivity, and specificity over microchemical or even other immunological assays such as immune precipitation or immunofluorescence. The development of a new RIA for peptides using well-established and adequately described methods (Berson and Yalow, 1973) is simplified when more than 25 mg of the peptide is available. Even when supplies of pure and even impure peptides are more limited, however, the development of a RIA may still be practical. Because the greatest losses of peptide generally occur in final stages of purification, it is desirable to use less purified peptides whenever possible in this development.

The requirements for RIA include a suitable antiserum, labeled antigen, appropriate standards, and some method for separation of antigen–antibody complexes from the free fraction, since these complexes are generally soluble under the usual conditions of RIA.

3.1. Production of Antisera

Since the presence of other immunological reactions does not interfere with the reaction between labeled hormone and specific antibody, it is possible to use relatively impure antigens for immunization. For example, immunization of guinea pigs with crude bovine parathyroid hormone ($\sim 5\%$ purity) or with crude porcine gastrin in antral extracts ($\sim 0.5\%$ purity) resulted in the production of antisera suitable for measurement in plasma of hPTH (mol.wt.~ 9000 daltons) or human gastrin (mol.wt.~ 2000 daltons), respectively (Berson et al., 1963; Yalow and Berson, 1970). Thus, relatively impure antigens of low molecular weight can serve as satisfactory immunogens. Immunogenicity of small molecules may be enhanced by coupling them to larger proteins such as bovine albumin. The protein impurities of the crude extracts could also serve as the larger protein to which the small peptide can be coupled. Coupling agents employed include polylysine (Talamo *et al.*, 1968) and carbodiimide (Goodfriend *et al.*, 1964).

For immunization against many human, bovine, or porcine peptide hormones we have generally found guinea pigs to be satisfactory. In some instances, rabbits have been equally suitable or even superior. Since antisera can frequently be used at dilutions of 10^5 or greater, a single bleeding from a single animal can provide for 10^5 or more assays. Since one cannot predict which animal will prove satisfactory, and since littermates are likely to show similar characteristics in antibody production, it has been our practice to use randomly bred, unrelated animals for immunization. Generally we initially immunize three guinea pigs and one rabbit. The immunization regimen we have employed is to inject about 50–200 μg of coupled or uncoupled peptide in a volume of about 1 ml homogenized with an equal volume of complete Freund's adjuvant (Difco) subcutaneously into the medial aspect of the thigh at 2–3 week intervals. If because of limited supplies of the peptide or because at that level its biological effect may prove toxic to the recipient, smaller immunizing doses may be attempted. After the second or third immunization, blood is taken by cardiac puncture from the guinea pig or from an ear vein of the rabbit and tested for the presence of antibody. Immunization is continued at irregular intervals if antibody concentrations continue to rise. Occasionally, an animal can maintain a steady and satisfactorily high concentration of antibody for many months without reimmunization. If on repeated immunizations the antibody concentration decreases, perhaps 6 months or more should elapse before reimmunization. We continue to maintain only those animals producing antisera satisfactory for our experimental purposes.

The antibody concentration or "titer" of antiserum is generally of secondary importance in RIA, provided it is sufficiently high to permit performance of a large number of assays from a reasonable volume of antiserum. The usefulness of an antiserum for measurement of low concentrations of peptide is dependent on the sensitivity or slope of the standard curve, i.e., the decrease in binding of labeled antigen to antibody with increase in peptide concentration. The sensitivity is primarily dependent on the energy of interaction of the peptide with the predominant order of antibody combining sites, and is usually characteristic of the response of a particular animal. It is frequently unrelated to changes in antibody concentration occurring during immunization. Antibody concentration and sensitivity appear to be unrelated in different animals immunized on the same schedule with the same antigen (Fig. 7).

3.2. Labeled Peptide

In RIA, it is desirable to use a concentration of labeled peptide small enough to be comparable to the lowest concentration of the peptide detectable at the appropriate dilution of the antiserum. Long-lived radioisotopes such as ^{14}C or ^3H are generally not satisfactory radiolabels in sensitive RIA systems in which the chemical amount of the tracer must be kept small. ^{125}I ($t_{1/2} = 60$ days) has proved to be the radioisotope most suitable for labeling peptides. The radioiodine generally

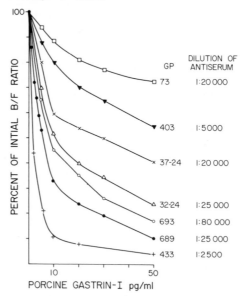

FIG. 7. Comparison of various antisera for sensitivity for measurement of gastrin. Initial B:F ratios between 0.5 and 1.0 in all cases. Reproduced from Berson and Yalow (1972).

substitutes with good efficiency onto the *ortho* position of tryosine in peptide linkage or with more difficulty onto histidyl. We have generally employed the chloramine T technique (Hunter and Greenwood, 1962), in which iodination is effected in the alkaline state. We recommend this method for the noncommercial laboratory in preference to the lactoperoxidase method (Thorell and Johansson, 1971) because of the potential radiation hazard associated with volatilization of I_2 during the iodination procedure, which takes place in an acid medium. Detailed discussion has been presented elsewhere (Berson and Yalow, 1973; Schneider *et al.,* 1976) concerning iodination of peptides by the chloramine T technique, including factors such as the extent of iodination, the distribution of iodine among the tyrosyl residues at a given level of iodination, problems introduced by more than one radioiodine atom per molecule (decay catastrophe), purification of and damage to labeled peptide, and other factors.

Of particular relevance is the question as to whether with limited amounts of peptide it is preferable to use the peptide in highly purified form for labeling or to label a peptide in a less pure state and then purify the labeled product. We would consider that preparations containing no more than 10% of the desired peptide can be quite satisfactory for labeling with radioiodine if sufficient is known about the chemistry of the peptide to choose the optimal methods for subsequent purification. Fifteen years ago, when available chemical purification methods were less sophisticated than now, human insulin of only about 25% purity was used for preparation of ^{131}I-human insulin suitable for RIA (Berson and Yalow, 1961b). Methods of purification following iodination might include those that separate on the basis of size, such as Sephadex and Bio-gel column fractionations, or electrophoretic methods, which separate on the basis of charge, or some combination of the two. In fact, the same purification methods used to free the labeled peptide from unreacted iodine and damaged moieties after iodination are equally suitable for freeing it from contaminants in the impure preparation used for labeling. It is important that the labeled preparation be highly purified to free it from contaminants that might be common to those in the impure peptide used for immunization.

RIA methodology does not require identical immunological behavior of the labeled peptide used as tracer and the unlabeled peptide to be assayed. Nevertheless, alterations in the configuration of the labeled peptide can affect the sensitivity of the assay. This was apparent in the secretin assay. Secretin contains an N-terminal histidyl and no tyrosyl residue. To facilitate iodination, it was considered desirable to prepare a synthetic peptide with tyrosine in place of phenylalanine in the sixth position (6-Tyr-secretin) (Guiducci, 1974). However, this substitution appears to produce a conformational alteration that renders 6-iodotyrosyl-secretin less immunoreactive than iodohistidyl secretin, thereby decreasing the sensitivity of the secretin RIA (Straus *et al.*, 1975).

If the peptide does not contain either a tyrosyl or histidyl residue, or if its configuration might be altered by iodination of such a residue, the peptide can be labeled by conjugation to a ^{125}I-containing moiety. A relatively simple method has been described (Bolton and Hunter, 1973) in which the peptide is treated with an ^{125}I-labeled acylating agent, which reacts with free amino groups in the peptide and attaches the labeled hydroxyphenylpropionic acid ester through amide bonds.

3.3. Preparation of Standards

For proper validation of RIA procedure, the unknown substance to be assayed must be immunochemically identical with the substance used as standard. There is no requirement for biological or chemical identity. It is quite unlikely that a highly purified and biologically characterized reference standard would be available for a newly discovered peptide. Crude tissue extracts or biological fluids containing a high concentration of the peptide can be used for evaluation of *relative* concentrations of the peptide in unknown samples.

It can be anticipated that CNS peptides may show the same heterogeneity of molecular forms as do many of the peptidal hormones. When immunochemical heterogeneity is found, as in the case of parathyroid hormone, it is necessary to fractionate standards or unknown or both to assure a properly validated RIA. If the various forms of the peptide are immunochemically similar but differ in biological potency, then one can anticipate discrepancies between bioassay and radioimmunoassay data.

3.4. Separation Techniques

In RIA, the labeled peptide–antibody complexes are generally soluble, and some technique must be employed to distinguish between them and the free labeled peptide. The most generally applicable and specific method is that depending on precipitation of the complex with an antibody directed against the antibody in the complex ("double-antibody"). However, this method is quite expensive because of the cost of the large amounts of second antibody required when thousands of samples are to be processed. If the antibody to the peptide was prepared in a guinea pig, then the carrier guinea pig plasma in the RIA incubation tubes should be about 0.1 or 0.2%. This percentage allows adequate precipitation of the complexes with a minimal amount of second antibody. It has been our experience that with the best of goat or rabbit anti-guinea-pig-γ-globulin sera available commercially or prepared in our laboratory, about 20–50 μl of undiluted antiserum is required to precipitate the labeled peptide–antibody complexes in 1-ml assay tubes. The potency of the second antibody should be tested using labeled γ-globulin of the appropriate species in place of the labeled peptide–antibody complexes.

The technique most commonly used for separation in RIA systems for small peptides depends on adsorption of the free peptide to a solid-phase material, although precipitation of antigen–antibody complexes by salting-out techniques or organic solvents has also been employed (Yalow and Berson, 1973b). Adsorption or complexing of antibody to solid-phase material is commonly used in commercial RIA kits, but is probably not advisable in designing new assays. Often there is some alteration of antibody in the coupling procedure, which results in less sensitivity than would otherwise obtain with the same antibody not complexed to the solid-phase material.

4. Conclusions and Conjectures

When a peptide or other substance of biological interest is highly purified, chemically characterized and available in relatively large quantities, i.e., perhaps microgram or milligram amounts, the need for a biological or immunological assay of its potency in that purified form may disappear. Such could be the case for some small peptides. At present, methods for peptide synthesis by solid-phase and classic approaches are sufficiently good for the production of highly purified synthetic peptides up to about 30 residues. Beyond this size, the classic approach is too difficult, and the solid-phase approach does not give a product sufficiently homogeneous for most applications. With further advances in methodology, it can be expected that the next decade will see an increase in the size of peptides of adequate purity and decrease in their cost. Nonetheless, determination as to which peptides are biologically relevant depends first on their purification from natural sources, followed by their characterization and eventual sequencing.

Measurement of potency of these peptides in body fluids and tissues will continue to require other than strictly chemical methodology. Bioassay and immunoassay have complementary roles in these studies.

The basic assumption required for validation of either assay is that standards and unknowns behave identically at all dilutions employed in the test system. It must be appreciated, however, that identical behavior in a single test system does not assure identity in other test systems. Earlier, we discussed the possible causes for discrepancy between bioassay and immunoassay concentrations: species-specificity of the immune reaction (Section 2.1), presence of peptidal precursors or metabolic fragments with different relative behavior in the two assay systems (Section 2.4), and other causes. We have also noted that RIA results obtained with different antisera reacting differently with different parts of the molecule may also appear discrepant (Section 2.4). Discrepancies similar to those among RIA systems and between RIA and bioassay systems occur among different bioassay systems as well. For instance, different tissues may exhibit quantitatively different responses to the complete hormone than to one of its peptide fragments. Thus, the N-terminal heptadecapeptide of ACTH exhibits full lipolytic activity on fat tissue *in vitro* (Hofman *et al.*, 1962), but has only about 5% of the steroidogenic potency of ACTH *in vivo* (Tanaka *et al.*, 1962). Even the route of administration is relevant to bioassay potency. Thus, when ACTH preparations were bioassayed using the adrenal ascorbic acid depletion method, two preparations giving identical responses to intravenous administration differed about 10-fold in potency when both were administered subcutaneously (Wolfson, 1953). Other examples of anomalous behavior with respect to biological specificity include the induction of 17-α-hydroxylase enzymes in the rabbit adrenal by administration of exogenous porcine (1–39) ACTH, but not under the action of the endogenous intermediate ACTH (see Coslovsky and Yalow, 1974); the persistence of plasma insulinlike activity (ILA) in the presence of severe hyperglycemia following total pancreatectomy (Schoeffling *et al.*, 1965); the cross-reaction of somatomedin in receptor assays for insulin, although not to the same extent with receptors from different tissues, despite the lack of chemical similarity between the two molecules (Van Wyk *et al.*, 1974); and other examples too numerous to mention.

In the past decade, our appreciation of the heterogeneity of peptide hormones has introduced problems in the interpretation of radioimmunoassay measurements, but it has also broadened our understanding of their nature, as well as of the paths of their synthesis and metabolism.

5. References

Assan, R., 1973, Gut glucagon, in: *Methods in Investigative and Diagnostic Endocrinology*, Part III (S.A. Berson and R.S. Yalow, eds.), pp. 888–901, North-Holland Publishing Co., Amsterdam.

Banting, F.G., and Best, C.H., 1921, Pancreatic extracts, *J. Lab. Clin. Med.* **7**:464–472.

Berson, S.A., and Yalow, R.S., 1961*a*, Immunochemical distinction between insulins with identical amino acid sequences from different mammalian species (pork and sperm whale insulins), *Nature London* **191**:1392, 1393.

Berson, S.A., and Yalow, R.S., 1961*b*, Preparation and purification of human insulin-I^{131}; binding to human insulin-binding antibodies, *J. Clin. Invest.* **40**:1803–1808.

Berson, S.A., and Yalow, R.S., 1966, Insulin in blood and insulin antibodies, *Amer. J. Med.* **40**:676–690.

Berson, S.A., and Yalow, R.S., 1968, Immunochemical heterogeneity of parathyroid hormone, *J. Clin. Endocrinol. Metab.* **28**:1037–1047.

Berson, S.A., and Yalow, R.S., 1972, Radioimmunoassay in gastroenterology, *Gastroenterology* **62:**1061–1084.

Berson, S.A., and Yalow, R.S., 1973, General radioimmunoassay, in: *Methods in Investigative and Diagnostic Endocrinology*, Part I (S.A. Berson and R.S. Yalow, eds.), pp. 84–120, North-Holland Publishing Co., Amsterdam.

Berson, S.A., Yalow, R.S., Aurbach, G.D., and Potts, J.T., Jr., 1963, Immunoassay of bovine and human parathyroid hormone, *Proc. Nat. Acad. Sci. U.S.A.* **49:**613–617.

Bolton, A.E., and Hunter, W.M., 1973, The labelling of proteins to high specific radioactivities by conjugation to a ^{125}I-containing acylating agent, *Biochem. J.* **133:**529–539.

Chretien, M., 1973, Lipotropins (LPH), in: *Methods in Investigative and Diagnostic Endocrinology* Part III (S.A. Berson and R.S. Yalow, eds.), pp. 617–636, North-Holland Publishing Co., Amsterdam.

Cohn, D.V., MacGregor, R.R., Chu, L.L.H., Kimmel, J.R., and Hamilton, J.W., 1972, Calcemic fraction-A: Biosynthetic peptide precursor of parathyroid hormone, *Proc. Nat. Acad. Sci. U.S.A.* **69:**1521–1525.

Coslovsky, R., and Yalow, R.S., 1974, Influence of the hormonal forms of ACTH on the pattern of corticosteroid secretion, *Biochem. Biophys. Res. Commun.* **60:**1351–1356.

Coslovsky, R., Schneider, B., and Yalow, R.S., 1975, Characterization of mouse ACTH in plasma and in extracts of pituitary and of adrenotropic pituitary tumor, *Endocrinology* **97:**1308–1315.

Fischer-Ferraro, C, Nahmod, V.E., Goldstein, D.J., and Finkielman, S., 1971, Angiotensin and renin in rat and dog brain, *J. Exp. Med.* **133:**353–361.

Ganten, D., Marquez-Julio, A., Granger, P., Hagduk, K., Karumky, K.P., Boucher, R., and Genest, J., 1971a, Renin in dog brain, *Amer. J. Physiol.* **221:**1733–1737.

Ganten, D., Minnich, V., Granger, P., Hagduk, K., Brecht, H.M., Barbeau, A., Boucher, R., and Genest, J., 1971b, Angiotensin-forming enzyme in brain tissue, *Science* **173:**64–65.

Glick, S.M., Roth, J., Yalow, R.S., and Berson, S.A., 1963, Immunoassay of human growth hormone in plasma, *Nature London* **199:**784–787.

Goodfriend, T.L., Levine, L., and Fasman, G.D., 1964, Antibodies to bradykinin and angiotensin: A use of carbodiimides in immunology, *Science* **144:**1344–1346.

Grimm-Jørgensen, Y., McKelvy, J.F., and Jackson, I.M.D., 1975, Immunoreactive thyrotropin releasing factor in gastropod circumoesophageal ganglia, *Nature London* **254:**620.

Guiducci, M., 1974, Solid phase synthesis of porcine secretin and 6-tyrosyl-secretin, in: *Endocrinology of the Gut* (W.Y. Chey and F.P. Brooks, eds.), pp. 103–106, Charles B. Slack, New Jersey.

Hofmann, K., Yajima, H., Liu, T., and Yanaihara, N., 1962, Studies on polypeptides XXIV. Synthesis and biological evaluation of a tricosapeptide prossessing essentially the full biological activity of ACTH, *J. Amer. Chem. Soc.* **84:**4475–4480.

Hunter, W.M., and Greenwood, F.C., 1962, Preparation of iodine-131 labeled human growth hormone of high specific activity, *Nature London* **194:**495, 496.

Knobil, E., and Hotchkiss, J., 1964, Growth hormone, *Annu. Rev. Physiol.* **26:**47–74.

Lee, T.H., 1973, Adrenocorticotropic hormone (ACTH)—Purification and biochemical characterization, in: *Methods in Investigative and Diagnostic Endocrinology*, Part II (S.A. Berson and R.S. Yalow, eds.), pp. 331–335, North-Holland Publishing Co., Amsterdam.

Li, C.H., 1960, Comparative biochemical endocrinology of pituitary growth hormone, *Acta Endocrinol. Suppl.* **50:**75–81.

Methods in Investigative and Diagnostic Endocrinology, 1973, Parts I–III (S.A. Berson and R.S. Yalow, eds.), North-Holland Publishing Co., Amsterdam.

Potts, J.T., Jr., 1973, Calcitonin—Extraction, purification and biochemical characterization, in: *Methods in Investigative and Diagnostic Endocrinology,* Part III (S.A. Berson and R.S. Yalow, eds.), pp. 991–998, North-Holland Publishing Co., Amsterdam.

Raben, M.S., and Hollenberg, C.H., 1960, Growth hormone and the mobilization of fatty acids, in: *Ciba Found. Colloq. Endocrinol. Proc.* **13:**89–105, Academic Press, New York.

Sanger, F., 1959, Chemistry of insulin (Nobel lecture), *Science* **129:**1340–1345.

Schneider, B.S., Straus, E., and Yalow, R.S., 1976, Some considerations in the preparation of radioiodoinsulin for radioimmunoassay and receptor assay, *Diabetes* **25**:260–267.

Schoeffling, K., Ditschuneit, H., Petzoldt, R., Beyer, J., Pfeiffer, E.F., Sirek, A., Geerling, H., and Sirek, O.V., 1965, Serum insulin-like activity in hypophysectomized and depancreatized (Houssay) dogs, *Diabetes* **14**:658–662.

Silverman, R., and Yalow, R.S., 1973, Heterogeneity of parathyroid hormone: Clinical and physiologic implications, *J. Clin. Invest.* **52**:1958–1971.

Smith, L.F., 1966, Species variation in the amino acid sequence of insulin, *Amer. J. Med.* **40**:662–666.

Steiner, D.F., Cunningham, D., Spigelman, L., and Aten, B., 1967, Insulin biosynthesis: Evidence for a precursor, *Science* **157**:697–700.

Straus, E., and Yalow, R.S., 1976, Artifacts in the radioimmunoassay of peptide hormones in gastric and duodenal secretions, *J. Lab. Clin. Med.* **87**:292–298.

Straus, E., Urbach, H.-J., and Yalow, R.S., 1975, Comparative reactivities of ^{125}I secretin and ^{125}I-6-tyrosyl secretin with guinea pig and rabbit anti-secretin sera, *Biochem. Biophys. Res. Commun.* **64**:1036–1040.

Talamo, R.C., Austen, K.F., and Haber, E., 1968, Effect of carrier and method of coupling on the immunogenicity of bradykinin, in: *Protein and Polypeptide Hormones* (M. Margoulies, ed.), pp. 93–98, Excerpta Medica Foundation, ICS No. 161, Amsterdam.

Tanaka, A., Pickering, B.T., and Li, C.H., 1962, Relationship of chemical structure to *in vitro* lipolytic activity of peptides occurring in adrenocorticotropic and melanocyte-stimulating hormones, *Arch. Biochem. Biophys.* **99**:294–298.

Tashjian, A.H., Jr., Levine, L., and Wilhelmi, A.E., 1968, Use of complement fixation for the quantitative estimation of growth hormone and as a method for examining its structure, in: *Growth Hormone,* (A. Pecile and E.E. Muller, eds.), pp. 70–83, Excerpta Medica Foundation, Milan.

Thorell, J.I., and Johansson, B.G., 1971, Enzymatic iodination of polypeptides with radioactive ^{125}I to high specific activity, *Biochim. Biophys. Acta* **251**:363–369.

Vanderhaegen, J.J., Signeau, J.C., and Gepts, W., 1975, New peptide in the vertebrate CNS reacting with antigastrin antibodies, *Nature London* **257**:604, 605.

Van Wyk, J.J., Underwood, L.E., Hintz, R.L., Clemmons, D.R., Voina, S.J., and Weaver, R.P., 1974, The somatomedins: A family of insulinlike hormones under growth hormone control, in: *Recent Prog. Horm. Res.* **30**:259–318, Academic Press, New York.

Winokur, A., and Utiger, R.D., 1974, Thyrotropin-releasing hormone. Regional distribution in rat brain, *Science* **185**:265–267.

Wolfson, W.Q., 1953, The three subtypes of pituitary adrenocorticotropin, *Arch. Intern. Med.* **92**:108–147.

Wong, E.T., and Lindall, A.W., 1971, Human parathyroid gland immunoreactive peptides—evidence for proparathormone, *J. Lab. Clin. Med.* **78**:825 (abstract).

Yalow, R.S., 1973, Radioimmunoassay: Practices and pitfalls, *Circ. Res.* **32**:I/116–I/128.

Yalow, R.S., 1974, Heterogeneity of peptide hormones, in: *Recent Prog. Horm. Res.* **30**:597–633, Academic Press, New York.

Yalow, R.S., and Berson, S.A., 1964, Reaction of fish insulins with human insulin antiserums: Potential value in the treatment of insulin resistance, *N. Engl. J. Med.* **270**:1171–1178.

Yalow, R.S., and Berson, S.A., 1970, Radioimmunoassay of gastrin, *Gastroenterology* **58**:1–14.

Yalow, R.S., and Berson, S.A., 1971a, Problems of validation of radioimmunoassays, in: *Principles of Competitive Protein-Binding Assays* (W.D. Odell and W.H. Daughaday, eds.), pp. 374–400, J.B., Lippincott Co., Philadelphia and Toronto.

Yalow, R.S., and Berson, S.A., 1971b, Further studies on the nature of immunoreactive gastrin in human plasma, *Gastroenterology* **60**:203–214.

Yalow, R.S., and Berson, S.A., 1973a, Characteristics of "big ACTH" in human plasma and pituitary extracts, *J. Clin. Endocrinol. Metab.* **36**: 415–423.

Yalow, R.S., and Berson, S.A., 1973b, Separation techniques—antigen adsorption, in: *Methods in Investigative and Diagnostic Endocrinology* Part I (S.A. Berson and R.S. Yalow, eds.), pp. 120–125, North-Holland Publishing Co., Amsterdam.

Yalow, R.S., Glick, S.M., Roth, J. and Berson, S.A., 1964, Radioimmunoassay of human plasma ACTH, *J. Clin. Endocrinol. Metab.* **24:**1219–1225.

Yong, H.-Y., and Neff, N.A., 1972, Distribution and properties of angiotensin converting enzyme of rat brain, *J. Neurochem.* **19:**2443–2450.

4

Immunocytochemistry of Neuropeptides and Their Receptors

Ludwig A. Sternberger

1. Introduction

Peripheral cells talk to each other via hormones. The message, although diffused into the bloodstream, is picked up selectively by those groups of cells that possess specific receptors for it. Neurons also talk to each other via hormones. By virtue of the axonal networks, they are able to direct a message to receptors on specific synapses with heightened selectivity and specificity. Thus, anatomical structure in the nervous system has added a relay system to the hormonal communication system found in the periphery. Immunocytochemistry provides simple and rapid means to differentiate axons as to the neurotransmitters carried, thereby permitting distinction among dopaminergic, serotonergic, and noradrenergic fibers.

As peripheral neurons synapse with muscles and organs, so do some specialized central neurons terminate on capillaries. Using peptides as their language, these neurons talk to specific receptors in peripheral cells and thus provide a physiological link between the brain and the peripheral endocrine system. Knowledge of the microanatomy of secretion of neuropeptides is largely due to immunocytochemistry.

Peripheral secretory cells can talk back to the sources from which they received the message. Usually, their reply is a feedback inhibition, such as the suppression of follicle-stimulating hormone (FSH) and luteinizing hormone (LH) secretion by the effects of estrogen on pituitary and hypothalamus. However, via still unknown mechanisms, an opposite reply can occur, such as the preovulatory LH surge evoked by accumulation of estrogen (Davidson, 1969; Smith and David-

Ludwig A. Sternberger · Immunology Branch, Edgewood Arsenal, Maryland.

son, 1974; Schönberg, 1975; Shin and Howitt, 1975). Such feedback and -forward effects can also be conceived of as providing a modulating effect on neurotransmission in pathways entirely confined to the CNS. It is still somewhat speculative to envoke neuropeptides in such a role, but nevertheless, suggestions of their widespread role in the CNS come forth with increasing frequency. Immunocytochemistry may be capable of describing such modulating pathways. By its ability to detect not only neuropeptides, but also their specific receptors, immunocytochemistry may conceivably disclose potential sites of action of neuropeptides even under those physiological conditions in which the neuropeptides themselves are not extensively secreted.

2. The Unlabeled Antibody Enzyme Method—Sensitivity of Immunocytochemistry

For years, it has been axiomatic that antibodies have to be labeled in order to provide a specific reagent for visualization of antigen in tissue by light and electron microscopy. Covalent bonding with fluorescein, ferritin, or peroxidase has, indeed, provided immunocytochemical methods. The relative sensitivities of these substances have apparently concealed the fact that orders of magnitude of even higher sensitivity could be achieved if the covalent bonding were entirely avoided. Covalent bonding decreases the efficiency of labeled antibody methods because it destroys some of the antibody and because it results, in all but exceptional applications, in reagents that consist of mixtures of labeled and unlabeled antibody. In such mixtures, unlabeled antibody competes, often preferentially, with the labeled antibody for the antigen to be detected. Also, labeling may conceivably impair the specificity of antibody as a result of reaction with polar groups (reviewed in Sternberger, 1974).

To avoid these difficulties, the unlabeled antibody enzyme method has been introduced. In this method, only immunological bonds are used to attach a detector molecule, such as horseradish peroxidase, to an antigen site in tissue (Sternberger *et al.*, 1970). Tissue antigen is first reacted with specific antibody (primary antiserum). This reaction is followed by an excess of a secondary antiserum produced in another species and specific for the immunoglobulin of the primary antiserum. One of the combining sites of the bivalent antibody in the secondary antiserum (antiimmunoglobulin) reacts with the bound antibody of the primary antiserum, while the other site remains free. As a third step, peroxidase antiperoxidase complex (PAP) is added. It becomes bound because the antibody in this soluble antigen–antibody complex reacts as antigen with the free combining site of the antiimmunoglobulin (Fig. 1a). Peroxidase is then reacted with its substrate, hydrogen peroxide, using diaminobenzidine tetrahydrochloride as electron-donor. The oxidized diaminobenzidine forms an insoluble, osmiophilic polymer, which is dark brown on light microscopy and becomes electron-opaque after treatment with osmium tetroxide.

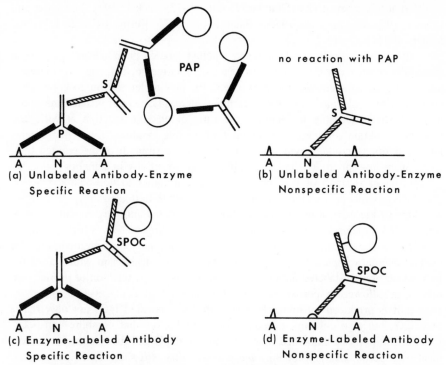

(a) Unlabeled Antibody-Enzyme
    Specific Reaction

(b) Unlabeled Antibody-Enzyme
    Nonspecific Reaction

(c) Enzyme-Labeled Antibody
    Specific Reaction

(d) Enzyme-Labeled Antibody
    Nonspecific Reaction

FIG. 1. In the unlabeled antibody enzyme method: (a) Tissue antigen (A) is localized by specific antibody in the primary antiserum (P). This localization is followed by reaction with antiimmunoglobulin in a secondary antiserum (S), peroxidase–antiperoxidase complex (PAP), and cytochemical reaction for peroxidase. (b) Nonspecific antigens in tissue (N) react with antibodies that contaminate the secondary antiserum. This nonspecific reaction remains undetected in the unlabeled antibody enzyme method, as contaminant antibodies cannot react with PAP. In the indirect enzyme-labeled antibody method: (c) Specific antibody in the primary antiserum (P) is followed by a secondary antiimmunoglobulin chemically conjugated with peroxidase (SPOC). (d) Antibodies that contaminate the secondary antiimmunoglobulin react with nonspecific antigens (N) in the tissue. Since these antibodies are labeled with peroxidase, a nonspecific background reaction ensues.

Optimally, the antiperoxidase component of PAP is of the same species as the primary antiserum. However, Erlandsen et al. (1975) were able to modify the method by using sheep or goat antirabbit immunoglobulin with rabbit PAP in the detection of primary antibodies of monkey and guinea pig. Marucci and Dougherty (1975) used rabbit antihuman immunoglobulin with baboon PAP in the detection of primary antibodies from human antisera.

One of the earliest findings with the unlabeled antibody enzyme method was a sensitivity that exceeded that of labeled antibodies, not by an expected factor of 3 or 4, but apparently by a factor of several orders of magnitude (Sternberger et al., 1970). Thus, it has become possible to detect antigens that by other methods were either undetectable (Taylor, 1974; Palmer et al., 1974; Bahr et al., 1975) or detected at lower sensitivity (Dougherty *et al.*, 1972; Böcker, 1974; Sotmoller and Cowan, 1974; Pickel *et al.*, 1975*a;* Baker *et al.*, 1975; Burns, 1975*a,b;* Elias and Miller, 1975).

All immunocytochemical methods, whether labeled or unlabeled, provide in theory the sensitivity for the detection of a single antigenic site. The ferritin detector could provide such sensitivity by itself, for in the electron microscope, each single ferritin molecule is readily discernible. Peroxidase could provide the sensitivity, for it is possible to lengthen the time of enzyme reaction at will, and thereby to obtain any desirable amount of visible product. Even fluorescein-labeled antibody could provide the sensitivity if the strength of the emitted light signal were intensified. The failure to obtain such sensitivity in actuality is not due to insufficient strength of the detector signal, but rather to nonspecific reaction with background, thus yielding a relatively low signal-to-noise ratio.

One of the earliest observations with the unlabeled antibody enzyme method was the low nonspecific background staining, providing thereby a high signal-to-noise ratio. The reason for the low background in the unlabeled antibody enzyme method becomes apparent if we compare its reaction mechanism with that of the indirect peroxidase-labeled antibody method. In the latter, reaction of primary antiserum is followed by peroxidase-labeled anti-IgG (Fig. 1c). Anti-IgG, however, even if it is produced by immunization with fairly purified IgG, not only reacts with IgG, but also contains antibodies or other factors that combine nonspecifically with the tissue examined. Both specific anti-IgG and the nonspecific factors react with peroxidase on labeling, and consequently will show up as background staining (Fig. 1d). In the unlabeled antibody enzyme method, the unlabeled anti-IgG contains nonspecifically reacting antibodies and other nonspecific factors just as the peroxidase-labeled anti-IgG does. However, these nonspecific factors will not be able to react with PAP, for PAP is a purified reagent, and since in any antibody both combining sites are identical, only components in antiserum to IgG specifically reacting with IgG will be able to cross-link with PAP (Fig. 1b). If, in fact, there existed a contaminant in PAP that would be shared with a tissue cross-reactive impurity in the IgG used for producing the anti-IgG, the contaminant, indeed, would be bound to the tissue. However, it would still not show up as nonspecific stain, for only antiperoxidase in PAP and not a contaminant can bind peroxidase.

Evidence for the mechanism underlying the sensitivity of the PAP method has recently been given by Marucci and Dougherty (1975). These workers were challenged by the need to detect human antibody to herpes simplex virus (HSV), while no human PAP was available. The conventional way to demonstrate these antibodies would be a four-layer sequence, such as human anti-HSV, rabbit anti-human immunoglobulin, guinea pig antirabbit immunoglobulin, and rabbit PAP. However, this four-layer sequence gave disturbing nonspecific background. When they introduced a new cross-reactive three-layer sequence (human anti-HSV, rabbit antihuman immunoglobulin, baboon PAP), the background staining had disappeared, and highly specific results were obtained. Closer examination showed that it was the rabbit antihuman immunoglobulin that caused the nonspecificity in the four-layer technique. This nonspecificity did not disappear even when that serum had been diluted 1:100. On the other hand, the reaction was specific in the three-layer sequence of Marucci and Dougherty, even though the same rabbit antihuman immunoglobulin had been used in it, and even when it had been used undiluted. Marucci and Dougherty explained the nonspecificity by a contaminant in the rabbit antihuman immunoglobulin that would be detected by sheep antirabbit immunoglobulin in the four-layer technique, but would not react with PAP in the three-layer technique. In fact, the four-layer technique provides a simulation of indirect labeled antibody techniques, using antiimmunoglobulin and PAP rather than label as a histochemical detector.

PAP had not been an original ingredient of the unlabeled antibody enzyme method. When it was first introduced (Sternberger, 1969), the method employed, instead of PAP, the application of specifically purified antibody to peroxidase, followed by peroxidase. A further simplification of this method (Mason et al., 1969) merely uses antiserum to peroxidase, followed by peroxidase. However, the serum immunoglobulin consists of a mixture of antiperoxidase with antibodies of other specificities. The latter compete with the binding of specific antiperoxidase. When the sensitivity of the method was expressed in staining intensity relative to dilution of primary antiserum, it had been found that the purified antibody–peroxidase sequence was 4 times more sensitive than the antiserum–peroxidase sequence, when the immunoglobulin in the antiserum contained 13% antiperoxidase (Petrali et al., 1974).

PAP is easily prepared from an immune precipitate obtained by reaction of peroxidase with antiperoxidase. The precipitate is dissociated at acid pH in a moderate excess of peroxidase. On neutralization, a solution consisting of free peroxidase and soluble peroxidase–antiperoxidase complex is obtained. The PAP is separated by precipitation with ammonium sulfate (Sternberger, 1974).

PAP differs from antigen–antibody complexes of other specificities in that it is obtainable at high yield with relatively small concentrations of peroxidase, in its stability in the absence of large amounts of free antigen, in its invariance of composition (three peroxidase and two antiperoxidase molecules), and in its cyclic, pentagonal structure. All these are properties that suggest a high stability in the bonds of peroxidase and antiperoxidase in the complex, a property probably resulting from its cyclicity.

Use of PAP in the unlabeled antibody enzyme method has been found to be 5 times more sensitive than purified antiperoxidase followed by peroxidase (Petrali et al., 1974), and 20–125 times more sensitive than antiserum to peroxidase followed by peroxidase (Baker et al., 1975; Böcker, 1974; Petrali et al., 1974; Vacca et al., 1975). The increase in sensitivity is explainable by the stability of the PAP complex. In the first and second steps of the unlabeled antibody enzyme method, antiserum is added to tissue-bound antigen. Among the heterogeneous antibodies in the antisera, those of the highest binding affinities are bound preferentially. PAP is also bound with high affinity because both combining sites of the bound anti-IgG are equal, and therefore of identical affinities. For the same reason, antibody to peroxidase is bound with high affinity (strong bonding between anti-IgG and antiperoxidase). However, when antiperoxidase and peroxidase are applied in sequence, no selection is made for binding of antiperoxidase with high affinity for peroxidase. Therefore, part of the peroxidase reacting with antiperoxidase in the subsequent step will be lost during washing. When, however, PAP is used, the cyclicity of the complex apparently prevents this loss, thus making the use of PAP more sensitive than that of antiperoxidase followed by peroxidase (Petrali *et al.*, 1974).

We have discussed two expressions of sensitivity accomplished by the unlabeled antibody enzyme method. One is a high signal-to-noise ratio; the other is high affinity of reaction in all steps, which makes it possible to use primary antisera at high dilutions. Many of the antisera available in endocrinology and neurology are of weak and sometimes unknown titer. Often, the activities are too low to give immunodiffusion lines with suitable antigens. Sometimes, the activities are so low as to be barely detectable by radioimmunoassay. Nevertheless, such sera are often suitable for immunocytochemistry even at dilutions in the 10^5 order of magnitude. If a primary antiserum is of unknown activity, a suitable starting dilution for immunocytochemistry is 1:1000 at 48 hr incubation.

The high signal-to-noise ratio of the unlabeled antibody enzyme method has probably been the main reason for its use in discriminating among nerve pathways, whether by antibodies to neuropeptides or to neurotransmitter synthetic enzymes.

3. Modifications of the Unlabeled Antibody Enzyme Method

Instead of PAP, complexes of other enzymes than peroxidase could conceivably be used in the unlabeled antibody enzyme method. However, much of the sensitivity would be lost if these complexes were not cyclic. A preliminary indication of stability of any antigen–antibody complex can be obtained even prior to establishment of its structure by the degree of excess of antigen required for its preparation. Only if the excess antigen added is moderate (such as 6 times equivalence, as in PAP) can one expect a stable complex of fairly uniform composition. Ferritin–antiferritin (FAF) complexes have been successfully used in electron microscopy by Marucci *et al.* (1974). Undiluted FAF was used, apparently because higher dilutions would lead to dissociation of this less stable, but otherwise useful, complex.

It has often been thought that complexes smaller than PAP would be preferable for preembedding-staining electron microscopy. Fab fragments of antiperoxidase would react in antigen excess with one molecule of peroxidase. However, these complexes may again be unstable because they are not cyclic and because Fab fragments may have lower affinity for antigen than whole antibody.

It may be tempting to conjugate peroxidase with nonspecific immunoglobulin by chemical means and use the resulting artificial complexes instead of PAP. Such an approach should be carried out with caution, for it is doubtful that the resulting complex would be very specific. As discussed above, anti-IgG contains components, presumably antibodies, that react nonspecifically with tissue. These nonspecific antibodies do not react with PAP, but they would react with artificial complexes of peroxidase and immunoglobulin, because the immunoglobulin used for immunization in the preparation of antiimmunoglobulin is of impurity comparable to that of the nonspecific immunoglobulin used for preparation of the artificial complex.

4. Immunocytochemical Staining of Nervous Tissue

Large serial sections are important in tracing nerve pathways at the light-microscope level. The easiest way to obtain such sections is by paraffin embedding. Frozen sections have long been the preferred mode of preparing tissue for light-microscopic immunocytochemistry because of best antigenic preservation. Optimal antigenic preservation is a necessity for relatively insensitive methods. However, the thickness of frozen sections makes it difficult to discriminate individual neurons and their projections. Although paraffin sections can be used in any light-microscopic immunocytochemical method (for a review, see Sternberger, 1974), it is only with the unlabeled antibody enzyme method that they have become the preferred mode (Halmi and Duello, 1976; Burns, 1975b). The reason probably stems from the sensitivity of the unlabeled antibody enzyme method. Although perhaps most antigenic sites become destroyed during preparation of tissue for paraffin embedding (fixation and dehydration), the number of sites remaining is sufficient for immunocytochemical localization by a sensitive method. It is becoming increasingly evident that the more sensitive an immunocytochemical method is, the less particular one has to be about tissue preparation. Blocks embedded in paraffin after routine fixation in formalin have been useful for immunocytochemical studies after years of storage (Halmi and Duello, 1976). Even stored sections stained with hematoxylin–eosin or with aldehyde–thionin–PAS–orange G (subjected to permanganate oxidation prior to initial staining) could be used after destaining. The wealth of material that has thus become available for immunocytochemical study has enabled Halmi and Duello (1976) to reappraise the classification of "acidophil" pituitary adenomas entirely on the basis of hormonal properties. Similarly, an immunocytochemical classification of gliomas of previously unknown nature has been accomplished by Deck et al. (1976).

Nerve tracts can be delineated by retrograde transport of injected peroxidase and histochemical visualization of the enzyme with diaminobenzidine and hydrogen peroxide (Nauta et al., 1974). Frozen sections are needed, because fixation and embedding destroys the enzymatic activity of peroxidase. However, Vacca et al. (1975) were able to increase sensitivity and range of applicability of the method by using paraffin sections and localizing the peroxidase as an antigen with antiperoxidase and the reagents of the unlabeled antibody enzyme method.

Since antibodies and even antibody fragments do not readily penetrate cellular and subcellular membranes, staining after embedding, i.e., directly on the ultrathin section, appears to be the most desirable mode of electron-microscopic immunocytochemistry. Staining after embedding has not been very successful prior to development of the unlabeled antibody enzyme method. Apparently, the sensitivity of older methods was insufficient to permit localization of antigens when most of their antigenic determinants had been destroyed as a result of embedding. When only few antigenic determinants persist, the effective concentration of antibodies available for staining is reduced.

Postembedding staining is usually done after fixation in aldehyde and embedding in conventional embedding media. To avoid nonspecific binding with the plastic of the section, it has been found necessary to pretreat the section with a protein that cannot be detected by the reagents of the unlabeled antibody enzyme method (Hardy *et al.*, 1970; Sternberger, 1972; Sternberger and Petrali, 1975). Postembedding staining has been used for the electron-microscopic localization of neuropeptides (Leclerc and Pelletier, 1974; Pelletier *et al.*, 1974*a,b*, 1975*a,b;* Silverman and Zimmerman, 1975; Goldsmith and Ganong, 1975; Goldsmith *et al.*, 1975; Castel and Hochman, 1976*b*), pituitary hormones (Moriarty and Halmi, 1972; Moriarty *et al.*, 1975; Parsons and Erlandsen, 1974; Doerr-Schott, 1974; Doerr-Schott and DuBois, 1973; Erlandsen *et al.*, 1975; Beauvillain *et al.*, 1975), steroid hormones (Gardner, 1975), insulin and glucagon (Erlandsen *et al.*, 1975), enzymes (Erlandsen *et al.*, 1974; Inoue *et al.*, 1976), bacterial (Short and Walker, 1975; Hardy *et al.*, 1970) and viral antigens (Wenkelschafer-Crabb *et al.*, 1976), and hormone receptors (Sternberger and Petrali, 1975; Sternberger *et al.*, 1976) and carriers (Pelletier *et al.*, 1974*b;* Silverman and Zimmerman, 1975). One of the disadvantages of postembedding staining is the need in most cases of aldehyde rather than osmium fixation. Osmium fixation would be necessary for preservation of membranes. The absence or poor preservation of membranes in aldehyde fixation gives the cell a ''washed-out'' appearance and makes it difficult to localize membrane-bound antigens. Also, some antigens seem to be entirely destroyed during embedding, and can therefore not be localized at all by postembedding staining. Postembedding staining is important in quantitative immunocytochemistry, for it permits comparison of experimental and semiadjacent control sections (Sternberger, 1974).

Preembedding staining has been difficult because even the fixed cell is penetrated by antibody fragments, but sluggishly. For years, it had been thought that size of the immunocytochemical reagents was the primary factor that determined penetration. Therefore, a search for ever smaller fragments as immunocytochemical reagents has been fostered (Kraehenbuhl *et al.*, 1971). Recently, however, excellent preembedding staining has been achieved with vibrotome sections by the technique of Pickel *et al.* (1975*b*), using the unlabeled antibody enzyme method. That a large PAP complex (420,000 daltons) was useful at all in preembedding staining suggests that size of reagents is not as critical a factor for preembedding staining as sensitivity of the method. Nevertheless, there is still a gradient in the intensity of preembedding staining, with areas close to the surface of the tissue showing better staining than deeper ones (Pickel *et al.*, 1975*b;* Joh *et al.*, *1975;* Rufener *et al.*, 1975).

Besides enabling identification of antigens that do not withstand plastic embedding, the preembedding procedure of Pickel *et al.* (1975*b*) offers the advantage of better structural preservation, thus making it possible to identify antigens that are bound on membranes or associated with microtubules. Another advantage, especially important in brain research, is the possibility of selecting an immunocytochemically stained area by light microscopy prior to processing it by plastic embedding for electron microscopy. Selection of specific brain areas in unstained material would be necessary for the postembedding-staining procedure.

5. Neurotransmitter Pathways—Catecholamines and Serotonin

Catecholamines can be converted to fluorescent substances by reaction with formaldehyde (Falck *et al.*, 1962). Maps of catecholamine fibers have been constructed by fluorescence microscopy (Ungerstedt, 1971). With specialized spectroscopy, noradrenergic fibers can be differentiated from dopaminergic fibers, but often the distinction appears ambiguous. Another approach involves immunocytochemistry of enzymes that are rate-limiting in the synthesis of catecholamines. Immunofluorescence has been successful when special precautions are taken to minimize background and to maximize sensitivity. To reduce nonspecific background, it has been essential in the work of Hartman (1973) to use frozen sections, employ specifically purified fluorescein-labeled antibody, and incorporate Triton X100 into the cytochemical reagents and wash solutions.

Paraffin and polyethylene glycol sections for light microscopy and vibrotome sections for electron microscopy have been used by Pickel and colleagues in specific mapping of dopaminergic, noradrenergic, and serotonergic neurons by the unlabeled antibody enzyme method (Pickel *et al.*, 1975*a,b*, 1976*a,b*; Joh *et al.*, 1975). Tyrosine hydroxylase is rate-limiting in the synthesis of norepinephrine and dopamine. Immunocytochemical staining with antibodies to tyrosine hydroxylase in the substantia nigra and in the nucleus locus coeruleus reveals intense staining in the axons and cytoplasm of cell bodies of dopaminergic neurons, while in noradrenergic neurons, only the cell bodies and the most proximal parts of cell processes are stained. This differential staining permits distinction of both neurotransmitters. Perikaryon staining is less intense in noradrenergic than in dopaminergic neurons. Furthermore, antiserum to dopamine-β-hydroxylase, an enzyme characteristic to the biosynthetic pathway of noradrenaline, localizes only noradrenergic neurons, but not dopaminergic neurons. Finally, antibodies to tryptophan hydroxylase, the rate-limiting enzyme in the synthesis of serotonin, localize only serotonergic neurons (Figs. 2–5).

Although it also has been possible to localize tyrosine hydroxylase in catecholamine-containing neurons by immunofluorescence or by the peroxidase-conjugated antibody method, the enzyme was less clearly cytoplasmic, the nerve processes were less evident, and, in the case of immunofluorescence, localization was obscured by nonspecific staining (Pickel *et al.*, 1975*a*).

The differing intensity of reaction of antiserum to tyrosine hydroxylase in

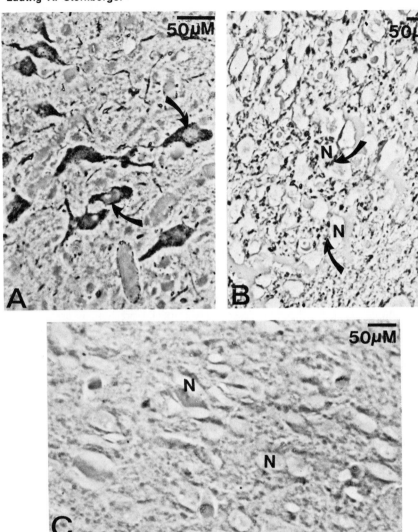

FIG. 2. Localization of tyrosine and tryptophan hydroxylase in the nucleus locus coeruleus of the rat brain. Unlabeled antibody enzyme method. (A) Phase-contrast photomicrograph of section incubated with tyrosine hydroxylase antiserum. Reaction product is present in the cytoplasm of the neuronal perikarya (arrows). (B) Phase-contrast photomicrograph of section incubated with tryptophan hydroxylase. Reaction product is in the processes (arrows) surrounding unstained neurons (N). (C) Phase contrast photomicrograph of a section incubated with normal rabbit serum replacing specific antiserum in staining procedure. No reaction product is present in either neurons (N) or surrounding neuropil. From the work of Virginia Pickel, Tong Joh, and Donald Reis.

noradrenergic and dopaminergic neurons is probably due to the presence of an aggregated form of tyrosine hydroxylase in noradrenergic neurons and a more disaggregated form in dopaminergic neurons (Joh and Reis, 1974; Pickel *et al.*,

1975*b*). The lower-molecular-weight form has been shown to react with higher affinity with antityrosine hydroxylase than did the higher-molecular-weight form.

Using vibrotome sections, stained by the unlabeled antibody enzyme method, postfixed in osmium tetroxide and embedded in Epon, Pickel *et al.* (1975*b*) were able to differentiate the manner of localization of tyrosine hydroxylase in cell bodies and processes. In the cell bodies, the enzyme was found on the membranes of

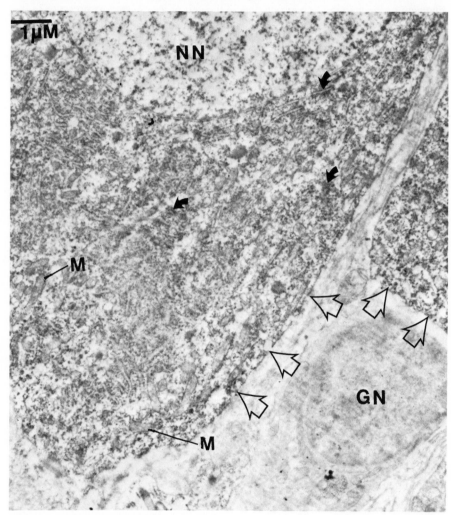

FIG. 3. Electron-microscopic immunocytochemical localization of tyrosine hydroxylase in neurons of the nucleus locus coeruleus by the unlabeled antibody enzyme method. Open arrows mark the outer boundaries of two neurons that have selective cytoplasmic staining for the enzyme. The reaction product is especially intense along membranes of the endoplasmic reticulum (solid arrows) and absent from mitochondria (M). The neuronal nucleus (NN) has a granular labeling not seen in glial nuclei (GN). This granular staining of the nucleus is also present in control sections. From the work of Virginia Pickel, Tong Joh, and Donald Reis.

the endoplasmic reticulum and the Golgi apparatus, as well as in more diffuse sites in the cytoplasm. Mitochondria and lysozomes were unstained (see Fig. 3). In the cell processes, the enzyme was confined to association with the neurotubules. Tyrosine hydroxylase was also seen in nerve terminals in the caudate, but only when the vibrotome sections had been pretreated with Triton (Pickel *et al.,* 1976*a*). Myelinated axons, glial cytoplasm, or nuclei were unstained. The confinement of the enzyme to neurotubules in the cell processes and its presence in the cytoplasm of the cell bodies may explain biochemical studies in which most of the enzyme isolated from neuronal terminals was bound to particulates, while that from cell bodies was soluble. The presence of the enzyme in the neurotubules suggests that it is rapidly transported from cell bodies to nerve terminals, and provides an explanation for the finding that most neurotransmitter norepinephrine is synthesized at the nerve terminal and rapidly replenished in this location on depletion.

With antibody to tryptophan hydroxylase, the rate-limiting enzyme in the synthesis of serotonin, Joh *et al.* (1975) were able to describe a similar ultrastructural distribution specific to serotonergic neurons (see Fig. 4).

Interestingly, dopamine-β-hydroxylase, which is not a rate-limiting enzyme, has been found only in the cytoplasm of noradrenergic neurons, and not in neurotubules (Pickel *et al.,* 1976*b*). The finding supports the conclusion of isolation studies, in which dopamine-β-hydroxylase has been found a readily soluble enzyme.

Staining of serial sections with antiserum to tyrosine, dopamine, and tryptophane hydroxylases, or for that matter with antiserum to neuropeptides, facilitates, among other things, an identification of the functionality of interneuronal synapses (Pickel *et al.,* 1976*a*). In the nucleus locus coeruleus, tyrosine hydroxylase was localized by light microscopy in the perikarya and the very proximal portions of nerve processes. Thus, noradrenergic neurons were identified (see Fig. 2). In adjacent sections, tryptophan hydroxylase was localized in a network of beaded fibers, presumably representing nerve terminals and preterminal axons of serotonergic fibers. By electron microscopy, tyrosine hydroxylase was seen in the cytoplasm with preferential association to the endoplasmic reticulum (see Fig. 3). Tryptophan hydroxylase, on the other hand, was restricted to unmyelinated neuronal processes, 0.1–1.4 μm in diameter, with the majority in the 0.3–0.8 μm range (see Fig. 4). Within these processes, most of the enzyme was associated with structures of 20–80 nm cross-sectional diameter. An analysis of the frequency distribution of these structures stained for tryptophan hydroxylase showed that they were of the appropriate size for the small, clear vesicles (40–60 nm) and the large vesicles reported for serotonergic terminals. The localization of tryptophan hydroxylase along microtubules was best seen in longitudinal section of what appear to be medium-sized particles in cross section (see Fig. 5). The fibers stained with antitryptophan hydroxylase were similar in size and distribution to the unstained axons and axon terminals, which in adjacent sections processed with antityrosine hydroxylase were associated with dendrites containing tyrosine hydroxylase. The catecholamine neurons in the nucleus locus coeruleus, therefore, appear to be innervated by serotonergic neurons.

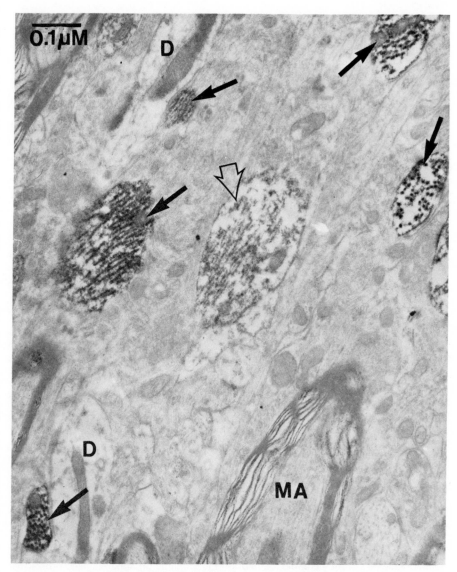

FIG. 4. Low-magnification electron micrograph of tryptophan hydroxylase immunocytochemically localized to neuronal processes within the nucleus locus coeruleus by the unlabeled antibody enzyme method. The labeled processes (solid-arrows) show considerable variability in both size and intensity of reaction. The large process (open arrow) lightly labeled for the enzyme may represent either a less extensively stained fiber or one that is at a greater depth in the section. Numerous unlabeled dendrites (D) are present in the neuropil adjacent to the unlabeled processes. Myelinated axons (MA) were always unlabeled. From the work of Virginia Pickel, Tong Joh and Donald Reis.

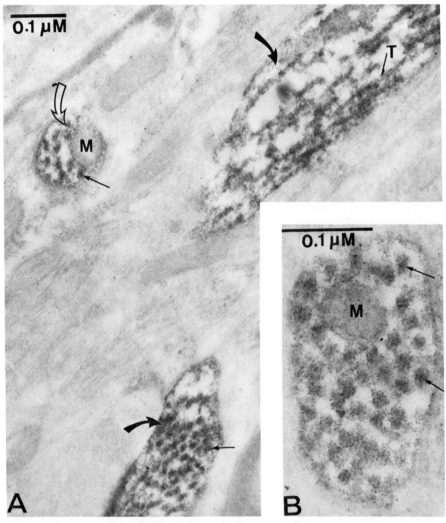

FIG. 5. Localization of tryptophan hydroxylase in processes in the nucleus locus coeruleus. Unlabeled antibody enzyme method. (A) Electron micrograph of labeled processes, including cross sections of small axon (open arrow) and longitudinal sections (solid arrows). Specific enzyme reaction product is evident in association with microtubules (T) and other subcellular organelles (small arrows). Mitochondria (M) were unlabeled. (B) Higher-magnification electron micrograph illustrating the labeled subcellular organelles (arrows) that are covered with small granules resembling PAP complexes. These organelles are 30–60 nm in cross-sectional diameter. From the work of Virginia Pickel, Tong Joh, and Donald Reis.

6. Hypothalamopituitary Pathways

6.1. Oxytocin, Vasopressin, and Angiotensin

Even though antiserum to Lys[8]-vasopressin has been available for many years (for a review, see Robinson and Frantz, 1973), it was not until the work of Leclerc and Pelletier (1974) that this nonapeptide had become localized immunocytochemically. Pelletier *et al.* (1975*b*) feel that the immunocytochemical detection of a molecule as small as a nonapeptide requires the availability of a sensitive technique, such as provided by the unlabeled antibody enzyme method. Neurophysins I and II are carriers for the nonapeptides oxytocin and vasopressin, respectively. Sokol *et al.* (1976) have demonstrated by quantitative immunocytochemistry that in the Brattleboro homozygous rat (a diabetes insipidus strain lacking vasopressin), total neurophysin in the neural lobe of the pituitary was decreased to 33% of normal, and in the heterozygous rat, to 75% of normal, presumably reflecting the absence and diminution, respectively, of neurophysin II, while neurophysin I was present at normal levels.

In the supraoptic and paraventricular nuclei, a dark and light cell population was distinguished in normal rats on the basis of immunocytochemical staining with antineurophysins (I and II) (Sokol *et al.*, 1976). The dark cells were more concentrated in the peripheral wings of the paraventricular nuclei and in the dorsal region of the supraoptic nucleus. In the Brattleboro rat, only the dark cells were stained with antineurophysin; the light cells remained unstained. Antioxytocin also stained the dark cells, but the light cells remained unstained. This staining was absorbed with oxytocin, but not with vasopressin. Antivasopressin gave weak staining in a few perikarya, but this staining was absorbed with oxytocin and not with vasopressin, suggesting that it was due to antioxytocin contaminating the antiserum to vasopressin. These data show that there exists a cell population that produces only vasopressin and its associated neurophysin (neurophysin II), and that this population is absent in the Brattleboro rat. Neurophysin from Brattleboro rats, therefore, can be used for absorption of antisera to nonpurified neurophysins, thereby making them specific for immunocytochemical localization of neurophysin II.

Even at the electron-microscopic level, there was complete absence of material reacting with anti-Lys[8]-vasopressin in the entire hypothalamic–hypophysial system of the Brattleboro rat (Leclerc and Pelletier, 1974; Pelletier *et al.*, 1975*b*).

Watkins (1975) also found that in the diabetes insipidus rat, less than half the cells in the paraventricular and supraoptic nuclei were stained for neurophysin. These cells presumably secreted neurophysin I and oxytocin. However, on staining with anti-Lys[8]-vasopressin, there still was some reaction in both nuclei. The finding was attributed to the sensitivity of the unlabeled antibody enzyme method, which is capable of detecting perhaps even minimal traces of antigen. However, Sokol *et al.* (1976) believe it could have been due to contamination of antivasopressin with antibodies cross-reactive with oxytocin.

In normal rats, all cells in the supraoptic nucleus and two-thirds of cells in the paraventricular nucleus react with antineurophysin (I plus II) (Sokol *et al.*, 1976). Half the stained cells were "dark" and half were "light." The dark cells corresponded in location to the only stained cells in the Brattleboro rat. Antioxytocin reacted only with the dark cells. Antivasopressin, however, reacted with both dark and light cells. Absorption with vasopressin reduced the staining, but absorption with oxytocin had no effect. It was concluded that there exist at least some cells that produce vasopressin as well as oxytocin (Sokol *et al.*, 1976). This conclusion may especially hold for the dark cells. However, the absence of any staining in the light cells of the Brattleboro rat suggests that in normal rats there exist many cells that produce vasopressin without producing oxytocin.

Vandesande and Dierickx (1975) were the first to stain two antigens in the same section by the unlabeled antibody enzyme method. Sections were first reacted with antioxytocin, anti-IgG, and PAP, and a brown color was obtained by using diaminobenzidine as electron-donor with hydrogen peroxide. Acidification and treatment with 10% dimethylformamide removed the immunoreagents, but not the polymerized enzyme reaction product, and repetition of staining, with antivasopressin, anti-IgG, and PAP and 4-Cl-1-naphthol as electron-donor with hydrogen peroxide, yielded a blue reaction product. Oxytocin and vasopressin were found in distinct perikarya in the magnocellular system of the rat; no cells producing both peptides have been detected. Serial sections, alternately stained for neurophysin I and II, confirmed and extended the finding to the bovine hypothalamus (DeMay *et al.*, 1975*a*; Vandesande *et al.*, 1975). Thus, the question whether there exist cells that can produce both hormones (Sokol *et al.*, 1976) still remains unresolved. The question is important, for it may challenge the widely cherished "one cell–one hormone," theory as well as the "single messenger" theory (see also Section 7). Unfortunately, the otherwise challenging double-staining technique of Vandesande and Dierickx (1975) suffers from the limitation that the brown reaction must precede the blue reaction, for the reaction product of 1-Cl-4-naphthol is not as insoluble as that of diaminobenzidine. Further, it has been found preferable to stain for oxytocin in the first sequence and for vasopressin in the second. Reversing the order of primary antisera and of color reactions would appear to be desirable for a fully controlled study.

Also by electron microscopy, cells secreting oxytocin can be distinquished from those secreting vasopressin (Leclerc and Pelletier, 1974; Pelletier *et al.*, 1975*c*). Whether antibodies to neurophysins or vasopressin or oxytocin are used, immunocytochemistry demonstrates these peptides and their carriers in the magnocellular neurosecretory hypothalamohypophysial system (Fig. 6) (Leclerc and Pelletier, 1974; Pelletier *et al.*, 1974*b*, 1975*c*; Vandesande *et al.*, 1974; Watkins, 1975; Zimmerman *et al.*, 1974; Kozlowski *et al.*, 1976*b*). Staining was intense in the perikarya of the supraoptic and paraventricular nuclei, and could be traced in coarsely staining nerve fibers that converged in the zona interna of the median eminence and terminated in the neural lobe. With their double-staining technique, Vandesande and Dierickx (1975) were able to distinguish vasopressin fibers entering the neural lobe of the rat as a medioventrally located tapering bundle from rostrodorsal parts mainly containing oxytocin. Mixed fibers were absent.

In addition, Vandesande *et al.* (1974) and Zimmerman (1976) observed a few stained perikarya in the suprachiasmic nuclei of rat and mouse. Finely stained fibers that possess small varicosities were also seen in the external zone of the median eminence of the rat (Vandesande *et al.,* 1974). These fibers were less prominent than the coarse fibers in the internal zone. They had a preferential orientation perpendicular to those in the internal zone, and terminated around the capillaries of the hypothalamic–pituitary portal circulation. At least in the rat, they appeared to originate in the suprachiasmic nuclei. That these fibers are part of a pathway separate from the supraoptic–paraventricular–posterior pituitary pathway has been demonstrated immunocytochemically by their increase after bilateral adrenalectomy and their disappearance after colchicine treatment. These procedures did not affect fibers in the internal zone. Adrenalectomy also increased the number of stained nerve fibers and perikarya in the suprachiasmic nucleus, while colchicine treatment decreased the number of stained fibers (it slightly increased the number of stained perikarya).

In the bovine hypothalamus, perikaryon staining is absent from the suprachiasmic nucleus, but is found in a few cells within the arcuate (bovine infundibular) nucleus (DeMey *et al.,* 1975b). Only neurophysin I, not neurophysin II, vasopressin, or oxytocin, could be detected.

Silverman (1975a,b) was unable to detect neurophysin in any parvicellular nucleus of the guinea pig, including the suprachiasmic, even after intensive colchicine treatment. However, strongly staining neurons, seemingly extending the magnocellular formations, were seen along the paraventricular fiber tract between the supraoptic and paraventricular nucleus and also associated with the paraventricular nucleus trailing all the way to the level of the median eminence. Neurophysin was absent from the suprachiasmic nucleus in monkey and human (Zimmerman, 1976).

In the developing guinea pig, immunocytochemically detectable neurophysin appears in the magnocellular perikarya and the external zone of the median eminence at about the same time, perhaps suggesting that in this species, neurophysin terminals on the portal vasculature originate in magnocellular nuclei (Silverman, 1975b).

By electron microscopy, neurophysin (using antisera reactive with neurophysin I and II) is seen in all the large secretory granules (larger than 130 nm) in axons, Herring bodies, and nerve terminals (Fig. 7) (Pelletier *et al.,* 1974b; Silverman and Zimmerman, 1975). Endothelial cells, pituicytes, and adjacent pars intermedia are negative. In the internal zone of the median eminence, in the axons that contain large neurosecretory granules, all the granules are stained (contain PAP molecules). Occasional expansions of the nerve fibers in this location, commonly known as ''Herring bodies,'' are also full of neurophysin-containing large secretory granules. In the supraoptic and paraventricular nuclei (Pelletier *et al.,* 1974b), most secretory neurons contain neurophysin in large secretory granules.

Neurophysins are also seen in the smaller secretory granules (90–100 nm) that belong to axons distinct from those possessing large secretory granules (Silverman and Zimmerman, 1975). Unlike axons with large granules that all stain

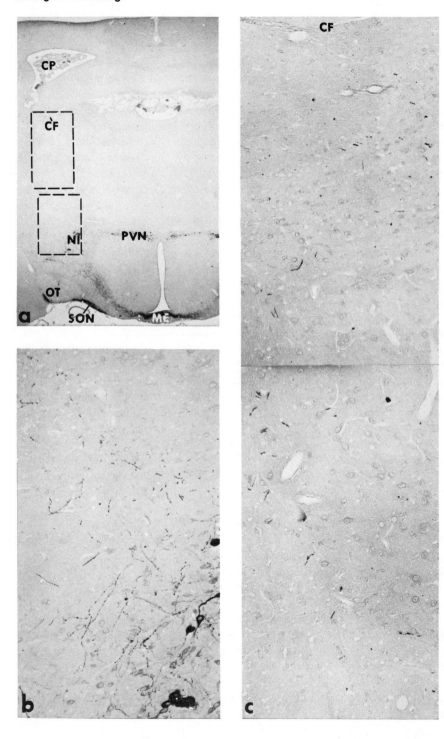

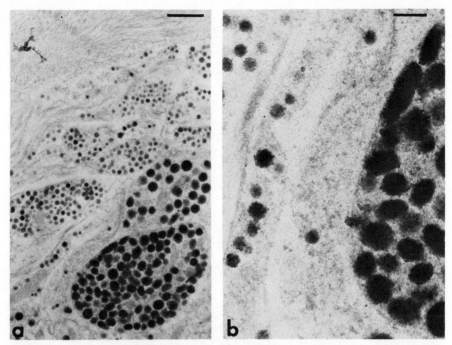

FIG. 7. Zona interna of guinea pig median eminence. Freeze substitution. Araldite section stained postembedding for neurophysin. Calibration bar in (a) is 0.59 μm; in (b), 0.18 μm. Axons with large and small granules stained and bearing PAP molecules. PAP also present in axoplasm near large granules. From Silverman and Zimmerman (1975).

strongly, the axons with small granules represent a heterogeneous population, of which only occasional ones bear neurophysin. Small granules are seen in the zona interna and externa of the median eminence, but not in the neurohypophysis. In the zona externa, they are seen in nerve endings close to or abutting on portal capillaries.

Vasopressin parallels in its entirety the distribution of neurophysin in the small secretory granules in the zona interna and externa of the median eminence (Fig. 8) (Silverman and Zimmerman, 1975). In the posterior pituitary, granules

---

FIG. 6. Rat hypothalamus stained with antineurophysin II and the unlabeled antibody enzyme method. (a) Low magnification of cross section showing neurophysin in neurons of the paraventricular nucleus (PVN), nucleus intermedius (NI), and supraoptic nucleus (SON). At this magnification, the majority of neurons from the PVN and NI are seen coursing ventromedially toward the median eminence (ME) as the supraopticohypophysial tract. However, at higher magnifications of areas outlined for (b) and (c), some neurophysin-containing fibers course dorsolaterally toward the choroidal fissure (CF) of the lateral ventricle. The choroid plexus (CP) of the lateral ventricle originates from the choroidal fissure. (OT) Optic tract. (b) Higher magnification of area of nucleus intermedius as outlined in (a). Neurophysin fibers originating from this area begin to course dorsally. (c) Higher magnification of area outlined in (a). Neurophysin fibers continue to course dorsally toward the area of the choroidal fissure (CF). From the work of Gerald Kozlowski.

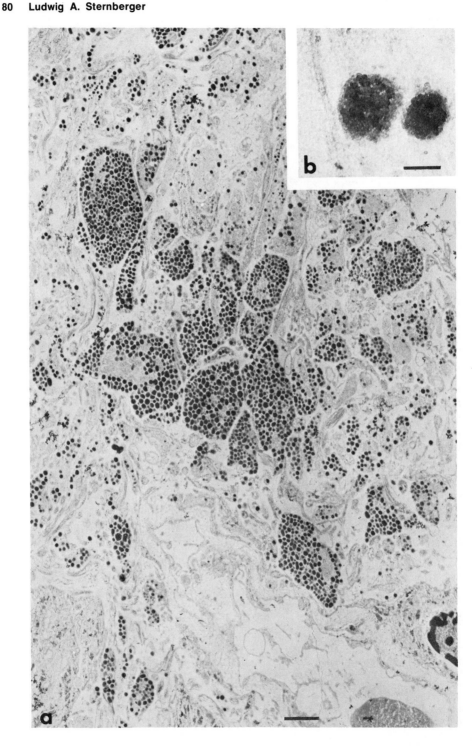

of all (Leclerc and Pelletier, 1974) or many (Silverman and Zimmerman, 1975) axons contain PAP molecules when stained for vasopressin.

The presence of neurophysin- and vasopressin-containing nerve endings in the external zone explains the finding of neurophysin by radioimmunoassay in the hypophysial portal blood (Zimmerman et al., 1973). Castel and Hochman (1976b) have recently demonstrated PAP molecules in the pars intermedia of the pituitary when staining with antivasopressin. Unlike the stain in the neurohypophysis and the median eminence, the pars intermedia staining was not absorbed with vasopressin. It may reflect a receptor for vasopressin in the pars intermedia (see Section 7). A relationship between corticotropin and vasopressin has long been postulated (see Section 6.2). This secretory pathway into the adenohypophysis suggests that the adenohypophysis may participate with the neurohypophysis in water and salt control.

The mammalian brain contains an endogenous renin–angiotensin system (Fischer-Ferraro et al., 1971). The neuropeptide angiotensin II causes release of antidiuretic hormone (Severs et al., 1970). Recently, Changaris et al. (1976) were able to demonstrate angiotensin I in the cytoplasm of pars intermedia cells, thus describing immunocytochemically a conceivable receptor site for this peptide.

In secretory neurons of the supraoptic and paraventricular nuclei, the secretory granules that are close to the Gogli apparatus, as well as those in the axonal processes, stain for neurophysins (Pelletier et al., 1974b). However, only the axonal granules stain for vasopressin. The findings suggest that the final step of formation of vasopressin (like that of other neurotransmitters, as mentioned in Section 5) is completed under normal conditions only during migration to the nerve terminals. Therefore, immunocytochemical localization of neurophysins (I or II) or other carriers of peptide hormones may be a more reliable indication for mapping potentially peptidergic neurons than localization of oxytocin or vasopressin itself, just as immunocytochemical localization of catecholamine- or serotonin-synthesizing enzymes may be a more reliable indicator for mapping of these fibers than the immunocytochemical or fluorescence staining of the neurotransmitters themselves.

With the electron microscope, Castel and Hochman (1976a,b) compared the distribution of vasopressin in xerotic species, such as the Dead Sea desert mouse, and a species of wild mouse with a mesoic species (the laboratory mouse). In the laboratory mouse, only a few granules in the magnocellular perikarya were stained, while in the desert mouse and wild mouse, stained granules were abundant. Apparently, in the xerotic species, which produce more vasopressin, assembly is started more proximally to the cell bodies than in the mesoic species. The limiting concentration of antivasopressin for perikaryon staining was 1:50, while that for staining in the median eminence and neurohypophysis was 1:2500. This

FIG. 8. (a) Posterior pituitary. Freeze substitution. Araldite section stained postembedding for vasopressin. Stain is over axons and nerve terminals. Calibration bar is 1.2 μm. (b) Higher magnification of (a). PAP molecules are only over the secretion granules. Calibration bar is 91 nm. From Silverman and Zimmerman (1975).

difference suggests a gradient in vasopressin concentration that increases with distance from cell body to nerve terminal.

It has commonly been observed in immunocytochemistry that specifically stained secretory granules are surrounded by a few PAP molecules in the adjacent cytoplasm. The scattering of discrete PAP molecules suggests that each represent only a single antigen site, and thus only a small amount of antigen compared with that found in the secretory granules. Until the work of Silverman and Zimmerman (1975), it has not been possible to decide whether the extragranular localization was real or the result of a diffusion artifact from tissue processing. However, in their work on neurophysin and vasopressin, Silverman and Zimmerman found cytoplasmic neurophysin only in the vicinity of the large granules, not in the vicinity of the small ones (see Fig. 7). This specificity of axoplasmic localization argues against a diffusion artifact. Furthermore, only neurophysin, not vasopressin (see Fig. 8), was found in the axoplasm in the posterior pituitary or median eminence, even though the smaller vasopressin is expected to diffuse more readily than the larger neurophysin.

The extragranular neurophysin may be of different physiological significance than the intragranular neurophysin. Neurophysin is present in the neurohypophysis of the Brattleboro rat (Zimmerman et al., 1975), even though the nerve terminals in these animals lack neurosecretory granules (Kalimo and Rinne, 1972).

Castel and Hochman (1976a,b) indeed found vasopressin in the extragranular cytoplasm in axonal dilations in the median eminence and in the neurohypophysis in the desert mouse. Apparently, in this xerotic animal, in which vasopressin is produced abundantly, the extragranular neurophysin described by Silverman and Zimmerman (1975) is utilized to carry vasopressin. Castel and Hochman (1976b) suggest that such extragranular vasopressin in nerve endings and Herring bodies mediates a tonic release, which is in contrast to a possibly more sporadic release that may accompany exocytosis.

Most immunocytochemical studies with the unlabeled antibody enzyme method employ conventional aldehyde fixation and conventional embedding media prior to immunocytochemical staining. Silverman and Zimmerman (1975), in their studies on the neurophysin–vasopressin secretory pathways, used a new freeze substitution method (Silverman et al., 1975) instead of fixation. This method of tissue preparation may well be superior in preservation of antigenic reactivity and avoidance of diffusion (Goldsmith and Ganong, 1975), and may even further increase the sensitivity of the technique by permitting use of still higher dilutions of primary antisera.

6.2. Luteinizing Hormone–Releasing Hormone, Somatostatin, and Corticotropin-Releasing Factor

The most anterior hypothalamic location in which the decapeptide luteinizing hormone–releasing hormone (LH–RH) has been immunocytochemically traced is the nervous tissue alongside blood capillaries of the organum vasculosum of the lamina terminalis (King and Gerall, 1976; King et al., 1974; Baker et al., 1975).

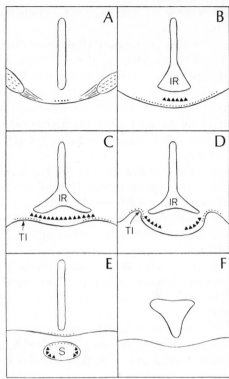

FIG. 9. Diagrammatic distribution of LH–RH fibers (circles) and somatostatin fibers (triangles) in coronal sections of the rat hypothalamus, rostral (A) to caudal (F). From King *et al.* (1975).

The finding provides the first clue to the possible function of this structure. Throughout the median eminence, LH–RH fibers are found in the external layer. At the most rostral end, they are clustered medially and pass ventral to the large anterior arcuate neurons (Fig. 9A). Traversing caudally, they soon spread to occupy the entire width of the external zone of the median eminence (Fig. 9B). The number of staining fibers appears to increase in the process. More caudally, the fibers disappear from the midline, and at the level at which the tuberoinfundibular sulci (which mark the lateral boundary of the median eminence) become deeper, LH–RH fibers cap these sulci in the palisade zone (Fig. 9C,D). As the tuberoinfundibular sulci deepen and as the lateral cephalic extensions of the pars tuberalis become more prominent, and when eventually stalk separation has occurred (Fig. 9E), the fibers appear to terminate near vascular elements. No LH–RH fibers are seen, at least in the rat, at the floor of the caudal portion of the mammillary recess, in the mammillary body, or in the pars nervosa or in the stalk caudal to the level of stalk separation (Fig. 9F). Interestingly, gonadotropic cells are found in specially high concentration in the extensions of the anterior pituitary into the tuberoinfundibular sulci.

The continuous increase in number of stained fibers caudally in successive coronal sections suggests that in the rat, LH–RH is found in long fibers of the preopticoinfundibular pathway, which are joined as they pass caudally by the short fibers of the tuberoinfundibular pathway that originate in the arcuate–ventromedial nuclei.

Within the fibers, PAP molecules are deposited on neurosecretory granules, 75–90 nm in diameter, grouped within neuronal processes of the palisade zone (Pelletier *et al.*, 1974c; Goldsmith and Ganong, 1975; Kozlowski *et al.*, 1976a). About 10–20% of nerve endings are stained. In every stained nerve ending, all granules contain LH–RH. This finding suggests that many of the fibers possess other functions than LH–RH neurotransmission. It decreases the likelihood, but does not exclude the possibility, that LH–RH fibers do not secrete anything but LH–RH.

In normal rats, no LH–RH has been detected in perikarya by specific immunocytochemical means. Perhaps the absence of LH–RH from perikarya is a reflection of nonribosomal synthesis, of derivation of LH–RH from an immunologically inactive precursor, of rapid transport along the axons, or of assembly of the peptide simultaneous with its transport from cell bodies toward nerve endings, so that a gradient in peptide concentration results. The distribution pattern of LH–RH seems to follow similar principles as that of vasopressin (Silverman and Zimmerman, 1975b; Castel and Hochman, 1976b) and of catecholamines and serotonin (Pickel *et al.*, 1975a,b; 1976a,b) discussed in Sections 5 and 6.1.

Interestingly, in the mouse, strong perikaryal staining was observed in the arcuate neurons by Hoffman *et al.* (1976). Also in the mouse, these investigators found, in addition to the main pathway in the external zone of the median eminence, less pronounced pathways in the internal zone. The finding is complementary to the localization of vasopressin and oxytocin, in which the bulk of coarse fibers is found in the internal zone, and a lesser amount of fine fibers in the external zone.

Treatment with monosodium glutamate resulted in loss of 80–90% of neurons in the arcuate nucleus (Hoffman *et al.*, 1976). This depletion did not decrease the number of fibers or perikarya stained for LH–RH. In fact, in the internal zone, the number of stained fibers or their intensity had increased. The results suggest that the deficit in reproductive capacity resulting from monosodium glutamate treatment is not due to damage to arcuate neurons, and that the increase in staining observed may be compensatory to damage elsewhere in the hypothalamo–pituitary–gonadal axis.

In the guinea pig, cell bodies were seen by immunofluorescence in the anterior and basal hypothalamus, and in some extrahypothalamic sites (Barry *et al.*, 1974). Extending these studies with the unlabeled antibody enzyme method, Silverman (1976) found LH–RH in the medial basal hypothalamus in cell bodies of the arcuate nucleus. In the anterior hypothalamus, LH–RH was seen in cell bodies of the bed nucleus of the stria terminalis, and of the medial preoptic, the anterior hypothalamic, and suprachiasmic nuclei. Only in the arcuate nucleus was the number of cells increased in castrates receiving colchicine. Surgical isolation of the medial basal hypothalamus caused only a slight decrease in immunocytochemical staining in parts of the median eminence, suggesting that most of the LH–RH that reaches the portal circulation originates in the arcuate nucleus.

The clear staining of perikarya permitted Silverman (1976) to define by immunocytochemical means the direction of neurotransmission in fibers. Six pathways (I–IV and two minor tracts) were described. Tract III, which originates in the arcuate nucleus and to some extent in the ventromedial nucleus, is the only pathway that terminates in the median eminence. Many of its fibers send off collaterals as they travel ventrad, some of which synapse with neurons devoid of LH–RH in the arcuate nucleus. Others travel toward the third ventricle. The other five pathways are discussed in Section 8.

In developing male and female rats, fibers staining for LH–RH are present at birth in the arcuate median eminence region (King and Gerall, 1976). The staining diminishes to a minimum on the 5th postnatal day. In the female, this diminution is followed by a rapid increase to reach adult-level staining by the 9th day. In the male, the number of stained fibers increases only slightly in this period.

The distribution of fibers carrying the quadridecapeptide somatostatin is roughly analogous to that of LH–RH (Pelletier et al., 1975b,c; King et al., 1975; Dube et al., 1975). Most of the fibers traverse the median eminence through the external zone. On occasion, however, some immunocytochemical reaction was also seen in the internal zone. The organum vasculosum of the lamina terminalis was found to contain somatostatin in addition to LH–RH. However, somatostatin was also traced in the subcommissural organ and the pineal gland (Pelletier et al., 1975b). Despite this superficial resemblance to LH–RH, somatostatin fibers were found to be anatomically distinct groups, rather than being interlaced with LH–RH fibers. Somatostatin fibers are caudal, dorsal, and medial to LH–RH fibers (King et al., 1975). Thus, in successive coronal sections, they make their appearance later than LH–RH fibers (see Fig. 9A) and are medial to them throughout the preinfundibular region of the median eminence (see Figs. 9B–D). Somatostatin fibers are more predominant in the postinfundibular median eminence and the distal stalk than LH–RH fibers (see Fig. 9E).

On staining with antisomatostatin, PAP molecules localize over the secretory granules of the nerve endings in the median eminence (Pelletier et al., 1974a). Somatostatin granules are larger (90–100 nm) than LH–RH granules. About 30% of nerve endings in the external zone stain for somatostatin. Apparently, both LH–RH and somatostatin are released from nerve endings directly into the fenestrated capillaries of the primary portal plexus before transport to their adenohypophysial sites of action.

Somatostatin is also found peripherally in the small intestine and the endocrine pancreas. Rufener et al. (1975) and Pelletier et al. (1975a) identified by immunocytochemistry the D-cell of the endocrine pancreas as the specific somatostatin-containing cell.

Let us return our attention to the neurophysin-containing fibers described by Vandesande et al. (1974) and Castel and Hochman (1976a,b) in the zona externa of the median eminence. Watkins et al. (1974) found that the increase in staining of these fibers observed after adrenalectomy can be further enhanced by administration of deoxycorticosterone acetate 2–3 weeks later. It was concluded from these observations that a neurophysin was stimulated that acts as a carrier for corticotropin-releasing factor (CRF) in a region close to or interspaced with LH–RH

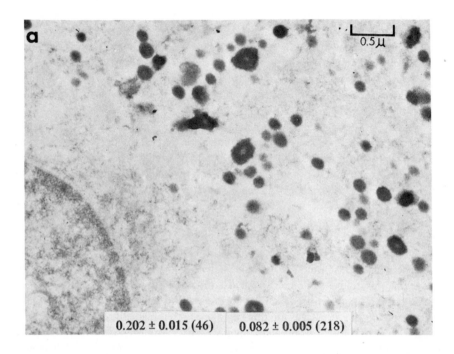

0.202 ± 0.015 (46)   0.082 ± 0.005 (218)

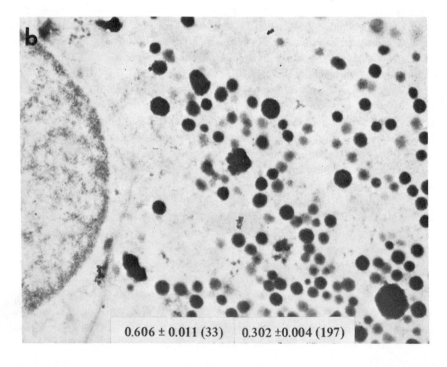

0.606 ± 0.011 (33)   0.302 ±0.004 (197)

and somatostatin terminals. Alternatively, vasopressin acts directly on receptors in the pars intermedia (Castel and Hochman, 1976*b*) (see Section 6.2).

7. Neuropeptide Receptors

It is of interest to localize not only the pathway of secretion of neuropeptides, but also their sites of action. Immunocytochemistry can accomplish this localization with antibodies to neuropeptides only if the reaction of the peptide with its receptor does not sterically hinder the peptide from reacting with the antibodies. Complete hindrance apparently does not occur, however, for antisera to LH–RH yielded a fair degree of localization in the secretion granules of gonadotropic cells (Fig. 10a). No other site in the pituitary became stained. Immunoabsorption would be the classic test for demonstrating whether this reaction was due to antibody to LH–RH held in the gonadotropic granules or to some other antibody conceivably formed in the various antisera to LH–RH used. Admixture of LH–RH with these antisera should suppress the reaction if it were due to antibody to LH–RH. However, even when an exceedingly large excess of LH–RH was used, the reaction became enhanced, rather than suppressed. This result suggested that the LH–RH reacted with the tissue, rather than with the antibody. Therefore, LH–RH was applied separately to the section, followed by washing and then by anti-LH–RH and the reagents of the unlabeled antibody enzyme method. This sequence led to strong enhancement of localization of LH–RH (Fig. 10b), suggesting reaction of LH–RH with a receptor in the gonadotropic secretion granules (Sternberger and Petrali, 1975). On quantitation, the enhancement was found to be as much as 23-fold over the reactivity without LH–RH pretreatment. The affinity of reaction appeared to be high, for as little as 4 pg LH–RH/μl gave significant enhancement of reactivity ($P < 0.005$).

The specificity of the reaction was demonstrated by the use of analogues, by solid-phase immunoabsorption, and with purified antibody to LH–RH. Des-Gly[1]-LH–RH, a peptide unable to react with LH–RH receptor, had no effect on the immunocytochemical reaction. Des-amide[10]-LH–RH, which is capable of reacting with receptor, inhibited the reaction. Apparently, only the *C*-terminal region of LH–RH is available for reaction with antibodies after binding with receptor, the remainder of the molecule being sterically hindered. Even though, on radioimmunoassay, apparently only 0.2% of anti-LH–RH reacts with the *C*-terminal amino acids of the peptide, this amount of antibody is sufficient for immunocytochemical reaction. In fact, anti-LH–RH gave significant ($P < 0.001$) staining of receptor sites to dilutions of 1:240,000.

FIG. 10. Gonadotropic granules from male rat pituitary semiadjacent araldite sections treated with Tris albumin (a) and LH–RH (b) prior to staining with anti-LH–RH and the unlabeled antibody enzyme method. Numbers at bottom of figures are average optical-density indices of secretion granules ± S.E.M. (number of granules measured in parentheses). The first index in each figure is that of the large granules (200–700 nm in diameter); the second index, that of the small granules (100–150 nm in diameter).

Pretreatment of sections with corticotropin or prolactin had no effect on reactivity with anti-LH–RH.

The data show that the enhancement reaction is LH–RH-specific. They demonstrate, therefore, that gonadotropic granules contain a free receptor for LH–RH, i.e., a receptor possessing a binding site for LH–RH that can react with LH–RH *in vitro* and be assayed immunocytochemically. But what is the nature of the weaker, gonadotropic granule–specific reaction that occurs on sections not pretreated with LH–RH? The elucidation of this reaction requires absorption of antisera, but at first this elucidation was difficult, because admixture of LH–RH with antisera results in enhancement rather than absorption of reactivity. We therefore prepared an immunoabsorbant by linking LH–RH onto cyanogen bromide–activated Sepharose via a ribonuclease spacer, and with this immunoabsorbant absorbed early and late bleedings of antisera produced by immunization with LH–RH-azo-peroxidase and LH–RH-azo-bovine serum albumin (Sternberger *et al.*, 1976). On absorption, all immunocytochemical reactivity had become abolished, with or without LH–RH pretreatment. This finding indicates that in the antisera, all the immunocytochemically reacting antibodies were specific to LH–RH. In addition, it shows that the reaction in gonadotropic granules not pretreated with LH–RH is also due to anti-LH–RH. It permits the conclusion that the gonadotropic granules contain not only a free receptor capable of reacting with LH–RH *in vitro,* but also a stable LH–RH–receptor complex derived from binding of endogenous LH–RH.

One of the most puzzling observations has been the effect of glycoprotein hormone pretreatment on staining with unabsorbed anti-LH–RH. Apparently, pretreatment with rat, human, and rabbit FSH, rat and human LH, and rat LH α-chains enhances a weak reaction of gonadotropic granules with anti-LH–RH, but inhibits a strong reaction. At first, two factors were postulated for this apparently diametrically opposing effect (Sternberger and Petrali, 1974), but more recent experiments suggested a single explanation. The evidence is based on immunoabsorbance and on specifically purified antibodies. Since immunoabsorption removed the glycoprotein enhancement effect, it must be related to LH–RH-specific antibodies. Purified anti-LH–RH was obtained in low yield, and purified LH–RH–anti-LH–RH complex in high yield, by absorption of early bleeding antiserum to LH–RH-azo-peroxidase onto LH–RH–ribonuclease–Sepharose and elution by acid in the absence or presence, respectively, of excess LH–RH. The reaction with purified antibody became enhanced by pretreatment with α-chains, while the presence of LH–RH inhibited any effect of α-chain or glycoprotein hormone treatment.

Since a stable LH–RH–receptor complex has been demonstrated in the gonadotropic granules, it is assumed that the glycoprotein hormones may be contaminated with traces of degradation products of this complex. Conceivably, the receptor portion of the LH–RH in this complex (the N-terminal end of the peptide) is less degraded than the free end of LH–RH (the C-terminal portion), which is the only portion that participates in the immunocytochemical detection of LH–RH–receptor complex. Consequently, the degradation product may have affinity for receptor equal to that of unaltered LH–RH, but may have lower immunoreactivity. If reaction of antiserum to LH–RH with gonadotropic granules is strong without pretreatment with LH–RH, i.e., if the receptor is largely occupied by endogenous LH–RH, addition of the degradation product competitively displaces some of the bound LH–RH and replaces it with a product of weaker immunoreactivity. The reaction intensity decreases. On the other hand, if the reaction of gonadotropic granules with anti-LH–RH in the absence of LH–RH pretreatment is weak, i.e., if most of the receptor is free, addition of the degradation product will lead to binding of weekly immunoreactive material without displacing any traces of LH–RH already there, and staining intensity increases.

Since membranes are absent in aldehyde-fixed and Araldite-embedded tissue, preembedding staining was used to demonstrate membrane-bound receptor (Sternberger and Petrali, 1974). The vibrotome technique of Pickel et al. (1975b) was used, and LH–RH receptor was demonstrated on the plasma membrane of the gonadotropic cell. However, pretreatment of the vibrotome section with LH–RH was necessary to reveal the staining. Thus, it appears that under the conditions of the experiment, the cell membranes contain only free receptor for LH–RH, while the secretion granules possess free and bound receptor.

The dual location of LH–RH receptor might be explained by the existence of two different kinds of receptors, each with its own function. However, if in fact there existed only one kind of receptor, the location at the cell membrane may be primary and that in the secretion granules the result of receptor translocation. The dual location of LH–RH receptor may not be too surprising in view of the five actions of the peptide: release and synthesis of LH as well as of FSH in the same cell (Shiino et al., 1972; Vilchez-Martinez et al., 1975; Moriarty et al., 1976) and stimulation of additional LH–RH receptor (Cooper et al., 1975). Edwardson and Filbert (1975) showed that hemipituitaries stimulated with constant bursts of LH–RH every ½ hr will secrete increasing amounts of LH with each successive stimulus, and this increase is abolished by cycloheximide. These authors suggested (personal communications) that perhaps a connection can be found between their observation and the dual location of LH–RH receptor. Conceivably, the formation of stable complexes of LH–RH receptor in the secretion granules may be of physiological significance.

The intracellular location of LH–RH receptor as demonstrated immunocytochemically is not unique to this neuropeptide. Indeed, previous findings by isolation techniques and autoradiography from the laboratory of Tixier-Vidal (Gourdji et al., 1973; Brunet et al., 1974) provide an antecedent observation of the intracellular location of thyrotropin-releasing hormone (TRH) receptor.

8. Nonpituitary Neurosecretory Peptide Pathways

Punch biopsy and regional dissection radioimmunoassay (Brownstein et al., 1975a,b; 1976) are important techniques that have predated and guided most immunocytochemical observations on neurosecretory pathways of hypothalamic peptides. At present, it seems that these techniques may also form a guide to the immunocytochemical search for pathways of neuropeptides beyond hypothala-mopituitary tracts. Somatostatin has already been described as possessing a wide distribution in the brain. Immunocytochemistry has not revealed a similar distribution so far. It may be possible that different forms of neurohormones are detected by radioimmunoassay and immunocytochemistry. Radioimmunoassay would be more effective in detecting free neuropeptides than any bound form. At least in the case of LH–RH, most of the free LH–RH is lost during tissue processing (Goldstein and Ganong, 1975). It is feasible that immunocytochemistry detects mainly a bound form of the peptides. The bound form could conceivably be attached to a still unknown carrier. If this were indeed the case, then antibodies to neuropeptides could never detect a wider distribution than antibodies to the carrier, for antibodies to the carrier would detect both bound and potentially bound peptide sites, but antibodies to the peptide would detect only bound sites. This may relate to the finding by Silverman and Zimmerman (1975) of neurophysin in juxtagranular cytoplasm as well as in secretion granules, while vasopressin was seen only in the granules.

Of the six LH–RH pathways discovered by Silverman (1976) in the guinea pig, only one is a hypothalamopituitary tract (tract III; see Section 6.2). Tracts I and II originate in the hypothalamus, but the cells of origin of tract IV and the two minor tracts have not been found.

Tract I originates mainly in the precommissural region at the level of the bed nucleus of the stria terminalis and the medial preoptic nucleus, and travels mainly ventrad along the precommissural side to terminate largely in the preoptic portion of the suprachiasmic nucleus. Tract II originates in the same area as tract I, but passes on the postcommissural side of the anterior commissure to terminate in the retrochiasmic portion of the suprachiasmic and paraventricular nucleus. Tract IV arises at the level of the mammillary bodies and proceeds caudally along the base of the brain to become indistinct at the most caudal portion of the interpenduncular nucleus. Minor tracts traverse the diagonal band of Broca into the organum vasculosum of the lamina terminalis. Other isolated fibers were seen in the posterior hypothalamus.

Pfaff (1973) and Moss and McCann (1975) have shown that LH–RH induces mating behavior in estrogen-primed rats even if they have been ovariectomized as well as hypophysectomized. A direct central action of LH–RH was suggested. Two distinct areas of sensitivity to estrogen stimulation of lordosis behavior exist in the guinea pig (Morin and Feder, 1974). One is in the arcuate nucleus, and may thus affect Silverman tract III. The other is in the preoptic anterior hypothalamic area, and may be related anatomically to tract I. Progesterone implanted in the posterior mammillary nucleus inhibits lordosis in response to estrogen injection. Tract IV may represent the anatomical correlate (Silverman, 1976).

Kozlowski et al. (1976b) has shown that in dehydrated rats, the ependymal cells of the choroid plexus of the lateral ventricles undergo changes similar to those undergone by cells of the renal collecting tubules. Therefore, choroid plexus was regarded as a target tissue for vasopressin. Indeed, at least two projections of extrahypothalamic neurophysin-containing fibers were observed immunocytochemically (See Fig. 6). Neurophysin-containing axons originating from the paraventricular and intermediate nuclei course dorsolaterad to enter the choroidal fissure of the superior horn. Neurophysin-containing axons originating from the supraoptic nucleus are diverted caudally and coarse ventrolaterad to the choroidal fissure of the anterior horn.

Any extrahypothalamic pathways for neuropeptides are in all likelihood not as active as the main secretory pathways under normal conditions, for otherwise they would have been detected by immunocytochemistry with equal ease. It is more likely that such pathways are not as frequently invoked as the acetylcholine, catecholamine, serotonin, and GABA pathways, and that perhaps they exert mainly a modulating influence on other pathways, conceivably forming secretory terminals in only fair approximation to them. These characteristics would lead to slower and more prolonged effects, in contrast to true synapses that engage in rapid transmission. Indirect support is lent to such a hypothesis by Brown and Vale (1975), who have shown that TRH increases the LD_{50} of nembutal, while somatostatin increases that of strychnine. Somatostatin alone, however, is not a depressant, and is indeed relatively free of clinical side effects. Apparently, neuro-hormone peptides are invoked only as a result of a physiological imbalance, such as nembutal or strychnine poisoning. Such a mechanism would resemble the feed-back commonly observed with peripheral hormones. Thus, neuropeptides could become pharmacological antagonists that are effective only when called for, but are relatively free of side effects in the absence of agonists. Since these neuropep-tides are not secreted all the time, we do not expect them to be localized easily by immunocytochemistry. However, we may expect a carrier for them to exist in spe-cific axons. Staining for carrier could conceivably be carried out in the same man-ner as staining for LH–RH receptor in the pituitary. Carrier could be revealed if deparaffinized sections were treated with neuropeptide prior to staining with an-tiserum to the peptide and the unlabeled antibody enzyme method.

In addition to their location in neurons and their projections, both neurophy-sin and LH–RH were on occasion seen in tanycytes of the median eminence in rat, mouse, and monkey (Zimmerman, 1976), but not in the guinea pig (Silverman, 1976). Several neuropeptides have been detected in the CSF by radioimmunoas-say. If tanycytes partake in channeling solutes between CSF and blood capillaries, the irregular detection of CSF contents in tanycytes by sensitive immunocy-tochemical technique may depend on physiological conditions yet to be discov-ered.

9. References

Bahr, G.F., Mikel, U., and Klein, G., 1975, Localization and quantitation of EBV-associated nuclear antigen (EBNA) in Raji cells, *Beitr. Pathol.* **155:**72–78.

Baker, B.L., Dermody, W.C., and Reel, J.R., 1975, Distribution of gonadotropin-releasing hormone in rat brain as observed with immunocytochemistry, *Endocrinology* **97:**125–135.

Barry, J., Dubois, M.P., and Carette, B., 1974, Immunofluorescence study of the preoptico-infundibular LRF neurosecretory pathway in the normal, castrated or testosterone-treated male guinea pig, *Endocrinology* **95:**1416–1423.

Beauvillain, J.C., Tramu, G., and Dubois, M.P., 1975, Characterization by different techniques of adrenal corticotropin- and gonadotropin-producing cells in lerot pituitary (*Eliomys quercinus*). A superimposition technique and an immunocytochemical technique, *Cell Tissue Res.* **153:**301–317.

Böcker, W., 1974, Use of triple layer enzyme method as an alternative to immunofluorescence for the detection of tissue antigens, *Beitr. Pathol.* **153:**410–414.

Brown, M., and Vale, W., 1975, Central nervous system effects of hypothalamic peptides, *Endocrinology* **96:**1333–1336.

Brownstein, M., Arimura, A., Sato, H., Schally, A.V., and Kizer, J.S., 1975a, The regional distribution of somatostatin in the rat brain, *Endocrinology* **96:**1456–1461.

Brownstein, M.J., Palkovits, M., and Kizer, J.S., 1975b, On the origin of luteinizing hormone–releasing hormone (LH–RH) in the supraoptic crest, *Life Sci.* **17:**679–682.

Brownstein, M.J., Arimura, A., Schally, A.B., Palkovits, M., and Kizer, J.S., 1976, The effect of surgical isolation of the hypothalamus on its luteinizing hormone–releasing hormone content, *Endocrinology* **98:**662–665.

Brunet, N., Gourdji, D., Tixier-Vidal, A., Pradelles, P., Morgat, J.L., and Fromegeot, P., 1974, Chemical evidence for associated TRF with subcellular fractions of intact rat prolactin cells (GH3) with ³H-labeled TRF, *Eur. J. Biochem.* **38:**129–133.

Burns, J., 1975a, Background staining and sensitivity of the unlabeled antibody enzyme (PAP) method. Comparison with peroxidase-labeled antibody sandwich method using formalin-fixed paraffin-embedded material, *Histochemistry* **43:**291–294.

Burns, J., 1975b, Immunoperoxidase localization of hepatitis B antigen (HB) in formalin–paraffin processed liver tissue, *Histochemistry* **44:**133–135.

Castel, M., and Hochman, J., 1976a, Immunohistochemistry of the hypothalamic neurohypophysial system in the common spiny mouse, *Acemys cahirius,* 7th International Symposium on Neurosecretion, Leningrad.

Castel, M., and Hochman, J., 1976b, Ultrastructural immunohistochemical localization of vasopressin in the hypothalamic–neurohypophysial system of three murids, *Cell Tissue Res.* in press.

Changaris, D.G., Demers, L.M., Keil, L.C. and Severs, L.M., 1976, Identification of angiotensin-I in rat and human pituitary gland, *Exp. Neurol.* **51:**699–704.

Cooper, K.J., Fawcett, C.P., and McCann, S.M., 1975, Augmentation of pituitary responsiveness to luteinizing hormone/follicle stimulating hormone–releasing factor (LH–RF) as a result of acute ovariectomy in the 4-day cycling rat, *Endocrinology* **96:**1123–1129.

Davidson, J.M., 1969, Feedback regulation of gonadotropic secretion, in: *Frontiers in Neuroendocrinology* (W.F. Ganong and L. Martini, eds.), pp. 343–388, Oxford University Press, New York.

Deck, J.H., Eng, L.F.and Bigbee, J., 1976, A preliminary study of glioma morphology using the peroxidase–antiperoxidase immunohistochemical method for glial fibrillar acidic protein, *J. Neuropath. Exp. Neurol.* **35:**362.

DeMey, J., Dierickx, K., and Vandesande, F., 1975a, Immunohistochemical demonstration of neurophysin I– and neurophysin II–containing nerve fibers in the external zone of the bovine median eminence, *Cell Tissue Res.* **157:**517–519.

DeMey, J., Dierickx, K., and Vandesande, F., 1975b, Identification of neurophysin-producing cells. III. Immunohistochemical demonstration of neurophysin I–producing neurons in the bovine infundibular nucleus, *Cell Tissue Res.* **161:**219–224.

Doerr-Schott, J., 1974, Localization submicroscopique par cyto-immunoenzymologie de differents principes hormonaux de l'hypophyse de *Rana tempora* L., *J. Micros.* **20:**151–164.

Doerr-Schott, J., and Dubois, M.P., 1973, Mise en evidence des hormones de l'hypophyse d'un Amphibien par la cyto-immuno-enzymologie au microscopie electronique, *C. R. Acad. Sci. Paris Ser. D* **256:**2179–2182.

Dougherty, R.M., Marucci, A.A., and DiStefano, H.S., 1972, Application of immunocytochemistry to the study of avian leukosis virus, *J. Gen. Virol.* **15:**149–162.

Dube, D., Leclerc, R., Pelletier, G., Arimura, A., and Schally, A. V., 1975, Immunohistochemical detection of growth hormone–release inhibiting hormone (somatostatin) in the guinea pig brain, *Cell Tissue Res.* **161:**385–392.

Edwardson, J.A., and Filbert, D., 1975, Sensitivity of a self-potentiating effect of luteinizing hormone-releasing hormone to cycloheximide, *Nature London* **225:**71.

Elias, J.M., and Miller, M., 1975, A comparison of the unlabeled enzyme method with immunofluorescence for the evaluation of human immunologic renal disease, *Amer. J. Clin. Pathol.* **64:**464–471.

Erlandsen, S.L., Parsons, J.A., and Taylor, T.D., 1974, Ultrastructural immunocytochemical localization of lysozyme in the Paneth cell of man, *J. Histochem. Cytochem.* **22:**516–541.

Erlandsen, S.L., Parsons, J.A., Burke, J.P., Redick., J.A., Van Orden, D.E., and Van Orden, L.S., 1975, A modification of the unlabeled antibody enzyme method using heterologous antisera for the light microscopic localization of insulin, glucagon and growth hormone, *J. Histochem. Cytochem.* **23:**666–677.

Falck, B., Hillarp, N.-A., Thieme, G., and Thorp, A., 1962, Fluorescence of catecholamine and related compounds condensed with formaldehyde, *J. Histochem. Cytochem.* **10:**348–354.

Fischer-Ferraro, C., Nahmed, V.E., Goldstein, D.J., and Finkleman, S., 1971, Angiotensin and renin in rat and dog brain, *J. Exp. Med.* **133:**353–361.

Gardner, P.J., 1975, Immunocytochemical localization of steroids in the testis; a preliminary study, Amer. Assn. Anatomists. 8th Ann. Session, pp. 354–360.

Goldsmith, P.C., and Ganong, W.F., 1975, Ultrastructural localization of luteinizing hormone–releasing hormone in the median eminence of the rat, *Brain Res.* **97:**181–193.

Goldsmith, P.C., Rose, J.C., Arimura, A., and Ganong, W.F., 1975, Ultrastructural localization of somatostatin in pancreatic islets of the rat, *Endocrinology* **97:**1061–1064.

Gourdji, D., Tixier-Vidal, A., Morin, A., Pradelles, P., Morgat, J.L., Fromegeot, P., and Kerdelhue, B., 1973, Binding of tritiated thyrotropin-releasing factor to a prolactin-secreting clonal cell line (GH3), *Exp. Cell Res.* **82:**39–46.

Halmi, N.S., and Duello, T., 1976, "Acidophil" pituitary tumors. A re-appraisal with differential staining and immunocytochemistry, *Arch. Pathol.* **100:**346–351.

Hardy, P.H., Petrali, J.P., and Sternberger, L.A., 1970, Postembedding staining for electron microscopy by the unlabeled antibody enzyme method, *J. Histochem. Cytochem.* **18:**678.

Hartman, B. K., 1973, Immunofluorescence of dopamine-β-hydroxylase: Application of improved methodology to the localization of the peripheral and central noradrenergic nervous system, *J. Histochem. Cytochem.* **21:**312–332.

Hoffman, G.E., Knigge, K.M., and Sladek, J.R., 1976, Luteinizing hormone–releasing hormone (LH–RH) localization in normal and monosodium glutamate treated mice, 9th Annual Winter Conference on Brain Research, p. 52.

Inoue, K., Tice, L., and Creveling, C.R., Immunohistochemical localization of catecholamine-*O*-methyltransferase, *J. Histochem. Cytochem.* in press.

Joh, T.H., and Reis, D.J., 1974, Different forms of tyrosine hydroxylase in noradrenergic and dopamenergic systems in brain, *Fed. Proc. Fed. Amer. Soc. Exp. Biol.* **33:**535.

Joh, T.H., Shikimi, T., Pickel, V.M., and Reis, D.J., 1975, Brain tryptophane hydroxylase: Purification of, production of antibodies to, and cellular and ultrastructural localization in serotonergic neurons of rat brain, *Proc. Nat. Acad. Sci. U.S.A.* **72:**3575–3579.

Kalimo, N., and Rinne, U.K., 1972, Ultrastructural studies on the hypothalamic neurosecretory neurons of the rat. II. The hypothalamoneurohypophysial system in rats with hereditary hypothalamic diabetes insipidus, *Z. Zellforsh.* **134:**205–225.

King, J.C., and Gerall, A.A., 1976, Localization of luteinizing hormone–releasing hormone, *J. Histochem. Cytochem.* **24:**829–845.

King, J.C., Parsons, J.A., Erlandsen, S.L., and Williams, T.H., 1974, Luteinizing hormone–releasing hormone (LH–RH). Pathway of the rat hypothalamus revealed by the unlabeled antibody peroxidase–antiperoxidase method, *Cell Tissue Res.* **153:**211–217.

King, J.C., Arimura, A., Gerall, A.H., Fishback, J.B., and Elkind, K.E., 1975, Growth hormone–release inhibiting hormone (GH–RIH) pathway of the rat hypothalamus revealed by the unlabeled antibody peroxidase–antiperoxidase method, *Cell Tissue Res.* **160:**423–430.

Kozlowski, G.P., Frank, S., McNeill, T.H., Scott, D.E., Hsu, K.C., and Zimmerman, E.A., 1976a, Light and electron microscopic localization of gonadotropin releasing hormone (Gn-RH) in rat median eminence, in *Immunoenzymatic Techniques*, p. 329, North Holland Publishing Co., Amsterdam.

Kozlowski, G.P., Brownfield, M.S., and Schultz, W.J., 1976b, Neurosecretory pathways to the choroid plexus, *Int. Res. Com. Sys.* **4**:299.

Kraehenbuhl, J.P., DeGrandi, P.B., and Campichi, M.A., 1971, Ultrastructural localization of intracellular antigens using enzyme-labeled antibody conjugates, *J. Cell Biol.* **50**:432–445.

Leclerc, R., and Pelletier, G., 1974, Electron microscopic immunohistochemical localization of vasopressin in hypothalamus and neurohypophysis of the normal and Brattleboro rat, *Amer. J. Anat.* **140**:583–588.

Marucci, A.A., DiStefano, H.S., and Dougherty, R.M., 1974, Preparation and use of soluble ferritin–antiferritin complexes as specific markers for immunoelectron microscopy, *J. Histochem. Cytochem.* **22**:35–39.

Marucci, A.A., and Dougherty, R.M., 1975, Use of unlabeled antibody immunohistochemical technique for the detection of human antibody, *J. Histochem. Cytochem.* **23**:618–623.

Mason, T.E., Phifer, R.F., Spicer, S.S., Swallow, R.A., and Dreskin, R.B., 1969, An immunoglobulin–enzyme bridge method for localizing tissue antigens, *J. Histochem. Cytochem.* **17**:563–569.

Moriarty, G.C., and Halmi, N.S., 1972, Electron microscopic localization of the adrenal corticotropin–producing cell with the use of unlabeled antibody and the peroxidase–antiperoxidase complex, *J. Histochem. Cytochem.* **10**:590–603.

Moriarty, G.C., Halmi, N.S., and Moriarty, C.M., 1975, The effect of stress on the cytology and immunocytochemistry of pars intermedia cells in the rat pituitary, *Endocrinology* **96**:1426–1436.

Moriarty, G., Garner, L., Gessman, R., and Tobin, R.B., 1976, Immunocytochemical studies of FSH and LH gonadotrophs in the cycling female rat, *Endocrine Society Program, 58th Annual Meeting*, p. 225, Endocrine Society, Bethesda, Maryland.

Morin, L.P., and Feder, N.H., 1974, Intracranial estradiol benzoate implants and lordosis behavior in ovariectomized guinea pigs, *Brain Res.* **70**:95–102.

Moss, R.L., and McCann, S.M., 1975, Action of luteinizing hormone–releasing factor (LH–RF) in the initiation of lordosis behavior in the estrone-primed ovariectomized female rat, *Neuroendocrinology* **17**:309–318.

Nauta, H.J.W., Printz, M.B., and Lasek, R.J., 1974, Afferents to the rat caudoputamen studied with horseradish peroxidase. An evaluation of a retrograde neuroanatomical research method, *Brain Res.* **76**:219–238.

Palmer, P.E., DeLellis, R.A., and Wolfe, H.J., 1974, Immunohistochemistry of liver in α-antitrypsin deficiency. A comparative study, *Amer. J. Clin. Pathol.* **62**:350–354.

Parsons, J.A., and Erlandsen, S.L., 1974, Ultrastructural immunocytochemical localization of prolactin in rat anterior pituitary by use of the unlabeled antibody enzyme method, *J. Histochem. Cytochem.* **22**:340–351.

Pfaff, D.W., 1973, Luteinizing hormone–releasing factor potentiates lordosis behavior in hypophysectomized ovariectomized female rats, *Science* **182**:1148–1149.

Pelletier, G., Labrie, F., Arimura, A., and Schally, A.V., 1974a, Electron microscopic immunohistochemical localization of growth hormone release inhibiting hormone (somatostatin) in the rat median eminence, *Amer. J. Anat.* **140**:445–450.

Pelletier, G., Leclerc, R., Labrie, F., and Puviani, R., 1974b, Electron microscopic immunohistochemical localization of neurophysin in the rat hypothalamus and pituitary, *Mol. Cell. Endocrinol.* **1**:157–166.

Pelletier, G., Labrie, F., Puviani, R., Arimura, A., and Shally, A.V., 1974c, Immunohistochemical localization of luteinizing hormone–releasing hormone in the rat median eminence, *Endocrinology* **95**:314–317.

Pelletier, G., Leclerc, R., Arimura, A., and Schally, A.V., 1975a, Immunohistochemical localization of somatostatin in the rat pancreas, *J. Histochem. Cytochem.* **23**:699–701.

Pelletier, G., Leclerc, R., Dubé, D., Labrie, F., Puviani, R., Arimura, A., and Schally, A.V., 1975b, Localization of growth hormone release–inhibiting hormone (somatostatin) in the rat brain, Amer. J. Anat. 142:397–401.

Pelletier, G., Leclerc, R., and Puviani, R., 1975c, Localization ultrastructurale d'hormones hypothamiques, Union Med. Can. 104:355–362.

Petrali, J.P., Hinton, D.M., Moriarty, G.C., and Sternberger, L.A., 1974, The unlabeled antibody enzyme method of immunocytochemistry. Quantitative comparison of sensitivities with and without peroxidase–antiperoxidase complex, J. Histochem Cytochem. 22:782–801.

Pickel, V.M., Joh, T.H., Field, P.M., Becker, C.G., and Reis, D.J., 1975a, Cellular localization of tyrosine hydroxylase by immunohistochemistry, J. Histochem. Cytochem. 23:1–12.

Pickel, V.M., Joh, T.H., and Reis, D.J., 1975b, Ultrastructural localization of tyrosine hydroxylase in noradrenergic neurons of brain, Proc. Nat. Acad. Sci. U.S.A. 72:659–663.

Pickel, V.M., Joh, T.H., and Reis, D.J., 1976a, Serotonergic innervation of noradrenergic neurons in nucleus locus coeruleus by immunocytochemical localization of transmitter specific enzymes, Brain Res. in press.

Pickel, V.M., Joh, T.H., and Reis, D.J., 1976b, Monamine-synthesizing enzymes in central dopaminergic, noradrenergic and serotonergic neurons, J. Histochem. Cytochem. 24:792–806.

Robinson, A.G., and Frantz, A.G., 1973, Radioimmunoassay of posterior pituitary peptides: A review, Metabolism 22:1047–1057.

Rufener, C., Amherdt, M., Dubois, M.P., and Orci, L., 1975, Ultrastructural immunocytochemical localization of somatostatin in rat pancreas monolayer cultures, J. Histochem. Cytochem. 23:866–869.

Schönberg, D.K., 1975, Dynamics of hypothalamo-pituitary function during puberty, Clin. Endocrinol. Metab. 4: 57–88.

Severs, W.B., Summy-Long, J., Taylor, J.S., and Connor, J.D., 1970, A central affect of angiotensin: Release of pituitary pressor material, J. Pharmacol. Exp. Ther. 174:27–34.

Shiino, M., Arimura, A., Schally, A.V., and Rennels, B.G., 1972, Ultrastructural observations of granule extrusion from rat anterior pituitary cell after injection of luteinizing hormone–releasing hormone, Z. Zellforsh. 128:152–161.

Shin, S.H., and Howitt, C., 1975, Effect of castration on luteinizing hormone and luteinizing hormone–releasing hormone in the male rat, J. Endocrinol. 65:447, 448.

Short, J.A., and Walker, P.D., 1975, The location of bacterial antigens on sections of Bacillus cereus by use of soluble peroxidase–antiperoxidase complex and unlabeled antibody, J. Gen. Microbiol. 89:93–101.

Silverman, A.J., 1975a, The hypothalamic magnocellular neurosecretory system of the guinea pig. I. Immunohistochemical localization of neurophysin in the adult, Amer. J. Anat. 144:433–443.

Silverman, A.J., 1975b, The hypothalamic magnocellular neurosecretory system of the guinea pig. II. Immunohistochemical localization of neurophysin and vasopressin in the foetus, Amer. J. Anat. 144:445–459.

Silverman, A.J., 1976, Distribution of luteinizing hormone–releasing hormone (LH–RH) in guinea pig brain, Endocrinology 99:30–41.

Silverman, A.J., and Zimmerman, E.A., 1975, Ultrastructural immunocytochemical localization of neurophysin and vasopressin in the median eminence and posterial pituitary of the guinea pig, Cell Tissue Res. 159:291–301.

Silverman, A.J., Knigge, K.M., and Zimmerman, E.A., 1975, Ultrastructural immunocytochemical localization of neurophysin in freeze-substituted neurohypophysis, Amer. J. Anat. 142:265–271.

Smith, E.R., and Davidson, J.M., 1974, Location of feedback receptors: Effects of intracranially implanted steroids on the plasma LH and LRF response, Endocrinology 95:1566–1573.

Sokol, H.W., Zimmerman, E.A., Sawyer, W.H., and Robinson, A.G., 1976, The hypothalamoneurohypophysial system of the rat: Localization and quantitation of neurophysin by light microscopic immunocytochemistry in normal rats and in Brattleboro rats deficient in vasopressin and neurophysin, Endocrinology in press.

Sternberger, L.A., 1969, Some new developments in immunocytochemistry, *Mikroscopie* **25**:346–361.

Sternberger, L.A., 1972, The unlabeled antibody peroxidase and the quantitative immunouranium methods in light and electron immunohistochemistry, *Tech. Biochem. Biophys. Morphol.* **1**:67–88.

Sternberger, L.A., 1974, *Immunocytochemistry*, Prentice-Hall, Englewood Cliffs, New Jersey.

Sternberger, L.A., and Petrali, J.P., 1974, Hormone receptors: Light and electron immunocytochemical localization of the target cell receptors for adrenocorticotropin (ACTH) and luteinizing hormone–releasing hormone (LH–RH), *J. Histochem. Cytochem.* **22**:296–297.

Sternberger, L.A., and Petrali, J.P., 1975, Quantitative immunocytochemistry of pituitary receptors for luteinizing hormone–releasing hormone, *Cell Tissue Res.* **162**:141–176.

Sternberger, L.A., Hardy, P.H., Jr., Cuculis, J.J., and Meyer, H.G., 1970, The unlabeled antibody enzyme method of immunohistochemistry. Preparation and properties of soluble antigen–antibody complex (horseradish peroxidase–antihorseradish peroxidase) and its use in identification of spirochetes, *J. Histochem. Cytochem.* **18**:315–333.

Sternberger, L.A., Petrali, J.P., Meyer, H.G., Mills, K.R., and Goding, S.L., 1976, Specificity of the immunocytochemical LH–RH receptor reaction, *Endocrine Society Program, 58th Annual Meeting*, p. 197, Endocrine Society, Bethesda, Maryland.

Sotmoller, P., and Cowan, K.M., 1974, The detection of foot and mouth disease virus antigen in injected cell cultures by immunoperoxidase techniques, *J. Gen. Virol.* **22**:287–291.

Taylor, C.R., 1974, The nature of Reed-Sternberg cells and other malignant "reticulum" cells, *Lancet*, Oct. 5, pp. 802–807.

Ungerstedt, U., 1971, The anatomy, pharmacology and function of the nigro-striated dopamine system, *Acta Physiol. Scand. Suppl.* **367**:1–48.

Vacca, L.L., Rosario, S.L., Zimmerman, E.A., Tomashevsky, P., Ng, P.–Y., and Hsu, K.C., 1975, Application of immunoperoxidase techniques to localize horseradish peroxidase-tracer in central nervous system, *J. Histochem. Cytochem.* **23**:216–234.

Vandesande, F., and Dierickx, K., 1975, Identification of the vasopressin producing and oxytocin producing neurons in the hypothalamic magnocellular neurosecretory system of the rat, *Cell Tissue Res.* **164**:153–162.

Vandesande, F., DeMey, J., and Dierickx, K., 1974, Identification of neurophysin-producing cells. I. The origin of the neurophysin-like substance–containing nerve fibers of the external region of the median eminence of the rat, *Cell Tissue Res.* **151**:187–200.

Vandesande, F., Dierickx, K., and DeMey, J., 1975, Identification of separate vasopressin-neurophysin II and oxytocin-neurophysin I containing nerve fibers in the external region of the bovine median eminence, *Cell Tissue Res.* **158**:509–516.

Vilchez-Martinez, J.A., Arimura, A., and Schally, A.V., 1975, Effect of actinomycin B on the pituitary response to LH–RH, *Fed. Proc. Fed. Amer. Soc. Exp. Biol.* **34**:240 (abstract).

Watkins, W.B., 1975, Presence of neurophysin and vasopressin in the hypothalamic magnocellular nuclei of rats homozygous and heterozygous for diabetes insipidus (Brattleboro strain), as revealed by immunoperoxidase histology, *Cell Tissue Res.* **157**:101–113.

Watkins, W.B., Schwabel, P., and Bock, R., 1974, Immunohistochemical demonstrations of a CRF-associated neurophysin in the external zone of the rat median eminence, *Cell Tissue Res.* **152**:411–421.

Wendelschafer-Crabb, T.W., Erlandsen, S.L., and Walker, D., Jr., 1976, Ultrastructural localization of viral antigens using the unlabeled antibody enzyme method, *J. Histochem. Cytochem.* **24**:517–526.

Zimmerman, E.A., 1976, Localization of hypothalamic hormones by immunocytochemical techniques, in: *Frontiers in Neuroendocrinology*, Vol. 4 (L. Martini and W.F. Ganong, ed.), pp. 25–61, Raven Press, New York.

Zimmerman, E.A., Carmel, P.W., Hussain, M.K., Ferrin, M., Tannenbaum, M., Frantz, A.G., and Robinson, A.G., 1973, Vasopressin and neurophysin: High concentrations in monkey hypophysial–portal blood, *Science* **182**:925–927.

Zimmerman, E.A., Hsu, K.C., Ferrin, M., and Kozlowski, G.P., 1974, Localization of gonadotropin-releasing hormone (Gn–Rh) in the hypothalmus of the mouse by immunoperoxidase technique, *Endocrinology* **95**:1–8.

Zimmerman, E.A., Defendi, R., Sokol, H.W., and Robinson, A.G., 1975, The distribution of neuro-physin-secreting pathways in the mammalian brain. Light microscopic studies using the im-munoperoxidase technique, *Ann. N. Y. Acad. Sci.* **248**:92–111.

5

Substance P and Neurotensin

Susan E. Leeman, Edmund A. Mroz, and Robert E. Carraway

1. Introduction

During the past 15 years, this laboratory has been involved in the isolation and chemical characterization of two neural peptides, Substance P and neurotensin. Substance P had been detected as early as 1931 (von Euler and Gaddum, 1931), but had resisted purification until the work of Chang and Leeman (1970). Neurotensin, another vasoactive peptide, was previously undiscovered, and is a new addition to the group of peptides of neural origin. It was detected in hypothalamic extracts during the course of purification of Substance P and its isolation reported by Carraway and Leeman (1973). This chapter will review general guidelines for the isolation and characterization of biologically active peptides, and then illustrate how they were applied to studies of Substance P and neurotensin, in the hope that this material will be of some value to those now considering similar problems for the first time.

In general, the study of biologically active peptides has involved the following sequence of steps: (1) detection of activity; (2) quantitation of activity; (3) establishment of the peptidic nature of the material; (4) development of isolation procedures; (5) determination of the amino acid sequence; and (6) synthesis of the peptide, including chemical and biological comparison of native and synthetic material.

Preparation of synthetic material has of course provided a great boost to the further study of the peptide in question, supplying far greater quantities of pure

Susan E. Leeman, Edmund A. Mroz, and Robert E. Carraway · Laboratory of Human Reproduction and Reproductive Biology and Department of Physiology, Harvard Medical School, Boston, Massachusetts.

material than could ever reasonably be isolated from natural sources. A great variety of further investigations has then become feasible in an effort to explore further the physiology of the peptide and its potential clinical usefulness. Some of the directions that such studies have taken include: (1) further characterization of the biological properties of the peptide; (2) determination of the biologically active region of the peptide; (3) development of a sensitive and specific radioimmunoassay, permitting studies on the distribution of the peptide in various body tissues and fluids and its regulation; and (4) preparation of specific antagonists, either by chemical synthesis of analogues or by production of neutralizing antisera.

2. Guidelines for the Isolation of Biologically Active Peptides

The biologically active neural peptides that have thus far been isolated are present in minuscule quantities in the tissues in which they have been found, perhaps as little as one ten-millionth the weight of the tissue or less. It seems reasonable that this will continue to be true for the as yet undiscovered biologically active peptides. Therefore, large amounts of tissue will probably have to be extracted to yield workable amounts of peptides and a sufficiently gentle isolation scheme developed that permits great purification without, one hopes, altering the chemical and biological properties of the substances sought.

2.1. Detection of Activity

There have been two general approaches taken in the search for biologically active neural peptides: (1) to suspect that a biological activity may be present in a given tissue on the basis of previous physiological evidence, as in, for example, the case of the releasing factors thyrotropin-releasing hormone (TRH) and luteinizing hormone–releasing hormone (LH–RH); or (2) to discover by chance, in extracts of a given tissue, a biological activity that looks interesting enough to warrant further investigation, as in, for example, the case of vasopressin, Substance P, and neurotensin. In both cases, the establishment of the peptidic nature of the substance came after the detection of the biological activity. It is, of course, possible to direct one's efforts toward isolating peptides in any given tissue using only chemical properties, such as staining characteristics, to monitor purification procedures, in an effort to discover one of biological interest. But this route seems to be a potentially more frustrating one. Thus, detection of a biological activity judged worth pursuing seems to us to be an important first step. However detection of activity may not be simple, and a considerable amount of preliminary work may be needed before an initial crude extract can be prepared containing the activity sought for or before an activity of interest is found. Some pitfalls to expect are: (1) A given extraction procedure may or may not solubilize the chemical responsible for the activity. (2) Even if the chemical is solubilized, there may be present in the extract other interfering chemicals that will mask the activity until they are separated away by appropriate fractionation steps. (3) The cooperation of two or more distinct molecules might be necessary for the activity to be observed, as has recently been suggested to be the case for corticotropin-releasing factor (CRF) (Pearlmutter *et al.*, 1975).

2.2. Quantitation of Activity

After detection, some method is needed for quantitating the biological activity to be pursued. This quantitation is important for several reasons: (1) During the subsequent isolation procedures, whatever they are, activity will usually be distributed over several fractions. Decisions must be made about which fraction to pool for further isolation steps. It is undesirable to discard very much of the active material, but it is also unproductive to retain much material from marginally active fractions because of their relatively low specific activity, reflecting the continued presence of contaminants. (2) It is necessary to monitor the overall progress of the purification process and to decide whether a chosen step has been beneficial. (3) It is necessary to assess the purity of the isolated material. Pure material must have a maximal, constant specific activity (i.e., units of biological activity per amount of peptide) after passage through several additional purification procedures, e.g., reelectrophoresis at several different pH values.

Desirable attributes of the quantitation procedure are, of course, that it be simple, precise, specific, and sensitive. A good choice for the assay system can be crucial; otherwise, the demands of the assay may jeopardize the success of the entire project. The workload must be tolerable for the laborious task of screening seemingly endless fractions, and the quantitation sufficiently sensitive that not all the extracted activity is consumed by the testing process.

2.3. Establishment of the Peptidic Nature of the Active Material

It seems important to determine, if possible, the general chemical nature of the active principle, since this nature will yield information valuable in the design of a purification scheme. More specifically, if the responsible chemical is thought to be a peptide, attempts should be made to verify this peptidic nature early in the project. Some distinguishing properties of the active material that can be examined using crude extracts are : (1) stability to enzymic and chemical treatments, (2) solubility in various solvents, and (3) size and functional group content.

2.3.1. Stability to Enzymic and Chemical Treatments

Various broad-spectrum proteolytic enzymes, such as pronase, are available that cleave peptide bonds between L-amino acids with little regard for the nature of side-chain groups. If the biologically active substance is a peptide, it will most likely be destroyed by incubation with such proteolytic enzymes. Enzymes with a higher order of specificity, such as trypsin, chymotrypsin, and carboxypeptidase, can give useful hints as to the nature of the amino acids present. However, some "proteolytic enzymes" are also "esterolytic," and might destroy the biological activity of a nonpeptidic substance; conversely, a true peptide (e.g., TRH with the sequence<Glu–His–Pro–NH$_2$) may be resistant to many proteolytic enzymes because of blocked amino- and carboxy-termini. The peptide bond is usually stable to boiling at neutral pH, but is cleaved during treatment at elevated temperatures with high concentrations of either HCl or Ba(OH)$_2$ (Bailey, 1967).

If at this stage the peptidic nature of the substance has not been established, various other steps can be performed to obtain information about the functional groups present in the active molecule.

2.3.2. Solubility in Various Solvents

Peptides are usually somewhat soluble in polar organic solvents such as methanol, ethanol, and acetone, but much less soluble in solvents such as benzene and petroleum ether; however, peptides with an unusually high content of hydrophobic amino acids and blocked amino- and carboxy-termini can display unusual behavior in this regard. Highly basic peptides are usually more soluble in acidic water, while acidic peptides are more soluble in base.

2.3.3. Size and Functional Group Content

A rapid estimate of the molecular size of the active material can be obtained by determining its apparent molecular weight during gel filtration through various preparations of Sephadex (Pharmacia, 1974).

Peptides often contain free carboxyl and amino groups, which titrate at pH values near 3 and 9, respectively. The presence (or absence) of these functional groups can usually be implicated by examining the behavior of the active substance in the presence of various ion-exchange resins (Bailey, 1967), or during electrophoresis at several pH values (Offord, 1966), or in both ways. Further information regarding the nature of the functional groups present in the active molecule might be obtained by examining the effects of various group-specific reagents on biological activity; 1-fluoro-2,4-dinitrobenzene, for example, will react with amino groups (Bailey, 1967); various other selective reagents can be used to implicate the presence of specific amino acids in the peptide (Bailey, 1967). It should be remembered, however, that certain functional groups present in the molecule may not be necessary for its biological action.

2.4. Development of Isolation Procedures

In designing a purification scheme, one attempts to minimize the total number of fractionation steps, choosing those that afford the greatest increase in specific activity and best yield. Harsh procedures that might alter the chemical structure of the peptide, perhaps without affecting biological activity, should be avoided. Yield is of prime importance for the successful isolation of biologically active neural peptides, since they are usually found in such low concentrations.

2.4.1. Choice of Starting Material

Most peptides will have a differential distribution in the body, or even within the nervous system. Choosing a starting tissue or region of the nervous system that has a high specific activity will simplify problems involved in the isolation. If a region of high specific activity is used as the source of active material, there will be a smaller percentage of other substances present in extracts. Furthermore, less tissue can be used to obtain the amount of pure peptide needed for structure determination, and logistical problems will also be minimized.

2.4.2. Initial Extraction Step

A careful choice of extraction method can itself serve as an important purification step. When dealing with small peptides, for example, extraction into organic solvents can be an effective means of separating them from larger proteins, which tend to be less soluble in nonaqueous media.

A crucial consideration in designing the extraction method is the economy and logistical simplicity of the method when applied to large quantities of tissue. In order to obtain sufficient pure material for the identification of the peptide, it is usually necessary to deal with kilogram quantities of tissue. The extraction procedure must be tractable and economically feasible when scaled up to this magnitude. Steps such as concentration of extracts by evaporation of organic solvents and lyophilization of water and volatile buffers are most easily performed on small volumes. Alternatively, a sequence of steps that precipitate the activity by treatment with solvent or salt, or that adsorb the activity of ion-exchange resins, may minimize logistical problems. Extraction methods that lead to small volumes of fluid or small quantities of a precipitate are thus desirable. But such concentration steps might lead to undesirable losses, and compromises between yield and ease of extraction must be made.

The initial steps in the extraction of active materials would tax the resources of most investigators' laboratories. However, access to the necessary equipment can be obtained at regional centers designed to assist investigators with such efforts, such as the New England Enzyme Center at Tufts Medical School in Boston, where much of the work reported here has been carried out.

2.4.3. Isolation Steps

Although the basic approach to the successful purification of peptides is rather general, the optimum purification scheme for a particular peptide can be determined only empirically. Even when purifying the same peptide from different sources, one must sometimes employ different procedures. The purification scheme should have a minimum number of fractionation steps, each affording the greatest increase in specific activity and the best yield. These steps may be based on a number of different principles, such as molecular size, solubility, charge, or specific affinity. Again, harsh procedures are to be avoided. The various steps should be ordered in a manner that minimizes time and effort, and takes advantage of the particular advantages afforded by each method and the peculiar problems of the specific isolation. For example, an extraction step yielding 100 liters of aqueous extract would better be followed by batch adsorption to an ion-exchange resin than by lyophilization and gel chromatography. Some procedures that have very limited capacity, such as high-voltage paper electrophoresis or paper chromatography, are most effectively employed after substantial purification has been performed. In this laboratory, the general plan has been first to subject the active extract to gel-exclusion chromatography, first on a large scale as a concentration and desalting procedure, then as a smaller-scale purification step. Volatile buffers are used so that they can be removed during lyophilization of active fractions, simplifying later steps. Then pilot experiments are performed with various ion-exchangers to determine the ion-exchange properties of the active principle in question; from these pilot studies, a rational choice is made about proper conditions of ion-exchange chromatography, again using volatile buffers, as further purification. After one or two ion-exchange steps, final purification is accomplished by preparative high-voltage paper electrophoresis.

2.4.4. Criteria of Purity

Multiple criteria must be satisfied before the purity of a peptide can be regarded as highly probable; each piece of evidence is a necessary but not sufficient condition to establish purity. Perhaps the most crucial evidence for the purity of a peptide is that the molar ratios of the constituent amino acids be integral and remain constant after further fractionation steps. The specific biological activity should be maximal and also remain constant throughout its running zone after refractionation.

A pure peptide should exhibit only one peptide-stainable spot on chromatography under different solvent conditions. Chemical or enzymatic determination of the amino- and carboxyl-termini should be consistent with a single residue in each position. The material will normally run as a single, well-defined band on high-voltage paper electrophoresis. The molecular weight predicted from the amino acid composition should be consistent with an independent estimate of the molecular weight. However, some of these latter criteria can be misleading. For example, an inhomogeneous mixture may appear homogeneous on chromatography or electrophoresis, or contaminants might not stain and thus remain undetected. On the other hand, a pure material may be changed during relatively harsh procedures such as paper electrophoresis, and smear as a result.

Even when these criteria are satisfied, however, one should still entertain the possibility of a heterogeneity with respect to content of non-amino-acid substituents, such as carbohydrate, lipid, phosphate, sulfate, and ammonia, or with respect to the optical configuration of the amino acids comprising the peptide giving rise to diastereomeric structures. Total enzymic digestion using enzymes specific for the L-configuration is often used as an argument for the "optical" purity of the isolated peptide (Keutmann et al., 1970). If total enzymic digestion and acid hydrolysis of the peptide give rise to the same amino acids, and if the molecular weight determined by amino acid analysis is in good agreement with that estimated by physical methods, the possibility of having non-amino-acid substituents is considered unlikely.

2.5. Sequencing

Certainly the biggest hurdle to the successful sequencing of neural peptides is the limited quanitity of pure material available, which sometimes strains even the most "ultrasensitive" methods. One is severely hampered by the inability to perform many trial experiments, and success is largely dependent on the structure and properties of the particular peptide. If the amino-terminus of the peptide is free, the manual or automated Edman degradation is applicable; however, losses on glassware and during extractions, often a problem with less polar peptides, can severely limit the procedure. Recent technical advances have reduced losses and increased the sensitivity of detection methods used for the Edman products (Niall, 1975), but at best, the number of successful cycles is less than 20 for nanomole quantities of peptide.

One approach to the sequencing of trace amounts of peptide involves the biosynthetic incorporation of radioactively labeled amino acids into the peptide prior to sequence analysis (Jacobs *et al.*, 1974). For example, Chu *et al.* (1975) have incubated slices of parathyroid glands with ^{3}H-labeled amino acids, introducing radioactivity into proparathyroid hormone. After isolation of the peptide from the tissue, it was possible to successfully perform more than 10 cycles of the sequential Edman degradation using less than 1 nmol total peptide; the radioactive phenylthiohydantoin amino acid derivatives obtained at each cycle could be identified in picomole amounts. The applicability of this technique to the study of brain peptides will depend most heavily on the degree of success obtained with the biosynthesizing system.

Sensitive, repetitive COOH-terminal sequencing procedures have not been developed as yet; thus, large peptides and those with "blocked" NH_2-termini must be cleaved to form more suitable fragments. There are numerous very specific chemical and enzymic methods of peptide fragmentation that can be used to generate overlapping sequences (Kasper, 1970), but, once again, losses of the trace quantities of peptide are a severely limiting factor. When small amounts of a pure peptide are chromatographed, a large portion of the peptide is lost by adsorption to chromatographic media and glassware; yields of less than 20% are not uncommon during paper electrophoresis of 10–50 nmol peptide (the amount of a neural peptide likely to be obtained from 10–50 kg tissue). Once generated and purified, however, the smaller peptide fragments (6 or fewer amino acids in length) should be amenable to sequence analysis using carboxypeptidase (Ambler, 1967) and the dansyl-Edman (Gray, 1967) or the subtractive Edman (Konigsberg, 1967) procedure.

The sensitivity of amino acid analysis using, for example, the Beckman 121

M machine, has advanced to the point where 100–500 pmol quantities of peptide can be rapidly and reliably analyzed; thus, sequencing methods employing amino acid analysis are highly favored. Total enzymic digestion of each peptide should be performed to establish the amide content of the acidic residues and to consider the possibility of other substituents. The structure of the "mother" peptide can be deduced by piecing together the information obtained from the various "daughter" peptides. All the data should be consistent with the known specificities of the various enzymes and reagents employed.

2.6. Synthesis

Chemical synthesis of the native peptide serves to support the proposed structure, and also provides the large amounts of peptide needed to characterize pharmacological properties and to study physiological function. Several methodologies are available for the construction of peptides from commercially available amino acid derivatives, but the Merrifield solid-phase technique (performed manually) is by far the easiest for the novice to learn, requires the fewest manipulations, and yields the desired product most rapidly. Although classic solution methods have withstood the test of time, they suffer from the need to perform time-consuming purifications on each of the intermediates. The basic principle of the solid-phase process is that anchorage of the growing peptide to an insoluble support allows removal of unreacted materials and by-products by simple filtration. All the operations are performed in a single vessel without need of transfer until the finished peptide is removed from the solid support. An excellent guide for the design and performance of solid-phase syntheses is available (Stewart and Young, 1969).

Since its introduction in 1962, several interesting variations on the Merrifield concept have been described, such as the Letsinger method and liquid-phase synthesis (Letsinger and Kornet, 1963). However, each of these methods suffers from the same major flaw; i.e., side-products accumulate on the peptide support, and must be separated from the desired peptide at the completion of the synthesis. Because of this cumulative error, the desired peptide (of greater than 10 amino acids) can be only a small portion of the product, even when each reaction is 99% successful (Bayer et al., 1970); removal of the contaminating peptides, which usually differ by 1 or 2 amino acids, is usually difficult or impossible. Another variation on the solid-phase concept involves the use of "solid-phase reactants" while the peptide is in solution (Fridkin et al., 1968); this procedure permits purification of the growing peptide to be performed at various stages during the synthesis. The side-products of coupling and deblocking failure, however, become increasingly difficult to remove as the peptide lengthens. A very recent innovation called *alternating liquid–solid phase peptide synthesis* (ALS) avoids the problem of cumulative error by having the peptide alternate from the liquid to the solid phase as it is successfully coupled and deblocked (Frank et al., 1975). Although ALS is only in its early stage of development, it has the potential for application in instances in which the more popular methods fail, particularly for larger peptides.

3. Guidelines for the Radioimmunoassay of Small Peptides

Radioimmunoassay (RIA) can provide a highly specific and reliable measure of neural peptides in the low concentrations in which they normally occur, provided great care is taken in the design, performance, and evaluation of the assays. Although Chapter 3 deals with this general topic, some comments, particularly relevant to RIA of small peptides, are included here.

3.1. General Approach

Since the binding site of an antibody molecule is of finite size (about 6–8 amino acids), antisera generated against peptides larger than this are often directed toward various small portions of the molecule, and thus display significant cross-reactivity toward fragments of the peptide and toward related peptides having these regions in common. It would seem that there might be two basic approaches to dealing with this specificity problem in RIA: (1) to use an antiserum that is heterogeneous with respect to specificity, and require that the cross-reactivity toward any particular fragment be only a small percentage of the total; or (2) to use multiple site-specific antisera and require that the unknown possess equal immunological potency toward all antisera. Since site-directed antisera offer other potential advantages, it would seem that the latter approach might be more beneficial in the long run. For example, armed with a battery of region-specific antisera, one can quickly characterize an immunoreactive substance and decide whether it is the entire peptide, a fragment of it, or a closely related peptide. In addition, antisera directed specifically toward the "biologically active" region of the peptide may neutralize its biological actions and be useful to induce and study a deficiency syndrome.

3.2. Generation of Antisera

Since small peptides are themselves seldom immunogenic, they are usually coupled to carrier proteins prior to immunization. Antisera raised toward hapten–protein conjugates are often directed primarily toward areas of the hapten most distal to the site of conjugation. Thus, when a small peptide hapten is attached to the carrier protein by its amino-terminus, antibody specificity is usually directed toward its carboxy-terminal region. In contrast, a heterogeneous peptide–protein conjugate with the peptide in various orientations usually gives rise to antisera with multiple specificities. Therefore, careful selection of the coupling procedure so as to minimize the number of possible reactions and generate as homogeneous a conjugate as possible can be of great benefit. Several coupling methods that are especially useful for peptides have been described (Abraham and Grover, 1971). An alternative means of directing antibody specificity that is especially useful for peptides of more than 10 amino acids involves the use of fragments of the molecule as haptenic determinants. Thus, antisera raised against a partial sequence of a larger peptide sometimes also bind to that portion of the entire molecule. This method is likely to be most successful with peptides having a random configuration in solution.

Another important consideration in the preparation of immunogen is whether to use the native or the synthetic material. While the synthetic peptide is more abundant, it is also more likely to be contaminated with similar peptides differing by 1 or 2 amino acids (see Section 2.6); immunization of animals with preparations containing several similar peptides is likely to give rise to heterogeneous nonspecific antisera. In contrast, the native material is not likely to be contaminated with similar sequences, and should not present this problem. However, even though the immunogen need not be absolutely pure, one seldom has sufficient highly purified native substance to immunize a colony of animals, especially if the peptide must be coupled to a larger protein to render it immunogenic. A sensible compromise might be to attempt both, doing fewer animals with native peptide.

3.3. Selection of Antisera

A simple method of assessing the specificity of an antiserum involves the determination of its relative cross-reactivity toward various partial sequences of the peptide. Amino-terminal fragments of the peptide can be obtained by treatment with specific enzymes, while carboxy-terminal partial sequences are most easily obtained during solid-phase synthesis of the peptide (see Section 2.6).

3.4. Performance and Interpretation

One should exercise caution in the interpretation of measurements of peptides made by RIA, especially in regard to measurements of tissue and plasma levels. Immunoreactivity cannot be equated with peptide content unless multiple criteria are rigorously satisfied, and even then, the exact chemical nature of the immunoreactive substance or substances is not established, and inferences should be made with reservation.

The validity of RIA measurements is a basic concern that is often treated too lightly. Multiple artifacts that interfere with the separation of bound and free antigen, or that alter the properties of the antibody or tracer, could register as false positives in the assay; such effects might even exhibit dose–response curves that run approximately parallel to that of the peptide. Large amounts of some proteins, for example, sometimes interfere nonspecifically with the apparent binding of tracer and antibody; serum proteins are notorious in this regard. Another artifact that plagues the investigator using RIA concerns the ability of tissue extracts to destroy the tracer. Any substances that alter the binding constant between antigen and antibody, such as protein denaturants, acids, bases, and salts, might register as false positives in the assay. The likelihood of obtaining these kinds of false measurements can be greatly reduced by using extraction methods that destroy proteolytic enzymes and by employing volatile buffers for chromatography; when other reagents are employed, appropriate controls should be included in the assays.

When one suspects that biological fluid or extract is giving a valid response in the RIA, one usually examines the physical–chemical, immunochemical, and biological character of the immunoreactive material in order to make comparisons with the "standard" peptide. Rapid comparisons of physical–chemical properties might be based on chromatographic and electrophoretic behavior, while more detailed studies could involve determinations of stability to various chemical and enzymic treatments. Since "precursors" and "breakdown products" of the peptide are likely candidates as cross-reactors in the RIA, information concerning the molecular size of the immunoreactivity is vital. One accepted criterion of similarity between test material and standard is that it gives a parallel dose–response curve; especially when employing site–specific antisera, equal immunological potency with several antisera differing in specificity would seem to be another sensible criterion. Information as to the biological activity of the immunoreactive material is of utmost interest, since biological tests are potentially very discriminating. It would seem that the ultimate test of similarity would involve chemical sequencing performed on the isolated immunoreactive substances. Such studies, if performed in at least one animal species, can greatly bolster one's confidence in RIA measurements.

4. Substance P

4.1. Detection of Activity

The original project undertaken by this laboratory, beginning in 1959, was not the isolation of either Substance P or neurotensin. Rather, the activity under examination was that of CRF. Leeman felt that a significant problem in the isolation of CRF was the lack of specificity in the assays then used to determine CRF activity. She had thus developed (Leeman, 1958) an assay for CRF activity that involved the injection of test materials into morphine-treated rats anesthetized with Nembutal. Adrenal steroid output in response to the injection was compared among normal and hypophysectomized animals. The rationale for this procedure was that morphine should inhibit endogenous release of CRF due to stress during the procedure; any CRF present in a test sample would lead to increased adrenal steroid output in a normal rat, but not in a hypophysectomized animal.

Initial fractionation of crude bovine hypothalamic extracts on Sephadex G-75 uncovered an unexpected activity. Figure 1 shows the OD_{280} profile of the eluate from such a column. Screening for CRF activity was carried out systematically over the whole column effluent, even in relative troughs of optical density. In such a trough, fractions were found that led to a copious production of fluid in the mouths of injected test animals (Leeman and Hammerschlag, 1967). Assay of this fluid from the mouth showed amylase activity, indicating that some component in these fractions led to salivation on intravenous injection. The sialogogic activity

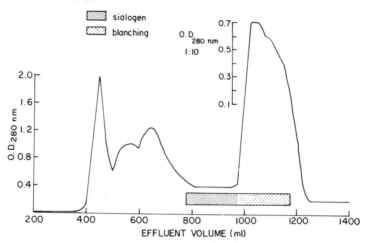

FIG. 1. Optical density profile of effluent from 1200 ml Sephadex G-75 column. Extract from about 200 bovine hypothalami was applied, and column was run in 0.1 M pyridine acetate, pH 2.8, at room temperature. Regions that elicited salivation (due to substance P) and cutaneous blanching (probably due to vasopressin) on injection into rats are indicated. After Leeman and Hammerschlag (1967).

was not affected by cholinergic or adrenergic blocking agents, and was observed in rats with surgically denervated salivary glands as well. However, incubation of these fractions with pepsin or pronase destroyed the sialogogic activity, while incubation without enzymes did not. This finding indicated that the responsible factor was a peptide. This activity was considered sufficiently interesting to be pursued, instead of CRF. After the sialogogic peptide had been isolated, its resemblance to the previously unpurified Substance P was noted, and Chang and Leeman (1970) then proved this identity. (Although for simplicity this peptide will be called Substance P below, the identity of the sialogogic peptide with Substance P was not suspected during the isolation.)

Thus, it was the sialogogic effect of Substance P that initially attracted sufficient interest in this laboratory to warrant its isolation. Two points about this activity should be noted: First, the main biological role of Substance P does not at this time seem to be to elicit salivation; however, this distinctive activity did allow a relatively simple and specific assay for use in isolation procedures.

Second, the activity cannot be seen very well in unfractionated crude extracts, at least in the extracts produced in this laboratory's work. Leeman and Hammerschlag (1967) showed that a material similar or identical to vasopressin interferes with sialogogic effect of Substance P. This interfering substance could be separated from Substance P during gel-exclusion chromatography, or could be destroyed in crude extracts by use of reducing agents that presumably broke disulfide bonds necessary for the action of the interfering substance. Thus, the detection of the Substance P activity depended on preliminary fractionation. It would have been detected only with great difficulty, if at all, had not the assay for CRF activity been performed systematically and with an open eye for unexpected findings.

4.2. Quantitation of Activity

The sialogogic effect of Substance P provided a simple and quantitative assay for its presence in test fractions, with sufficient sensitivity for routine use in large-scale isolation. For the assay, 80–100 g rats were anesthetized with Nembutal, which suppresses salivation. Test material was injected into the tail vein, the saliva produced in response was collected with a Pasteur pipette into a small tube, and the volume of saliva was measured with glass microliter pipettes. Over the lower range of the dose–response curve, the amount of saliva produced was proportional to the amount of active material injected. A "sialogogic dose" of material was defined as the amount needed to elicit approximately 50 μl saliva. Although fresh rats were routinely used for each test, it was not necessary to do so, because rats could respond identically to repeated injections of active material.

4.3. Identification as a Peptide

The identification of the material as a peptide can influence the choice of further isolation steps, so an early test for the peptide nature of the active principle was important. In the initial report of the sialogogic activity, Leeman and Hammerschlag (1967) showed that incubation of active material with pepsin or pronase destroyed the sialogogic activity, while similar incubation without enzyme did not. Similar tests were performed by Chang and Leeman (1970) on the sialogogic peptide after its isolation; pepsin and chymotrypsin inactivated the material, while incubation with denatured pepsin or chymotrypsin did not. These tests, of course, did not rule out the possible importance of carbohydrate or other moieties in the active chemical, but they did show the presence of a crucial peptide component.

4.4. Isolation Procedures

4.4.1. Choice of Starting Material

The initial choice of starting material for these isolations was a result of the initial project undertaken by the laboratory, the isolation of CRF. Leeman chose bovine hypothalamic fragments as the tissue from which to extract CRF activity. This choice was somewhat unorthodox at the time. Other groups involved in the isolation of releasing factors in the late 50's and early 60's had been using pituitary powders as the source of starting material (Schally et al., 1958). Such powders were readily available; investigators apparently hoped that CRF might be transported toward the pituitary in analogy to oxytocin and vasopressin, or perhaps thought that vasopressin itself was CRF. In any event, Leeman thought that hypothalamic fragments might afford a better chance of success, since the hypothalamus was the apparent source of CRF. The hypothalamic fragments used were rather large, extending from the optic chiasm to the mammillary bodies, and weighed 8–10 g each. They are thus considerably larger than the median eminences, which have been used more recently as a source of releasing factors for isolation (Schally et al., 1966).

At this time, we cannot determine the wisdom of this choice of starting material as a source of CRF, but the choice was quite felicitous as a source of Substance P. The hypothalamus is one of the regions of brain most highly concentrated in Substance P activity. Strangely enough, the median eminence alone seems to be relatively poor in it, according to recent data from radioimmunoassays (see Section 4.7).

4.4.2. Initial Extraction Step

When work first began on the isolation of CRF, an extraction method was chosen for preliminary studies that minimized the handling of large volumes of extracts by precipitating activity out of the liquid phase. Frozen hypothalamic fragments were homogenized in 4 vol of a mixture of acetone, water, and HCl (2000:250:50) and spun, and the supernatant was added to 6 vol cold acetone. The precipitate was collected, washed with acetone, and dried. In this manner, 20 kg hypothalami (some 2000 in number) were reduced to about 25 g of powder, which contained CRF activity and exhibited sialogogic activity after gel-exlusion chromatography in 0.1 M acetic acid.

Although this procedure was successful in the initial demonstration of sialogogic activity in extracts of bovine hypothalami, it turned out to be too wasteful for purposes of isolation. Several other methods were tried. Extraction of acetone powders with 2 M acetic acid, or homogenization of tissue in distilled water followed by acidification, could yield 5–10 times as much sialogogic activity as the method first used by Leeman and Hammerschlag (Chang, 1970). However, these methods also produced volumes of aqueous extracts some 4–5 times the volume of the starting tissue. Such extracts had to be concentrated by lyophilization, but lyophilizing 100 liters of extract is somewhat unwieldy.

A method was finally chosen for extraction of Substance P (and used with minor modifications for extraction of neurotensin) that afforded high yield with minimal volumes of extracts remaining for lyophilization. First, the tissue was homogenized in a mixture of acetone and 1 M HCl (100:3, vol/vol) in a volume 5 times that of the initial tissue. This mixture was stirred overnight in the cold, and was then filtered by suction through filter paper in Buchner funnels. The filtrate was saved, and the residue was reextracted with a mixture of acetone and 0.01 M HCl (80:20, vol/vol), with a volume 2 times that of the initial volume of the tissue. This second mixture was filtered as described above, and the filtrates were pooled. A volume of petroleum ether equal to the volume of the combined filtrates was taken and used in 5 aliquots to extract the combined filtrates. This extraction with petroleum ether removed lipids and the bulk of the acetone in the initial extracts. A volume of fluid less than twice that of the starting tissue was left, which was rotary-evaporated to remove traces of organic solvents, and then lyophilized to remove water. The initial extraction with acid acetone solubilized Substance P and precipitated most large proteins, while the petroleum ether extraction significantly reduced the volume of the extract, but left the Substance P in the aqueous phase.

4.4.3. Isolation Steps

Given the peptidic nature of the sialogogic material, it was decided to pursue its isolation using a combination of gel-exclusion chromatography, ion-exchange chromatography, and high-voltage paper electrophoresis. Sephadex G-25 was found to be more appropriate as a medium for gel-exclusion chromatography than was the G-75 first used by Leeman and Hammerschlag. A typical preparative isolation would start with some 20 kg of hypothalamic fragments, extracted as described above. The lyophilized extract was resuspended in 0.1 M acetic acid, and applied to a 20-liter column of G-25 at the New England Enzyme Center. This first step was most important as a desalting procedure, although it did afford some separation of the material (especially removing from the Substance P the vasopressin, which interferes with the sialogogic assay). Fractions of peak activity were pooled and lyophilized to remove water and acetic acid, then redissolved and applied to a smaller 5-liter column in this laboratory. This second gel-exclusion step provided further purification and concentration of the active material.

After this initial fractionation, it was necessary to run a variety of pilot studies to determine the general ion-exchange properties of the peptide. This approach allowed reasonable choices of conditions for purification on the basis of the peptide's ion-exchange properties. Active fractions were mixed with anion- and cation-exchangers at various levels of pH and ionic strength, and the supernatants were assayed for activity. Substance P was found to be quite basic in character. At no pH tried would it adsorb to anion-exchange resins, while cation-exchangers readily removed sialogogic activity from extracts. However, the Substance P that adsorbed to cation-exchangers could be recovered by increasing the pH or ionic strength. The strongly basic character of the peptide suggested cation-exchange chromatography under fairly strong conditions as a step to remove the sialogogic activity from other materials, most of which would be expected to be considerably less basic.

Thus, the peak fractions from the second G-25 column were applied to a 50-ml column of the strong cation-exchanger Sulfoethyl (SE) Sephadex C-25. (SE sephadex has been replaced by Sulfopropyl Sephadex in recent years.) The column had been equilibrated at pH 5.5 with 0.5 M pyridine acetate, a buffer chosen for its volatility. The peak fractions from the G-25 column were redissolved in this same buffer and washed into the column; then the column was washed with 170 ml more of the equilibration buffer. The eluent was then changed to 0.75 M pyridine acetate at the same pH. Sialogogic activity was eluted after some 200 ml of this latter buffer had passed through the column.

A second ion-exchange step was then used for further purification of Substance P. It was found to adsorb to the weak cation-exchanger carboxymethyl cellulose even in 5 mM ammonia. Active fractions from the SE Sephadex column were applied to a 25-ml column of carboxymethyl cellulose in 5 mM ammonia, and 150 ml more of this ammonia was washed through the column, removing all but the most basic substances from it. Then 40 mM acetic acid was applied as an eluent. This buffer titrated the carboxymethyl group on the column, releasing the

TABLE 1. Scheme for Purification of Substance P from Bovine Hypothalami (20 kg)

Purification step	Total protein (mg)	Total sialogogic doses	Specific activity (doses/mg)
Initial extraction	100,000	30,000	0.3[a]
First gel filtration on Sephadex G-25	2,000	25,000	12[a]
Second gel filtration on Sephadex G-25	1,000	20,000	20[a]
Chromatography on sulfoethyl Sephadex	5	8,000	1,300[a]
Chromatography on Cm-cellulose	1	5,000	5,000[b]
Paper electrophoresis	0.150	2,000	13,000[b]

[a] Protein was determined by the method of Lowry *et al.* (1951).
[b] Protein was calculated from quantitative amino acid analyses.

Substance P. At this point, the material was deemed sufficiently pure for preparative high-voltage paper electrophoresis. When the active fractions from the carboxymethyl cellulose column were lyophilized and subjected to electrophoresis at pH 1.9 or 3.5, a major band of protein and activity was found, which showed only one component on reelectrophoresis. Table 1 presents a summary of the purification scheme for Substance P.

4.4.4. Criteria of Purity

Although there was only a single stained spot after electrophoresis, this result was far from adequate demonstration of the purity of the product. Another important but also inadequate criterion is the presence of a single amino-terminus. The material showed only arginine at the amino-terminus by dansylation, but it was noted that the inhomogeneous material applied to electrophoresis also showed only an arginyl amino-terminus. This result demonstrates the inadequacy of this criterion of purity. Similarly, it is important to show homogeneity of the carboxyl-terminus. In the case of Substance P, neither carboxypeptidase A or B released any amino acids from the peptide; more direct demonstration of the homogeneity of the carboxyl-terminus was not immediately possible.

The most stringent criterion of purity is a constant, integral ratio of the component amino acids on electrophoresis of the isolated material at several different values of pH. The molecular weight as calculated from these amino acid ratios must be consonant with other determinations of the molecular weight. In the case of Substance P, the isolated peptide could be rerun on electrophoresis at either pH 1.9 or 3.5 with a constant amino acid composition; the molecular weight predicted from the amino acid composition was 1340 (Table 2). Gel exclusion is often used to obtain the necessary independent estimate of molecular weight. It was found, however, that the pure material had a tendency to adsorb to Sephadex if run in an eluent of low ionic strength. There are a small number of negative charges present on Sephadex, and these charges apparently interacted with the strongly basic pep-

TABLE 2. Amino Acid Composition of Substance P after Electrophoresis

| | Electrophoresis performed at: | | | | |
| Amino acid | pH 3.5 | | pH 1.9 | | Assumed residues per mol peptide |
	Composition (nmol)	Molar ratio	Composition (nmol)	Molar ratio	
Lys	30	1.0	6.0	1.0	1
Arg	33	1.1	6.0	1.0	1
Ser	5	0.2	0.7	0.1	0
Gln	62	2.1	11.6	1.9	2
Pro	56	1.9	12.3	2.0	2
Gly	35	1.1	6.1	1.0	1
Ala	3	0.1	trace	trace	0
Met	23	0.8	3.5	0.6	1
Leu	35	1.2	5.6	0.9	1
Phe	62	2.1	10.8	1.8	2
TOTAL					11
Assumed nmol peptide	32		5.9		
Sialogogic doses applied	650		125		
Doses/nmol peptide	20.3		21.2		

tide unless the eluent contained a sufficient amount of salt. Thus, gel-exclusion columns were run with 200 mM NaCl to provide a sufficiently high ionic strength; under these conditions, the estimated molecular weight was consistent with the value of 1340 from the amino acid composition. Although the data from electrophoresis at pH 1.9 and 3.5 seemed to indicate homogeneity, if the material were run at a higher pH of 6.5 or 8.9, it formed a broad smear rather than a well-defined band. It was possible to demonstrate that the material was still pure, despite the smearing. The amino acid composition of the material was unchanged under these conditions (though more methionine appeared as the sulfone on amino acid analysis), and material taken from the rear of the smear did run in a well-defined band on reelectrophoresis, though with a lower mobility than the original product. Chang and Leeman concluded that the relatively harsh procedure of high-voltage paper electrophoresis at higher levels of pH led to some irreversible change in the molecule, accounting for the observations. The increased levels of methionine sulfone after this treatment suggested that oxidation of the methionine may play a role in this behavior, but this suggestion has not been demonstrated directly. This problem in the electrophoresis of Substance P at higher levels of pH point up another problem with high-voltage electrophoresis: not only may an apparently homogeneous band be composed of more than one peptide, but an apparently ill-defined band may be due to homogeneous material.

One further step was necessary in the determination of the composition of the peptide: spectrophotometric studies to test for the presence of tryptophan or tryosine, which are destroyed under usual conditions of acid hydrolysis. Neither was found to be present.

4.4.5. Characterization of the Sialogogic Peptide as Substance P

The identity of the sialogogic peptide and Substance P was not suspected in this laboratory until Chang began to search the literature for previous reports of other neural peptides. Substance P had first been noted by von Euler and Gaddum (1931); it was characterized by its ability to contract isolated intestinal smooth muscle and to lower the blood pressure of test animals, even after pretreatment with atropine. It had not been completely purified, so the only definition of Substance P was based on its pharmacological activities. Since 1931, other hypotensive and gut-contracting peptides had been found, notably bradykinin and its relatives (Rocha e Silva *et al.*, 1949), but the kinins relaxed the rat duodenum, while Substance P contracted it (Horton, 1959). The sialogogic peptide met all these criteria of Substance P (Chang, 1970). Furthermore, Substance P had been shown to be sialogogic (Vogler *et al.*, 1963; Lembeck and Starke, 1968). The sialogogic peptide shared many of the chemical and physical characteristics reported for partially purified preparations of Substance P; the one notable exception, the sensitivity of Substance P to trypsin but a lack of effect of trypsin on the sialogogic peptide, was laid to probable contamination with chymotrypsin of the trypsin used on Substance P (Chang, 1970).

4.4.6. Comparison with Previous Isolation Attempts

It is interesting to note some of the differences between this successful isolation of Substance P and some of the previous incomplete purification attempts. (Lembeck and Zetler, 1971, provide a useful summary of isolation attempts prior to 1970.) It is impossible to say precisely why Chang and Leeman had better success than other investigatiors, but several differences in approach may conceivably have made a difference.

The simplicity and specificity of the sialogogic assay may have been one of the most important factors contributing to this successful isolation. Previous attempts used the hypotensive or gut-contracting properties of Substance P, the activities first noted by von Euler and Gaddum, as assays for screening fractions. Various smooth muscle preparations *in vitro* were used as assays; these preparations had varying degrees of specificity and sensitivity. The isolated guinea pig ileum was considered the most specific of these preparations (Cleugh *et al.*, 1964), but this preparation can also contract in response to kinins, albeit with a different time-course (Pernow and Rocha e Silva, 1955). Differentiation of Substance P from kinins can be performed on the rat duodenum as indicated above, but this preparation is much less sensitive than the guinea pig ileum. Many of the other muscle bath preparations would show adenine and uridine nucleotides, histamine, and serotonin as false positives for Substance P, unless the muscle preparation were pretreated with a variety of blocking or desensitizing agents, or the extract were treated so that none of these contaminants was present (Cleugh *et al.*, 1964). Furthermore, most of these preparations show some tachyphylaxis, so that several minutes must elapse between tests. During the course of screening the effluent from a column, each sample must be tested, the response recorded, the bath washed out thoroughly, and the preparation left to sit long enough that tachyphylaxis presents no problem. In the course of screening a column effluent, this procedure must be repeated scores of times. In this laboratory, at least, it has been found difficult to avoid gradual deterioration of such isolated smooth muscle preparations during such a process.

Although Substance P was reported in 1968 to be the only material from a mammalian source to have a sialogogic effect not blocked by adrenergic or cholinergic blocking agents (Lembeck and Starke, 1968), this more specific sialogogic activity had not been used in any previous attempts to isolate Substance P. The sialogogic assay for Substance P is considerably less sensitive than the various muscle bath preparations, but the scale of the isolation was sufficient that the sensitivity of the assay did not present a significant problem. The specificity of the sialogogic assay prevented false positives from leading the isolation astray, and the relative simplicity of the assay enabled effort and energy to be expended on the isolation procedures themselves.

The choice of starting material also might have helped. Intestine was often used by others as a starting material. This tissue has a fair specific activity of Substance P, and an isolation of Substance P from equine intestine has been accomplished (Studer *et al.*, 1973). However, experience in this laboratory with intestinal extracts has given us the impression that there may be more small peptides present in the intestine than in the hypothalamus, and that intestine may pose more of a problem with peptidase activity than does hypothalamus. Other attempts at isolating Substance P had used whole bovine brain, in one attempt 1000 kg of such material (Zuber, 1966). Unfortunately, cerebral and cerebellar cortex, which form the bulk of such brains, has a much lower concentration of Substance P than do hypothalamus and brainstem. The preponderance of cortical tissue in whole bovine brain could have diluted out the Substance P, present mostly in subcortical regions, to such an extent as to make isolation difficult.

The extraction method for Substance P used by von Euler and Gaddum (1931) involved homogenization of tissue in acid and extraction of activity by ethanol. This method did form the basis for some isolation attempts. Chang tried to extract Substance P by this classic method from the classic source of Substance P, equine intestine, and found extremely low yields in comparison with that obtained by acid–acetone extraction of hypothalami. Other methods of extraction (Pernow, 1953) involved repeated precipitation steps, which may have led to significant loss of activity.

An isolation step often used in attempts to isolate Substance P is adsorption and desorption on aluminum oxide columns. Lembeck and Zelter (1971) list 14 reports of this approach. In retrospect, this procedure seems to have been an unfortunate choice. Chang (1970) found that purified Substance P is very difficult to recover from aluminum oxide, and does not elute from it in a well-defined way. As much as 70% of applied sialogogic activity could not be recovered from an aluminum oxide column eluted with varying concentrations of methanol, a step used in various permutations by some previous workers.

Countercurrent distribution was also used as an isolation method. This method, based on differential solubility, did afford substantial purification, but did not yield homogeneous material (Meinardi and Craig, 1966).

Zuber (1966) used a general approach to the purification of Substance P not unlike the one that succeeded for Chang and Leeman: G-25 Sephadex, Carboxymethyl Sephadex, and preparative high-voltage paper electrophoresis. Their product was apparently the purest available until the complete purification by Chang and Leeman, but it was inhomogeneous. Again, we cannot say exactly what the sources of difficulty were, but a few relatively minor variations may account for the difference in final result. As mentioned above, the starting material was whole bovine brain instead of hypothalamus; thus, these workers began with material of lower specific activity than did Chang and Leeman. Their extraction procedure involved acetic acid extraction of an ammonium sulfate precipitate, followed by further precipitation of the active material by ether. The several precipitation steps could have led to substantial losses. Also, Zuber carried out electrophoresis at pH 9.5, at which pH Substance P probably does not run in a well-defined band.

We cannot be sure that any of the above-mentioned differences in the approach of Chang and Leeman from that of previous workers actually account for the difference in success, but they at least serve to demonstrate some of the issues that must be considered in performing an isolation.

4.5. Amino Acid Sequence

The first sequence studies that led to a proposed structure for Substance P were performed in collaboration with Hugh D. Niall at the Massachusetts General Endocrine Unit (Chang *et al.*, 1971) on material isolated from bovine hypothalami and superior and inferior colliculi (Chang and Leeman, 1970). The results of sequential Edman degradation on the intact peptide clearly established the amino-terminal sequence, H–Arg–Pro–Lys–Pro–. Other studies on the unseparated mixture of peptides produced by chymotryptic cleavage, when taken with carboxy-terminal analyses, argued for the carboxy-terminal sequence –Phe–Phe–Gly–Leu–Met–X, in which X was assumed to be NH_2. Two Gln residues, shown to be amidated by total enzymic digestion, were placed in the middle of the peptide by difference.

The structure of Substance P of equine intestinal origin was investigated by Studer *et al.* (1973). These workers also experienced some difficulty in directly positioning the Gln residues; however, their results argued for the same sequence proposed for the hypothalamic peptide. Their work, also based on chymotryptic treatment of the peptide, argued for a major cleavage by this enzyme at a place different from that reported by Chang *et al.* (1971).

Recently, Carraway and Leeman (unpublished results) have confirmed by a

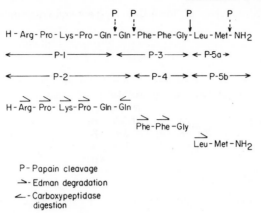

P - Papain cleavage

→ - Edman degradation

↙ - Carboxypeptidase digestion

FIG. 2. Complete amino acid sequence of Substance P and the alignment of the fragments obtained by papain cleavage. (P) Papain; ($P_{1-4,5a,b}$) papain peptides.

more rigorous analysis the proposed structure that is summarized in Fig. 2. The Substance P was obtained as a side-product during the isolation of neurotensin from bovine hypothalami, and was judged pure by multiple criteria. The pure peptide was treated with papain, and the 6 peptides indicated in Fig. 2 were isolated by high-voltage paper electrophoresis. The structure of each peptide was determined by amino-terminal and carboxy-terminal analyses as shown in Fig. 2 for 3 of the peptides. The carboxy-terminal Met–NH_2 was identified directly by high-voltage paper electrophoresis. Total enzymic digestion of each peptide established that the two glutamic acids were amidated, and argued against the presence of substituents. In total, the results unambiguously established the placement of each of the amino acids, including the glutamine residues, and defined the nature of the carboxy-terminal blocking group. Thus, Substance P from several different tissue sources has been found to have the same amino acid sequence.

The structure of Substance P is closely related to that of several other peptides (Table 3): physalaemin, isolated from the skin of a South American amphibian (Erspamer *et al.*, 1964); eledoisin, isolated from the salivary glands of a cephalopod (Erspamer and Anastasi, 1962); and uperolein, isolated from the skin of an Australian amphibian (Anastasi *et al.*, 1975). These undecapeptides differ somewhat in their amino-terminal regions, but their carboxy-terminal regions are very similar; indeed, the last three residues are identical in all these peptides. Physalaemin and eledoisin have biological properties in common with Substance P, suggesting that the biological activity resides in the carboxy-terminus. Much work on fragments and analogues of these peptides confirms this view (Schröder and Lübke, 1966). Bergmann *et al.* (1974) report that carboxy-terminal fragments

TABLE 3. Structures of Substance P and Related Peptides

Substance P	Arg–Pro–Lys–Pro–Gln–Gln–Phe–Phe–Gly–Leu–Met–NH_2
Physalaemin	<Glu–Ala–Asp–Pro–Asn–Lys–Phe–Tyr–Gly–Leu–Met–NH_2
Eledoisin	<Glu–Pro–Ser–Lys–Asp–Ala–Phe–Ile–Gly–Leu–Met–NH_2
Uperolein	<Glu–Pro–Asp–Pro–Asn–Ala–Phe–Tyr–Gly–Leu–Met–NH_2

TABLE 4. Relative Activities of Some Carboxy-Terminal Fragments of Substance P

Fragment	Sialogogic activity $(\%)^a$	Immunologic activity $(\%)^b$
H–Arg–Pro–Lys–Pro–Gln–Gln–Phe–Phe–Gly–Leu–Met–NH₂ (Substance P)	100	100
H–Pro–Gln–Gln–Phe–Phe–Gly–Leu–Met–NH₂	12	47
<Glu–Phe–Phe–Gly–Leu–Met–NH₂	15	0.01
H–Phe–Gly–Leu–Met–NH₂	≈1	<0.00004
H–Leu–Met–NH₂	<0.04	<0.00004

^{a} Based on the intravenous dose needed to elicit 50 μl saliva from a 100-g rat anesthetized with Nembutal; 100% = 40 pmol.
^{b} Based on the dose needed to displace 50% of bound counts under the conditions described in the text. 100% = 8 fmol.

of Substance P, especially the octapeptide, have even greater activity than the intact peptide when tested on isolated guinea pig ileal tissue. In our laboratory, however, the fragments of Substance P that we have tested using the sialogogic assay do not have complete biological activity (Table 4).

4.6. Synthesis

Tregear *et al.* (1971) performed the first solid-phase synthesis of a peptide according to the structure determined for Substance P. Since Substance P has a C-terminal methioninamide, the synthesis was performed using a benzhydrylamine resin as support. Cleavage of the final product from this resin with hydrofluoric acid directly yielded a molecule with an amidated carboxy-terminus. The purified synthetic product was found to be chemically and biologically indistinguishable from native Substance P. A comparison of the specific activity of native and synthetic material using the sialogogic assay is shown in Fig. 3.

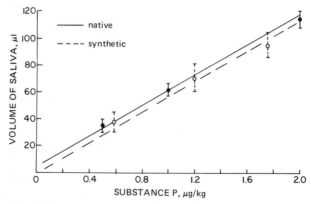

FIG. 3. Comparison of the sialogogic activity of native Substance P (solid line) and synthetic Substance P (dashed line). Each point represents the mean ± S.D. for 3 determinations. All rats weighed between 90 and 120 g. From Tregear *et al.* (1971).

4.7. Radioimmunoassay

As discussed above, biological assays for Substance P were difficult to perform properly, and were plagued with problems from interfering substances. The fairly specific sialogogic assay was extremely insensitive, requiring some 30–40 pmol to produce a reasonable response of 50 μl saliva in a 100-g test rat. The various gut-contraction assays were more sensitive, but less specific for the undecapeptide isolated by Chang and Leeman. The availability of large quantities of pure synthetic material suggested that a RIA for Substance P be developed. This was done, again in collaboration with the Endocrine Unit of the Massachusetts General Hospital.

Substance P presented two difficulties in the development of a RIA: First, it is a fairly small molecule, and thus would not be expected to be immunogenic. Thus, the peptide was coupled to larger carrier molecules, such as bovine γ-globulin or thyroglobulin, and such conjugates were used for the production of antisera, using the method of Vaitukaitis *et al.* (1971). The second problem was the lack of a tyrosine in the native molecule, so that Substance P could not be iodinated directly in order to produce tracer for the assay. Several analogues of Substance P containing tyrosine were tested for biological activity. One, containing tyrosine instead of phenylalanine in position 8 (in analogy to the related peptide physalaemin), was found to have sialogogic activity equal to that of native Substance P, and thus was chosen as the most appropriate Substance P analogue for purposes of iodination.

The initial report of the RIA (Powell *et al.*, 1973) contained several demonstrations of the validity of the assay: The closely related peptides physalaemin and eledoisin were not detected by the assay. The dose–response curves of the RIA for standard synthetic Substance P and for a variety of tissue extracts were parallel. Over a wide range of purification of such extracts, the Substance P concentration determined by the sialogogic assay and that detected by the RIA were identical. Thus, the immunoassay was quite specific for Substance P. Unfortunately, the RIA reported in 1973 was only marginally more sensitive than the more sensitive bioassay techniques, requiring some 100 fmol Substance P in the incubation volume of 0.5 ml for significant displacement of iodinated tracer. Since that time, two improvements have been made, increasing the sensitivity of the assay by almost 2 orders of magnitude. First and most important, another antiserum was obtained that had higher affinity for Substance P, thus improving the detection limit, under the same conditions, to 10–15 fmol Substance P. Further increases in sensitivity have been made by using highly purified monoiodinated 8-tyrosyl Substance P (sp act $\approx$ 2000 Ci/mmol) as tracer, adding fewer counts to the incubation mixture, but counting for a longer time. Under these conditions, the assay can now detect 1.5–2 fmol Substance P.

Use of such low concentrations of Substance P has demonstrated another technical problem in dealing with Substance P: its tendency to adsorb to glass and plastic surfaces. This tendency was shown even for impure preparations by Cleugh and Gaddum (1963). Such behavior is seen with most peptides and proteins in dilute solutions, but Substance P seems to be especially troublesome in this regard. Precautions such as adding gelatin at 1 mg/ml help, but such solutions stored in a refrigerator overnight lose much of their recoverable Substance P. Serial dilutions of Substance P standards or samples, though reproducible, tend to lead to erroneous results, apparently due to adsorption to pipette tips. For this reason, solutions are kept in the liquid state for as little time as possible, and standards and samples are prepared to be as close as possible in their final concentrations of Substance P, and are delivered by individual pipettings in identical manner.

After incubation of standards or samples, antiserum, and tracer together for 4 days, separation of antibody-bound from free tracer is performed with dextran-treated charcoal. Data are analyzed with the logit-log linearization method. This method has the advantage of allowing quantitative statistical comparison of sample and standard dose–response curves as a continuing check on the validity of the assay when applied to samples from new sources.

Of course, any measurements of Substance P made with RIA are really measurements of Substance P–like immunoreactivity. Table 4 shows that the present assay will sense the carboxy-terminal octapeptide in addition to the intact undecapeptide. Thus, the possibility that fragments of Substance P, or larger molecules with similar immunological determinants, will be included in such values cannot be immediately ruled out, and data must be considered with this caution in mind.

Early RIA studies essentially confirmed the tissue distribution of Substance P and its distribution in subcellular fractions of brain homogenates that had been determined previously by bioassay. It is found in highest concentrations in the hypothalamus, preoptic area, and brainstem, with olfactory bulb, cerebral, and cerebellar cortex having a much lower concentration. Substance P is concentrated in the ''synaptosomal'' fractions of brain homogenates. The increased sensitivity and simplicity of the present RIA has enabled us to analyze Substance P levels in much smaller regions of brain than had previously been possible. In collaboration with Dr. Michael Brownstein and co-workers (Brownstein *et al.*, 1976), levels of Substance P in microdissected regions of rat brain have been analyzed, and 10-fold differences in levels of Substance P even within regions such as the mesencephalon have been found. Biossay studies had shown that the substantia nigra has the highest concentration of Substance P in the brain. This finding has been confirmed by immunoassay, and a differential distribution within the substantia nigra has been shown with levels in the zona reticulata (11.4 pmol/mg protein) being 4 times those in the zona compacta (2.98 pmol/mg protein.). Levels in the red nucleus (1.34 pmol/mg protein) and the colliculi (1.77 pmol/mg superior, 1.17 inferior) are comparatively unremarkable. Within the hypothalamus, the level in the paraventricular nucleus is twice that in the supraoptic (3.06 vs. 1.58 pmol/mg protein), while the median eminence is relatively poor (1.00 pmol/mg protein.)

The sensitivity of the RIA has enabled us to investigate the possible existence of Substance P–containing tracts in the CNS by examining the effects of restricted lesions on Substance P levels in regions receiving endings from axons arising in or passing through the lesioned area. Although the Substance P level in the superior colliculus is unaffected by unilateral enucleation of the eye in the white rat (Lam, Leeman, and Mroz, unpublished data), lesions of the habenula lead to a 75% reduction in Substance P in the interpeduncular nucleus, and results of several studies point to the presence of Substance P–containing fibers in the internal capsule passing into the reticularis of the substantia nigra, accounting for the high concentration of Substance P in that region. (Mroz *et al.*, 1976, 1977).

Substance P–like immunoreactivity has been reported in plasma by Nilsson *et al.* (1975*b*). We have also found immunoreactive material in unextracted plasma, at about the same level as reported by these workers (approximately 10^{-10} mol/liter), but preliminary results indicate that about half this apparent immunoreactivity does not disappear on prolonged incubation of the plasma under conditions where added Substance P is quantitatively destroyed. Furthermore, comparison of values obtained from ethanol-extracted with unextracted plasma samples has led us to question the validity of RIA for Substance P in unextracted plasma samples. This question can be resolved only by determination of the chemical nature of the material or materials leading to apparent immunoreactivity in unextracted plasma.

4.8. Immunocytochemical Studies

Although this laboratory has not undertaken studies involving immunocytochemical localization of Substance P, at least two groups have now begun such work. A Swedish group has reported studies of the distribution of Substance P–like immunoreactivity throughout the body, using fluorescent antibody methods (Nilsson *et al.*, 1974, 1975*a*; Hökfelt *et al.*, 1975*a*), and Pearse and Polak (1975) have studied the distribution in the gastrointestinal tract. The presence of Substance P in the nervous system has been shown to be concentrated in nerve terminals, as had been expected. This approach has also been used to show the disappearance of Substance P from the dorsal horn of the cat spinal cord following transaction of the dorsal root (Hökfelt *et al.*, 1975*b*).

Work on the gastrointestinal tract (Nilsson *et al.*, 1975*a*; Pearse and Polak, 1975) has produced a surprising finding of substance P–like immunoreactivity in endocrinelike cells in the intestinal mucosa, suggesting a possible hormonal role for the peptide.

4.9. Substance P as a Neurotransmitter

The result of most relevance to neurobiology obtained with Substance P since its isolation and availability in pure synthetic form has been its identification as the motoneuron-depolarizing peptide detected by Otsuka and his associates. These workers had begun a search for a new transmitter candidate. They had developed an assay for their transmitter candidate, an isolated preparation of frog spinal cord. The test samples were applied into the tissue bath, and the electrical response of the motoneurons was recorded from the ventral root. As a source of tissue, they chose the region of bovine nervous tissue most likely to contain a high concentration of the suspected transmitter, the dorsal sensory ganglion. Chromatographic fractions of crude extracts of this tissue did show depolarizing activity, while similar samples prepared from ventral roots, as expected, showed much less. These fractions also showed hypotensive and gut-contracting properties, but all these effects were destroyed by chymotrypsin, demonstrating the peptide nature of the active principle or principles (Otsuka *et al.*, 1972). With this information in hand, Otsuka and his colleagues set out to isolate and identify the motoneuron-depolarizing material, using the isolated frog spinal cord as the assay for screening fractions.

During the course of their work, they examined the effects of various known peptides on this bioassay system. They found that physalaemin (Konishi and Otsuka, 1971) and eledoisin both had depolarizing effects in their system, but these peptides were chemically distinguishable from the motoneuron-depolarizing peptide extracted from bovine tissue. Otsuka, Leeman, and their associates began to compare Substance P, which was just then in synthetic form, with the motoneuron-depolarizing peptide.

Although the motoneuron-depolarizing peptide has not yet been isolated in pure form, a wealth of evidence shows that it is probably identical to Substance P (Takahashi *et al.*, 1974). The two peptides share the chymotrypsin-sensitive ability to depolarize motoneurons in the frog spinal cord assay, to contract guinea pig ileum, and to lower blood pressure. Although physalaemin and eledoisin share the same qualitative effects, the quantitative responses are most consistent with the depolarizing peptide's being Substance P, rather than either of the other peptides. The molecular weights of the depolarizing peptide and Substance P, as determined on a Sephadex LH-20 column, are identical within experimental error. The electrophoretic mobilities of Substance P and the depolarizing peptide are identical at either pH 2.0 or 6.2, whereas physalaemin and eledoisin differ substantially from the depolarizing peptide in mobility at either pH.

Furthermore, in a RIA using an antiserum raised against Substance P that did not respond to either physalaemin or eledoisin, it was shown that the motoneuron-depolarizing peptide reacted identically to synthetic Substance P. This chemical, pharmacological, and immunological evidence leaves little doubt that the motoneuron-depolarizing peptide of Otsuka and his colleagues is identical to Substance P. These findings substantiated the earlier suggestion of Lembeck (1953), based on the distribution of Substance P determined by biosassay, that Substance P might be a transmitter of primary sensory neurons. Otsuka's group has continued studying the effects of Substance P, and has more recently begun to use a mammalian system to investigate its properties. They use the hemisected spinal cord of a newborn rat as an assay system (Otsuka and Konishi, 1974), and are continuing to characterize the effects of Substance P upon this system. It has been shown to be considerably more potent than glutamate, the previous transmitter candidate, in depolarizing motoneurons. Konishi and Otsuka (1974) also showed that Substance P exerts its depolarizing action in the presence of low Ca^{2+} and high Mg^{2+}, indicating that its effects are probably postsynaptic in nature. Furthermore, Lioresal, a GABA derivative, blocks the monosynaptic reflex in this preparation and blocks the response to bath-applied Substance P, but has comparatively little effect on the response to glutamate (Saito *et al.,* 1975).

Henry *et al.* (1975), however, have shown little direct effect of Substance P when they applied it to motoneurons, though responses have been observed from interneurons in the spinal cord and higher in the CNS (Krnjevic and Morris, 1974; Phillis and Limacher, 1974). Such responses to iontophoretically applied Substance P have been of a comparatively long-latency, long-lasting sort, in contrast to the effect expected for a primary sensory transmitter, especially in the monosynaptic reflex pathway.

Another crucial gap in the argument that Substance P might be a transmitter has been the apparent lack of success thus far in demonstrating its release under appropriate conditions from neural tissue. At this point, however, Substance P still seems to be the best candidate for the transmitter of primary sensory neurons. (The peptide released from cerebral cortex in the study of Shaw and Ramwell, 1968, contracted rat uterus, and thus was not Substance P.)

5. Neurotensin

5.1. Detection of Activity

During the course of work on the isolation of Substance P, a second unexpected biological activity was found. When testing fractions for sialogogic activity after passage of a hypothalamic extract over sulfoethyl Sephadex, Leeman detected the presence of material in a region of the column effluent, distinct from the sialogogic region, that caused a marked vasodilatation in the exposed cutaneous areas of anesthetized rats; larger doses of this vasodilatory factor caused the rats to become severely cyanotic. Later, it was found that the vasodilatation was associated with a transient hypotension, and that these effects were labile to proteolytic digestion. After the peptide was isolated (Carraway and Leeman, 1973), it was named *neurotensin* (NT), because of its presence in neural tissue and its ability to affect blood pressure.

5.2. Quantitation of Activity

The vasodilatory effect of NT can be seen easily and graded visually; therefore, this simple response was used to locate and quantitate NT throughout the rest of the purification steps. Fractions were roughly quantitated by determining the dilution of the sample at which the vasodilatation could no longer be detected.

The presence of NT was masked in the initial crude extract by other vasoactive factors, e.g., Substance P and vasopressin. Thus, the marker used to obtain the fractions containing NT during the initial G-25 Sephadex steps was the sialogogic activity of the cochromatographing Substance P. After ion-exchange chromatography on SE-Sephadex, which separated Substance P and NT, the vasodilatory response was used routinely to locate NT.

Once it was established that the vasodilatory response to NT is associated with a hypotensive effect, a more precise estimate of the biological activity of NT became possible. A unit of NT activity was defined as that amount of material necessary to cause a fall in systemic blood pressure of about 35 mm Hg in a test rat weighing approximately 250 g. This measure was important, of course, in assessing the specific activity of NT during its purification. However, since the procedure is rather cumbersome, and since the response exhibits acute tachyphylaxis, the blood pressure assay was not used routinely for the screening of fractions. Instead, the active region was usually located using the vasodilatory response, and the activity in the pool was determined using the blood pressure assay.

Although it was precise, the blood pressure assay was known not to be specific for NT; therefore, it was also required that the active material elicit cyanosis when given at higher doses to anesthetized rats. The cyanotic effect of NT may be unique to the family of NT-like peptides since, of the numerous substances tested (Carraway, 1972), only the related peptide xenopsin has been shown to display this activity (see Section 5.8).

5.3. Identification as a Peptide

Incubation of the nonsialogogic, vasoactive material obtained after ion-exchange chromatography in SE-Sephadex with two rather nonspecific proteolytic enzymes, pronase and papain, as well as enzymes with high substrate specificity such as trypsin and chymotrypsin, destroyed its biological activity, as measured by the vasodilatory response and its hypotensive action (Carraway and Leeman, 1973).

5.4. Isolation Procedures

5.4.1. Isolation Steps

The purification scheme used for the isolation of NT from hypothalamic tissue was based on four basic techniques: extraction into aqueous acetone, gel filtration, ion-exchange, and paper electrophoresis (Carraway and Leeman, 1973). Overall, these procedures effected a several million-fold purification with respect to protein (Table 5). Concern that the lengthy flash-evaporation step that previously served to remove the acetone from the initial acidic extract might be harsh enough to alter the peptide slightly prompted us to search for a different concentrating procedure. Batch adsorption of sulfopropyl–Sephadex, which could be performed in the presence of as much as 50% acetone, proved to be a good substitute, and is employed now. Whereas the adsorption and elution could be performed at 4°C within 1 day, the flash evaporation was performed at 35–45°C and required 2–3 days for completion. Furthermore, the eluate from the SP-Sephadex (1 liter) could be subjected to gel chromatography immediately, while the aqueous residue of flash evaporation (40 liters) was usually lyophilized, which took a week's time.

TABLE 5. Scheme for Purification of Neurotensin (NT) from Bovine Hypothalami (45 kg)

Purification step	Total Protein (mg)	Total NT (doses)	Yield of NT (%)	Specific activity (doses/mg)
80% Acetone extraction	270,000[a]	15,000[b]	100	0.06
Gel chromatography on Sephadex G-25				
First	10,000[a]			
Second	1,500[a]	11,000	75	7.0
Chromatography on SE-Sephadex				
First	250[a]	7,500	50	30.0
Second	10[c]	6,000	40	600.0
Paper electrophoresis at pH 3.5	0.3[c]	3,100	21	10,300.0

[a] Protein is expressed as absorbance units at 280 nm.
[b] Calculated by assuming the same yield of NT as for Substance P through the two initial gel-chromatography stages.
[c] Protein was calculated from quantitative amino acid analyses.

We recently accomplished the isolation of immunoreactive NT from intestinal tissue (Kitabgi *et al.*, 1976), in which it is also present in high concentration as will be discussed later. The availability of antibody directed toward NT prompted us to add an affinity-chromatography step to our original procedure. An antibody–Sepharose conjugate with a high capacity for NT was prepared and used to specifically adsorb immunoreactive NT; after extensive washing of the conjugate, the adsorbed materials were eluted and collected. This process was repeated as many times as necessary to purify all the immunoreactivity, which eluted from the immunoadsorbent in a near pure state. However, the final product obtained after paper electrophoresis appeared to contain not only NT, but also some of another peptide with the amino acid composition of des-Leu[13]–NT (possibly a carboxypeptidase degradation product of NT, although its structure has not been determined). This other peptide not only registered in the "specific" RIA for NT, but also was carried along through the antibody-affinity chromatography step. These results demonstrate the usefulness of immunoadsorbents, but also clearly illustrate that antibodies are usually only relatively specific for their respective antigens, and should be employed with this in mind.

5.4.2. Criteria of Purity

The purity of the peptide obtained from bovine hypothalami was evidenced by the fact that acid hydrolysis gave integral molar ratios of the constituent amino acids, and reelectrophoresis on paper at three different pH values did not alter the amino acid composition or the biological specific activity (Table 6). When stained for peptide using ninhydrin or Cl_2-O-tolidine, each of these paper electrophoregrams displayed a single spot that was adjacent to the region of biological activity. Other evidence for homogeneity stemmed from amino- and carboxy-terminal analyses. The amino-terminus of the peptide was blocked, since it was unreactive toward dansyl-chloride and the Edman reagent; however, the carboxy-

TABLE 6. Molar Ratios of Amino Acids in Neurotensin after Electrophoresis

Amino Acid	Electrophoresis at pH 3.5	Electrophoresis at pH 3.5, then at pH 6.5	Electrophoresis at pH 3.5, then at pH 8.9	Assumed residues per mol peptide
Lysine	1.0	1.2	1.1	1
Arginine	2.1	1.9	2.0	2
Aspartic acid	1.2	1.1	1.1	1
Glutamic acid	1.8	2.1	2.0	2
Proline	2.1	1.9	1.8	2
Isoleucine	1.0	1.0	1.0	1
Leucine	2.0	1.9	2.0	2
Tyrosine	2.1 (2.2)[a]	2.0 (2.0)[a]	1.8	2
TOTAL				13
Specific activity (doses/mg)	10,500	10,000	10,500	

[a] The value in parentheses is that of tyrosine determined spectrophotometrically.

terminus was free. Treatments with carboxypeptidase A&B released sequentially in time 1.0 mol of Leu, Ile, and Tyr per mole of the peptide, arguing strongly for the presence of a single peptide with a COOH-terminal sequence, –Tyr–Ile–Leu–OH.

5.5. Amino Acid Sequence

The amino acid sequence of NT was deduced from information obtained on the intact peptide, as well as on its tryptic, chymotryptic, and papain-generated fragments (Carraway and Leeman, 1975a). The structure of NT and the sites of cleavage by the various enzymes are shown in Fig. 4; the results of sequence studies on the papain-generated fragments, P-1, P-2, and P-3, are given in the lower part of the figure. Similar studies were performed on the other fragments, and all information obtained was consistent with the structure of NT shown. The results of total enzymic digestions indicated that all amino acids were in the L-configuration, that there were no non-amino-acid substitutes attached to the molecule, and that the aspartic acid was in the amidated form. Also consistent with this structure were the results of charge measurements for each of the peptides at pH 1.9 and 6.5 by a technique involving paper electrophoresis (Offord, 1966). The presence of an amino-terminal pyrrolidone carboxylic acid residue (< Glu) was established by treatment of NT and its fragments with the specific enzyme pyrrolidonecarboxylyl peptidase (<GLUase). The COOH-terminal amino acid sequence was determined by kinetic experiments using carboxypeptidase A&B.

Comparison of the amino acid sequence of NT with the sequences of other hypothalamic peptides indicated that there was a very distant relatedness among NT, vasopressin, and LH–RH (Fig. 5). When aligned from their amino-termini, 5 of the 9 amino acids in vasopressin and 5 of the 10 amino acids in LH–RH were either identical or similar to the corresponding ones in NT. Fewer similarities were

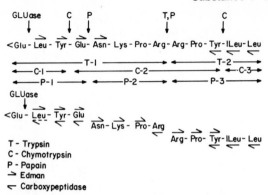

FIG. 4. Complete amino acid sequence of NT and the alignment of the fragments obtained by enzymic cleavage. (<GLUase) Pyrrolidonecarboxylyl peptidase; (T-1, 2) tryptic peptides; (C-1–3) chymotryptic peptides; (P-1–3) papain peptides.

noted for all other peptides examined. These data suggest that NT, vasopressin, and LH–RH may have derived from a common ancestor that underwent a gene-duplication event; at this time, however, the relationship is rather weak.

5.6. Synthesis

The first synthesis of NT was accomplished in our laboratory using the Merrifield solid-phase procedure, performed manually (Carraway and Leeman, 1975b). Our approach was to assemble the molecule in the Gln^1-form and to cyclize this to the $<Glu^1$-form, reasoning that purification from some of the possi-

NEUROTENSIN	< Glu - Leu - Tyr - Glu - Asn - Lys - Pro - Arg - Arg - Pro - Tyr - Ileu - Leu-OH
VASOPRESSIN	H- Cys · Tyr - Phe Gln - Asn - Cys - Pro Arg - Gly - NH₂
LH – RH	<Glu - His - Trp Ser - Tyr - Gly - Leu - Arg - Pro Gly - NH₂
OXYTOCIN	H- Cys · Tyr - Ileu Gln Asn - Cys - Pro Leu-Gly - NH₂
TRH	<Glu - His - Pro- NH₂
SOMATOSTATIN	H- Ala - Gly - Cys - Lys Asn Phe- Phe- Trp - Lys Thr Phe Thr – Ser - Cys - OH
SUBSTANCE P	H - Arg - Pro - Lys - Pro Gln Gln - Phe - Phe - Gly - Leu - Met - NH₂

[a] Shifting the Arg - Gly - NH₂ segment of vasopressin and the Arg-Pro- Gly- NH₂ segment of LH - RH one position to the right enhances their relatedness to NEUROTENSIN.

FIG. 5. Comparison of the amino acid sequence of NT with the sequences of several hypothalamic peptides of mammalian origin. Residues of the other peptides that are identical with the corresponding ones of NT are enclosed in rectangles; those that are related through highly favored codon substitutions are enclosed in circles. <Glu was taken to be equivalent to glutamine. From Carraway and Leeman (1975a).

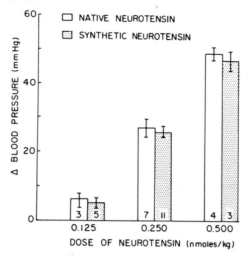

FIG. 6. Comparison of the hypotensive effect of native and synthetic NT. Plotted is the fall in carotid blood pressure (mean ± S.E.M.) measured 30 sec after injection of a test sample into the femoral vein of rats weighing about 250 g. Each injection was done in a separate rat. From Carraway and Leeman (1975*b*).

ble side-products might be facilitated in this manner. The final product, obtained in a 38% overall yield after purification, appeared to be homogeneous during chromatography and electrophoresis, and possessed the desired amino acid composition and sequence. Furthermore, it gave rise to homogeneous peptide fragments when treated with various specific enzymes. When compared with native NT, the synthetic product was indistinguishable by multiple chemical and biological criteria, e.g., the hypotensive effect shown in Fig. 6.

5.7. Biological Properties

Since its isolation, identification, and synthesis, NT has been shown to be capable of a number of biological actions (Table 7); however, whether it exerts any of these effects physiologically is not yet known. NT can be classified as a

TABLE 7. Some Biological Actions of Neurotensin

In vivo (anesthetized rat)	Vasodilatation and hypotension
	Increased vascular permeability
	Stasis and cyanosis
	Hyperglycemia
	Increased corticosterone secretion
	Increased LH and FSH Secretion without effect on TSH and GH
In vitro	Contraction of guinea pig ileum
	Relaxation of rat duodenum
	Chemotaxis of human leukocytes

"kinin," since it is a hypotensive peptide that stimulates the contraction of the isolated guinea pig ileum, while it relaxes the rat duodenum in an organ bath. However, NT also possesses other biological properties that may not be shared by the "kinin family," the most striking of which is its ability to induce cyanosis in the anesthetized rat. Circumstantial evidence suggests that this response is likely to be the result of a stasis of blood in the peripheral tissues, rather than an effect on respiration or oxygen exchange; i.e., it is not associated with a change in the partial pressure of oxygen or carbon dioxide in arterial blood; however, it is accompanied by a dramatic increase in vascular permeability to proteins and a marked hemoconcentration. The increase in vascular permeability induced by NT can be demonstrated easily in the rat or guinea pig that has received an intravenous injection of Evan's blue dye. The dye has a high affinity for albumin, and remains within the vascular compartment until the animal is challenged with NT. Within minutes after an intravenous (tail vein) injection of NT (>1 nmol/kg), the dye is seen to be extruded from the circulation into various tissues which appear blue. It is interesting that while most tissues are "blued," especially the skin and abdominal organs, the brain and the lungs appear not to be affected. When NT is administered intradermally on the shaved backs of dye-injected animals,

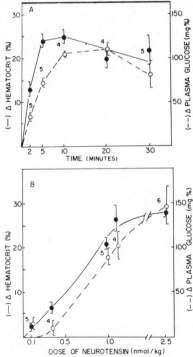

FIG. 7. Time-course and dose–response relationships for the effects of NT on hematocrit and on plasma glucose levels. Plotted are the increments observed (mean ± S.E.M.) for the numbers of animals indicated. (A) Time-course of response to a single dose of 1 nmol/kg; (B) dose-response obtained at 15 min.

blue spots appear at the sites of injection; the size of the blue spot and the amount of dye extractable from it are related to the dose of NT given. The hemoconcentration induced by NT is easily demonstrated by measuring hematocrit; within minutes after an intravenous injection of NT into rats, hematocrit increases dramatically (Fig. 7), probably the result of a movement of plasma protein and water from the circulation into the extravascular space.

NT is a potent hypotensive agent in the anesthetized rat, the threshold intravenous dose for a measurable response being about 0.1 nmol/kg (see Fig. 6). This effect is likely to represent a direct action of NT on the vasculature, as it is not altered by prior hypophysectomy or adrenalectomy, or by pretreatment of the test animal with atropine, phenozybenzamine, or propranolol. The hypotensive effect of NT exhibits acute tachyphylaxis; i.e., a second equal dose given 1, 10, or 60 min after an active first dose produces no effect. However, a second dose given several hours later is effective. The mechanism of this effect is not yet clear; however, present results are consistent with the notion that tachyphylaxis involves a tight binding of NT to its "vascular receptors," so as to maintain them in an inactive form until dissociation takes place. At any rate, NT, whether given as a bolus or as a constant infusion, does not appear to be able to maintain its hypotensive action over a prolonged period.

In addition to its effects on the circulation, NT also stimulates liver glycogenolysis in the anesthetized, fed rat, rapidly bringing about a marked hyperglycemia (Fig. 8). The increment observed in plasma glucose is related to the decrement in liver glycogen, which is also roughly sufficient to account for the glucose units appearing in the blood. A dose-dependent activation of liver glycogen phosphorylase has been noted to precede the development of the hyperglycemia; however, the precise site of action of NT in promoting this effect is not yet clear. Since NT has been shown to be ineffective in a liver slice preparation shown to respond to

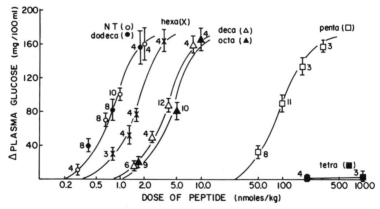

FIG. 8. Hyperglycemic effect of NT and its carboxy-terminal partial sequences. Plotted is the increment in plasma glucose levels measured 15 min after tail vein injection of test samples into rats weighing 250–300 g, as described in the text. From Carraway and Leeman (1975c).

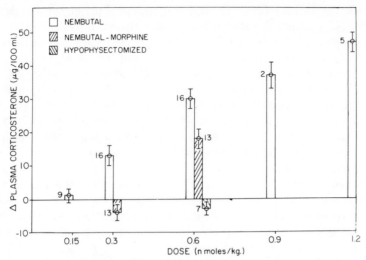

FIG. 9. Dose–response relationships for the steroidogenic effect of NT in various treated rats. The increment in plasma corticosterone levels above saline controls obtained 15 min after injection is plotted against the dose of NT given. From Carraway (1972).

glucagon, a mediator of its effect on the liver has been sought. The hyperglycemic response occurs in hypophysectomized as well as in adrenalectomized animals, indicating that the constituents of these tissues are not required for the effect. In addition, the response is not prevented by pretreatment of the test animal with the neural blocking agents phenoxybenzamine, propranolol, and morphine, nor with the catecholamine-depleting agent reserpine. There is, however, some suggestion that pancreatic glucagon might mediate this action of NT. Although we have been unable to demonstrate an elevation of glucagon levels in either peripheral or hepatic portal vein blood after injection of NT, Brown and Vale (1976) have recently reported evidence of a rapid, short-lived burst in glucagon secretion following NT. However, the dose of NT employed by this group, $2\,\mu g$/animal, was rather high, being about 5 times that required for half-maximal hyperglycemia; it is possible that glucagon is involved in the response to high doses, but does not mediate the effect at lower doses.

Intravenous injection of NT into rats did not raise the plasma levels of growth hormone (GH) or thyroid-stimulating hormone (TSH); however, it was found to promote the release of ACTH from the pituitary gland, as evidenced by the resultant increase in plasma corticosterone levels (Carraway, 1972). Figure 9 shows the dose–response relationship obtained 15 min after injection of NT into untreated, morphine-pretreated, and hypophysectomized rats under pentobarbital anesthesia. The results show that the NT-induced increase in plasma corticosterone is abolished by prior hypophysectomy, and significantly diminished ($P < 0.01$) by morphine pretreatment. That the presence of the pituitary gland is essential for the response suggests that ACTH is mediating the steroidogenic effect. It can be concluded from the morphine inhibition that NT is acting primarily through the CNS to promote the ACTH release. The response elicited at the higher dose probably represents NT overcoming the morphine block; however, it might also be attributable to a direct effect on the pituitary. In total, the data show that the most sensitive pathway (threshold 0.2–0.3 nmol/kg) for stimulation of ACTH secretion by NT involves excitation of the CNS and release of endogenous CRF or CRFs.

NT has also been shown to bring about increased secretion of pituitary luteinizing hormone (LH) and follicle-stimulating hormone (FSH) when injected intravenously into estrogen/progesterone primed ovariectomized rats (Makino *et al.*, 1973). It is likely that this response is mediated via release of hypothalamic LH–RH, since it is relatively slow (30 min) and is associated with a fall in hypothalamic LH–RH content.

It is possible that all the actions of NT *in vivo* described above are secondary to the hypovolemia that results from its effects on the vasculature. This notion is supported by the fact that the threshold doses for these effects are similar, and by the finding that the partial sequences and analogues of NT thus far examined display a similar potency for each response.

Recently, NT has been shown to be chemotactic for human neutrophils (Carraway *et al.*, 1974), and to interact with these cells to desensitize them to further chemotactic stimuli. Not much is known about the significance of this effect; however, that NT has this ability suggested to us that it might play a role in inflammation (see Section 5.9).

5.8. Biologically Active Region

An investigation of the structural requirements for the biological activity of NT indicated that the participants in the biological action of this peptide reside primarily in its carboxy-terminal region (Carraway and Leeman, 1975c). While amidation of the carboxy-terminus, digestion with carboxypeptidase, and substitution of the carboxy-terminal amino acid all destroyed biological activity ($<0.1\%$), alterations at the NH_2-terminus did not reduce intrinsic activity. Furthermore, NH_2-terminal partial sequences were inactive, while COOH-terminal partial sequences of 5 or more amino acids in length were gut-contracting, produced hypotension and hyperglycemia in the rat, and increased cutaneous vascular permeability in the guinea pig. Figure 8 shows dose–response relationships obtained for several COOH-terminal partial sequences of NT when assayed for ability to induce hyperglycemia in anesthetized rats; rather similar results were obtained in the other bioassays. The carboxy-terminal pentapeptide H–Arg–Pro–Tyr–Ile–Leu–OH had reduced binding affinity (about 1% that of NT), but expressed full intrinsic biological activity, indicating that it possessed all the structural features necessary for these actions. Affinity for the receptor jumped dramatically to about 50% with addition of another Arg residue to give H–Arg–Arg–Pro–Tyr–Ile–Leu–OH, but did not change much with further additions to the amino-terminus.

In keeping with these results, an antiserum (HC-8; see Section 5.9) shown to be directed specifically toward the carboxy-terminal half of NT neutralizes the biological actions of NT (Carraway and Leeman, unpublished results). It is also interesting to note that a nonmammalian peptide, xenopsin, <Glu–Gly–Lys–Arg–Pro–Trp–Ile–Leu–OH (Araki *et al.*, 1973), bears a striking resemblance to the carboxy-terminal region of NT, and although it lacks a corresponding amino-terminal segment, it possesses the biological properties of NT.

5.9. Radioimmunoassay

5.9.1. General Approach

In developing RIA for NT, an attempt was made to obtain multiple, site-directed antisera, some of which would preferentially bind to the "biologically active" carboxy-terminal region of NT. Although several methods of coupling to carrier protein were investigated, one in particular yielded the best antisera when rabbits were immunized with conjugates (Carraway and Leeman, 1974). This coupling procedure linked NT specifically through its lysine side chain, as shown in Fig. 10. Briefly, the coupling method involved the following steps: (1) esterification of the carboxyl-groups in NT; (2) addition of the NT-diester to the carbodiimide-activated, carboxyl-enriched protein[succinylated hemacyanin (HC), succinylated thyroglobulin (TG), or polyglutamic acid–lysine (PGL)]; and (3) deesterfication and dialysis of the NT–protein conjugate. The carboxy-terminus of the coupled NT was shown to be free (>90%) by treatments with carboxypeptidase A. Using these conjugates, several interesting antisera were obtained, three of which displayed very different specificities toward NT (Table 8). Antiserum PGL-4 appeared to bind exclusively to the carboxy-terminal region of NT, cross-reacting nearly 100% with all carboxy-terminal partial sequences of NT that were 4 amino acids or greater in length. Anti-serum HC-8 was also carboxy-terminally directed, but required about 8 amino acids for full binding affinity. In contrast, the binding of antiserum PGL-6 was sensitive to the removal of NH_2-terminal residues as well, reacting only 9% with des-<Glu[1]-NT. A diagrammatic representation of antibody-binding sites that is consistent with these data is illustrated in the schematic of Fig. 10.

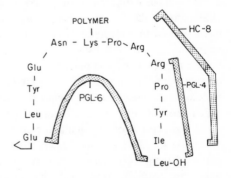

FIG. 10. Diagrammatic representation of the binding sites determined for several antisera toward NT. The antigenic conjugate of NT to protein (polymer) is shown, and the binding regions for the antisera are in brackets.

TABLE 8. Comparative Immunoreactivity of Neurotensin and Its Partial Sequences Using Various Antisera

Peptide	Immunoreactivity (%)[a]		
	PGL-4	HC-8	PGL-6
<Glu–Leu–Tyr–Glu–Asn–Lys–Pro–Arg–Arg–Pro–Tyr–Ile–Leu–OH	100	100	100
H–Leu–Tyr–Glu–Asn–Lys–Pro–Arg–Arg–Pro–Tyr–Ile–Leu–OH	100	100	9
H–Lys–Pro–Arg–Arg–Pro–Tyr–Ile–Leu–OH	100	100	9
H–Arg–Arg–Pro–Tyr–Ile–Leu–OH	100	50	9
H–Arg–Pro–Tyr–Ile–Leu–OH	75	2	3
H–Pro–Tyr–Ile–Leu–OH	440	<0.1	<0.1
H–Tyr–Ile–Leu–OH	4	<0.1	<0.1
H–Ile–Leu–OH	<0.1	<0.1	<0.1
<Glu–Leu–Tyr–Glu–Asn–Lys–Pro–Arg–Arg–Pro–Tyr–Ile–Leu–NH$_2$	<0.5	<0.5	<0.5
<Glu–Leu–Tyr–Glu–Asn–Lys–Pro–Arg–Arg–Pro–Tyr–OH	<0.1	<0.1	<0.1

[a] As determined by the amount of peptide needed to inhibit by 50% the binding of ^{125}I-NT to antiserum.

5.9.2. Native vs. Synthetic NT

Isolated native bovine NT and synthetic NT gave superimposable displacement curves in assays employing each of these antisera, indicating that native and synthetic NT, in addition to being chemically and biologically indistinguishable, are immunochemically equivalent.

5.9.3. Measurements in Hypothalami

In consideration of the limitations inherent in RIA (see Section 3), it was decided that although antiserum PGL-6 appeared to be relatively specific for NT, extracts would be examined with all three antisera, and equal immunological potency would be used as a criterion for similarity of unknown to NT. Crude acid–acetone extracts of bovine hypothalami were found to produce displacement curves closely paralleling that of synthetic NT; furthermore, measurements with the three antisera were in good agreement. This agreement is nicely illustrated in Fig. 11, which shows the profile of radioimmunossayable NT (R-NT) obtained when such an extract was subjected to gel chromatography on Sephadex G-25. One major peak of immunoreactivity, displaying equal potency with the three antisera, eluted in the region of NT. In order to prove that R-NT was, indeed, NT, and not a mixture of substances, the immunoreactivity in an extract of about 50 kg bovine hypothalami was purified to homogeneity and shown to be attributable to a peptide with the same amino acid composition and sequence as NT. These findings were taken as strong evidence that this RIA was giving a valid measure of the NT content of acid–acetone extracts of bovine hypothalami. Further studies indicated that extracts of hypothalami from mice, rats, rabbits, guinea pigs, and calves contained R-NT, and that the concentration equivalents of NT determined to be extracted from the tissues were very similar (40–70 pmol NT/g wet wt), irrespective of the antibody used for the measurement. In summary, the conclusions from

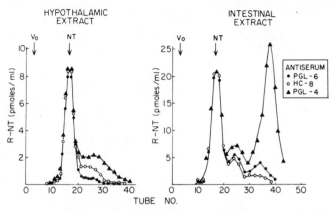

FIG. 11. Gel chromatography of bovine hypothalamic extract and rat intestinal extract on Sephadex G-25. (Left) Sample: acid–acetone extract of about 2 g bovine hypothalamic tissue; column size: 1.6 × 91 cm; fraction size: 75 ml initially, then 2.8 ml/tube; eluent: 0.1 M acetic acid. (Right) Sample: acid–acetone extract of about 100 g rat intestinal tissue; column size: 4 × 95 cm, with 1-liter volume; fraction size: 400 ml initially, then 12 ml/tube; eluent: 0.1 M acetic acid. Immunoreactivity was measured using the three antisera indicated.

this study were that: (1) the structure of NT in these various species was likely to be very similar; (2) the concentration of hypothalamic R-NT in these species was similar; and (3) the RIA was likely to be giving a valid measure of the NT content of this tissue. Assuming that breakdown of NT occurs at or near its receptor or receptors, giving rise to significant amounts of some fragments, then the fact that large amounts of such fragments were not present in the hypothalamus might lead one to conjecture that this tissue is primarily a storage site for NT.

5.9.4. Distribution in Brain

A study of the differential distribution of R-NT in the CNS of the adult rat (Fig. 12) indicated that even though NT was originally isolated from hypothalamic tissue, only about 30% of the total brain content of R-NT was found in this region, the bulk of the remainder being located in the midbrain and brainstem. The highest concentrations of R-NT, however, were found in the region of the hypothalamus and in the pituitary gland, and the lowest in the cerebellum. It is not yet known whether these measurements reflect the presence of NT or of similar peptides that cross-react in the RIA; however, crude extracts of extrahypothalamic tissue have been found to give discrepant measurements using the three antisera previously described. It is possible that NT is confined to the hypothalamus and that these inconsistent measurements in other regions indicate the presence of other biologically active, NT-related peptides which comprise the "NT family"; however, high concentrations of carboxy-terminal breakdown products of NT might also give rise to these results. Current studies are aimed at distinguishing between these two possibilities.

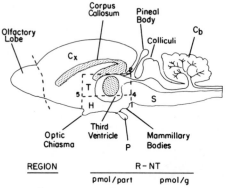

REGION	R – NT	
	pmol/part	pmol/g
H	4.3	60.0
S	4.3	12.9
T	1.3	16.1
C_x	2.2	2.0
C_b	0.2	0.8
P	0.2	28.7

FIG. 12. Regional distribution of immunoreactive NT in adult rat brain determined by RIA (antiserum PGL-6) of acid–acetone extracted tissue. (H) Hypothalamus; (S) brainstem; (T) thalamus; (Cx) cerebral cortex; (Cb) cerebellum; (P) pituitary gland.

5.9.5. Distribution in Gastrointestinal Tissues

Recent measurements show that R-NT is also present in tissue other than the brain, particularly in the gastrointestinal region. The initial studies, performed with the 1-week-old rat, indicated that while only about 10% of the R-NT in the rat was found in the head, about 90% was found in extracts of the rest of the body, and most of this was present in the intestine. In adult animals, more than 10 times as much R-NT was found in extracts of small intestine as in extracts of brain. The concentration of R-NT was highest in the jejunoileal section of the intestine (about 50 pmol/g), but it was also detected in the esophagus, stomach, duodenum, and large intestine (about 1–8 pmol/g). Concentrations in other tissues, such as skeletal muscle, liver, kidney, lung, and pancreas, were much lower (<0.2 pmol/g). These measurements were obtained using only one antiserum (PGL-6), and it is not yet known whether they indicate the presence of NT in crude extracts of each of these tissues, since results with other antisera are in disparity with these. However, some results do suggest the presence of NT along with some smaller cross-reacting materials in extracts of rat small intestine. Figure 11 shows the profile of immunoreactivity obtained when such an extract was subjected to gel chromatography on Sephadex G-25; a major peak of immunoreactivity eluted from the column in the region of NT and exhibited equal potency with the three antisera examined. The least specific antiserum, PGL-4, also detected the presence of smaller, cross-reacting substances that eluted later. In order to determine whether intestinal R-NT has the same structure as hypothalamic NT, an isolation and characterization of R-NT from about 50 kg bovine small intestine has been attempted (Kitabgi *et al.*, 1976). The results to date indicate that a peptide with chromatographic immunochemical, and biological properties similar to those of NT can be isolated, and that it has the amino acid composition of NT. It can be concluded, then, that by far the largest amount of R-NT is found in the gastrointestinal region of the rat, and that only about 3% of the total body content is in the hypothalamus; the structure of bovine and rat intestinal R-NT is likely to be similar to that of NT from bovine hypothalami.

5.9.6. Measurements of Plasma

R-NT has been detected in human, bovine, rabbit, and rat plasma, and its normal concentration estimated to be about 50 fmol/ml (Carraway and Leeman, unpublished results). Similar measurements were obtained on both unextracted and extracted plasmas using two of the antisera, PGL-6 and HC-8, suggesting that this activity is due to the presence of NT. Furthermore, the activity was found to elute in the region of synthetic NT when it was subjected to gel chromatography on Sephadex G-25. The PGL-4 antiserum, however, gave plasma measurements that were 10–40 times those obtained with PGL-6 and HC-8. This PGL-4 immunoreactivity was also extractable into acid–acetone and acid–ethanol, suggesting that it represents a small peptide, possibly a carboxy-terminal breakdown product of NT. Although it is possible that several peptides contribute to this PGL-4 immunoreactivity, size estimates suggest that the tryptic product, H–Arg–Pro–Tyr–Ile–Leu–OH, is a likely possibility.

In summary, plasma appears to contain a peptide that is chromatographically and immunochemically similar to NT, as well as much larger amounts of a substance with properties similar to the carboxy-terminal tryptic breakdown product of NT. Whether R-NT activity appears in the blood as a result of the active secretion of this peptide from a storage site, such as the hypothalamus, en route to a target tissue, or whether it merely accumulates in blood after removal from a target tissue or as a result of tissue breakdown, is not yet clear. However, the finding that its concentration is rather constant in the various animals examined suggests that its release may be regulated.

5.9.7. Measurements of Synovial Tissue and Fluid

Although NT was originally isolated from central nervous tissue, it possesses biological properties that might qualify it as a mediator of the inflammatory response; i.e., NT has the ability to induce vasodilation, increased vascular permeability, stasis, edema, and pain, as well as the ability to attract leukocytes. In order to examine the possibility that NT participates in acute or chronic inflammation, an investigation of an involvement of this peptide in human synovitis was undertaken (Carraway *et al.*, 1974). Synovial fluid samples from knee joints of patients with diverse arthropathies were found to register in the RIA for NT and to give dose–response curves that were parallel to that of NT when antiserum PGL-4 was employed. Very high levels of immunoreactivity appeared to be associated

**TABLE 9. Comparison of Some Physical and Chemical Properties
of Neurotensin (NT) and Neurotensin-Related Antigen (NRA)**

Property	Value NT	NRA
Molecular weight	1692	75,000
Precipitable by 50% $(NH_4)_2SO_4$	No	Yes
Soluble in 80% acid–acetone	Yes	No
Binds to DEAE-Sephadex at pH 6.5	No	Yes
Isoelectric pH	10	5.5
Dialyzable in saline	Yes	No
Dialyzable in 8 M urea	Yes	No
Immunoactivity stable to boiling for 10 min	Yes	Yes
Immunoactivity destroyed by carboxypeptidase A	Yes	Yes

with states of acute inflammation and erosive synovitis, the highest levels (0.2–1.2 nmol/g) being found in synovial fluids and extracts of synovial tissue obtained from rheumatoid arthritics. However, when these samples were examined using the other antisera (PGL-6 and HC-8), discrepant measurements were obtained that were less than 1% of those obtained with the carboxy-terminal-directed antiserum, PGL-4. This discrepancy suggested that the immunoreactive substance or substances were different from NT, and further studies indicated that a number of the physical and chemical properties of NT-related antigen (NRA) of synovial fluid were strikingly different from those of NT (Table 9).

Since NRA is nondialyzable in 8 M urea or in saline, the immunoreactive sequence is likely to be covalently attached to the large protein. The immunoactivity of NRA is also stable to boiling for 10 min, again supporting the notion that the immunoreactivity resides in the primary structure of the protein, and is not dependent on the pattern of folding. However, the immunoreactivity is destroyed by treatment with carboxypeptidase, suggesting that the NT-like sequence is located at the carboxy-terminus of the protein. It is interesting that highly purified NRA, like NT, was found to exhibit chemotactic activity toward human leukocytes (Carraway *et al.,* 1975), and that this activity was destroyed by treatment of NRA with carboxypeptidase A. These data argue strongly that NRA possesses the immunoreactive sequence at its carboxy-terminus, and that it is likely to have at a minimum the biologically active portion of NT, –Arg–Pro–Tyr–Ile–Leu–OH. It is not yet clear how NRA relates to NT; however, these findings illustrate the usefulness of multiple site-directed antisera in characterizing immunoreactive substances (see Section 3). An exciting possibility is that NRA serves as a precursor to NT or another peptide in the "NT family."

Note Added in Proof. At the Nobel Symposium on substance P (June, 1976) Otsuka reported the release of substance P from superfused spinal cord *in vitro,* and our laboratory reported work of Schenker, Mroz, and Leeman on the release of substance P from subcortical synaptosomes. Both studies demonstrated calcium-dependent release of substance P in response to depolarization by high K^+.

6. References

Abraham, G.E., and Grover, P.K., 1971, Covalent linkage of hormonal haptens to protein carriers for use in radioimmunoassay, in: *Principles of Competitive Protein Binding Assays* (W.D. Odell and W.H. Daughaday, eds.), pp. 134–139, J.B. Lippincott Co., Philadelphia.

Ambler, R.P., 1967, Enzymic hydrolysis with carboxypeptidase, in: *Methods in Enzymology,* Vol. XI (C.H.W. Hirs, ed.), pp. 155–166, Academic Press, New York.

Anastasi, A., Erspamer, V., and Endean, R., 1975, Structure of uperolein, a physalaemin-like endecapeptide occurring in the skin of *Uperoleia rugosa* and *Uperoleia marmorata, Experientia* **31:**394, 395.

Araki, K., Tachibana, S., Uchiyana, M., Nakajima, T., and Yashuhara, T., 1973, Isolation and structure of a new active peptide, xenopsin, from the skin of *Xenopus laevis, Chem. Pharm. Bull.* **21:**2801–2804.

Bailey, J.L., 1967, *Techniques in Protein Chemistry,* Elsevier Publishing Co., New York.

Bayer, E.H., Eckstein, K., Hägele, W.A., Konig, W., Brüning, H., and Hagenmaier, W.P., 1970, Failure sequences in solid phase synthesis of polypeptides, *J. Amer. Chem. Soc.* **92:**1735–1738.

Bergmann, J., Bienert, M., Niedrich, H., Mehlis, B., and Oehme, P., 1974, Über den Einfluss der Kettenläge bei C-terminalen Sequentzen der Substanz P—im Vergleich mit Analogen Physalaemin- und Eledoisin-peptiden—auf die Wirksamkeit am Meerschweinchenileum, *Experientia* **30:**401–403.

Brown, M.R., and Vale, W.W., 1976, Glucoregulatory effects of neurotensin (NT) and substance P (SP), *Clin. Res.* **24:**154.

Brownstein, M.J., Mroz, E.A., Kizer, J.S., Palkovits, M., and Leeman, S.E., 1976, Regional distribution of Substance P in the brain of the rat, *Brain Res.,* in press.

Carraway, R.E., 1972, The Isolation, Chemical and Pharmacological Characterization and the Synthesis of a New Hypothalamic Peptide, Neurotensin, Ph.D. Thesis, Brandeis University, available as University Microfilms 72-32,090, Ann Arbor, Michigan.

Carraway, R.E., and Leeman, S.E., 1973, The isolation of a new hypotensive peptide, neurotensin, from bovine hypothalami, *J. Biol. Chem.* **248:**6854–6861.

Carraway, R.E., and Leeman, S.E., 1974, The amino acid sequence, chemical synthesis, and radioimmunoassay of neurotensin, *Fed. Proc. Fed. Amer. Soc. Exp. Biol.* **33:**548.

Carraway, R.E., and Leeman, S.E., 1975*a,* The amino acid sequence of a hypothalamic peptide, neurotensin, *J. Biol. Chem.* **250:**1907–1911.

Carraway, R.E., and Leeman, S.E., 1975*b,* The synthesis of neurotensin, *J. Biol. Chem.* **250:**1912–1918.

Carraway, R.E., and Leeman, S.E., 1975*c,* Structural requirements for the biological activity of neurotensin, a new vasoactive peptide, in: *Peptides: Chemistry, Structure and Biology* (R. Walter and J. Meienhofer, eds.), pp. 679–685, Ann Arbor Science Publishers, Ann Arbor, Michigan.

Carraway, R.E., Demers, L., and Leeman, S.E., 1973, Hyperglycemic effect of a hypothalamic peptide, *Fed. Proc. Fed. Amer. Soc. Exp. Biol.* **32:**211.

Carraway, R.E., Goetzl, E.J., and Leeman, S.E., 1974, Detection of a neurotensin-related antigen in human synovial fluid, Paper presented at the 6th Pan American Congress on Rheumatic Diseases.

Carraway, R.E., Goetzl, E.J., and Leeman, S.E., 1975, Mononuclear leucocyte chemotactic activity associated with neurotensin-related antigens (NRA) in rheumatoid synovitis, Paper presented at the Annual Meeting of the American Rheumatism Association.

Chang, M.M., 1970, The Isolation of a Sialogogic Peptide from Bovine Hypothalamic Tissue and Its Characterization As Substance P, Ph.D. Thesis, Brandeis University.

Chang, M.M., and Leeman, S.E., 1970, Isolation of a sialogogic peptide from bovine hypothalamic tissue and its characterization as substance P, *J. Biol. Chem.* **245:**4784–4790.

Chang, M.M., Leeman, S.E., and Niall, H.D., 1971, Amino acid sequence of substance P, *Nature London New Biol.* **232:**86, 87.

Chu, L.L.H., Huang, W.Y., Littledike, E.T., Hamilton, J.W., and Cohn, D.V., 1975, Porcine proparathyroid hormone. Identification, biosynthesis, and partial amino acid sequence, *Biochemistry* **14**:3631–3635.

Cleugh, J., and Gaddum, J.H., 1963, The stability of purified preparations of substance P, *Experientia* **19**:72, 73.

Cleugh, J., Gaddum, J.H., Mitchell, A.A., Smith, M.W., and Whittaker, V.P., 1964, Substance P in brain extracts, *J. Physiol. London* **170**:69–85.

Erspamer, V., and Anastasi, A., 1962, Structure and pharmacological actions of eledoisin, the active undecapeptide of the posterior salivary glands of eledone, *Experientia* **18**:58, 59.

Erspamer, V., Anastasi, A., Bertaccini, G., and Cei, J.M., 1964, Structure and pharmacological actions of physalaemin, the main active polypeptide of the skin of *Physalaemus fuscumaculatus*, *Experientia* **20**:489, 490.

Frank, H., Meyer, H., and Hagenmaier, H., 1975, Investigation of solid supports and carboxyl-protecting groups in alternating liquid–solid phase peptide synthesis, in: *Peptides: Chemistry, Structure, and Biology* (R. Walter and J. Meienhofer, eds.), pp. 439–445, Ann Arbor Science Publishers, Ann Arbor, Michigan.

Fridkin, M.A., Patchornick, E., and Katchalski, E., 1968, Use of polymers as chemical reagents. II. Synthesis of bradykinin, *J. Amer. Chem. Soc.* **90**:2953–2957.

Gray, W.R., 1967, Sequential degredation plus dansylation, in: *Methods in Enzymology*, Vol. XI (C.H.W. Hirs, ed.), pp. 469–475, Academic Press, New York.

Henry, J.L., Krnjevic, K., and Morris, M.E., 1975, Substance P and spinal neurons, *Can. J. Physiol. Pharmacol.* **53**:423–432.

Hökfelt, T., Kellerth, J.-O., Nilsson, G., and Pernow, B., 1975a, Substance P: Localization in the central nervous system and in some primary sensory neurons, *Science* **190**:889, 890.

Hökfelt, T., Kellerth, J.-O., Nilsson, G., and Pernow, B., 1975b, Experimental immunohistochemical studies on the localization and distribution of substance P in cat primary sensory neurons, *Brain Res.* **100**:235–252.

Horton, E.W., 1959, Human urinary kinin excretion, *Br. J. Pharmacol. Chemother.* **14**:125–132.

Jacobs, J.W., Kemper, B., Niall, H.D., Habener, J.F., and Potts, J.T., Jr., 1974, Structural analysis of human proparathyroid hormone by a new microsequencing approach, *Nature London* **249**:155–157.

Kasper, C.B., 1970, Fragmentation of proteins for sequence studies and separation of peptide mixtures, in: *Protein Sequence Determination* (S.B. Needleman, ed.), pp. 137–184, Springer-Verlag, New York.

Keutmann, H.T., Parsons, J.A., Potts, J.T., Jr., and Schleuter, R.J., 1970, Isolation and chemical properties of two calcitonins from salmon ultimobranchial glands, *J. Biol. Chem.* **245**:1491–1496.

Kitabgi, P., Carraway, R., and Leeman, S.E., 1976, Isolation of a tridecapeptide from bovine intestinal tissue and its partial characterization as neurotensin, *J. Biol. Chem.*, in press.

Konigsberg, W., 1967, Subtractive Edman degradation, in: *Methods in Enzymology*, Vol. XI (C.H.W. Hirs, ed.), pp. 461–469, Academic Press, New York.

Konishi, S., and Otsuka, M., 1971, Actions of certain polypeptides on frog spinal neurons, *Jpn. J. Pharmacol.* **21**:685–687.

Konishi, S., and Otsuka, M., 1974, Excitatory action of hypothalamic substance P on spinal motoneurones of newborn rats, *Nature London* **252**:734, 735.

Krnjevic, K., and Morris, M.E., 1974, An excitatory action of substance P on cuneate neurons, *Can. J. Physiol. Pharmacol.* **52**:736–744.

Leeman, S.E., 1958, The Problem of Neurohormonal Stimulation of ACTH Secretion, Ph.D. Thesis, Radcliffe College.

Leeman, S.E., and Hammerschlag, R., 1967, Stimulation of salivary secretion by a factor extracted from hypothalamic tissue, *Endocrinology* **81**:803–810.

Lembeck, F., 1953, Zur Frage der zentralen Übertragung afferenter Impulse. III. Das Vorkommen und die Bedeutung der Substanz P in den dorsalen Wurseln des Rückenmarks, *Naunyn-Schmiedebergs Arch. Exp. Pathol. Pharmakol.* **219**:197–213.

Lembeck, F., and Starke, K., 1968, Substanz P und Speichelsekretion, *Naunyn-Schmiedebergs Arch. Pharmakol. Exp. Pathol.* **259**:375–385.

Lembeck, F., and Zetler, G., 1971, Substance P, in: *International Encyclopedia of Pharmacology and Therapeutics*, Sec. 72, Vol. 1 (J.M. Walker, ed.), pp. 29–71, Pergamon Press, Oxford.

Letsinger, R.L., and Kornet, M.J., 1963, Popcorn polymer as a support in multistep syntheses, *J. Amer. Chem. Soc.* **85:**3045, 3046.

Lowry, O.H., Rosenbrough, N.J., Farr, A.L., and Randall, R.J., 1951, Protein measurement with the Folin phenol reagent, *J. Biol. Chem.* **193:**265–275.

Makino, T., Carraway, R.E., Leeman, S.E., and Greep, R.O., 1973, *In vitro* and *in vivo* effects of newly purified hypothalamic tridecapeptide on rat LH and FSH release, Abstracts of papers presented at meeting of Society for the Study of Reproduction, p. 26.

Meinardi, H., and Craig, L.C., 1966, Studies of substance P, in: *Hypotensive Peptides* (E.G. Erdös, N. Back, and F. Sicuteri, eds.), pp. 594–607, Springer-Verlag, New York.

Mroz, E.A., Brownstein, M.J., and Leeman, S.E., 1976, Evidence for substance P in the habenulo-interpeduncular tract, *Brain Res.* **113:**597–599.

Mroz, E.A., Brownstein, M.J., and Leeman, S.E., 1977, Evidence for substance P in the striato-nigral tract, *Brain Res.*, in press.

Niall, H.D., 1975, Advances in ultrasensitive sequence analysis of peptides, in: *Peptides: Chemistry, Structure and Biology* (R. Walter and J. Meienhofer, eds.), pp. 975–984, Ann Arbor Science Publishers, Ann Arbor, Michigan.

Nilsson, G., Hökfelt, T., and Pernow, B., 1974, Distribution of substance P-like immunoreactivity in the rat central nervous system as revealed by immunohistochemistry, *Med. Biol.* **52:**424–427.

Nilsson, G., Larsson, L.-I., Håkanson, R., Brodin, E., Pernow, B., and Sundler, F., 1975*a*, Localization of substance P-like immunoreactivity in mouse gut, *Histochemistry* **43:**97–99.

Nilsson, G., Pernow, B., Fischer, G.H., and Folkers, K., 1975*b*, Presence of substance P-like immunoreactivity in plasma from man and dog, *Acta Physiol. Scand.* **94:**542–544.

Offord, R.E., 1966, Electrophoretic mobilities of peptides on paper and their use in the determination of amide groups, *Nature London* **211:**591–593.

Otsuka, M., and Konishi, S., 1974, Electrophysiology of mammalian spinal cord *in vitro*, *Nature London* **252:**733, 734.

Otsuka, M., Konishi, S., and Takahashi, T., 1972, The presence of a motoneuron-depolarizing peptide in bovine dorsal roots of spinal nerves, *Proc. Jpn. Acad.* **48:**342–346.

Pearlmutter, A.F., Rapino, E., and Saffran, M., 1975, The ACTH-releasing hormone of the hypothalamus requires a co-factor, *Endocrinology* **97:**1336–1339.

Pearse, A.G.E., and Polak, J.M., 1975, Immunocytochemical localization of substance P in mammalian intestine, *Histochemistry* **41:**373–375.

Pernow, B., 1953, Studies on substance P. Purification, occurrence and biological effects, *Acta. Physiol. Scand.* **29** (Suppl. 105): 1–90.

Pernow, B., and Rocha e Silva, M., 1955, A comparative study of bradykinin and substance P, *Acta Physiol. Scand.* **34:**59–66.

Pharmacia Fine Chemicals, 1974, *Sephadex Gel Filtration in Theory and Practice*, Upplands Grafiska AB, Sweden.

Phillis, J.W., and Limacher, J.J., 1974, Substance P excitation of cerebral cortical Betz cells, *Brain Res.* **69:**158–163.

Powell, D., Leeman, S., Tregear, G.W., Niall, H.D., and Potts, J.T., Jr., 1973, Radioimmunoassay for substance P, *Nature London New Biol.* **241:**252–254.

Rocha e Silva, M., Beraldo, W.T., and Rosenfeld, G., 1949, Bradykinin, a hypotensive and smooth muscle stimulating factor released from plasma globulin by snake venoms and by trypsin, *Amer. J. Physiol.* **156:**261–273.

Saito, K., Konishi, S., and Otsuka, M., 1975, Antagonism between Lioresal and substance P in rat spinal cord, *Brain Res.* **97:**177–180.

Schally, A.V., Saffran, M., and Zimmerman, B., 1958, A corticotropin-releasing factor: Partial purification and amino acid composition, *Biochem. J.* **70:**97–103.

Schally, A.V., Bowers, C.Y., Redding, T.W., and Barrett, J.F., 1966, Isolation of thyrotropin releasing factor (TRF) from porcine hypothalamus, *Biochem. Biophys. Res. Commun.* **25:**165–169.

Schröder, E., and Lübke, K., 1966, *The Peptides,* Vol. II, *Synthesis, Occurrence and Action of Bio-*

logically Active Polypeptides (translated by E. Gross), pp. 127–156, Academic Press, New York.

Shaw, J.E., and Ramwell, P.W., 1968, Release of a substance P polypeptide from the cerebral cortex, *Amer. J. Physiol.* **215**:262–267.

Stewart, J.M., and Young, J.D., 1969, *Solid Phase Peptide Synthesis*, W.H. Freeman and Co., San Francisco.

Studer, R.O., Trzeciak, H., and Lergier, W., 1973, Isolierung und Aminosäure-sequenz von Substanz P aus Pferdedarm, *Helv. Chim. Acta* **56**:860–866.

Takahashi, T., Konishi, S., Powell, D., Leeman, S.E., and Otsuka, M., 1974, Identification of the motoneuron-depolarizing peptide in bovine dorsal root as hypothalamic substance P, *Brain Res.* **73**:59–69.

Tregear, G.W., Niall, H.D., Potts, J.T., Jr., Leeman, S.E., and Chang, M.M., 1971, Synthesis of substance P, *Nature London New Biol.* **232**:87–89.

Vaitukaitis, J., Robbins, J.B., Nieschlag, E., and Ross, G.T., 1971, A method for producing specific antisera with small doses of immunogen, *J. Clin. Endocrinol. Metab.* **33**:988–991.

Vogler, K., Haefely, W., Hürlimann, A., Studer, R.O., Lergier, W., Strässle, R., and Bemeis, K.H., 1963, A new purification procedure and biological properties of substance P, *Ann. N. Y. Acad. Sci.* **104**:378–389.

von Euler, U.S., and Gaddum, J.H., 1931, An unidentified depressor substance in certain tissue extracts, *J. Physiol. London* **72**:74–87.

Zuber, H., 1966, Purification of substance P, in: *Hypotensive Peptides* (E.G. Erdös, N. Back, and F. Sicuteri, eds.), pp. 584–593, Springer-Verlag, New York.

6

Biologically Active Peptides in the Mammalian Central Nervous System

Michael J. Brownstein

1. Introduction

The ancient Greeks believed that the brain functioned, among other ways, by secreting proteinaceous material (*pituita*) via the hypophyseal stalk into the nasal cavity (Singer and Underwood, 1962; Abel, 1924). Hundreds of years passed before it was demonstrated conclusively that certain central neurons—neurosecretory neutrons—do make small proteins (peptides) and release them into the periphery, where they act on such structures as the pituitary. Between the Greeks' time and our own, people who were trying to understand the nature and function of peptides secreted by the CNS made many false starts and collected a great deal of uninterpretable data. Thus, it is difficult to summarize their work in a neat and concise way.

2. Historical Perspective

One of the earliest demonstrations of the importance of the brain for normal endocrine function was that of Aschner, who reported in 1912 that dogs with hypothalamic lesions developed atrophy of the genitalia. Eight years later, this finding was confirmed by Camus and Roussy (1920), who also found that

Michael J. Brownstein · Section on Pharmacology, Laboratory of Clinical Science, National Institute of Mental Health, Bethesda, Maryland.

lesions of the hypothalamus were associated with diabetes insipidus. More than a decade later, Richter (1933) described changes in the estrous cycle that followed transection of the hypophyseal stalk.

The hypothalamus was found to be involved in regulating the thyroid gland as well as the gonads. Grafe and Grünthal (1929) reported that hypothalamic lesions altered metabolic rates, and Cahane and Cahane (1936) found that the destruction of medial basal hypothalamic structures changed the histological appearance of the thyroid.

The proposal that neurons might secrete hormonal products was made by Speidel (1919), after he noted that giant neurons in the posterior spinal cord of elasmobranchs had morphological features in common with glandular cells. Ernst Scharrer (1928) discovered similar neurosecretory cells in the brains of vertebrates, initially in the hypothalami of fish.

As the result of the work of Gomori, the studies of endocrinologists and morphologists were brought together. In the course of examining the cellular makeup of the pancreatic islets, Gomori (1941) discovered a stain that seemed specific for secretory material. Gradually, the Gomori staining technique (chrom–alum–hematoxylin, aldehyde–fuchsin) began to be applied to nervous tissue, and in the late 1940's and the 1950's, cells that appeared to be secretory in nature were demonstrated in the CNS of both vertebrates and invertebrates. In vertebrates, large neurons of the hypothalamic supraoptic and paraventricular nuclei were found to contain granular elements in their cytoplasm that were stained selectively. Cutting the pituitary stalk resulted in a proximal buildup of the neurosecretory material. Thus, this material seemed to be transported from the cell bodies along axons to nerve terminals in the posterior pituitary. On the basis of the observations described above, Scharrer and Scharrer (1954) and Bargmann (1954, 1966, 1968) suggested that certain hormones of the posterior pituitary were of hypothalamic origin. This hypothesis was confirmed when oxytocin and vasopressin were purified in the mid-1950's (du Vigneaud et al., 1953a,b), and were shown to be present in the hypothalamus—just as Abel had said in 1924 they would be—specifically, in cells of the supraoptic and paraventricular nuclei.

As a consequence of the localization and purification of oxytocin and vasopressin, the role of the brain and pituitary in regulating the ejection of milk, the contraction of the uterus, and the conservation of water could be studied in detail. The mechanism by which the brain controlled the anterior pituitary remained a mystery, however.

It has been known for at least two and a half centuries that there is no direct connection between the brain and the anterior pituitary: "Although the infundibulum seems to go straight down, as if into the gland below, and although the anterior part seems to be inserted into it, this is certainly not so. For where it touches the anterior gland, it curves back caudally . . ." (G.D. Santorini, 1724; see Heller, 1963). But despite the absence of a neuronal input to the anterior pituitary from the brain, the two are linked. In 1930, Popa and Fielding (1930) described a system of vessels that connected the hypothalamus with the gland.

Wislocki and King (1936) showed that the blood in these vessels flowed from the hypothalamus toward the anterior pituitary.

Since the neurons of the hypothalamus were known to be capable of secreting hormones, and since the hypothalamus had been implicated in regulating the activity of the anterior pituitary, Taubenhaus and Soskin (1941) and Harris (1955, 1960) argued that hypothalamic control over the anterior lobe was effected by means of the release of chemical agents into the portal vessels. The presence of such agents, or *factors,* as they were called, in extracts of hypothalamic tissue was soon demonstrated. Hume and Wittenstein (1950) showed that a factor existed in the hypothalamus that produced eosinophilia when injected into animals. Corticotropin-releasing activity was shown to be present in hypothalamic extracts by Guillemin and Rosenberg (1955), and by Saffran *et al.* (1955); a year later, Porter and Jones (1956) found corticotropin-releasing factor (CRF) in portal blood. Even though CRF was the first hypothalamic hormone to be discovered, it has not yet been purified sufficiently for its structure to be established.

The first compelling evidence for the existence of a factor that stimulates the release of thyrotropin was provided by Shibusawa *et al.* (1956) and by Scheiber *et al.* (1961). This factor (TRH) has been identified as pyroglutamyl–histidyl–prolineamide (Burgus *et al.,* 1969; Bøler *et al.,* 1969). In addition to stimulating the release of thyrotropin, this peptide also causes prolactin to be released (Vale *et al.,* 1973).

TRH has a large number of behavioral and pharmacological effects. Shortly after its discovery, the tripeptide was said to be of potential value in treating patients suffering from depression (Prange and Wilson, 1972; Prange *et al.,* 1972; Kastin *et al.,* 1972). Recently, however, its therapeutic efficacy has been questioned (Mountjoy *et al.,* 1974; Copper *et al.,* 1974).

In animals, TRH has been shown to increase spontaneous motor activity and to potentiate the hyperactivity produced by monoamine oxidase inhibitors plus either L-dopa (Plotnikoff *et al.,* 1972) or trytophan (Green and Grahame-Smith, 1974). Similarly, TRH reverses the narcosis induced by either ethanol (Breese *et al.,* 1974) or pentobarbital (Prange *et al.,* 1974); in fact, it reduces the toxicity of the latter (Brown and Vale, 1975). Overall, TRH appears to be a central stimulant, but its central actions are not confined to this area. TRH may be involved in temperature regulation by the brain as well. When it is injected into the CNS, it causes an intense tremor (Schenkel-Hulliger *et al.,* 1974; Wei *et al.,* 1975). This shaking or shivering is associated with the development of hypothermia (Metcalf, 1974). In fact, TRH is more potent in reducing body temperature than any agent heretofore studied (Metcalf, 1974).

McCann *et al.* (1960) showed that the hypothalamus has luteotropin-releasing activity. Luteinizing hormone–releasing hormone (LH–RH) has been isolated, and its structure is (pyro)Glu–His–Tryp–Ser–Tyr–Gly–Leu–Arg–Pro–Gly–NH$_2$ (Baba *et al.,* 1971; Matsuo *et al.,* 1971). This decapeptide promotes the release of both luteinizing hormone (LH) and follicle-stimulating hormone (FSH). Furthermore, it induces mating behavior in animals (Moss, 1973; Moss and McCann, 1975).

A third hypothalamic hormone, somatostatin (somatotropin release–inhibiting hormone, SRIH), has been obtained in pure form (Brazeau *et al.*, 1973; Ling *et al.*, 1973; Burgus *et al.*, 1973). Its amino acid composition is:

$$H-Ala-Gly-Cys-Lys-Asn-Phe-Phe-Trp-Lys-Thr-Phe-Thr-Ser-Cys-OH$$

Somatostatin inhibits the release of growth hormone (GH) and, under certain circumstances, of thyrotropin and prolactin from cells of the anterior pituitary (Vale *et al.*, 1974*a*). In addition, it appears to have a depressant action on the CNS. It decreases the duration of seizures caused by strychnine and reduces the toxicity of the drug; conversely, it enhances the toxicity of pentobarbital (Brown and Vale, 1975).

Synthetic TRH, LH–RH, and somatostatin are commercially available. Antibodies against the synthetic peptides have been raised in laboratory animals, and have been used for purposes of radioimmunoassay and immunohistochemistry (see Chapters 3 and 4).

Besides the hormones mentioned above, there are probably others that alter the function of the anterior pituitary. FSH-releasing activity, prolactin release–inhibiting activity, melanocyte-stimulating hormone (MSH)–releasing activity, MSH release–inhibiting activity, and GH-releasing activity have been found in extracts of hypothalamic tissue. Peptides other than those that are directly involved in endocrine regulation are also present in the brain. Among these peptides are Substance P (SP), neurotensin, enkephalin, and carnosine.

3. Indirect Methods for Locating Neurosecretory Cells

Due to their high neurophysin levels, the magnocellular neurons of the paraventricular and supraoptic nuclei were stained by Gomori reagents, and these particular neurosecretory neurons were easy to locate. The parvicellular neurons of the hypothalamus, which were said by Szentágothai *et al.* (1968) to be good candidates for the role of producing releasing factors and release-inhibiting factors, were found to be "Gomori negative." Thus, there seemed to be no available way to stain peptidergic cells in the hypothalamus and elsewhere, and methods other than histochemical ones had to be relied on for information about their locations.

Prior to the development of sensitive and specific bioassays and radioimmunoassays for releasing and release–inhibiting hormones, a number of indirect techniques were used to map areas of the brain that were involved in the production and secretion of these substances. It is useful to consider briefly data obtained by means of these indirect methods, since they complement those obtained by more direct approaches.

3.1. Lesions

Physicians have realized for more than a century that tumors within or pressing on the hypothalamus cause endocrine abnormalities. It was not until the beginning of this century, however, that attempts were made to find the neuroanatomical substrate of these endocrine changes. As mentioned earlier, Aschner (1912) reported that hypothalamic lesions result in gonadal atrophy. Camus and Roussy (1920), Bailey and Bremer (1921), and Smith (1927) confirmed Aschner's findings, and Dott (1923), Mahoney and Sheehan (1936), Harris (1937), Hinsey (1937), and Westman and Jacobsohn (1938) have shown that pituitary stalk sectioning also causes atrophy of the gonads, so long as care is taken that the portal vessels do not regenerate (Harris, 1949, 1950; Harris and Johnson, 1950). Dey (1943) found that animals with hypothalamic lesions at the level of the caudal third of the optic chiasm no longer ovulate.

Cahane and Cahane (1936) observed changes in the histological appearance of the thyroid after lesions were made in the infundibular area. Soulairac *et al.* (1954) and D'Angelo and Traum (1956) showed that the secretion of thyroid-stimulating hormone (TSH) is derpressed after destruction of the anterior median eminence, while Ganong *et al.* (1955), Reichlin (1960), and Halász *et al.* (1963*a*) described a deterioration in the function of the thyroid after lesions of the anterior hypothalamus.

Endröczi and his colleagues (1957) have made electrolytic lesions in the anterior hypothalamus of weaning rats, and have found that the growth of these animals is retarded subsequently. A number of workers have reproduced these findings, and generally attribute them to a reduced elaboration of GH.

Lesions of the medial basal hypothalamus result in a decrease in corticoid secretion and cause adrenal atrophy (Laqueur *et al.*, 1955). Furthermore, the release of ACTH by the pituitary following stress is abolished or depressed in animals with hypothalamic lesions, especially lesions in the area of the median eminence (Laqueur *et al.*, 1955; Hume and Nelson, 1955).

By studying the effects of large lesions at first, and of smaller, more discrete lesions later, neuroendocrinologists have tried to infer the locations of neurosecretory neurons. The problem with making such inferences is that it is not possible to determine whether the deficits produced by any given lesion result from injury to cell bodies or to axons. Moreover, it is not always possible to judge whether a lesion has destroyed the final neuron in a polysynaptic pathway—in this case, the neurosecretory neuron—or a neuron that regulates the activity of the final one. In the following sections, indirect methods, which have similar shortcomings, will be described.

3.2. Pituitary Grafts

If the anterior pituitary is transplanted to an area of the body remote from the sella turcica, such as the anterior chamber of the eye or the space beneath the renal capsule, its cells appear to dedifferentiate (Siperstein and Greer, 1956). Similarly, if the pituitary is separated from the hypothalamus by sectioning the stalk, the basophilic and eosinophilic cells disappear from its anterior lobe (Fortier *et al.*, 1957). Halász reasoned that pituitary grafts that were implanted into releasing hormone–rich areas of the brain might not undergo cellular degeneration (Halász *et al.*, 1962). In fact, hypophyseal tissue that was introduced into the medial ventral part of the hypothalamus was quite different in appearance from tissue that was placed elsewhere. The former grafts had nearly normal numbers of basophils. The area of the hypothalamus that could support normal cellular differentiation of the pituitary was called the *hypophysiotrophic area* by Halász (Halász *et al.*, 1962). Healthy grafts in the hypophysiotrophic area secrete considerable amounts of ACTH, TSH, and LH. Animals with such grafts have larger thyroids, adrenals, ovaries, and testes than hypophysectomized rats; the histology of these glands is reasonably normal (Halász *et al.*, 1962, 1963*b*; Halász and Pupp, 1965; Flament-Durand, 1965). Hypophysectomized rats with grafts in the hypophysiotrophic area grow better than hypophysectomized controls (Halász *et al.*, 1963*b*).

Although it is undoubtedly true that the levels of releasing hormones in the medial basal hypothalamus are sufficiently high to maintain grafts of pituitary, it may not necessarily be true that the nerve cells that synthesize these hormones are present there. In fact, if the hormones tend to accumulate in axons and their terminals, and not to accumulate in perikarya (see Chapter 4), more insight may be gained into the topography of peptidergic axons than into the topography of cell bodies by studying hypophyseal grafts.

3.3. Electrical Stimulation

Excitation of specific areas of the brain may produce alterations in the function of the endocrine organs by a variety of mechanisms. In the simplest case, stimulation of a population of neurosecretory cells or their axons might cause a releasing hormone or release-inhibiting hormone to be secreted. On the other hand, stimulation of two populations of neurosecretory neurons or their axons might result in the coincident secretion of a releasing hormone and a release-inhibiting hormone. The mutually antagonistic effects of these two hormones might cancel one another. Electrical stimulation of cells other than neurosecretory neurons might ultimately cause the latter to fire more quickly or more slowly than normal via a transsynaptic mechanism.

The excitation of an area of the brain that plays absolutely no role in regulating neuroendocrine processes should, of course, have no neuroendocrine consequences.

Harris (1937) found that electrical stimulation of the tuber cinereum, the posterior hypothalamus, or the pituitary of rabbits caused them to ovulate. Markee *et al.* (1946) and Harris (1948) then demonstrated that lower currents were required to produce ovulation when the tuber cinereum was stimulated than when the pituitary was stimulated. Thus, the effect of stimulation of the tuber cinereum was not a result of spread of current to the pituitary. Critchlow (1958) has shown that stimulation of the hypothalamus of the rat in the neighborhood of the median eminence mobilizes LH. Stimulation of the suprachiasmatic area, the medial preoptic area, the medial amygdala, or the septum pellucidum must also act to release LH; in rats with constant estrus, stimulation of these areas causes ovulation (Everett, 1961; Bunn and Everett, 1957; Koikegami *et al.*, 1954).

In addition to affecting LH release, stimulation of the anterior hypothalamus and the tuberal region affects TSH release. Electrical excitation of these regions causes a decrease in TSH in the pituitary and an increase of TSH in the blood (D'Angelo and Snyder, 1963; D'Angelo *et al.*, 1964).

Stimulation of various hypothalamic areas results in the discharge of ACTH (Endröczi and Lissak, 1963; Snyder and D'Angelo, 1963). Acute lowering of the temperature of the preoptic area, but not of adjacent structures, evokes a rise in plasma cortisol concentrations (Chowers *et al.*, 1964); stimulation of the amygdala causes ACTH to increase in the blood as well (Porter, 1954). Conversely, the electrical excitation of several forebrain and rhinencephalic structures (hippocampus, septum, anterior and lateral hypothalamus, and olfactory tracts) has been reported to inhibit adrenocortical function and to prevent the stress-induced release of ACTH (Endröczi and Lissak, 1963; Endröczi *et al.*, 1959; Mason, 1958).

3.4. Hypothalamic Deafferentation

Halász devised a technique for localizing neurosecretory cells that circumvents some of the problems inherent in studying the effects of electrolytic lesions. Using a special knife, he isolated the hypothalamus from the remainder of the brain (Halász and Pupp, 1965). The "neural island" survives the surgical procedure because the medial basal hypothalamus receives its blood supply from vessels located on its ventral surface. Halász's islands extended from the back of the optic chiasm to the back of the mammillary body. They contained the median eminence and the arcuate nucleus, and parts of the ventromedial and medial mammillary nuclei.

Animals with islands of this sort no longer ovulate, but neither do they exhibit ovarian involution (Halász and Pupp, 1965; Halász and Gorski, 1967). Knife cuts that separate the medial hypothalamus from the preoptic area have the same effect as complete deafferentation (Halász and Gorski, 1967). When large islands that include both the medial hypothalamus and the medial preoptic area are made, the rats persist in ovulating.

The basal secretion of ACTH did not decrease after complete deafferentation; in fact, in the rat, it appeared to increase (Halász *et al.*, 1967*b*). TSH and GH secretion seem to be somewhat depressed after the surgical procedure, but certainly continue at moderate rates (Halász *et al.*, 1967*a;* Szentágothai *et al.*, 1968).

Apparently the hypophysiotrophic area is capable of maintaining a certain level of endocrine function in the absence of any neural input. But animals with hypothalamic islands clearly exhibit deficiencies in regulation of the endocrine system. It is likely, given that the basal secretions of LH, TSH, and ACTH are not grossly impaired by hypothalamic deafferentation, that some LH–RH, TRH, and CRF must be made in and released by cells within the hypothalamic islands.

3.5. Electrophysiological Approaches

A new technique for mapping neurons with an input into the neuroendocrine system involves stimulating the median eminence or other medial hypothalamic structures while recording from cells elsewhere (Novin *et al.*, 1970; Koizumi and Yamashita, 1972). It has not yet proved possible to determine whether the cells that have been found in this manner are peptidergic. Perhaps combining this electrophysiological approach with immunocytochemical staining of "positive" cells will permit neurosecretory cells to be located clearly.

Another electrophysiological approach to the study of central neuroendocrine phenomena is to correlate hypothalamic neuronal activity with such events as ovulation or the onset of sexual behavior (Terasawa and Sawyer, 1969; Sawyer, 1970). Presumably, cells that respond to endocrine events must be either directly or indirectly involved in regulating secretion by the anterior pituitary.

4. Regional Distribution of Selected Peptides

The concentration of a substance that is found in any given area of the brain should be related to the density of cells that make or store it there. Thus, there is a good, pragmatic reason for wanting to know about the regional distribution of peptides in the brain: it is much easier to undertake biochemical, anatomical, and physiological studies in regions that are rich in peptidergic neurons, their processes, and their peptides than in areas from which they are virtually absent. In addition to using data on the regional localization of peptides to plan future experiments, the data can be used, cautiously, to make inferences about the functional organization of the brain. For example, by looking for areas that contain high levels of LH–RH, it might be possible to map those regions that are intimately involved in regulating gonadal function. Alternatively, by studying the distribution of a compound the role of which is unclear, one might obtain some hint about its activity in the CNS. Thus, one would guess that if a peptide were found only in the occipital cortex, it might be involved in vision.

Unfortunately, that the level of a peptide is relatively low somewhere does not mean that it is unimportant there. Indeed, when low levels of a compound are found, it sometimes indicates that the release of the material is so rapid that it outstrips the synthetic capacity of the nerve to make it. Consequently, a peptide can be considered to be functionless in a specific region of the brain only if there is none of the peptide in that region. Since some biologically active peptides are present in the CSF, they must be considered potentially capable of acting everywhere in the CNS.

By assaying peptides in homogenates of tissue, one learns nothing about the structures, neurons or other, that normally contain the peptides. In order to decide whether the peptides are in neuronal perikarya, or axons of passage, or terminals one can study the effect of lesions on peptide levels. At first, the largest possible lesion that separates one region from another should be used (e.g., hypothalamic deafferentation). Subsequently, smaller, more discrete lesions can be employed. Obviously, visualizing peptidergic neurons is the best method for mapping such neurons, but this visualization has not always proved possible.

In the sections that follow, the distributions of peptides that have known roles will be detailed first. TRH, LH–RH, and somatostatin all act by being released into the hypophyseal portal vessels to regulate the anterior pituitary. Interestingly, these peptides are present outside the medial basal hypothalamus, in addition to being present inside it. They seem to act on central neurons as well as on cells of the pituitary; their electrophysiological properties are described elsewhere (see Chapter 11). Thus, the "hypothalamic" hormones may be released at synapses distant from the median eminence and produce the behavioral effects mentioned earlier. These effects often appear to be tied in with the endocrine roles of the peptides; LH–RH, for example, induces both release of LH and lordosis in female animals. These effects provide an elegant orchestration of reproductive activities.

After "hypothalamic" hormones are considered, peptides without known functions in the brain will be discussed briefly. Among these are Substance P, neurotensin, carnosine, and gastrin.

Substance P and neurotensin are found in high concentrations in areas of the brain that are involved in the regulation of vegetative and endocrine mechanisms (hypothalamus, preoptic area), and Substance P is also concentrated in the mesencephalon, specifically in the substantia nigra, where it may be involved in the control of movement.

On the basis of their distributions, it is safe to say that carnosine and "gastrin" may play important parts in olfaction and cortical integrative mechanisms, respectively.

4.1. Luteinizing Hormone–Releasing Hormone (Tables 1 and 2)

After McCann discovered LH-releasing activity in the hypothalamus in 1960 (McCann *et al.*, 1960), he set out to find more precisely where it was located. At first, he prepared extracts of various hypothalamic regions—stalk–median eminence, ventral hypothalamus, dorsal hypothalamus, caudal hypothalamus, lateral hypothalamus, and suprachiasmatic hypothalamus—and tested the ability of these extracts to release LH. Most of the releasing factor seemed to reside in the stalk and median eminence; some activity was also found in the anterior basal hypothalamus immediately overlying the median eminence (McCann, 1962).

Next, McCann and his colleagues cut serial sections of the hypothalamus and measured LH–RH in each section by bioassay. They showed that it was present in a region that extended from the preoptic area through the suprachiasmatic area to the area of the median eminence and arcuate nuclei, where the bulk of the activity was found (Schneider *et al.*, 1969; Crighton *et al.*, 1970). Subsequently Wheaton *et al.* (1975) have measured LH–RH in slices of brain by means of radioimmunoassay. The distribution of LH–RH measured in this way agrees well with the distribution demonstrated earlier.

Instead of assaying LH–RH in brain slices, Palkovits *et al.* (1974) assayed

TABLE 1. Luteinizing Hormone–Releasing Hormone in the
Brain of the Rat [a]

Brain region	LH–RH (ng/mg protein)
Hypothalamus[b] (total content ≈ 2.7 ng)	
Median eminence	22
Arcuate nucleus	3
Ventromedial nucleus (lateral part)	0.6
Suprachiasmatic nucleus	trace (<0.1)
Supraoptic nucleus	trace (<0.1)
Posterior hypothalamic nucleus	trace (<0.1)
Ventromedial nucleus (medical part)	trace (<0.1)
Preoptic area (total content ≈ 0.2 ng)	
Supraoptic crest	14
Tissue surrounding the supraoptic crest	0.3
Medial preoptic nucleus	0.15
Circumventricular organs	
Subfornical organ	4.2
Subcommissural organ	5.9
Area postrema	10.2

[a] From Palkovits et al. (1974), Wheaton et al. (1975), and Kizer et al. (1976a).
[b] LH–RH could not be detected in the periventricular, anterior hypothalamic,
paraventricular, dorsomedial, perifornical, dorsal premamillary, or ventral
premamillary nuclei, or in the medial forebrain bundle.

TABLE 2. Luteinizing Hormone–Releasing Hormone and
Thyrotropin-Releasing Hormone in the Bovine Median
Eminence [a]

Subdivision	LH–RH (ng/mg protein)	TRH (ng/mg protein)
Rostral	2.2	3.5
Anterior external	2.6	7.1
Anterior internal	4.0	5.2
Middle external medial	1.8	25.3
Middle external lateral	4.1	22.8
Middle internal medial	2.5	11.2
Middle internal lateral	3.3	4.2
Caudal	2.1	1.9

[a] From Kizer et al. (1976b)

the peptide in discrete hypothalamic nuclei, which were punched out of frozen
frontal sections of the hypothalamus of the rat. For this purpose, hollow stainless
steel needles or "punches" were used (Palkovits, 1973). The amount of LH–RH
in each sample was determined by radioimmunoassay. A very high concentration
of LH–RH was present in the median eminence; less was found in the arcuate
nucleus, where LH–RH seems to be in cells or processes other than dopaminergic
ones (Kizer et al., 1975), and still less was present in the suprachiasmatic area.

The preoptic region contained some LH–RH, as reported by McCann. Kizer *et al.* (1976*a*) and Wheaton *et al.* (1975) have concluded that most of the LH–RH in this region is in the supraoptic crest, a vascular structure that forms part of the rostral tip of the third ventricle. The supraoptic crest is one of several circumventricular organs in the brain; the others are the median eminence, the subfornical organ, the subcommissural organ, and the area postrema. These organs are characterized by a vascular and cellular architecture considerably different from that of neighboring structures. They are all permeable to vital dyes, rich in specialized ependymal cells (tanycytes), and endowed with a portal circulation. Furthermore, they have been found to be rich in LH–RH (Kizer *et al.*, 1976*a*). The reason they are is unclear at present.

LH-releasing activity in the medial basal hypothalamus was reported to fall after lesions were made in the suprachiasmatic region (Martini *et al.*, 1968; Schneider *et al.*, 1969). Isolation of the hypothalamus from the remainder of the brain caused LH–RH to fall by 70–90% in the medial basal hypothalamus (Brownstein *et al.*, 1976*a;* Weiner *et al.*, 1975), but did not cause the level of LH–RH to change in the supraoptic crest (Brownstein *et al.*, 1975*b;* Weiner *et al.*, 1975). Knife cuts made behind the optic chiasm that separated the posterior two-thirds of the hypothalamus from the suprachiasmatic and preoptic areas also produced marked decreases in LH–RH in the median eminence and arcuate nucleus. Apparently much of, but not all, the LH–RH in the medial basal hypothalamus is synthesized by or under the tropic influence of cells that are rostral to this area. Furthermore, the LH–RH that is present in the supraoptic crest must be in cells other than those that comprise the hypothalamic islands or those that are destroyed in the course of the surgical procedure, i.e., cells of the arcuate, ventromedial, dorsomedial, and premammillary nuclei. The LH–RH that is endogenous to the medial basal hypothalamus may be released tonically and may support basal gonadal function, while the LH–RH that is made by cells rostal to the medial basal hypothalamus may be released phasically and may drive the estrus cycle.

Studies based on quantitation of LH–RH in the brains of normal and lesioned animals are complemented by immunocytochemical data. Workers in a number of laboratories have succeeded in visualizing LH–RH in the median eminence, where it is present mainly in the zona externa (Barry *et al.*, 1973; Baker *et al.*, 1974; King *et al.*, 1974; Kordon *et al.*, 1974; Hökfelt *et al.*, 1975*b;* Kozlowski and Zimmerman, 1974; Sétáló *et al.*, 1975; Naik, 1975*a*).* At the electron-micro-scopic level, LH–RH has been shown to be in small granules in axon terminals (Goldsmith and Ganong, 1974; Naik, 1975*b,* Pelletier *et al.*, 1974*b*).

Although LH–RH has been seen only in axons in the median eminence of the rat and sheep, the peptide has been found in axons and tanycytes in the median eminence of the mouse (Kozlowski and Zimmerman, 1974). The presence of LH–RH in tanycytes and in CSF (Joseph *et al.*, 1974) has provided support for the suggestion made by Ben-Jonathan and her colleagues (1974) and by Knigge

---

* In the bovine median eminence, which can be dissected into several subdivisions, the highest concentration of LH–RH was measured in the zona externa (Kizer *et al.*, 1976*b*).

(1974) that LH–RH might be secreted into the CSF of the third ventricle, taken up by tanycytes in the median eminence, and tranported by them to the portal blood.

Most investigators who have sought to demonstrate LH–RH (and other peptides) in neuronal perikarya by application of immunocytochemical techniques have failed to do so. The reason may be that the cell bodies can store only small amounts of the peptides that they synthesize. Alternatively, the loss of peptides from nervous tissue may be so considerable that only areas that are especially rich in them can be stained; Goldsmith and Ganong (1975) have found that 98% of the immunoreactive LH–RH was lost from hypothalamic tissue while it was being dehydrated and embedded for electron microscopy. To make matters worse, it may also be that cell bodies are more prone than axons and terminals to lose their peptides while the tissue is being processed.

Despite the problems inherent in doing so, workers in several laboratories claim to have visualized LH–RH in cells of the arcuate nucleus, preoptic area, septum, and parolfactory region of the rat, mouse, and guinea pig (Zimmerman et al., 1974; Naik, 1975a; Barry et al., 1973, 1974; Kozlowski and Zimmerman, 1974). These findings are consistent with the results of studies of lesions summarized above, but continuing effort should be expended in confirming them.

4.2. Thyrotropin-Releasing Hormone (Tables 2–4)

TRH, measured by radioimmunoassay, is not restricted to the hypothalamus; only about ¼ of the TRH in the brain is found there (Jackson and Reichlin, 1974; C. Oliver et al., 1974; Winokur and Utiger, 1974). Fairly high TRH concentrations are present in the preoptic area and septum; lower concentrations are found in brainstem, mesencephalon, basal ganglia, and cerebral cortex.

Table 3. Regional Distribution of Thyrotropin-Releasing Hormone, Somatostatin, and Substance P in the Rat Brain [a]

Brain region	Concentration (ng/g wet wt)			Regional content (ng)			% of total in brain		
	TRH	SRIH	SP	TRH	SRIH	SP	TRH	SRIH	SP
Olfactory bulb	—	20	27	—	1.0	1.6	—	0.7	1.3
Striatum	—	50	126	—	3.2	4.6	—	2.3	3.7
Septum and preoptic area	9[b]	640	211	3.5	24.7	8.2	25.6	17.7	6.7
Hypothalamus	129	2120	268	4.1	39.3	5.8	31.2	28.2	4.8
Thalamus	9	150	84	1.9	17.5	6.6	13.8	12.6	5.4
Midbrain	—	60	243	—	9.5	33.9	—	6.8	27.8
Brainstem	12	50	201	2.1	9.8	42.1	16.9	7.0	34.5
Cerebellum	1	20	3	0.26	4.5	0.6	2.1	3.2	0.6
Cortex	2[c]	30	17	1.3	30.0	18.2	10.6	21.5	15
TOTAL				12.3	139.5	121.7			

[a] From Winokur and Utiger (1974) and Brownstein et al. (1975a, 1976b).
[b] Sample included frontal cortex.
[c] Sample did not include frontal cortex.

**TABLE 4. Thyrotropin-Releasing Hormone in the
Hypothalamus, Preoptic Area, Septum, and Circumventricular
Organs of the Rat** [a]

Brain region	Concentration (ng/mg protein)
Hypothalamus	
Periventricular nucleus	4.3
Suprachiasmatic nucleus	1.8
Supraoptic nucleus	0.9
Anterior hypothalamic nucleus	0.8
Medial forebrain bundle (anterior)	0.7
Paraventricular nucleus	2.6
Arcuate nucleus	3.9
Ventromedial nucleus (medial)	9.2
Ventromedial nucleus (lateral)	3.0
Dorsomedial nucleus	4.0
Perifornical nucleus	2.0
Medial forebrain bundle (posterior)	1.2
Posterior hypothalamic nucleus	1.8
Dorsal premamillary nucleus	1.5
Ventral premamillary nucleus	1.3
Median eminence	38.4
Mamillary body	0.3
Preoptic area	
Medial preoptic nucleus	2.0
Supraoptic crest	1.8
Septum	
Medial nucleus	0.4
Dorsal nucleus	1.9
Dorsal nucleus (intermediate)	0.5
Fimbrial nucleus	0.6
Triangular nucleus	0.5
Lateral nucleus	30.0
Circumventricular organs	
Subfornical organ	0.7
Subcommissural organ	1.1
Area postrema	0.9

[a] From Brownstein *et al.* (1975c) and Kizer *et al.* (1976a).

Within the hypothalamus, high levels of TRH are encountered in the medial part of the ventromedial nucleus, the periventricular nucleus, the arcuate nucleus, and the dorsomedial nucleus. The median eminence, of course, contains a very large amount of TRH (Brownstein *et al.*, 1974). Krulich and his co-workers (1974), who used a bioassay for TRH, observed that it was distributed essentially as described above.

At first, it was hypothesized that the neurosecretory cells that synthesized TRH were in the basal hypothalamus, and that they sent their axons both to the median eminence and to remote areas of the brain. To test this hypothesis, animals with hypothalamic islands were prepared, and TRH was measured both in the neural islands and in selected regions of the brain outside the islands. Of the TRH that was normally found in the medial basal hypothalamus, 75% disappeared after the surgical procedure, but the level of TRH did not change elsewhere in the brain (Brownstein *et al.*, 1975*c*). Clearly, the TRH that is located outside the hypophysiotrophic area is not in axons originating in cells of that area. Furthermore, it seems likely that a large part of the TRH that can be detected in the medial basal hypothalamus is in axons that are provided by other regions of the brain. On the other hand, it is possible that an important input to hypothalamic neurosecretory cells was eliminated by means of the operation. As a consequence, the neurosecretory cells might not afterward make and store as much TRH as usual. This possibility was also invoked to explain the drop in LH–RH after isolation of the hypothalamus.

Knife cuts rostral to the medial basal hypothalamus also result in substantial decreases in hypothalamic TRH; thus, the cells that are synthesizing TRH or regulating its synthesis must either be rostral to the hypothalamus or send their processes into this region from a rostral direction.

TRH has been visualized in axons and terminals in the spinal cord, brainstem, mesencephalon, and hypothalamus (Hökfelt *et al.*, 1975*c;* Hökfelt *et al.*, 1976). It has not yet been seen in specific populations of neuronal cell bodies. Presumably, efforts to visualize TRH immunocytochemically have been hampered by the small size of the molecule and by its blocked *N*-terminal end. Because of these two factors, TRH is (1) a poor antigen, against which it has been difficult to prepare antisera of high titer; (2) a fairly soluble molecule that diffuses out of tissue quickly; and (3) a compound that, in the absence of a free amino group, cannot easily be bound to tissue components by fixatives.

4.3. Growth Hormone Release–Inhibiting Hormone (Somatostatin) (Tables 3 and 5)

No matter whether it is measured by bioassay (Vale *et al.*, 1974*b*) or radioimmunoassay (Brownstein *et al.*, 1975*a*), somatostatin, like TRH, is found throughout the brain. Furthermore, it has been visualized in cells of the pancreas and gastrointestinal tract (Hökfelt *et al.*, 1975*a*). Its role in regulating glucagon and insulin release is being investigated by basic scientists and clinicians. It may play some part in the pathophysiology of diabetes mellitus, and its analogues may prove helpful in treating this disease.

The hypothalamus and preoptic area have higher levels of somatostatin than do other regions of the brain; the median eminence is especially rich in this peptide (Brownstein *et al,* 1975*a*). The arcuate, periventricular, ventral premammillary, and ventromedial nuclei have more somatostatin than the remainder of the hypothalamic nuclei (Brownstein *et al.*, 1975*a*). The subdivisions of the ventromedial nucleus all contain about the same concentrations of somatostatin. The subdivisions of the arcuate nucleus also contain roughly equal amounts of somatostatin (Palkovits *et al.*, 1976).

TABLE 5. Somatostatin in Nuclei of the Hypothalamus and in the Circumventricular Organs [a]

Brain region	Concentration (ng/mg protein)
Hypothalamus	
Medial preoptic nucleus	10.4
Periventricular nucleus	23.7
Suprachiasmatic nucleus	8.0
Supraoptic nucleus	3.2
Anterior hypothalamic nucleus	8.6
Medial forebrain bundle (anterior)	4.9
Paraventricular nucleus	4.4
Arcuate nucleus	44.6
I	64.3
II	74.6
III	72.2
IV	80.9
V	26.6
Ventromedial nucleus	14.6
Anterior	27.6
Anterior medial	33.2
Anterior lateral	34.8
Posterior medial	25.2
Posterior lateral	43.4
Dorsomedial nucleus	5.4
Perifornical nucleus	3.8
Medial forebrain bundle (posterior)	3.5
Ventral premamillary nucleus	17.3
Dorsal premamillary nucleus	4.3
Posterior hypothalamic nucleus	3.8
Median eminence	309.1
Mamillary body	2.7
Circumventricular organs	
Supraoptic crest	21
Subfornical organ	9.8
Subcommissural organ	17.7
Area postrema	14.7

[a] From Brownstein et al. (1975a) and Palkovits et al. (1976).

Compared with other parts of the CNS, the circumventricular organs are all endowed with moderately high somatostatin levels (Palkovits et al., 1976).

Somatostatin has consistently been visualized close to portal capillaries in the zona externa of the median eminence (Hökfelt et al., 1975b; Alpert et al., 1976; Pelletier et al., 1974a, 1975). Hökfelt has seen somatostin in structures that appear to be axons in the ventromedial and periventricular nuclei of the hypothalamus, as well as in the basolateral portion of this region. Moreover, he has reported seeing the peptide in scattered fibers of the amygdala.

Recently, Johansson (see Hökfelt et al., 1975a) and Alpert and his colleagues

(1976) have discovered a population of neurons in the preoptic area and anterior periventricular hypothalamus that appear to contain somatostatin. These findings remain to be extended and confirmed, and other central somatostatinergic neurons have yet to be demonstrated.

It is probable that TRH and somatostatin have functions in the brain other than those neuroendocrine ones that they perform at the level of the median eminence. They may act as neurotransmitters or neuromodulators at specific synapses in the CNS. When the neurons that synthesize and release these peptides are located, physiologists, pharmacologists, and psychologists should be able to study what they do more simply.

4.4. Vasopressin, Oxytocin, and the Neurophysins (Tables 6 and 7)

In 1895, G. Oliver and Schafer showed that extracts of the pituitary increased systemic blood pressure when injected intravenously. Subsequently, Farini (1913) and von den Velden (1913) found that the pituitary also contained a water-extractable substance that reduced the amount of urine made by the kidneys of healthy subjects and of patients with diabetes insipidus, and Dale (1906) reported that extracts of the posterior lobe of the pituitary induced uterine contractions. Van Dyke and his co-workers (1941) isolated a protein with a molecular weight of about 30,000 that had pressor, antidiuretic, and oxytocic activities. Soon it was recognized that this large molecule consisted of more than one component. It is now known that the octapeptide hormones vasopressin (du Vigneaud *et al.*, 1953*b*) and oxytocin (du Vigneaud *et al.*, 1953*a*) are often extracted from the posterior lobe loosely bound to carrier molecules—neurophysins. Neurophysins are proteins with molecular weights on the order of 10,000 daltons (Walter *et al.*, 1971; Capra and Walter, 1975). In some species, there appear to be two separate but closely related neurophysins, one for vasopressin and one for oxytocin. In other species, there seems to be only one neurophysin for both hormones.

TABLE 6. Vasopressin and Oxytocin in the Hypothalamus of the Rat[a]

Hypothalamic region	Vasopressin (ng/mg protein)	Oxytocin (ng/mg protein)
Supraoptic nucleus	40	60
Paraventricular nucleus	24	57
Retrochiasmatic area	72	10
Arcuate nucleus	4	23
Median eminence	441	223
Anterior hypothalamic nucleus	<0.1	8
Medial preoptic nucleus	<0.1	3
Suprachiasmatic nucleus	23	<0.1

[a] From George and Jacobowitz (1975) and George and Marks (1976).

TABLE 7. Vasopressin and Oxytocin in the Magnocellular Nuclei of the Hypothalamus [a]

Nucleus	Sheep (hormone content, mU)		Rat (hormone content, ng)[b]	
	Vasopressin	Oxytocin	Vasopressin	Oxytocin
Supraoptic	75	28	3.5	5.2
Paraventricular	19	30	1.7	4.1

[a] From Lederis (1962), George and Jacobowitz (1975), and George and Marks (1976).
[b] Based on the concentrations in Table 6 and on the following assumptions: volume of paraventricular nucleus: 0.72 μl; volume of supraoptic nucleus: 0.87 μl; oxytocin and vasopressin are distributed equally throughout the nuclei; protein makes up about 10% of the net weight of the nuclei.

When the hypothalamic localization of vasopressin and oxytocin were first determined by bioassay, the hormones were found in both the supraoptic nucleus and the paraventricular nucleus (Lederis, 1961; van Dyke *et al.*, 1957; Adamsons *et al.*, 1956). Initially, workers in the field were careful to point out the preponderance of oxytocin over vasopressin in the paraventricular nucleus and the preponderance of vasopressin over oxytocin in the supraoptic nucleus (Heller, 1963). The idea that each one of the magnocellular nuclei is principally involved in synthesizing one hormone, and that as a consequence the two nuclei function differently, is no longer in vogue. Currently, the similarities between the hormonal contents of the magnocellular nuclei are emphasized more than the differences, and both nuclei are felt to be involved in the synthesis and secretion of the two peptides (George and Jacobowitz, 1975; George and Marks, 1976; Zimmerman *et al.*, 1974).

The distribution of vasopressin and oxytocin among nuclei of the hypothalamus is depicted in Table 6. The peptides are present in relatively large amounts in the supraoptic and paraventricular nuclei, where they have been visualized immunohistochemically. The two hormones seem usually to be made in different cells. Whether this is always the case is currently a matter of debate (see Chapter 4). In the ox, which has two neurophysins, one neurophysin (I) is related to oxytocin, and the other (II) is related to vasopressin (Zimmerman, 1976).

As much vasopressin is present in the suprachiasmatic nucleus as in the paraventricular nucleus, and Zimmerman has reported that there is a population of parvicellular neurons in the dorsal pole of the suprachiasmatic nucleus that appear to have vasopressin in them (Zimmerman, 1976). The role of these neurons and the place where their axons terminate is not known.

It is interesting that the anterior hypothalamic nucleus and the medial preoptic nucleus contain some oxytocin (George and Marks, 1975). Which cells in these nuclei have stores of oxytocin remains to be seen.

4.5. Substance P and Neurotensin (Table 3)

These peptides are present in high concentrations in specific areas of mammalian CNS. Their localization is dealt with in Chapter 5.

4.6. β-Lipotropin, ACTH, β-MSH, and Enkephalin (Endorphin)

Enkephalin is the name given to two pentapeptides that resemble morphine in their actions on central and peripheral structures (Hughes *et al.*, 1975). It is tempting to speculate that one of these peptides, H–Tyr–Gly–Gly–Phe–Met–OH, might be derived from a larger molecule, β-lipotropin, which contains the sequence of amino acids in methionine–enkephalin as residues 61–65 (Li *et al.*, 1965; Gráf *et al.*, 1971; Cseh *et al.*, 1972). β-Lipotropin, a protein with a molecular weight of 9600, is present in the pituitary gland. Part of this protein bears a strong resemblance to β-MSH and ACTH.

Using a radioimmunoassay, an ACTH-like material has recently been found to be a normal constituent of the brain as well as the anterior pituitary, and the "ACTH" in the brain does not disappear from the brain after the pituitary is removed (Krieger and Brownstein, in preparation). Whether or not β-lipotropin exists in the CNS is unknown, but it may well be that this protein and its peptide products are made and stored in certain cells of the brain. If so, it will be interesting to see how the distribution of β-lipotropin relates to that of methionine–enkephalin, β-MSH, and ACTH. It may be that several different classes of central neurons make a single precursor molecule, such as β-lipotropin, and then generate their own specific neurosecretory products by employing unique sets of peptidases.

4.7. Carnosine

The only region of the brain that contains high levels of carnosine (β-alanylhistidine) is the olfactory bulb (Margolis, 1974). This worker has shown that the carnosine there goes away when the nasal mucosa is ablated. He has suggested that carnosine may act as the neurotransmitter released by the sensory neurons that innervate the bulb, and he is currently engaged in electrophysiological studies designed to prove his hypothesis.

4.8. Gastrin

Using a radioimmunoassay for gastrin, Van der Hagghen and his collaborators (1975) discovered a substance (or substances) in the brains of humans, dogs, pigeons, trouts, and frogs that reacted with antigastrin antibodies. No activity was found in liver, skeletal muscle, myocardium, lung, or kidney. In the brain, the concentration of the gastrinlike like material was much higher in the cerebral cortex than elsewhere.

The material appeared to be a peptide; the immunoreactivity disappeared after incubation with Pronase E. Like gastrin 2–17, the substance was fairly resistant to digestion by trypsin. The peptide did not seem to be identical to gastrin, however. Judging from Sephadex G-25 chromatography, it was smaller than gastrin 2–17, and its affinity for the antiserum used in the immunoassay was greater than its affinity for gastrin 2–17.

A number of neurotransmitters and biologically active peptides in peripheral organs of vertebrates and invertebrates also seem to be present in the mammalian CNS. Among these substances are serotonin, epinephrine, histamine, Substance P, and somatostatin. Since nature often acts conservatively, it may be useful to look in the brain for biologically active peptides that exist in other parts of the body. Special attention should be given to compounds like gastrin that are present in the gastrointestinal tract and pancreas, since the peptide-producing cells there may have their origin in the neural crest (Pearse and Polak, 1971).

5. Conclusion

By combining microdissection procedures with microassay methods, it is possible to determine the distribution within the brain of substances that cannot be mapped in any other way.

Information about the location of peptidergic cell bodies can be derived from studies of brains with surgical or chemical lesions, and the magnitude of the contribution of peptide by one area of the nervous system to another can also be studied by making lesions. Data obtained by use of the approaches outlined above complement those obtained by use of histochemical techniques, and serve as controls for them.

When the locations of groups of peptidergic neurons are known, the synthesis, transport, storage, and release of the peptides can begin to be determined. Ultimately, it will be possible to study the role of other cells in regulating the metabolism of peptidergic neurons in an unambiguous way, and the intricate and subtle central mechanisms mediated by neurosensory cells may come to be appreciated.

6. References

Abel, J.J., 1924, Physiological, chemical and clinical studies on pituitary principles, *Bull. Johns Hopkins Hosp.* **35**:305–328.

Adamsons, K., Jr., Engel, S.L., van Dyke, H.B., Schmidt-Nielson, B., and Schmidt-Nielsen, K., 1956, The distribution of oxytocin and vasopressin (antidiuretic hormone) in the neurohypophysis of the camel, *Endocrinology* **58**:272–278.

Alpert, L.C., Brawer, J.R., Patel, Y.C., and Reichlin, S., 1976, Somatostatinergic neurons in anterior hypothalamus: Immunohistochemical localization, *Endocrinology* **98**:255–258.

Aschner, B., 1912, Über die Funktion der Hypophyse, *Pfluegers Arch. Gesamte Physiol. Menschen Tiere* **146**:1–146.

Baba, Y., Matsuo, H., and Schally, A.V., 1971, Structure of the porcine LH- and FSH-releasing hormone. II. Confirmation of the proposed structure by conventional sequential analyses, *Biochem. Biophys. Res. Commun.* **44**:459–463.

Bailey, P. and Bremer, F., 1921, Experimental diabetes insipidus, *Arch. Intern. Med.* **28**:773–803.

Baker, B.L., Dermody, W.C., and Reel, J.R., 1974, Localization of luteinizing hormone–releasing hormone in the mammalian hypothalamus, *Amer. J. Anat.* **139**:129–134.

Bargmann, W., 1954, *Das Zwischenhirn-Hypophysensystem,* Springer-Verlag, Berlin.

Bargmann, W., 1966, Neurosecretion, *Intern. Rev. Cytol.* **19**:183–201.

Bargmann, W., 1968, Neurohypophysis: Structure and function, in: *Handbook of Experimental Phar-*

macology, Vol. 23, *Neurophypopseal Hormones and Similar Polypeptides* (B. Berde, ed.), pp. 1–39, Springer-Verlag, New York.

Barry, J., Dubois, M.P., and Poulain, P., 1973, LRF producing cells of the mammalian hypothalamus, *Z. Zellforsch.* **146:**351–366.

Barry, J., Dubois, M.P., and Carette, B., 1974, Immunofluorescence study of the preoptico-infundibular LRF neurosecretory pathway in the normal, castrated, or testosterone-treated male guinea pig, *Endocrinology* **95:**1416–1423.

Ben-Jonathan, N., Michal, R.S., and Porter, J.C., 1974, Transport of LRF from CSF to hypophysial portal and systemic blood and the release of LH, *Endocrinology* **95:**18–25.

Bøler, J., Enzmann, F., Folkers, K., Bowers, C.Y., and Schally, A.V., 1969, The identity of chemical or hormonal properties of the thyrotropin releasing hormone and pyroglutamyl–histidyl–proline amide, *Biochem. Biophys. Res. Commun.* **37:**705–710.

Brazeau, P., Vale, W., Burgus, R., Ling, N., Butcher, M., Rivier, J., and Guillemin, R., 1973, Hypothalamic polypeptide that inhibits the secretion of immunoreactive pituitary growth hormone, *Science* **179:**77–79.

Breese, G.R., Cott, J.M., Cooper, B.R., Prange, A.J., and Lipton, M.A., 1974, Antagonism of ethanol narcosis by thyrotropin releasing hormone, *Life Sci.* **14:**1053–1063.

Brown, M., and Vale, W., 1975, Central nervous effects of hypothalamic peptides, *Endocrinology* **96:**1333–1336.

Brownstein, M.J., Palkovits, M., Saavedra, J., Bassiri, R., and Utiger, R.D., 1974, Thyrotropin-releasing hormone in specific nuclei of rat brain, *Science* **185:**267–269.

Brownstein, M., Arimura, A., Sato, H., Schally, A.V., and Kizer, J.S., 1975*a*, The regional distribution of somatostatin in the rat brain, *Endocrinology* **96:**1456–1461.

Brownstein, M.J., Palkovits, M., and Kizer, J.S., 1975*b*, On the origin of luteinizing hormone–releasing hormone (LH–RH) in the supraoptic crest, *Life Sci.* **17:**679–682.

Brownstein, M., Utiger, R. Palkovits, M., and Kizer, J.S., 1975*c*, Effect of hypothalamic deafferentation on thyrotropin releasing hormone levels in rat brain, *Proc. Nat. Acad. Sci. U.S.A.* **72:**4177–4179.

Brownstein, M., Arimura, A., Schally, A.V., Palkovits, M., and Kizer, J.S., 1976*a*, The effect of surgical isolation of the hypothalamus on its luteinizing hormone–releasing hormone content, *Endocrinology* **98:**662–665.

Brownstein, M., Mroz, E., Kizer, J.S., Palkovits, M., and Leeman, S., 1976*b*, Regional distribution of Substance P in the rat brain, *Brain Res.* in press.

Bunn, J.P., and Everett, J.W., 1957, Ovulation in persistent-estrous rats after electrical stimulation of the brain, *Proc. Soc. Exp. Biol. Med.* **96:**369–371.

Burgus, R., Dunn, T.F., Desiderio, D., and Guillemin, R., 1969, Structure moléculaire du facteur hypothalamique hypophysiotrope TRF d'origine ovine: Mise en évidence par spectrométrie de masse de la séquence PCA–His–Pro-NH₂, *C. R. Acad. Sci. Paris Ser. D* **269:**1870–1873.

Burgus, R., Ling, N., Butcher, M., and Guillemin, R., 1973, Primary structure of somatostatin, a hypothalamic peptide that inhibits the secretion of pituitary growth hormone, *Proc. Nat. Acad. Sci. U.S.A.* **70:**684–688.

Cahane, M., and Cahane, T., 1936, Sur certaine modifications des glands endocrine aprés une lésion diencéphalique, *Rev. Fr. Endocrinol.* **14:**472–487.

Camus, J., and Roussy, G., 1920, Experimental researches on the pituitary body. Diabetes insipidus, glycosuria, and those dystrophies considered as hypophysial in origin, *Endocrinology* **4:**507–522.

Capra, J.D., and Walter, R., 1975, Primary structure and evolution of neurophysins, *Ann. N.Y. Acad. Sci.* **248:**92–111.

Chowers, I., Hammel, H.T., Stromme, S.B., and McCann, S.M., 1964, Comparison of effect of environmental and preoptic cooling on plasma cortisol levels, *Amer. J. Physiol.* **207:**577–583.

Copper, A., Peet, M., Montgomery, S., and Bailey, J. 1974, Thyrotropin-releasing hormone in the treatment of depression, *Lancet* **2:**433.

Crighton, D.B., Schneider, H.P.G., and McCann, S.M., 1970, Localization of LH-releasing factor in the hypothalamus and neurohypophysis as determined by *in vitro* method, *Endocrinology* **87:**323–329.

Critchlow, B.V., 1958, Ovulation induced by hypothalamic stimulation in the anesthetized rat, *Amer. J. Physiol.* **195**:171–174.

Cseh, G., Barát, E., Patthy, A., and Gráf, L., 1972, Studies on the primary structure of human β-lipotropic hormone, *FEBS Lett.* **21**:344–346.

Dale, H.H., 1906, On some physiological actions of ergot, *J. Physiol. London* **34**:163–206.

D'Angelo, S.A., and Snyder, J., 1963, Electrical stimulation of the hypothalamus and TSH secretion in the rat, *Endocrinology* **73**:75–80.

D'Angelo, S.A.,and Traum, R.E., 1956, Pituitary–thyroid function in rats with hypothalamic lesions, *Endocrinology* **59**:593–596.

D'Angelo, S.A., Snyder, J., and Grodin, J.M., 1964, Electrical stimulation of the hypothalamus: Simultaneous effects on the pituitary, adrenal, and thyroid system of the rat, *Endocrinology* **75**:417–427.

Dey, F.L., 1943, Evidence of hypothalamic control of hypophyseal gonadotrophic function in the female guinea-pig, *Endocrinology* **33**:75–82.

Dott, N.M., 1923, An investigation into the functions of the pituitary and thyroid glands. Part I. Technique of their experimental surgery and summary of results, *Q. J. Exp. Physiol.* **13**:241–282.

du Vigneaud, V., Ressler, C., Swan, J.M., Roberts, C.W., Katsoyannis, P.G., and Gordon, S., 1953a, Synthesis of an octapeptide amide with the hormonal activity of oxytocin, *J. Amer. Chem. Soc.* **75**:4879, 4880.

du Vigneaud, V., Lawler, H.C., and Popenoe, A., 1953b, Enzymatic cleavage of glycinamide from vasopressin and a proposed structure for the pressor–antidiuretic hormone of the posterior pituitary, *J. Amer. Chem. Sco.* **75**:4880, 4881.

Endröczi, E., and Lissak, K., 1963, Effect of hypothalamic and brain stem structure stimulation on pituitary–adrenocortical function, *Acta Physiol. Acad. Sci. Hung.* **24**:67–77.

Endröczi, E., Kovács, S., and Szalay, Gy., 1957, Einfluss von Hypothalamus Läsionen auf die Entwicklung des Körpers und verschiedener Organe bei neugeborenen Tieren, *Endokrinologie* **34**:168–175.

Endröczi, E., Lissak, K., Bohus, B., and Kovacs, S., 1959, The inhibitory influence of archicortical structures on pituitary–adrenal function, *Acta Physiol. Acad. Sci. Hung.* **16**:17–22.

Everett, J.S., 1961, The preoptic region of the brain and its relation to ovulation, in: *Control of Ovulation* (C.A. Villee, ed.), pp. 101–112, Pergamon Press, New York.

Farini, F., 1913, Diabete insipido ed opoterapia, *Gazz. Osped. Clin.* **34**:1135–1139.

Flament-Durand, J., 1965, Observations on pituitary transplants into the hypothalamus of the rat, *Endocrinology* **77**:446–454.

Fortier, C., Harris, G.W., and McDonald, I.R., 1957, The effect of pituitary stalk section on the adrenocortical response to stress in the rabbit, *J. Physiol. London* **136**:344–363.

Ganong, W.F., Frederickson, D.S., and Hume, M.D., 1955, The effect of hypothalamic lesions on thyroid function in the dog, *Endocrinology* **57**:355–363.

George, J.M., and Jacobowitz, D., 1975, Localization of vasopressin in discrete areas of the rat hypothalamus, *Brain Res.* **93**:363–366.

George, J., and Marks, B., 1976, Oxytocin content of microdissected areas of rat hypothalamus, *Endocrinology* in press.

Goldsmith, P.C., and Ganong, W.F., 1974, Ultrastructural localization of luteinizing hormone–releasing hormone in the rat hypothalamus, *Program, Abstracts of the Fourth Annual Meeting of the Society of Neuroscience* (abstract A250).

Goldsmith, P.C., and Ganong, W.F., 1975, Ultrastructural localization of luteinizing hormone–releasing hormone in the median eminence of the rat, *Brain Res.* in press.

Gomori, G., 1941, Observations with differential stains on human islets of Langerhans, *Amer. J. Pathol.* **17**:395–406.

Gráf, L., Barát, E., Cseh, G., and Sajgó, M., 1971, Amino acid sequence of porcine β-lipotrophic hormone, *Biochim.Biophys. Acta* **229**:276–278.

Grafe, E., and Grünthal, E., 1929, Uber isolierte Bee influssung des Gesamtstoffwechsels vom Zwischenhirn aus, *Klin. Wochenschr.* **8**:1013–1016.

Green, A.R., and Grahame-Smith, D.G., 1974, TRH potentiates behavioral changes following increased brain serotonin accumulation in rats, *Nature London* **251**:524–526.

Guillemin, R., and Rosenberg, B., 1955, Humoral hypothalamic control of anterior pituitary: A study with combined tissue cultures, *Endocrinology* **57**:599–607.

Halász, B., and Gorski, R.A., 1967, Gronadotrophic hormone secretion in female rats after partial or total interruption of neural afferents to the medial basal hypothalamus, *Endocrinology* **80**:608–622.

Halász, B., and Pupp, L., 1965, Hormone secretion of the anterior pituitary gland after physical interruption of all nervous pathways to the hypophysiotrophic area, *Endocrinology* **77**:553–562.

Halász, B., Pupp, L., and Uhlarik, S., 1962, Hypophysiotrophic area in the hypothalamus, *J. Endocrinol.* **25**:147–154.

Halász, B., Pupp, L., and Uhlarik, S., 1963a, Changes in the pituitary–target gland system following electrolytic lesion of median eminence and hypophyseal stalk in male rats, *Acta Morphol. Acad. Sci. Hung.* **12**:23–31.

Halász, B., Pupp, L., Uhlarik, S., and Tima, L., 1963b, Growth of hypophysectomized rats bearing pituitary transplants in the hypothalamus, *Acta Physiol. Acad. Sci. Hung.* **23**:287–292.

Halász, B., Florsheim, W.H., Corcorran, N.L., and Gorski, R.A., 1967a, Thyrotrotrophic hormone secretion in rats after partial or total interruption of neural afferents to the medial basal hypothalamus, *Endocrinology* **80**:1075–1082.

Halász, B., Slusher, M.A., and Gorski, 1967b, Adrenocorticotrophic hormone secretion in rats after partial or total deafferentation of the medial basal hypothalamus, *Neuroendocrinology* **2**:43–55.

Harris, G.W., 1937, The induction of ovulation in the rabbit by electrical stimulation of the hypothalamo–hypophysial mechanism, *Proc. R. Soc. London Ser. B* **122**:374–394.

Harris, G.W., 1948, Electrical stimulation of the hypothalamus and the mechanism of neural control of the adenohypophysis, *J. Physiol. London* **107**:418–429.

Harris, G.W., 1949, Regeneration of the hypophysial–portal vessels, *Nature London* **163**:70.

Harris, G.W., 1950, Oestrous rhythm. Pseudopregnancy and the pituitary stalk in the rat, *J. Physiol. London* **111**:347–360.

Harris, G.W., 1955, *Neural Control of the Pituitary Gland,* Arnold, London.

Harris, G.W., 1960, Central control of pituitary secretion, in: *Handbook of Physiology, Section 1: Neurophysiology,* Vol. 11 (American Physiological Society, J. Field, Editor-in-Chief), pp. 1007–1038, Williams and Wilkins Co., Baltimore.

Harris, G.W., and Johnson, D., 1950, Regeneration in the hypophysial portal vessels, after section of the hypophysial stalk in the monkey (*Macacus rhesus*) *Nature London* **165**:819, 820.

Heller, H., 1963, Neurophypophyseal hormones, in: *Comparative Endocrinology* (U.S. von Euler and H. Heller, eds.), Vol. 1, pp. 25–72, Academic Press, New York.

Hinsey, J.C., 1937, The relation of the nervous system to ovulation and other phenomena of the female reproductive tract, *Cold Spring Harbor Symp. Quant. Biol.* **5**:269–279.

Hökfelt, T., Efendic, S., Hellerstrom, C., Johansson, O., Luft, R., and Arimura, A., 1975a, Cellular localization of somatostatin in endocrine-like cells and neurons of the rat with special references to the A_1-cells of the pancreatic islets and to the hypothalamus, *Acta Endocrinol.* **80**(Suppl.):200.

Hökfelt, T., Fuxe, K., Goldstein, M., Johansson, O., Fraser, H., and Jeffcoate, S.L., 1975b, Immunofluorescence mapping of central monoamine and releasing hormone (LRH) systems, in: *Anatomical Neuroendocrinology* (W.E. Stumpf and L.D. Grant, eds.), Karger, Basel.

Hökfelt, T., Fuxe, K., Johansson, O., Jeffcoate, S., and White, N., 1975c, Thyrotropin releasing hormone (TRH)–containing nerve terminals in certain brain stem nuclei and in the spinal cord, *Neurosci. Lett.* **1**:133–139.

Hökfelt, T., Fuxe, K., Johansson, O., Jeffcoate, S., and White, N., 1976, Distribution of thyrotropin releasing hormone in the central nervous system as revealed with immunohistochemistry. *Eur. J. Pharmacol.* in press.

Hughes, J., Smith, T.W., Kosterlitz, H.W., Fothergill, L.A., Morgan, B.A., and Morris, H.R., 1975, Identification of two related pentapeptides from the brain with potent opiate agonist activity, *Nature London* **258**:577–579.

Hume, D.M., and Nelson, D.H., 1955, Effect of hypothalamic lesions on blood ACTH levels and 17-hydroxycorticosteroid secretion following trauma in the dog, *J. Clin. Endocrinol.* **15**:839, 840.

Hume, D.M., and Wittenstein, G.J., 1950, The relationship of the hypothalamus to pituitary–

adrenocortical function, in: *Proceedings of the First Clinical ACTH Conference* (J.R. Mote, ed.), pp. 134–146, Blakiston, Philadelphia.

Jackson, I.M.D., and Reichlin, S., 1974, Thyrotropin-releasing hormone (TRH): Distribution in hypothalamic and extrahypothalamic brain tissues of mammalian and sub-mammalian chordates, *Endocrinology* **95:**854–862.

Joseph, S.A., Sorrentino, S., Jr., and Sundberg, D.K., 1974, Releasing hormones, LRF, and TRF in the cerebrospinal fluid of the third ventricle, in: *Brain Endocrine Interaction. II. The Ventricular System*, 2nd International Symposium, Tokyo, 1974 (K.M. Knigge, D.E. Scott, H. Kobayashi, and S. Ishli, eds.), pp. 306–312, Karger, Basel.

Kastin, A.J., Schalch, D.S., Ehrensing, R.H., and Anderson, M.S., 1972, Improvement in mental depression with decreased thyrotropin response after administration of thyrotropin releasing hormone, *Lancet* **2:**470.

King, J.C., Parsons, J.A., Erlandsen, S.L., and Williams, T.H., 1974, Luteinizing hormone–releasing hormone (LH–RH) pathway of the rat hypothalamus revealed by the unlabeled antibody peroxidase–antiperoxidase method, *Cell Tissue Res.* **153:**211–217.

Kizer, J.S., Arimura, A., Schally, A.V., and Brownstein, M.J., 1975, Absence of luteinizing hormone–releasing hormone (LH–RH) from catecholaminergic neurons, *Endocrinology* **96:**523–525.

Kizer, J.S., Palkovits, M., and Brownstein, M., 1976a, Releasing factors in the circumventricular organs of the rat brain, *Endocrinology* **98:**309–315.

Kizer, J.S., Palkovits, M., Tappaz, M., Kebabian, J., and Brownstein, M., 1976b, Distribution of releasing factors, biogenic amines, and related enzymes in the bovine median eminence, *Endocrinology* **98:**649–659.

Knigge, K.M., 1974, Role of the ventricular system in neuroendocrine processes: Initial studies on the role of catecholamines in transport of thyrotropin in releasing factor, in: *Frontiers in Neurology and Neuroscience Research*, (P. Seeman and G.M. Brown, eds.), pp. 40–47, University of Toronto Press, Toronto.

Koikegami, H., Yamada, T., and Usei, K., 1954, Stimulation of the amygdaloid nuclei and periamygdaloid cortex with special reference to its effects on uterine movements and ovulation, *Folia Psychiatr. Neurol. Jpn.* **8:**7–31.

Koizumi, K., and Yamashita, H., 1972, Studies of antidromically identified neurosecretory cells of the hypothalamus by intracellular and extracellular recordings, *J. Physiol. London* **211:**683–705.

Kordon, C., Kerdelhue, B., Pattou, E., and Jutisz, M., 1974, Immunocytochemical localization of LH–RH in axons and nerve terminals in the rat median eminence, *Proc. Soc. Exp. Biol. Med.* **147:**122–127.

Kozlowski, G.P., and Zimmerman, E.A., 1974, Localization of gonadotropin-releasing hormone (Gn-RH) in sheep and mouse brain, *Anat. Rec.* **178:**396.

Krulich, L., Quijada, M., Hefco, E., and Sundberg, D.K., 1974, Localization of thyrotropin-releasing factor (TRF) in the hypothalamus of the rat, *Endocrinology* **95:**9–17.

Laqueur, C.L., McCann, S.M., Schreiner, L.H., Rosemberg, E., Rioch, D.Mck., and Anderson, E., 1955, Alterations of adrenal cortical and ovarian activity following hypothalamic lesions, *Endocrinology* **57:**44–54.

Lederis, M.K., 1961, Vasopressin and oxytocin in the mammalian hypothalamus, *Gen. Comp. Endocrinol.* **1:**80–89.

Lederis, K., 1962, The distribution of vasopressin and oxytocin in hypothalamic nuclei, in: *Neurosecretion* (H. Heller and R.B. Clark, eds.), pp. 227–236, Academic Press, New York.

Li, C.H., Barnafi, L., Chretien, M., and Chung, D., 1965, Isolation and amino-acid sequence of β-LPH from sheep pituitary glands, *Nature London* **208:**1093, 1094.

Ling, N., Burgus, R., Rivier, J., Vale, W., and Brazeau, P., 1973, The use of mass spectrometry in deducing the sequence of somatostatin—a hypothalamic polypeptide that inhibits the secretion of growth hormone, *Biochem. Biophys. Res. Commun.* **52:**786–791.

Mahoney, W., and Sheehan, D., 1936, The pituitary–hypothalamic mechanism: Experimental occlusion of the pituitary stalk, *Brain* **59:**61–75.

Margolis, F.L., 1974, Carnosine in the primary olfactory pathway, *Science* **184:**909–911.

Markee, J.E., Sawyer, C.H., and Hollinshead, W.H., 1946, Activation of anterior hypophysis by electrical stimulation in the rabbit, *Endocrinology* **38**:345–357.

Martini, L., Fraschini, F., and Motta, M., 1968, Neural control of anterior pituitary functions, *Recent Prog. Horm. Res.* **24**:439–485.

Mason, J.W., 1958, Plasma 17-hydroxycorticosteroid response to hypothalamic stimulation in the conscious rhesus monkey, *Endocrinology* **63**:403–411.

Matsuo, H., Baba, Y., Nair, R.M.G., Arimura, A., and Schally, A.V., 1971, Structure of the porcine LH- and FSH-releasing hormone. I. The proposed amino acid sequence, *Biochem. Biophys. Res. Commun.* **43**:1334–1339.

McCann, S.M., 1962, A hypothalamic luteinizing-hormone-releasing-factor, *Amer. J. Physiol.* **202**:395–400.

McCann, S.M., Taleisnik, S., and Friedman, H.M., 1960, LH-releasing activity in hypothalamic extracts, *Proc. Soc. Exp. Biol. Med.* **104**:432–434.

Metcalf, G., 1974, TRH: A possible mediator of thermoregulation, *Nature London* **252**:310, 311.

Moss, R.L., 1973, Induction of mating behavior in rats by luteinizing hormone–releasing factor, *Science* **181**:177–179.

Moss, R.L., and McCann, S.M., 1975, Action of luteinizing hormone–releasing factor (LRF) in the initiation of lordosis behavior in the estrone-primed ovariectomized female rat, *Neuroendocrinology* **17**:309–318.

Mountjoy, C.Q., Wheller, M., Hall, R., Price, J.S., Hunter, P., and Dewar, J.H., 1974, A double-blind crossover sequential trial of oral thyrotropin-releasing hormone in depression, *Lancet* **1**:958–960.

Naik, D.V., 1975*a*, Immunoreactive LH–RH neurons in the hypothalamus identified by light and fluorescent microscopy, *Cell Tissue Res.* **157**:423–436.

Naik, D.V., 1975*b*, Immuno-electron microscopic localization of luteinizing hormone–releasing hormone in the arcuate nuclei and median eminence of the rat, *Cell Tissue Res.* **157**:437–455.

Novin, D., Sundsten, J.W., and Cross, B.A., 1970, Some properties of antidromically activated units in the paraventricular nucleus of the hypothalamus, *Exp. Neurol.* **26**:330–341.

Oliver, C., Eskay, R.L., Ben-Jonathan, N., and Porter, J.C., 1974, Distribution and concentration of TRH in the rat brain, *Endocrinology* **95**:540–553.

Oliver, G., and Schafer, E.A., 1895, On the physiological actions of extracts of the pituitary body and certain other glandular organs, *J. Physiol. London* **18**:277–279.

Palkovits, M., 1973, Isolated removal of hypothalamic or other brain nuclei of the rat, *Brain Res.* **59**:449, 450.

Palkovits, M., Arimura, A., Brownstein, M.J., Schally, A.V., and Saavedra, J.M., 1974, Luteinizing hormone–releasing hormone (LH–RH) content of the hypothalalmic nuclei in rat, *Endocrinology* **96**:554–558.

Palkovits, M., Brownstein, M., Arimura, A., Sato, H., Schally, A.V., and Kizer, J.S., 1976, Somatostatin content of the hypothalamic ventromedial and arcuate nuclei and the circumventricular organs in the rat, *Brain Res.* **109**:430–434.

Pearse, A.G.E., and Polak, J.M., 1971, Neural crest origin of the endocrine polypeptide (APUD) cells of the gastrointestinal tract and pancreas, *Gut* **12**:783–788.

Pelletier, G., Labrie, F., Arimura, A., and Schally, A.V., 1974*a*, Electron microscopic immunohistochemical localization of growth hormone–release inhibiting hormone (somatostatin) in the rat median eminence, *Amer. J. Anat.* **140**:445–450.

Pelletier, G., Labrie, F., Puviani, R., Arimura, A., and Schally, A.V., 1974*b*, Electron microscopic localization of luteinizing hormone–releasing hormone in rat median eminence, *Endocrinology* **95**:314, 315.

Pelletier, G., LeClare, R., Dube, D., Labrie, F., Puviani, R., Arimura, A., and Schally, A.V., 1975, Localization of growth hormone–release inhibiting hormone (somatostatin) in the rat brain, *Amer. J. Anat.* **142**:397–400.

Plotnikoff, N.P., Prange, A.J., Breeze, G.R., Anders, M.S., and Wilson, I.C., 1972, Thyrotropin releasing hormone: Enhancement of dopa activity by a hypothalamic hormone, *Science* **178**:417.

Popa, G.T., and Fielding, V., 1930, A portal circulation from the pituitary to the hypothalamic region, *J. Anat. London* **65**:88–91.

Porter, R.W., 1954, The central nervous system and stress-induced eosinophilia, *Recent Prog. Horm. Res.* **10**:1–27.

Porter, J.C., and Jones, S.C., 1956, Effect of plasma from hypophyseal–portal vessel blood on adrenal ascorbic acid, *Endocrinology* **58**:62–67.

Prange, A.J., and Wilson, E.C., 1972, Thyrotropin-releasing hormone (TRH) for the immediate relief of depression: A preliminary report, *Psychopharmacologia* **26**(Suppl.):82.

Prange, A.S., Lara, P.P., Wilson, I.C., Alltop, L.B., and Breese, G.R., 1972, Effects of thyrotropin-releasing hormone in depression, *Lancet* **2**:999.

Prange, A.J., Breese, G.R., Cott, J.M., Martin, B.R., Cooper, B.R., Wilson, I.C., and Plotnikoff, N.P., 1974, TRH: Antagonism of pentobarbital in rodents, *Life Sci.* **14**:447–455.

Reichlin, S., 1960, Thyroid response to partial thyroidectomy, thyroxine, and 2,4-dinitrophenol in rats with hypothalamic lesions, *Endocrinology* **66**:340–354.

Richter, C.P., 1933, Cyclic phenomena produced in rats by section of the pituitary stalk and their possible relation to pseudopregnancy, *Amer. J. Physiol.* **106**:80–90.

Saffran, M., Schally, A.V., and Benfey, B.G., 1955, Stimulation of release of corticotropin from the adenohypophysis by a neurophyophyseal factor, *Endocrinology* **57**:439–444.

Sawyer, C.H., 1970, Some endocrine applications of electrophysiology, in: *The Hypothalamus* (L. Martini, M. Motta, and F. Fraschini, eds.), pp. 83–101, Academic Press, New York.

Scharrer, E., 1928, Die Lichtempfindlichkeit blinder Elritzen I. Untersuchungen über das Zwischenhirn der Fische, *Z. Vergleich. Physiol.* **7**:1–38.

Scharrer, E., and Scharrer, B., 1954, Hormones produced by neurosecretory cells, *Recent Prog. Horm. Res.* **10**:183–240.

Schenkel-Hulliger, L., Koella, W.P., Hartmann, A., and Maître, L., 1974, Tremorogenic effect of TRH in rats, *Experientia* **30**:1168–1170.

Schneider, H.P.B., Crighton, D.B., and McCann, S.M., 1969, Suprachiasmatic LH-releasing factor, *Neuroendocrinology* **5**:271–280.

Schreiber, V., Eckertova, A., Franz, Z., Koci, J., Rybau, M., and Kmentova, V., 1961, Effect of a fraction of bovine hypothalamic extract on release of TSH by rat adenohypophysis *in vitro, Experientia Basel* **17**:264, 265.

Sétáló, G., Vigh, S., Schally, A.V., Arimura, A., and Flerko, B., 1975, LH–RH containing neural elements in the rat hypothalamus, *Endocrinology* **96**:135–142.

Shibusawa, K., Saito, S., Nishi, K., Yamamoto, T., Tomizawa, K., and Abe, C., 1956, The hypothalamic control of the thyrotroph–thyroidal function, *Endocrinol. Jpn.* **3**:116–124.

Singer, C., and Underwood, E.A., 1962, *A Short History of Medicine,* Oxford University Press, New York.

Siperstein, E.R., and Greer, M.A., 1956, Observation on the morphology and histochemistry of the mouse pituitary implanted in the anterior eye chamber, *J. Nat. Cancer Inst.* **17**:569–599.

Smith, P.E., 1927, The disabilities caused by hypophysectomy and their repair. The tuberal (hypothalamic) syndrome in the rat, *J. Amer. Med. Assoc.* **88**:158–161.

Snyder, J., and D'Angelo, S.A., 1963, Hypothalamic stimulation and ascorbic acid content of endocrine glands of the albino rat, *Proc. Soc. Exp. Biol. Med.* **112**:1–4.

Soulairac, A., Desclaux, P., and Teysseyre, J., 1954, Modifications endocriniennes apres lesions hypothalamiques experimentales chex le rat, *C. R. Assoc. Anat. Paris* **40**:363–373.

Speidel, C.C., 1919, Gland cells of internal secretion in the spinal cord of the skates, *Carnegie Inst. Washington Publ. No.* 13, 1–31.

Szentágothai, J., Flerkó, B., Mess, B., and Halász, B., 1968, *Hypothalamic Control of the Anterior Pituitary,* Akademiai Kiado, Budapest.

Taubenhaus, M., and Soskin, S., 1941, Release of luteinizing hormone from anterior hypophysis by an acetylcholine-like substance from the hypothalamic region, *Endocrinology* **29**:958–964.

Terasawa, E., and Sawyer, C.H., 1969, Changes in electrical activity in the rat hypothalamus related to electrochemical stimulation of adenohypophyseal function, *Endocrinology* **85**:143–149.

Vale, W., Blackwell, R., Grant, G., and Guillemin, R., 1973, TRF and thyroid hormones on prolactin secretion by rat anterior pituitary cells *in vitro, Endocrinology* **93**:26–33.

Vale, W., Rivier, C., Brazeau, P., and Guillemin, R., 1974*a*, Effects of somatostatin on the secretion of thyrotropin and prolactin, *Endocrinology* **95**:968–977.

Vale, W., Rivier, C., Palkovits, M., Saavedra, J.M., and Brownstein, M.J., 1974*b*, Ubiquitous brain distribution of inhibitors of adenohypophyseal secretion, *Endocrinology* **94**:A128.

Van der Hagghen, J.J., Signeau, J.C., and Gepts, W., 1975, New peptides in the vertebrate CNS reacting with antigastrin antibodies, *Nature London* **257**:604–605.

van Dyke, H.B., Chow, B.F., Greep, R.D., and Rothen, A., 1941, The isolation of a protein from the pars neuralis of the ox pituitary with constant oxytocic, pressor, and diuresis inhibiting effects, *J. Pharmacol. Exp. Ther.* **74**:190–209.

van Dyke, H.B., Adamson, K., Jr., and Engel, S.L., 1957, The storage and liberation of neurohypophyseal hormones, in: *The Neurohypophysis* (H. Heller, ed), pp. 65–76, Academic Press, New York.

von den Velden, R., 1913, Beitrage zur WirKung von Hypophysenextrakten, *Berl. Klin. Wochenschr.* **50**: 1969.

Walter, R., Schlesinger, D.H., Schwarts, I.L., and Capra, J.D., 1971, Complete amino acid sequence of bovine neurophysin II, *Biochem. Biophys. Res. Commun.* **44**:293–298.

Wei, E., Sigel, S., Loh, H., and Way, E.L., 1975, Thyrotropin-releasing hormone and shaking behavior in rat, *Nature London* **253**:739, 740.

Weiner, R.I., Pattou, E., Kerdelhue, B., and Kordon, C., 1975, Differential effects of hypothalamic deafferentation upon luteinizing hormone–releasing hormone in the median eminence and organum vasculosum of the lamina terminalis, *Endocrinology* **97**:1597–1600.

Westman, A., and Jacobsohn, C., 1938, Endocrinologische Untersuchungen an Ratten mit durchtrenntem Hypophysenstiel, 4. Mitteilung: Die Genitalveränderungen der Rattenmannchen, *Acta Pathol. Microbiol. Scand.* **15**:301–306.

Wheaton, J.E., Krulich, L., and McCann, S.M., 1975, Localization of luteinizing hormone–releasing hormone in the preoptic area and hypothalamus of the rat using radioimmunoassay, *Endocrinology* **97**:30–38.

Winokur, A., and Utiger, R.D., 1974, Thyrotropin-releasing hormone: Regional distribution in rat brain, *Science* **185**:265–267.

Wislocki, G.B., and King, L.S., 1936, The permeability of the hypophysis and hypothalamus to vital dyes, with a study of the hypophyseal vascular supply, *Amer. J. Anat.* **58**:431–472.

Zimmerman, E.A., 1976, Localization of hypothalamic hormones by immunocytochemical techniques, in: *Frontiers in Neuroendocrinology* (L. Martini and W.F. Gangon, eds.), Vol. 4, pp. 25–62, Raven Press, New York.

Zimmerman, E.A., Robinson, A.G., Husain, M.K., Acosta, M., Frantz, A.G., and Sawyer, W.H., 1974, Neurohypophyseal peptides in the bovine hypothalamus: The relationship of neurophysin I to oxytocin, and neurophysin II to vasopressin in supraoptic and paraventricular regions, *Endocrinology* **95**:931–936.

7

Peptides Containing Probable Transmitter Candidates in the Central Nervous System

K.L. Reichelt and P.D. Edminson

1. General Properties of CNS Peptides

Peptides in the CNS are characterized by their (1) low overall concentrations, (2) high potency, (3) marked specificity, (4) both transmitter and hormonal-like properties, (5) content of potential transmitter candidates, and (6) apparent low rate of *de novo* synthesis *in vitro*.

Even in hypothalamic tissue, the overall concentrations of peptides are extremely low (Schally *et al.*, 1973). This also applies to γ-glutamyl peptides (Sano *et al.*, 1966; Reichelt, 1970), and to peptides such as homocarnosine and carnosine (Pisano *et al.*, 1961; Margolis, 1974). The recently discovered and characterized opiate agonist enkephalin also occurs in very low concentrations (Hughes, 1975; Hughes *et al.*, 1975). This seems to be true for all peptides presently known, although the concentrations of certain peptides may be fairly high when related to specific isolated structures, or when sequestered in certain synaptosomal populations (Fink *et al.*, 1972; Barnea *et al.*, 1975). The most abundant peptide, acetyl–Asp–Glu, is present in a quantity of 0.2 μmol/g fresh whole brain (Curatolo *et al.*, 1965).

Peptides in the CNS with transmitterlike or hormonal effects are characterized by their extreme potency, and are active in picomole concentrations. Metabolically active peptides given to intact animals are often active in nanomole

K.L. Reichelt and P.D. Edminson · Department of Neurochemistry, The University Psychiatric Clinic, Vinderen, Oslo, Norway.

doses. Substance P is several hundred times more active than Glu, on a molar basis, in dorsal root neurons (Otsuka *et al.*, 1975). Enkephalin is likewise studied in nanomole concentrations (Hughes *et al.*, 1975), as is the humoral peptidic sleep factor of CNS origin, delta (Monnier and Schoenbergen, 1974). A major stumbling block in peptide research seems to be the low CNS concentrations, which are probably due to extremely high potency.

Specificity of structure is apparent for thyrotropin-releasing hormone (TRH) (Schally *et al.*, 1973), and in the case of enkephalin, the substitution of C-terminal methionine with leucine causes a 3–5-fold decrease in potency (Hughes *et al.*, 1975). This also seems to be the case for peptides that modulate bursting pacemaker activity in single molluscan ganglion cells (Ifshin *et al.*, 1975). The effect on the extinction of escape-avoidance behavior by vasopressin is severely restricted as to the amino acid composition vasopressin (Van Wimersma-Greidanus *et al.*, 1975). The same is the case for ACTH fragments (Garred *et al.*, 1974).

Bifunctional properties are common to many of the known CNS peptides, which often act as possible transmitters and, simultaneously, may show hormonal properties. Vasopressin has a local hormonal effect on kidney tubules, but also inhibits extinction of active-avoidance learning (de Wied *et al.*, 1975). TRH and luteinizing hormone–releasing hormone (LH–RH) act on the hypophysis via the portal vessels of the pituitary stalk, but a direct depressant effect on central neurons has been described (Renaud *et al.*, 1975). It is interesting that TRH and other releasing factors are also found in measurable quantities outside the hypothalamus (Winokur and Utiger, 1974; Hökfelt *et al.*, 1975). Specific mnemic peptides have also been reported (Ungar *et al.*, 1972; Ungar, 1974).

A surprising number of the oligopeptides in the CNS (Table 1) contain possible neurotransmitters, or transmitterlike amino acids. A series of γ-glutamyl peptides has been found *in vivo* (Sano *et al.*, 1966; Reichelt, 1970), as well as acetyl–Asp–Glu (Curatolo *et al.*, 1965), all of which contain Glu. Another series contains GABA and β-Ala, such as GABA–His, GABA–Lys, β-Ala–His, and β-Ala–Lys (Pisano *et al.*, 1961). N-Acetyl-homocarnosine and N-acetyl-carnosine have also been found (Sobue *et al.*, 1975), and, in addition, GABA–CH_3His and β-Ala–CH_3His (Kumon *et al.*, 1970). The synthesis of acetyl–Asp peptides containing Glu, Asp, Gly, and Tau has been described for an *in vitro* system (Reichelt and Kvamme, 1973), and also for peptides containing GABA (Reichelt and Edminson, 1974). Many of the other amino acids readily incorporated into peptides, such as Pro, Ser, and cysteic acid, also have membrane effects.

CNS peptides are synthesized in surprisingly low yields in organ cultures (McKelvy *et al.*, 1975), in tissue cultures or in *in vitro* systems. Most data show nanomole yields per gram fresh tissue. Thus, the nonribosomal synthesis of TRH found in supernatants of hypothalamic extracts is characterized by very low yields (Mitnick and Reichlin, 1972). Likewise, formation of LH–RH in a crude mitochondrial fraction is very low (Johansson *et al.*, 1972). Work in our laboratory (to be published) on TRH also shows very low yields, as do the data for homocarnosine and carnosine synthetase (Skaper *et al.*, 1973). These low yields obtained in very similar incubation mixtures may reflect either a very complicated

TABLE 1. Oligopeptides of the CNS Found *in Vivo*

γ-Glutamyl	Other peptides	Releasing hormones
Glu–Glu	GABA–His	TRH
Glu–Gln	GABA–Lys	FSH–RH
Glu–Ser	β-Ala–Lys	LH–RH
Glu–Ala	β-Ala–His	MIH (MRIH)
		GHRH
Glu–Val	GABA–CH₃His	GHRIH
Glu–Ile	β-Ala–CH₃His	MSHRH
Glu–Cys	Asp–Ser	PRLRH
Glu–S–CH₃Cys	Ala–Phe	PRLRIH
Glu–Cys–Gly	acetyl-Asp–Glu	LHIH
Glu–Ala–Gly	acetyl-β-Ala–His	CRH
Glu–S–CH₃Cys–Gly	acetyl-GABA–His	
Glu–β-AIB [a]	Substance P	
Glu–β-AIB–Gly	neurotensin	
	enkephalin	
	(Tyr–Gly–Gly–Phe–Met)	
	vasopressin	
	oxytocin	

[a] AIB is the abbreviation for aminoisobutyric acid. The following amino acids have been reported to show excitatory effects: Glu, Asp, Cys, homocysteate, N–CH₃–Asp, and Pro. The following have, in various connections, been shown to have inhibitory properties: Tau, Gly, β-Ala, Ala, β-AIB, GABA, and betaine (Reichelt, 1972).

control of the synthetic process, as indicated by the considerable dependence on intact cellular structures (McKelvy *et al.*, 1975), or difficulties in blocking the numerous general and specific peptidases present in nervous tissue (Marks, 1971; Griffiths *et al.*, 1975*b*).

2. Peptide and Peptidoamine Synthesis with *N*-Terminal Acetyl-Asp

In homogenates of cortical tissue from mice, with monoamine oxidase (EC 1.4.3.4) inhibited with nialamide, and in the presence of ATP and the amino acids acetyl–Asp, Glu, Gly, Ser, Ala, and GABA or Tau in 1 mM concentrations, histamine (1 mM) was able to induce the synthesis of acetyl–Asp peptides during 30–60 min incubations (Reichelt and Kvamme, 1973). The natures of the peptides isolated by removing amino acids and non-*N*-substituted peptides by cation exchange (Dowex 50) and subsequent fractionation on an anion exchanger (Dowex 1), followed by TLC and P2 gel filtration, are illustrated in Table 2. The structures are tentative, since rearrangements during partial hydrolysis to remove acetyl-Asp cannot be excluded. Protein synthesis inhibitors did not prevent this peptide formation (Reichelt and Kvamme, 1973), and the level of each peptide formed was low, with a maximum of 100 nmol/g fresh tissue for acetyl-Asp–Glu. A certain amine specificity was also found when comparing dopamine and noradrenaline

TABLE 2. Peptidic Compounds Formed in the Presence of Histamine [a]

Column	Elution volume (ml)	Tentative structure
Dow. 1	44	acetyl-Asp–Glu–Ser–Gly
	110	acetyl-Asp–Glu–Gly (Cys^2, Ser^2, Ala)
	126	acetyl-Asp–Glu (Gly^4, Cys^2), Tau
	126	acetyl-Asp–Gly–Ser
	126	acetyl-Asp–Glu
	136	acetyl-Asp–Glu–Asp
	170	acetyl-Asp–Tau
P2 gel	$V_0 + 34$–38	acetyl-Asp (Ser^2, Glu^2, Gly^2, Ala^1, histamine^1)
	$V_0 + 34$–40	acetyl-Asp (Ser^1, Glu^1, Gly^1, Ala^1, histamine^1)
	$V_0 + 34$–40	acetyl-Asp (Ser^1, Glu^2, Gly^1, GABA^2, histamine^1)
	$V_0 + 34$–40	acetyl-Asp (Ser^2, Glu^2, Gly^2, Ala^1, GABA^1, histamine^1)

[a] Representative compounds isolated by anion and cation exchange. Thin-layer and paper chromatography are shown (Reichelt and Kvamme, 1973). The overall amino acid composition was determined by hydrolysis in glass distilled HCl for 24 hr. Structural studies were done by partial hydrolysis under nitrogen and recovery of acetyl-Asp and histamine as such, and histamine also as dansyl-histamine (Reichelt et al., 1976). The structure must be tentative due to the possibility of rearrangement during splitting of the acetyl group of acetyl-Asp.

with histamine, and the overall numbers and quantities of acetyl-Asp peptides found were much smaller (Reichelt and Edminson, 1974). Stimulation of the peptide synthesis by 3',5'-cAMP was found, using 0.1 mM 3',5'-cAMP.

The passage of the deproteinized incubation mixtures through Dowex 50 cationic-exchange resin in the H$^+$ form at pH 6 led to the retention of all free amino-terminal peptides, and also of peptides containing histidine or histamine. To avoid this loss, P2 gel filtration in 0.5 M acetic acid or formic acid was employed (Reichelt et al., 1976). We were then able to isolate the peptidoamine compounds shown in Table 2 by anion-exchange, Sephadex G10 filtration, and thin-layer chromatography. We have not, however, been able to exclude the action of transglutaminase (Ginsburg et al., 1963).

The salient feature of these peptides and peptidoamines is that transmitter candidates like histamine, Tau, Asp, Glu, and Gly (Table 2) are all included. It should be stressed that labeled precursors were recovered as Asp, acetyl-Asp, and histamine after partial hydrolysis. Also, the dependence on an ATP source makes degradative processes improbable (Reichelt et al., 1976). Similar data have been found for dopamine (manuscript in preparation). Recently, nanomole per gram fresh tissue quantities of acetyl-Asp-peptides bound to serotonin have also been found (Edminson, 1975). Following intracranial injection of acetyl-Asp in vivo, trace quantities of acetyl-Asp-labeled peptides have been found (Kvamme et al., 1975). It should be noted that peptidases, and also aryl-amidase, have not been inhibited.

3. Factors That Affect the Levels and Release of Peptides in the CNS

3.1. Regulation of Secretion of Releasing Factors

The release of hypothalamic releasing factors into the portal blood system to the hypophysis is apparently regulated by a considerable number of different inputs. Evidence is accumulating for the control of LH–RH release by catecholamines and other amines, dopamine and noradrenalin having a stimulatory effect, serotonin an inhibitory effect. Histamine is also implicated in the release of LH–RH (for a review, see McCann and Moss, 1975). Circulating levels of steroid hormones also affect the release of LH, both through the hypothalamus and through alterations in hypophyseal sensitivity to LH–RH (Mahesh et al., 1975). Other compounds that have effects on LH–RH release include stimulatory effects of prostaglandins and possibly GABA (Ojeda et al., 1975; Ondo, 1974). Another complicating factor is the isolation of a factor from hypothalamus with LHIH activity (Johansson et al., 1975). Stimulatory effects of metal ions of hypothalamic origin on the release of hypophyseal hormones have also been reported (LaBella et al., 1973). Similar results also apply to the regulation of release of TRH, and the control of growth hormone (GH) and prolactin release (Garver et. al., 1975; McCann and Moss, 1975). Thus, the hypothalamus may be considered to be a "mixing box" in which multisignal integration results in the formation and release of peptides.

3.2. Role of Peptidases in the CNS

Brain tissue contains a large number of peptidases with varying specificities, including acid and neutral proteinases, di- and tripeptidases, peptidases inactivating oxytocin, Substance P, and releasing factors from hypothalamus (Marks et al., 1974; Benuck and Marks, 1975; Griffiths and Hooper, 1974; Griffiths et al., 1973; Griffiths et al., 1975a,b). The peptidase that inactivates LH–RH is thus found both in the hypothalamus and in other regions of the brain (Griffiths et al., 1975a). The hypothalamic peptidases that use oxytocin or LH–RH as substrate are found in soluble and particulate forms, the activities and degree of solubilization varying with sex, castration, and treatment with gonadal steroids, indicating a feedback mechanism (Griffiths and Hooper, 1974; Griffiths et al., 1975c). This type of feedback regulation and intracellular distribution may play an important role in the regulation of levels of peptides, both in the hypothalamus and in other parts of the brain.

4. A Working Hypothesis of Peptides as the Final Common Pathway of Multisignal Integration

From the foregoing data the following seems established:

1. Peptides are present in the CNS, and act as possible neurotransmitters, hormones, and modulators of electrophysiological and secretory processes in neurons.

2. Many CNS oligopeptides contain neurotransmitter candidates.

3. An apparatus for peptide synthesis is present that can sustain *de novo* synthesis either by a ribosome-independent synthesis, or by ribosomal synthesis followed by peptidase-dependent splitting off of active components (vasopressin and oxytocin).

4. There is extensive apparatus for peptide breakdown (peptidases) in the CNS that is partially under specific and extensive feedback control.

5. Peptides have behavioral, mnemic, and extensive metabolic effects. (For a detailed discussion, see Reichelt and Edminson, 1976.)

6. Peptides such as releasing factors are known to be under multifactorial control. In fact, as noted in Section 3.1, the hypothalamus may be considered a "mixing box" in which multiple inputs are integrated. The releasing hormones may therefore be considered as the final common pathway in a multisignal-regulated servomechanism. As several of these peptides are also detectable in brain regions other than hypothalamus (Winokur and Utiger, 1974; Hökfelt *et al.*, 1975), they may therefore be suspected of having similar relationships.

7. Peptides have considerable informational capacity. Thus, 15 nucleotide bases, including initiation and terminating codons, are necessary to relay the information carried by a tripeptide (Reichelt and Edminson, 1976). Amidation and transamination (Ginsburg *et al.*, 1963) further increase the information capacity, and a very flexible coding possibility is apparent.

As a consequence of the facts presented above, we feel that it is reasonable to propose, as a *working hypothesis,* that peptides: (1) may be the integrated response to multisignal inputs to key cells, such as ganglion cells and neurosecretory cells in the CNS; and (2) possibly reflect the temporal and geometrical sequence of events in the neuron by their composition.

We think that the considerable number of signals impinging on dendrites, perikarya, and axons must somehow be *integrated* into meaningful chemical signals that directly represent series of neuronal events. The millisecond time scale of electrophysiological changes, and the minutes-to-hours time scale of chemical responses in neural tissue, entail a paramount requirement for an intermediate, stable registering system with a large and flexible information potential. In this connection, it is important that catechol–estrogen has been reported in hypothalamus (Fishman and Norton, 1975), which points to such an endocrine and neuronal integration as being possible. Our hypothesis would possibly explain the findings of Ungar *et al.* (1972), and retain sufficient plasticity and specificity to meet the great variety of possible inputs to a neuron.

5. Specific Examples of the Working Hypothesis

Our model would have two different forms: one in which the *release* of peptide is the result of converging pathways (Fig. 1), and another in which peptide synthesis is caused by the confluence of inputs impinging on the cell (Fig. 2). Figure 1 is typical of the first situation, and is an inferential model of the present knowledge on GH-releasing factor control (Garver *et al.*, 1975). Dopamine, serotonin, noradrenalin, stress, estrogens, glucocorticoids, free fatty acids, and GH all influence the release of this peptide. As the last synthesized transmitter is usually first released, this might be a special case of the second form of our hypothesis.

In the second form (Fig. 2), the synthesis of peptide is caused by the converging neuronal pathways, and is regulated by and dependent on the transmitter release. This model is entirely *speculative*. In Dendrite A (Fig. 2), the incoming inputs are B, then A, giving rise to a peptide such as acetyl-Asp–B–A (NAA–B–A). In Dendrite B, the sequence of firing is A, then B, and acetyl-Asp–A–B is formed. If the transmitter as such is not incorporated into the peptidic compounds, but causes amino acids such as X and Y, representing each specific transmitter, to be bound, a compound such as acetyl-Asp–X–Y or acetyl-Asp–Y–X might be formed respectively. Here, X depends on the release of B and Y on A.

Certain amines seem intimately related to the reward system in the CNS (Wise *et al.*, 1973; Goodman *et al.*, 1975), and the incorporation of amines such as those indicated in Fig. 2 by MA could constitute the go-ahead signal for further processing. Such peptidoamine compounds have been found (Reichelt *et al.*, 1976). Similarly, if the information should be blocked from further processing by

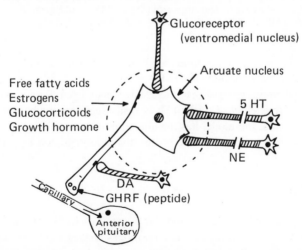

FIG. 1. Inferential model of the present knowledge of the converging controls pertaining to GH-releasing factor release (Garver *et al.*, 1975). Thickening of the membrane represents receptors (specific). Clearly, the secretory cell represents a multiple pathway integration of its release of peptide. It is noteworthy that in molluscan neurosecretory cell, the "bursting pacemaker activity" is peptide-regulated (Barker and Gainer, 1974).

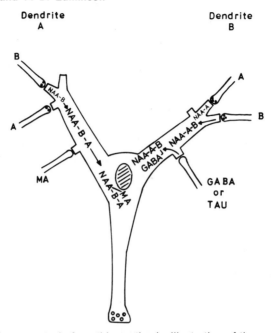

FIG. 2. Multipathway control of peptide synthesis. Illustration of the way in which signal integration into synthetic peptides of acetyl-Asp could be formed and would contain the transmitters sequentially. Peptide transmitters would also be incorporated. Transmitters not themselves incorporated may still give rise to the incorporation of corresponding amino acid from the cytoplasm. A specific amino acid thus represents this transmitter. In Dendrite A, the sequence of firing—B, then A—gives rise to the peptide acetyl-Asp-B-A. In Dendrite B, the firing sequence is A, then B, then GABA, and would cause the formation of acetyl-Asp–A–B–GABA. The aminergic input (MA) could give rise to a complex such as acetyl-Asp/MA–B–A or acetyl-Asp–B–A–MA. (NAA) Acetyl-Asp.

the protein-synthesis apparatus, inhibitory transmitters such as GABA (or Tau or 5-HT) might form acetyl-Asp–A–B–GABA (or Tau) compounds, as shown in Fig. 2. Peptides containing these transmitters have also been found (Reichelt and Kvamme, 1973; Reichelt *et al.*, 1976). It is easily seen that the permutations possible are endless, and, the model has the attraction of taking the geometric pattern of the brain and the sequential nature of information integration into account.

6. Conclusion

The high levels of peptidases and the potency of CNS peptides as possible neurotransmitters, neurohormones, and hormones makes this one of the prime fields of investigation. Peptides as postulated multisignal integrators would agree with Sherrington's (1906) concept of the neurons as the "final common pathway of a multichannel system."

7. References

Barker, L., and Gainer, H., 1974, Peptide regulation of bursting pacemaker activity in a molluscan neurosecretory cell, *Science* **184**:1371–1373.

Barnea, A., Ben-Jonathan, N., Colston, C., Johnston, J.M., and Porter, J.C., 1975, Differential subcellular compartmentalization of thyrotropin releasing hormone (TRH) and gonadotropin releasing hormone (LRH) in hypothalamic tissue, *Proc. Nat. Acad. Sci. U.S.A.*, **72**:3153–3157.

Benuck, M., and Marks, N., 1975, Enzymatic inactivation of substance P by a partially purified enzyme from rat brain, *Biochem. Biophys. Res. Commun.* **65**:153–160.

Curatolo, A., D'Arcangelo, P., Lino, A., and Brancati, A., 1965, Distribution of N-acetyl-aspartic and N-acetyl-aspartyl-glutamic acids in nervous tissue, *J. Neurochem.* **12**:339–342.

De Wied, D., Bohus, B., and Van Wimersma-Greidanus, Tj.B., 1975, Memory deficit in rats with hereditary diabetes insipidus, *Brain Res.* **85**:152–156.

Edminson, P.D., 1975, Tentative identification of N-acetyl-aspartyl–peptide complexes containing serotonin, or a serotonin derivative, formed *in vitro* in brain tissue extracts, *10th Meeting of the FEBS Paris* (abstract 1291).

Fink, G., Smith, G.G., Tibballs, J., and Lee, V.W.K., 1972, LRF and CRF release in subcellular fractions of bovine median eminence, *Nature London New Biol.* **239**:57–59.

Fishman, J., and Norton, B., 1975, Catechol–estrogen formation in the central nervous system of the rat, *Endocrinology* **96**:1054–1059.

Garred, P., Gray, J.A., and De Wied, D., 1974, Pituitary–adrenal hormones and extinction of reversal behaviour in the rat, *Physiol. Behav.* **12**:109–119.

Garver, D.L., Ghanshyam, N.P., Haroutune, D., and Deleon-Jones, F., 1975, Growth hormone and catecholamines in affective disease, *Amer. J. Psychiatry* **132**:1149–1154.

Ginsburg, M., Wajda, I., and Waelsch, H., 1963, Transglutaminase and histamine incorporation *in vivo*, *Biochem. Pharmacol.* **12**:251–264.

Goodman, R.H., Flexner, J.B., and Flexner, L.B., 1975, The effects of acetooxycyclohexemide on rate of accumulation of cerebral catecholamines from circulating tyrosine as related to its effect on memory, *Proc. Nat. Acad. Sci. U.S.A.* **72**:479–482.

Griffiths, E.C., and Hooper, K.C., 1974, Peptidase activity in different areas of the rat hypothalamus, *Acta Endocrinol.* **77**:10–18.

Griffiths, E.C., Hooper, K.C., Hopkinson, C.R.N., 1973, Evidence for an enzymic component in the rat hypothalamus capable of inactivating luteinizing hormone-releasing factor (LRF), *Acta Endocrinol.* **74**:49–55.

Griffiths, E.C., Hooper, K.C., Jeffcoate, S.L., and Holland, D.T., 1975a, Peptidases in different areas of the rat brain hypothalamus inactivating luteinizing hormone–releasing hormone (LHRH), *Brain Res.* **85**:161–164.

Griffiths, E.C., Hooper, K.C., Jeffcoate, S.L., and Holland, D.T., 1975b, The effects of gonadectomy and gonadal steroids on the activity of hypothalamic peptidases inactivating luteinizing hormone–releasing hormone (LH–RH), *Brain Res.* **88**:384–388.

Griffiths, E.C., Hooper, K.C., Jeffcoate, S.L., and White, N., 1975c, Peptidases in the rat hypothalamus inactivating thyrotropin-releasing hormone (TRH), *Acta Endocrinol.* **79**: 209–216.

Hökfelt, T., Fuxe, K., Johansson, O., Jeffcoate, S., and White, W., 1975, Thyrotropin releasing hormone (TRH), containing nerve terminals in certain brain stem nuclei in the spinal cord, *Neurosci. Lett.* **1**:133–140.

Hughes, J., 1975, Isolation of an endogenous compound from brain with pharmacological properties similar to morphine, *Brain Res.* **88**:295–308.

Hughes, J., Smith, T.W., Kosterlitz, H.W., Fothergill, L.A., Morgan, B.A., and Morris, H.R., 1975, Identification of two related pentapeptides from the brain with potent opiate agonist activity, *Nature London* **258**:577–579.

Ifshin, M.S., Gainer, A., and Barker, J.C., 1975, Peptide factor extracted from milluscan ganglia that modulates bursting pacemaker activity, *Nature London* **254**:72, 73.

Johansson, K.N.-G., Hooper, F., Sievertsson, H., Currie, B.L., Folkers, K., and Bowers, C.Y., 1972, Biosynthesis *in vitro* of the luteinizing releasing hormone by hypothalamic tissue, *Biochem. Biophys. Res. Commun.* **49**:656–660.

Johansson, K.N.-G., Greibrokk, T., Currie, B.L., Hansen, J., Folkers, K., and Bowers, C.Y., 1975, Factor C-LHIH which inhibits the luteinizing hormone from basal release and from synthetic LHRH and studies on purification of FSHRH, *Biochem. Biophys. Res. Commun.* **63**:62–68.

Kumon, A., Matsuoka, Y., Nakajima, T., Kakimoto, T., Imaoka, N., and Sano, I., 1970, Isolation and identification of N^{α}-(β-alanyl) lysine and N^{α}-(γ-aminobutyryl) lysine from bovine brain, *Biochem. Biophys. Acta* **200**:170, 171.

Kvamme, E., Reichelt, K.L., Edminson, P.D., Sinichkin, A., Sterri, S., and Svenneby, G., 1975, *N*-Substituted peptides in brain, *Proceedings of the 10th FEBS Meeting*, pp. 127–136.

LaBella, F., Dular, R., Vivian, S., and Queen, G., 1973, Pituitary hormone releasing or inhibiting activity of metal ions present in hypothalamic extracts, *Biochem. Biophys. Res. Commun.* **52**:786–791.

Mahesh, V.B., Muldoon, T.G., Eldridge, J.C., and Korach, K.S., 1975, The role of steroid hormones in the regulation of gonadotropin secretion, *J. Steroid Biochem.* **6**:1025–1036.

Margolis, F.L., 1974, Carnosine in the primary olfactory pathway, *Science* **184**:909–911.

Marks, N., 1971, Peptide hydrolases, in: *Handbook of Neurochemistry* (A. Lajtha, ed.), Vol. 3, pp. 133–171, Plenum Press, New York.

Marks, N., Galoyan, A., Grynbaum, A., and Lajtha, A., 1974, Protein and peptide hydrolases of the rat hypothalamus and pituitary, *J. Neurochem.* **22**:735–739.

McCann, S.M., and Moss, R.L., 1975, Putative neurotransmitters involved in discharging gonadotropin-releasing neurohormones and the action of LH-releasing hormone on the CNS, *Life Sci.* **16**:833–852.

McKelvy, J.F., Sheridan, M., Joseph, S., Phelphs, C.H., and Perrie, S., 1975, Biosynthesis of thyrotropin-releasing hormone in organ cultures of the guinea pig median eminence, *Endocrinology* **97**:908–918.

Mitnick, M., and Reichlin, S., 1972, Enzymatic synthesis of thyrotropin-releasing hormone (TRH) by hypothalamic TRH synthetase, *Endocrinology* **91**:1145–1153.

Monnier, M., and Schoenbergen, G.A.S., 1974, Neuro-humoral coding of sleep by the physiological sleep factor delta, in: *Neurochemical Coding of Brain Function* (R.D. Myers and R.R. Drucker, eds.), Odin, New York.

Ojeda, S.L., Wheaton, J.E., and McCann, S.M., 1975, Prostaglandin E_2 induced release of luteinizing hormone–releasing factor (LRF), *Neuroendocrinology* **17**:283–287.

Ondo, J.G., 1974, Gamma-aminobutyric acid effects on pituitary gonadotropin secretion, *Science* **186**:738, 739.

Otsuka, M., Konishi, S., and Takahashi, T., 1975, Hypothalamic substance P as a candidate for transmitter of primary afferent neurons, *Fed. Proc. Fed. Amer. Soc. Exp. Biol.* **34**:1922–1928.

Pisano, J.J., Wilson, J.D., Cohen, L., Abraham, D., and Udenfriend, S., 1961, Isolation of γ-aminobutyryl–histidine (homocarnosine) from brain, *J. Biol. Chem.* **236**:499–502.

Reichelt, K.L., 1970, The isolation of gamma-glutamyl peptides from monkey brain, *J. Neurochem.* **17**:19–25.

Reichelt, K.L., 1972, *An Investigation into the Nature and Function of Anionic Acids, Peptides, and Acid-Soluble Proteins from Brain Tissue,* pp. 1–42, Oslo Univ. Press, Oslo.

Reichelt, K.L., and Edminson, P.D., 1974, Biogenic amine specificity of cortical peptide synthesis in monkey brain, *FEBS Lett.* **47**:185–189.

Reichelt, K.L., and Edminson, P.D., 1976, Transmitter dependent peptide synthesis in the CNS, in: *Advances in Biochemical Neuropharmacology,* pp. 211–223, Raven Press, New York.

Reichelt, K.L., and Kvamme, E., 1973, Histamine-dependent formation of *N*-acetyl–aspartyl peptides in mouse brain, *J. Neurochem.* **21**:849–859.

Reichelt, K.L., Edminson, P.D., and Kvamme, E., 1976, The formation of peptido-amines from constituent amino acids and histamine in nervous tissue, *J. Neurochem.* **26**:811–815.

Renaud, L.P., Martin, J.B., and Brazeau, P., 1975, Depressant action of TRH, LH–RH and somatostatin on activity of central neurones, *Nature* **255**:233–235.

Sano, I., Kakimoto, Y., Kanazawa, A., Nakajima, T., and Schimizu, M., 1966, Identifizierung einiger Glutamyl Peptide aus Gehirn, *J. Neurochem.* **13**:711–719.

Schally, A.V., Arimura, A., and Kastin, A.J., 1973, Hypothalamic regulatory hormones, *Science* **179**:341–350.

Sherrington, C.S., 1906, *The Integrative Action of the Nervous System*, Yale Univ. Press, New Haven.

Skaper, S.D., Das, S., and Marshall, F.D., 1973, Some properties of a homocarnosine–carnosine synthetase isolated from rat brain, *J. Neurochem.* **21**:1429–1445.

Sobue, K., Konishi, H., and Nakayima, T., 1975, Isolation and identification of *N*-acetyl-homocarnosine and *N*-acetyl-carnosine from brain and muscle, *J. Neurochem.* **24**:1261, 1262.

Ungar, G., 1974, Molecular coding of memory, *Life Sciences* **14**:595–604.

Ungar, G., Desiderio, D.M., and Parr, W., 1972, Isolation, identification of a specific behaviour inducing peptide, *Nature London* **238**:198–202.

Van Wimersma-Greidanus, Tj.B., Bohus, B., and De Wied, D., 1975, The role of vasopressin in memory consolidation, *J. Endocrinol.* **164**:30, 31.

Winokur, A., and Utiger, R.D., 1974, Thyrotropin-releasing hormone: Regional distribution in rat brain, *Science* **185**:265–267.

Wise, C.D., Berger, B.D., and Stein, L., 1973, Evidence for α-noradrenergic reward receptors and serotonergic punishment receptors in rat brain, *Biol. Psychiatry* **6**:3–21.

8

Biosynthesis of Neuronal Peptides

Harold Gainer, Y. Peng Loh, and Yosef Sarne

1. Introduction

The general acceptance of the idea that peptides represent a new class of intercellular messengers in the nervous system (i.e., neurotransmitters and neuromodulators) naturally raises the question whether the biosynthetic mechanisms for peptides in "peptidergic" neurons (i.e., neurons that synthesize peptides for release as intercellular messengers) confer distinct properties on these neurons. One of the central issues is whether neuronal peptides are synthesized by ribosomal mechanisms followed by posttranslational cleavage, or by enzymatic ("synthetase") processes similar to those of the conventional neurotransmitters (e.g., acetylcholine, γ-aminobutyric acid, biogenic amines). The answer to this question has significance with regard to the cell biology and "functional–morphological" organization of these neurons. A ribosomal mechanism would imply that the biosynthetic process would be restricted to the neuronal perikaryon, whereas an enzymatic mechanism would allow biosynthesis and its regulation to occur at the neuron's site of release in the axon terminal. In either case, axonal transport mechanisms would be intimately involved. In the former case, however, the transported material would be the presynthesized peptides, whereas in the latter, the biosynthetic enzyme would be transported.

The best evidence for nonribosomal biosynthesis of polypeptides can be found in studies on the antibiotic peptides (e.g., gramicidin S, tyrocidines, and polymyxin-group antibiotics) in prokaryotes (Lipmann, 1971; Kurahashi, 1974).

Harold Gainer and Y. Peng Loh · Section on Functional Neurochemistry, Behavioral Biology Branch, National Institute of Child Health and Human Development, National Institutes of Health, Bethesda, Maryland. **Yosef Sarne** · Department of Behavioral Biology, Technion Medical School, Haifa, Israel.

This novel biosynthesis process, which is referred to as the "protein template" mechanism (Kurahashi, 1974), has not yet been reported in eukaryotic systems, and has been characterized as a possible evolutionary intermediate between fatty acid synthesis and the more complex ribosomal protein synthesis mechanisms (Lipmann, 1971). The sequence specificity of this mechanism is determined by specific enzymes, and it is postulated that the "heavy" enzyme (270,000 daltons) in gramicidin S–synthetase would possess 18 or 19 specific catalytic functions (Laland and Zimmer, 1973). In general, protein template mechanisms have been found to be less reliable than ribosomal (or "mRNA template") mechanisms, since a given strain of bacteria using protein template mechanisms may produce various peptide analogues in which some constituent amino acids are substituted by other structurally related ones (Laland and Zimmer, 1973; Kurahashi, 1974).

In eukaryotes, nonribosomal synthesis has been reported for several dipeptides and tripeptides (e.g., carnosine, glutathione, ophthalmic acid). The various γ-glutamyl peptides extracted from nervous tissue (Sano, 1970) probably derive from the γ-glutamyl cycle described by Meister (1973). The γ-aminobutyryl, β-alanyl, and N-acetylated di- and tripeptides also appear to be synthesized by "synthetase" enzymes with broad specificities (Sano, 1970; see also Chapter 7).

Recently, enzymatic (synthetase) mechanisms have been proposed for the synthesis of thyrotropin-releasing hormone (TRH) (Mitnick and Reichlin, 1971; Reichlin et al., 1972; Reichlin and Mitnick, 1973), growth hormone–releasing factor (GH–RF) (Reichlin and Mitnick, 1972), and prolactin-releasing factor (PRF) (Mitnick et al., 1973). However, a number of conflicting findings and objections to the proposal that TRH is synthesized via an enzymic mechanism have led Reichlin et al. (1976) to write in a recent review: "In light of these conflicting findings, including those from our own laboratory, we feel that a final judgment as to the mechanism of TRH biosynthesis cannot be made at this time." Many physiologically active peptides in eukaryotes appear to be derived from protein precursors by the action of specific proteinases. The production of the decapeptide angiotensin I by the proteolytic action of renin on angiotensinogen, and of angiotensin II by the subsequent cleaving of a dipeptide from angiotensin I by the converting enzyme, as well as the generation of kallidin I (bradykinin) and II by the limited proteolysis of kininogen in the blood are well known. It is the purpose of this chapter to review briefly the current status of the precursor-protein concept of polypeptide and peptide biosynthesis, and to show that this mechanism is relevant to "peptidergic neurons."

2. The Precursor-Protein (Prohormone) Concept

Since the discovery that insulin is synthesized in a larger precursor form as proinsulin (Steiner and Oyer, 1967; Steiner et al., 1967), substantial evidence has accumulated that this is a common mode of biosynthesis of eukaryotic proteins and peptides that are destined for secretion (Tager and Steiner, 1974). Following the translation of the messenger RNA into the precursor protein, the precursor is then converted into smaller proteins or peptides or both by posttranslational cleavage (i.e., by specific proteolytic enzymes). A detailed study of this mechanism and its morphological correlates has been described for the conversion of proinsulin to insulin and the C-peptide (Steiner *et al.*, 1974).

Precursors (or prohormones) have been reported for several other hormones, including glucagon (Noe and Bauer, 1971; Trakatellis *et al.*, 1975), calcitonin (Moya *et al.*, 1975), erythropoietin (Peschle and Condorelli, 1975), growth hormone (GH) (Stachura and Frohman, 1974, 1975; Sussman *et al.*, 1976), parathyroid hormone (PTH) (Sherwood *et al.*, 1970; Kemper *et al.*, 1972, 1974), and ACTH (Yalow and Berson, 1973; Gewirtz and Yalow, 1974; Hirata *et al.*, 1975). In the case of melanocyte-stimulating hormone (MSH), comparisons of amino acid sequences have suggested that ACTH is the precursor of γ-MSH and that β-lipotropin is the precursor of β-MSH (Scott *et al.*, 1973; Lowry and Scott, 1975). Similar arguments have been put forth for the proposal that β-lipotropin is the precursor for the morphinelike peptide enkephalin (see Chapters 9 and 13).

Such precursor–product relationships are not restricted to the endocrine system, and appear to be relevant to other secretory cells as well. Precursors involved in biosynthesis have also been reported for serum albumin (Judah *et al.*, 1973; Urban and Schreiber, 1975; Quinn *et al.*, 1975), collagen (Goldberg and Scherr, 1973; Bornstein, 1974; Davidson *et al.*, 1975), and immunoglobulin light chains (Milstein *et al.*, 1972; Schmeckpeper *et al.*, 1974).

3. Strategy for the Study of Peptide Biosynthesis in Neurons

3.1. Intact Systems

If it is anticipated that neuronal peptides are synthesized via a ribosomal mechanism, then attention must be paid to the segregation of biosynthetic and storage mechanisms in neurons. Regions of the nervous system that contain high concentrations of the specific peptide may reflect storage sites in axon terminals that are not capable of synthesis of the peptide. The biosynthetic site may be at a considerable distance from the storage region, and may contain low levels of the peptide. This problem of the regional localization of peptide biosynthesis and storage is discussed in Chapter 6. Clearly, biosynthetic studies must be conducted using the region occupied by the neuron perikaryon. However, axonal transport processes must also be accounted for in quantitative studies on "turnover."

The initial step is to demonstrate that radioactive amino acid precursors are incorporated into the peptide of interest by the tissue under study. The labeled amino acid used should obviously be of high specific activity and found in abundance in the peptide. As many detection techniques as are available (i.e., biochemical, immunological, and bioassay techniques) should be used to prove that the incorporated label is indeed in the peptide. Protein synthesis inhibitors (in particular, those that block initiation and elongation; see Grollman and Huang, 1973) are useful to provide supporting evidence that ribosomally directed protein synthesis is required for incorporation of the labeled amino acid into the peptide. Determination of the conditions and minimum concentrations of the inhibitor necessary to block *de novo* protein synthesis will also be of value in the pulse–chase experiments described below.

The critical issue is whether biosynthetic evidence can be obtained for a precursor–product relationship. The minimal requirement is to detect during the pulse, in pulse–label and –chase experiments, a labeled peptide (or polypeptide) of higher molecular weight, which then with time (in the chase) declines in radioactivity concurrently with the emergence of the label in the peptide product. Such experiments are best done in isolated, intact neuronal systems (e.g., dissected ganglia) in culture media, since it is desirable during the chase period to prevent additional incorporation of labeled amino acids by the addition of a large amount of unlabeled amino acids or a protein synthesis inhibitor or both to the culture medium. Control of the chase conditions is more difficult when the experiment is done *in vivo* (i.e., by intraventricular injection of the labeled amino acid or by microinjection in a restricted brain region). The major problem in these experiments is again one of detection. Extraction and separation procedures shown to be adequate for the peptide product may not be appropriate for the precursor. A variety of chromatographic and polyacrylamide-gel electrophoretic (PAGE) techniques should be employed to determine the optimum conditions for separation and resolution. Since the precursor may be bound to membranes or may tend to aggregate, the use of chaotropic agents or detergents may be necessary in extraction and separation procedures.

Probably the greatest difficulty in these experiments is to detect the "signal" (labeled precursor) over the "noise" (the rest of the labeled protein in the tissue). The "signal-to-noise" ratio will depend on (1) the extent of biosynthesis of precursor in the peptidergic neuron (i.e., the percentage of labeled precursor *vis-à-vis* the rest of the labeled protein in the cell) and (2) the degree to which the tissue is "contaminated" with other cell types (i.e., neurons and glia) that are not involved in the biosynthesis of the peptide. Clearly, in order to reduce this problem, the analysis should be done on biological material that most closely approximates the isolated peptidergic neurons alone. Hence, attention must be paid to the morphological composition of the tissue to be studied.

It should be pointed out that while the detection of a putative precursor using the pulse–chase paradigm described above in intact systems is a *necessary* step, it is not *sufficient* to prove a precursor–product relationship. The detection of higher-molecular-weight immunoreactive biologically active forms of the peptide (see Chapter 3) *per se* also represents suggestive but not compelling evidence that these forms are precursors. However, these immunological detection methods, used in combination with the biosynthesis studies described above, greatly strengthen the identification of a putative precursor.

3.2. Identification of a Precursor

Following the detection of a labeled polypeptide that behaves as a precursor in pulse–chase experiments, the labeled precursor must be isolated for further analysis. The use of affinity chromatography or immunoprecipitation, when the specific antibody to the peptide product or products that cross-react with the precursor is available, greatly facilitates this process, and in addition provides immunological evidence for some sequence homology. In the absence of this possibility, various conventional chromatographic and PAGE techniques can be used to obtain purified precursor material.

Tryptic peptide maps of the purified precursor should be compared with similar maps for the peptide product or products, in order to determine whether there is correspondence in tryptic peptides. In some cases, as for insulin (Steiner *et al.*, 1974), trypsin converts the precursor to a peptide of interest without further degradation. In many cases, however, there is more degradation of the peptides by trypsin than would occur *in situ,* and a careful matching of the total map is necessary. In order to improve detection of the tryptic peptides by autoradiography, a number of different radioactive amino acids should be used to label the precursor and peptide product or products. In addition, cyanogen bromide cleavage as well as tryptic digestion could be used in mapping studies. The ultimate identification of a precursor would, of course, depend on determining the amino acid sequence of the putative precursor and showing that it contains the sequences found for the peptide products. It is likely, however, that only very small amounts of neuronal peptide precursor material will be available, thus precluding this approach despite new "microsequencing" methods (Jacobs *et al.*, 1974). In recent years, precursors of secretory proteins have been obtained by direct translation of messenger RNA in cell-free systems (e.g., the wheat germ system). This approach has shown that prohormones may themselves be products of degradation of larger precursors called *preprohormones* (see Kemper *et al.*, 1974; Devillers-Thiery *et al.*, 1975; Sussman *et al.*, 1976). This elegant and powerful approach requires the isolation of the specific mRNA from the tissue, and would be extremely difficult in view of the heterogeneity of the nervous system.

4. Peptidergic Neurons in *Aplysia* as Model Systems

4.1. Peptidergic and Nonpeptidergic Identified Neurons

The identified neurons in the abdominal ganglion of a marine mollusk, *Aplysia californica,* have been extensively studied morphologically, electrophysiologically, and biochemically (Frazier *et al.*, 1967; Kandel and Kupfermann, 1970). This experimental preparation is unique in that individual specific neurons with defined properties can be repeatedly identified in each animal, and can be isolated for biochemical analysis. In addition, several topographical maps of the ganglion that catalogue the individual neurons have been prepared (Frazier *et al.*, 1967; Kandel and Wachtel, 1968), and in many cases, specific functions have been associated with specific neurons. The identified neurons R_2, L_{10}, and L_{11} have been characterized as cholinergic (Giller and Schwartz, 1971; McCaman and Dewhurst, 1971), whereas R_{15}, R_{3-14}, and the bag cells have been classified as "neurosecretory" cells (Frazier *et al.*, 1967; Coggeshall *et al.*, 1966; Coggeshall, 1967, 1970). Studies on the protein metabolism of these cells have demonstrated that the "neurosecretory" cells invariably synthesized large amounts of specific low-molecular-weight proteins, whereas the cholinergic neuron, R_2, did not (Arch, 1972*b;* Gainer and Wollberg, 1974; Loh and Peterson, 1974; Wilson, 1974). The "neurosecretory" cells appeared to be specialized in the synthesis of small polypeptides, and contained "neurosecretory granules." This finding indicated that they were "peptidergic" neurons. In the case of the bag cells, it could be shown that these polypeptides were indeed released from the nerve terminals (Arch, 1972*a;* Loh *et al.*, 1975). Only for R_{15} and the bag cells has a function been proposed. A peptide factor extracted from the neuron perikaryon of R_{15} appears to regulate salt or water balance, or both, in the animal (Kupfermann and Weiss, 1976), and a low-molecular-weight protein extracted from the bag cells causes egg-laying when injected into *Aplysia* (Kupfermann, 1967, 1970, 1972; Strumwasser *et al.*, 1969; Toevs and Brackenbury, 1969; see also Chapter 10).

Several protein biosythesis studies on the peptidergic neurons of *Aplysia* have indicated that the polypeptides released from the nerve terminals were derived from larger precursor molecules (Arch 1972*a,b;* Loh and Gainer, 1975*a,b;* Loh *et al.*, 1975), and we have focused on the R_{15} and bag cell neurons as models for the cellular analysis of neuronal peptide biosynthesis.

4.2. Pulse-and-Chase Experiments in Neuron R15

In a typical experiment (Gainer and Wollberg, 1974; Loh and Gainer, 1975*a,b*), intact abdominal ganglia are removed from the animal, and pulsed for 3 hr in *Aplysia* culture media containing [^{3}H]leucine. Following the "pulse" incubation period, the ganglia are either immersed in ice-cold saline to terminate the incubation, or placed in fresh culture medium containing a high concentration of only unlabeled leucine or a protein synthesis inhibitor, anisomycin, or of both, for chase incubation. The "chase" incubations are terminated as described above for the pulse, and dissection of the specific neuron somata is carried out. Each neuron soma is homogenized in an appropriate buffer, and the extracted polypeptides are separated by PAGE. The polyacrylamide gels are sliced and counted by conventional techniques.

Figure 1A depicts the results of a typical pulse–chase experiment in R_{15}, in which the proteins are separated according to molecular weight by PAGE in sodium dodecyl sulfate (SDS). Immediately after the pulse (Fig. 1A, 0 h chase), the R_{15} labeling profile was dominated by a major peak around 12,000 daltons and a relatively minor peak at 6–9000 daltons. After 3 hr of chase, there was a decrease in the 12,000-dalton peak, an increase in the 6–9000 dalton peak, and an appearance of a smaller peak at about 1500 daltons. At 6 hr after the chase, the labeling

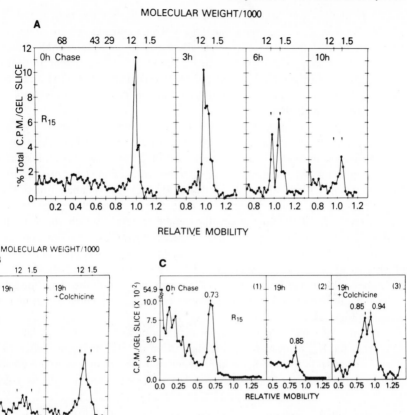

FIG. 1 (A) Molecular-weight distributions on 11% SDS gels of labeled proteins extracted from R_{15} neurons after a 3-hr "pulse" incubation in [³H]leucine, followed by 0, 3, 6, and 10 hr of "chase" incubations in physiological salines containing unlabeled leucine (at 15°C). The radioactivity (counts per minute minus background) in each slice of the gel, expressed as a percentage of total counts per minute on the gel, is plotted on the ordinate. The mobility of the proteins relative to cytochrome c is plotted on the lower abscissa. In the "chase" incubations, only data for the low-molecular-weight regions of the gel are shown (i.e., at relative mobilities >0.8), since no significant changes were observed at the lower relative mobilities. The upper abscissa shows the relative mobilities of a set of internal marker proteins of known molecular weight. The arrows represent extensions of the 12,000- and 1500-dalton marker protein positions. (B) Labeling patterns on SDS gels of low-molecular-weight proteins extracted from R_{15} neurons after a 3-hr "pulse" incubation in [³H]leucine at 15°C, and subsequent chase incubations for 19 hr in unlabeled leucine at 22°C in the absence and presence of colchicine (1 mg/ml). See (A) for explanation of axes. (C) Acid–urea gel patterns of labeled proteins from R_{15} neurons that were: (1) not chased; (2) chased for 19 hr; and (3) chased for 19 hr in the presence of 1 mg colchicine/ml at 22°C. The cpm/gel slice is shown on the ordinate, and the mobility relative to cytochrome c (R_f) is shown on the abscissa. The arrows indicate the R_f values of the specific peaks; data for R_f values less than 0.5 were omitted in C(2) and C(3) for the reasons discussed above. From Loh and Gainer (1975b).

profile is qualitatively similar to that seen at 3 hr, but the radioactivity of the 12,000-dalton peak was less than that of the 6–9000 molecular weight component. At 10 hr, the 12,000-dalton peak was virtually gone, whereas the 6–9000 dalton peak was present, although greatly diminished. In addition to the apparent transformation of the 12,000-dalton protein into a smaller (6–9000 dalton) protein in R_{15}, there was a substantial selective decrease in the total radioactivity of all the lower-molecular-weight proteins with time of chase.

The selective disappearance from the R_{15} soma of the newly synthesized, low-molecular-weight proteins after 10–19 hr of chase could have been due to (1) selective degradation of these protein species, or (2) their selective transport out of the neuron soma. We interpret these data to be a reflection of axoplasmic transport for two reasons. First, when the somata of R_{15} neurons incubated in [³H]leucine are separated from their axons and then chased in isolation, there is no decrease in the radioactivity of the low-molecular-weight proteins, which suggests that the continuity between the soma and axon is necessary for this phenomenon. Second, in intact ganglia, when agents (i.e., colchicine and vinblastine) that specifically block axonal transport (Dahlstrom, 1968, 1971; Karlson and Sjöstrand, 1969; Kreutzberg, 1969; McGregor *et al.*, 1973) were included in the chase medium, the low-molecular-weight proteins in R_{15} were retained. The latter experiment is depicted in Fig. 1B.

In Fig. 1B, the molecular-weight distributions on SDS gels of labeled proteins from the R_{15} neurons following incubation in [³H]leucine and after 19 hr of chase incubation in the absence and presence of 1 mg colchicine/ml are illustrated. In these experiments, the incubation in [³H]leucine was done at 15°C as usual, but the chase incubation was done at 22°C in order to increase the rate of axoplasmic transport, which is known to be sensitive to temperature (Gross, 1973). After 19 hr of chase, most of the newly synthesized low-molecular-weight proteins were absent. In the chase incubations in the presence of colchicine, however, the labeled low-molecular-weight proteins were still in the soma, although transformed—i.e., the pulsed cell (Fig. 1A, 0 h chase) showed a dominant label at 12,000 daltons, whereas the cells chased in colchicine (Fig. 1B) had dominant labeled peaks at 6000 and about 1500 daltons. After 3 hr of pulse label, 28% of the total gel label can be found in the 12,000-dalton peak, whereas after 19 hr of chase at 22°C, there was about 8% of the total gel counts per minute in the low-molecular-weight (12,000 and less) proteins. If colchicine was present, however, about 30% of the total gel counts per minute was found in the low-molecular-weight proteins. Therefore, colchicine, a classic blocker of axoplasmic transport, prevented the disappearance of the labeled low-molecular-weight proteins during the chase, but did not inhibit their processing or conversion into lower-molecular-weight forms.

Similar results are obtained from experiments in which the labeled proteins from R_{15} neurons, treated as described above, were separated by charge and size on acid–urea gels (Fig. 1C). Immediately after the 3-hr pulse, the labeling profile is dominated by a peak at $R_f = 0.73$ (Fig. 1C); after 19 hr of chase at 22°C, this peak is absent, and only a small peak at $R_f = 0.85$ is present [Fig. 1C(2)]. When

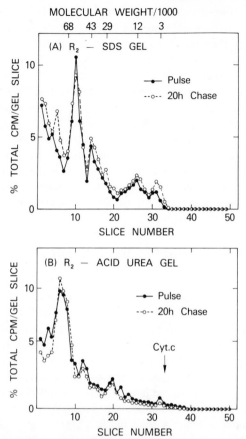

FIG. 2. Labeling patterns on SDS gels (A) and acid–urea gels (B) of proteins extracted from R_2 neurons after a 3-hr pulse incubation in [³H]leucine at 15°C, and subsequent chase incubations for 20 hr in unlabeled leucine at 22°C. The radioactivity (counts per minute minus background) in each slice of the gel, expressed as a percentage of total counts per minute on the gel, is plotted on the ordinate. The lower abscissa shows the gel slice number, and the upper abscissa of the SDS gels shows gel-slice positions of a set of internal marker proteins of known molecular weight. Cytochrome c (arrow) was used as an internal marker for mobility in the acid–urea gels. From Loh *et al.* (1977).

colchicine was present in the chase medium [Fig. 1C(3)], the 0.73 R_f peak was again missing, but comparable amounts of label were then found in two new peaks ($R_f = 0.85$ and 0.94). Therefore, the R_{15} neuron appeared to be processing the 12,000-dalton (acid–urea gel: $R_f = 0.73$) peak into two smaller proteins at about 6000 and 1500 daltons (acid–urea gel: $R_f = 0.85$ and 0.94), and one or all of these proteins appeared to be transported out of the soma.

In contrast to these studies on the peptidergic neuron R_{15} are the results of similar experiments depicted in Fig. 2 for the nonpeptidergic (cholinergic) neuron, R_2, in *Aplysia*. When the proteins extracted from R_2 neurons that had been pulsed for 3 hr in [³H]leucine, or pulsed and then chased for 20 hr at 22°C, were examined by either SDS (Fig. 2A) or acid–urea (Fig. 2B) PAGE, no evidence of significant synthesis of low-molecular-weight proteins or posttranslational processing could be observed. Therefore, several significant differences between the peptidergic neuron, R_{15}, and the nonpeptidergic neuron, R_2, are apparent: (1) The synthetic profile of proteins in R_{15} shows dominance of a low-molecular-weight protein, while the profile of R_2 does not. (2) There is evidence of posttranslational cleavage of the protein in R_{15}, and no such obvious phenomenon in R_2. (3) A large quantity of low-molecular-weight proteins (peptides) is rapidly transported out of the soma of R_{15}, whereas no such dramatic transport is seen in the soma of R_2. In the *Aplysia* neurons, one has a unique situation, in which the biosynthetic processes in the cell *soma* can be studied in the absence of extraneous cellular tissue. Whether the phenomena described above are characteristic of "peptidergic" neurons in general will depend on the results of comparable studies in other systems (see the discussion in Section 5).

4.3. Biosynthetic and Subcellular Fractionation Studies on the Bag Cells

Pulse-and-chase experiments similar to those described above for R_{15} and R_2 were done on the bag cells. Previous reports by Arch (1972*a*,*b*) showed that labeled proteins with molecular weights (around 6000 daltons) corresponding to that of the egg-laying hormone found in these cells appeared to be derived from a 25–29,000 dalton precursor. The data illustrated in Fig. 3A confirm these observations. Following a pulse incubation of the bag cells in [^{3}H]leucine (15°C) for 1 hr (Fig. 3A, open circles) most of the label is found in a 29,000-dalton peak on SDS gels. If one extends this pulse-incubation period to 3 hr (Fig. 3A, solid circles), one finds a substantial amount of the label in a 6000-dalton peak, and to a lesser extent a 12,000-dalton peak. These data are in agreement with the interpretation by Arch (1972*a*,*b*) of his extensive pulse–chase experiments on the bag cells, which was that the 25–29,000 dalton precursor was processed into 12,000- and 6000-dalton peaks. However, when these pulsed neurons are chased for longer periods of time, two additional events occur. In addition to the complete disappearance of label in the 29,000-dalton peak, there is: (1) a large decrease in label in all the low-molecular-weight peaks in the bag cells, which can be blocked by the presence of colchicine in the medium (Fig. 3B); and (2) a redistribution of label in the low-molecular-weight peaks, so that the major peak is about 3000 daltons or less, with minor peaks at 6000 and 12,000 daltons (Fig. 3B, solid circles). Therefore, the initial posttranslational events described by Arch (1972*a*,*b*) are followed by further processing to even smaller peptides. It is of interest to note here that the proteins released from the bag cell terminals have been shown to reflect the distribution illustrated in Fig. 3B (Loh *et al.*, 1975).

A clearer illustration of this point and a consideration of the subcellular localization of these processes is presented in Fig. 4. In these experiments, bag cells were pulsed in [^{3}H]leucine and also pulse–chased (in the presence of colchicine to prevent axonal flow) as described above. The cells were homogenized in

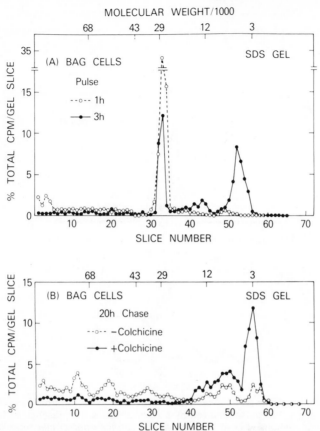

FIG. 3. Molecular-weight distributions on 11% SDS gels of labeled proteins extracted from bag cells after a 1-hr pulse (A, open circles) and a 3-hr pulse (A, solid circles) in [³H]leucine at 15°C. Note that some of the 29,000-dalton "precursor" seen after the 1-hr pulse has been converted into 12,000- and 6000-dalton proteins after the longer 3-hr pulse. (B) Chase incubations (20 hr at 15°C) of bag cells following a 3-hr pulse in the absence (open circles) and presence (solid circles) of 1 mM colchicine. Note the appearance of a new major peak at 3000 daltons. The descriptions of the coordinates of the graphs is the same as described for the SDS gels in Fig. 2. Adapted from Loh *et al.* (1975).

isotonic sucrose buffered with Tris at pH 7.8, and a fraction of the homogenate was subjected to differential centrifugation. The P₃ fraction (i.e., the 30,000*g* × 60-min pellet of the postmitchondrial supernatant fraction, which is enriched in neurosecretory granules) was extracted for proteins, which were then separated by acid–urea PAGE. Figure 4A compares the labeling profile of the P₃ fraction vs. the unfractionated cells (total cells) after a 3-hr pulse; Fig. 4B represents a similar comparison after the long chase in colchicine. Although four distinct peaks of radioactivity are detected in the bag cells by acid–urea PAGE after the pulse (Fig. 4A, open circles), only the most rapidly migrating two peaks are detected in the "granule" fraction (Fig. 4A, solid circles). Following the chase incubation (Fig.

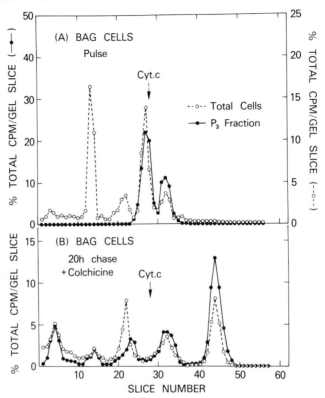

FIG. 4. Labeling patterns on acid–urea gels of proteins extracted from bag cells after a 3-hr pulse (A) and a 3-hr pulse followed by a 20-hr chase in the presence of 1 mM colchicine (B). All incubations were at 15°C. In (A) and (B), the open circles represent proteins extracted from the entire bag cell perikarya, whereas the solid circles represent proteins extracted from a subcellular fraction, P_3 (enriched in neurosecretory granules, see the text), derived from the bag cell perikarya. Note that after the 3-hr pulse, the "precursor" (slice 14) is not found in the P_3 fraction (A), whereas all the "final products" are in the P_3 fraction (B). The description of the coordinates of the graphs is the same as in Fig. 2. From Loh *et al.* (1977).

4B), most of the label is in a *new* very rapidly migrating peak, and all the labeled peaks in the cells are now found in the P_3 fraction.

We interpret these data as indicating that the two stages of posttranslational cleavage in the bag cells are occurring at separate intracellular locations. The two slowly migrating peaks in the total-cell labeling profile after the pulse (Fig. 4A) correspond to the 29,000- and 12,000-dalton peaks, while the two rapidly migrating peaks found in the P_3 fraction appear to be around 6000 daltons when reelectrophoresed on SDS gels. The new rapidly migrating peak after the 20-hr chase in colchicine (Fig. 4B) appears to correspond to the 3000-dalton protein on SDS gels (see Fig. 3B). Because of the molecular-weight relationships described above and the absence of the 29,000-dalton precursor and 12,000-dalton product in the putative granule fraction (P_3), we have tentatively concluded that the 29,000-to-12,000 dalton conversion occurs extragranularly (possibly in the rough endoplasmic reticulum), while the 6000-to-3000 dalton conversion occurs within the neurosecretory granule. Conclusive evidence for this hypothesis must await future electron-microscopic autoradiographic analysis in combination with correlative biochemical data (Neale *et al.*, in preparation).

4.4. A Model of Neuronal Peptide Biosynthesis and Transport

The results of studies on *Aplysia* peptidergic neurons described above support the concept of ribosomally dependent synthesis of a precursor, with subsequent transformation to smaller peptides. The experimental results with the bag cells (Fig. 4) raise the issue whether some of the processing (i.e., the 6000-to-3000 dalton protein conversion) is occurring in an intracellular compartment (i.e., the secretory granules) that is destined for axonal transport to the secretory site in the axon terminal. If this is the case, then the mechanism for processing, i.e., the specific proteolytic enzyme, must also be in the granule. In this regard, it is of interest that the conversion of proinsulin to insulin and C-peptide in the pancreatic β-cell appears to occur in the secretory granule (Kemmler *et al.*, 1973). If post-translational cleavage is occurring intragranularly in neurons, then it is likely that at least some of this phenomenon may be occurring in the axon during axonal transport.

A hypothetical model depicting the organization of peptide biosynthesis and transport is shown in Fig. 5. Several of our findings in the peptidergic neuron, R_{15}, provide support for this model. First, when R_{15} is pulsed in [^{3}H]leucine at 22°C, and then chased at 22°C, one finds, in contrast to pulse–chase experiments in R_{15} at 15°C (see Fig. 1A), that the 12,000-dalton precursor protein synthesized after the pulse rapidly disappears from the neuron soma during the chase, with little evidence of conversion to the lower-molecular-weight forms seen at 15°C (Gainer and Barker, 1975). However, if colchicine is present during the 22°C chase, then complete conversion in the soma is evident (see Fig. 1B). These data suggest that increases in temperature appear to increase the rate of axonal transport of the compartment containing the 12,000-dalton protein (i.e., the secretory granule) more than the conversion rate of precursor to product (see Loh and Gainer, 1975b), and therefore the conversion at 22°C was occurring almost exclusively in the axon. Recent autoradiographic experiments in collaboration with Dr. E. Neale indicate that the radioactivity in R_{15} following a pulse at 22°C and a chase at 0°C to prevent axonal transport is associated primarily with neurosecretory granules in the soma (Fig. 6). Thus, the low-molecular-weight proteins in R_{15} appear to be located in the granules, providing further evidence for the model described in Fig. 5.

Similar models have been described elsewhere for *Aplysia* peptidergic neurons (Loh *et al.*, 1975; Gainer, 1976), and for peptidergic neurons in vertebrates (Sachs and Takabatake, 1964; Sachs *et al.*, 1969; McKelvy, 1975; Cross *et al.*, 1975). In the following section, the applicability of this model to the hypothalamo-neurohypophyseal system in the mammalian brain will be considered.

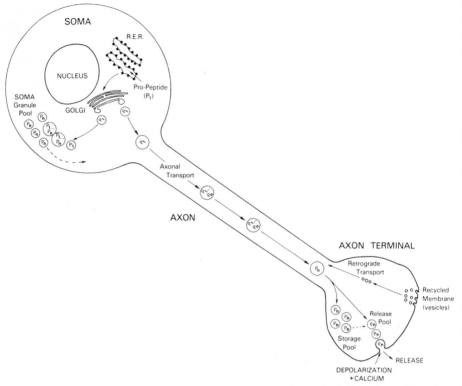

FIG. 5. Hypothetical model of biosynthesis, translocation, processing, and release of peptides in a peptidergic neuron. (R.E.R.) Rough endoplasmic reticulum; (P_1) propeptide or precursor molecule; ($P_1 \ldots P_n$) intermediates between P_1 and P_n; (P_n) final peptide products of processing. From Gainer *et al.* (1977b).

5. Biosynthesis of Neurohypophyseal Peptides and Neurophysin

5.1. Historical Background

The neurohypophyseal peptides and their associated neurophysins are synthesized by neurons in hypothalamic nuclei and transported to axon terminals in the posterior pituitary (Bargmann, 1968). Excellent reviews describing the history and current status of this system (Lederis, 1974; Heller, 1974), as well as the chemistry of the neurohypophyseal peptides (Sawyer, 1971; Acher, 1974, Hope and Pickup, 1974; Breslow, 1974), are available, and these issues will not be extensively discussed here. In mammals, both vasopressin and oxytocin are sythesized in two bilateral nuclei in the hypothalamus—the supraoptic nuclei (SON) and the paraventricular nuclei (PVN) (see Chapters 4 and 6). Evidence for the synthesis of these peptides in separate neurons has been reviewed elsewhere (Valtin *et al.*, 1974, 1975; see also Chapter 4).

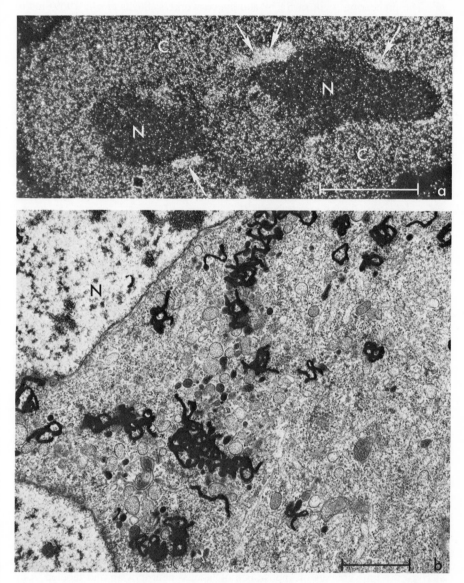

FIG. 6. Radioautograms of cell R_{15} pulse-labeled with [^{3}H]leucine for 3 hr at 22°C and chased for 19 hr at 0°C. (a) Light-microscope radioautogram photographed with dark-field illumination. The lightly labeled nucleus (N) can be distinguished from the more heavily labeled cytoplasm (C). Note areas of very high grain density (arrows) adjacent to the nucleus. Calibration bar: 100 μm × 340. (b) Electron-microscope radioautogram of the cell shown above. The highest silver grain density is associated with large numbers of neurosecretory granules frequently clustered near the nucleus (N). Fewer silver grains are seen over cytoplasm, where granules are scant. Calibration bar: 1 μm × 23,000. From E. Neale (unpublished data).

The pioneering work of Sachs and his co-workers (Sachs *et al.*, 1969) led to the view that vasopressin and neurophysin were formed from a common precursor protein synthesized by conventional ribosomal mechanisms. A similar mechanism for the synthesis of oxytocin has recently been proposed (Pickering and Jones 1971). The evidence in support of this hypothesis is as follows:

1. The synthesis of vasopressin occurs only in the hypothalamus, where the neuron perikarya are located, while the neurohypophysis containing the axons and terminals of the neurons is incapable of vasopressin synthesis (Sachs, 1960; Sachs and Takabatake, 1964; Sachs *et al.*, 1969).

2. Experiments both *in vivo* and *in vitro* showed that the incorporation of [^{35}S]cysteine into vasopressin underwent a considerable time lag (>1.5 hr in the dog) after exposure of the tissue to the isotope, and application of a protein-synthesis inhibitor (puromycin) during the chase period, when labeled vasopressin did appear, did not inhibit incorporation of [^{35}S]cysteine into the peptide (Sachs and Takabatake, 1964; Takabatake and Sachs, 1964; Sachs *et al.*, 1969). However, if puromycin was present during the 1.5 hr pulse period, no labeled vasopressin was formed during the subsequent chase period. From these data, Sachs and Takabatake (1964) concluded that the incorporation of [^{35}S]cysteine was initially into a precursor via conventional protein synthesis mechanisms, but that the formation of labeled vasopressin from the precursor did not require protein synthesis. This proposal of a prohormone for vasopressin preceded the discovery of proinsulin (Steiner and Oyer, 1967; Steiner *et al.*, 1967).

3. Additional but less direct evidence for a ribosomal mechanism is the observation that the increase of vasopressin biosynthesis induced by prolonged dehydration of the animal is correlated with increases in the ribosome and RNA content of SON neurons (Edstrom *et al.*, 1961; Osinchak, 1964; Zambrano and De Robertis, 1966; Eneström and Hamberger, 1968).

4. The ratio of vasopressin to oxytocin is higher in the hypothalamus than in the neurohypophysis (Vogt, 1953; Lederis and Jayasena, 1970), consistent with the notion that processing of the respective precursors of these peptides occurs at different rates and en route during axonal transport. Similarly, Sachs (1963) found that the vasopressin content of neurosecretory granules isolated from the hypothalamus–median eminence of dogs was about 5-fold less than in granules isolated from the neural lobe.

5. The proposal for a common precursor for both vasopressin and neurophysin stems from several diverse observations. Following a pulse with [^{35}S]cysteine *in vivo,* the time-courses of appearance of labeled neurophysin and labeled vasopressin during an *in vitro* chase were similar (Sachs *et al.*, 1971). Inhibition of neurophysin synthesis by employing analogues of amino acids (e.g., methyleucine, γ-methyl-methionine, and histidinol) present in neurophysin, but not in vasopressin, also inhibited the synthesis of vasopressin (Sachs *et al.*, 1969). Consistent with these biosynthetic data are the findings that rats with genetic defects in vasopressin synthesis (i.e., the Brattleboro strain) are also deficient specifically in the vasopressin-associated neurophysin (Valtin *et al.*, 1975; Pickering *et al.*, 1975).

6. The physical–chemical properties of neurophysin suggest that this protein is not in its originally synthesized form. Chaiken *et al.* (1975) have pointed out that the marked susceptibility of neurophysin to disulfide interchange (see also Breslow, 1974), in comparison to its relative stabilization by bound peptide ligand, is analogous to the behavior of ribonuclease-*S*-(21–124) alone vs. its behavior with bound ribonuclease-*S*-(1–20). That is, in both cases, the peptide ligand confers an increase in the thermodynamic stability of the macromolecule. Since the forms of ribonuclease-*S* named above are derived by enzymatic conversion of ribonuclease-*A* (Richards and Vithayathil, 1959), it has been suggested by analogy that neurophysins and their complementary neurohypophyseal peptides derive similarly from a common precursor (Chaiken *et al.*, 1975).

Although compelling evidence for the existence of a precursor of neurophysin and vasopressin has been available for some time, efforts to isolate or to directly identify (in biosynthetic experiments) this postulated precursor have been unsuccessful. In the following sections, recent experimental work in our laboratory relevant to this issue will be described.

5.2. The Experimental System

A diagrammatic representation of the hypothalamo-neurohypophyseal system of the rat is depicted in Fig. 7. In our experiments, 1 μl of [^{35}S]cysteine in 0.9% NaCl was injected bilaterally into the SON of ether-anesthetized rats. The rats were allowed to recover from the anesthetic, and were killed at various time intervals postinjection. Their brains and pituitaries were rapidly removed and frozen until analysis. Representative regions of the neuron perikarya (SON) and axons (median eminence) were isolated from frozen sections by the Palkovitz punch technique (Palkovitz, 1973), and the posterior pituitary was isolated as a sample of the nerve terminal region (see Fig. 7). The isolated tissues were then homogenized in 0.1N HCl in order to destroy any degradative enzymes (Dean *et al.*, 1967), and stored at -70°C. These experiments were done in collaboration with Dr. M. Brownstein at the National Institutes of Health.

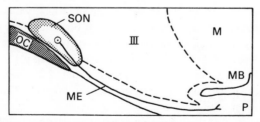

FIG. 7. Diagrammatic representation of a saggital section through the hypothalamus showing the course of axons from the supraoptic nucleus to the posterior pituitary. The broken line indicates the border of the third ventricle. The supraoptic nucleus would actually be lateral to the plane of this section. (SON) Supraoptic nucleus; (OC) optic chiasm; (ME) median eminence; (P) posterior pituitary; (MB) mamillary body; (M) mesencephalon; (III) third ventricle.

5.3. Time-Course of Synthesis and Transport of Protein

The time-course of appearance of ^{35}S-labeled protein (i.e., TCA-precipitable counts per minute) in the various regions of the hypothalamo-neurohypophyseal system is shown in Fig. 8. The data in Fig. 8A (solid circles, solid line) show that after 30 min postinjection, about 30% of the total radioactivity in the punched SON region was TCA-precipitable, and that this value did not change significantly until about 6 hr postinjection. Thus, the incorporation of [^{35}S]cysteine into protein by the SON was very rapid and approximated a pulse incubation.

Labeled proteins did not appear at the level of the median eminence (Fig. 8B, solid line) until 1 hr postinjection, and rose rapidly between 1 and 2 hr, reaching a maximum at about 12 hr. In the posterior pituitary (Fig. 8C, solid line), labeled proteins were first detected at 2 hr, and rose rapidly to a maximum value at 12 hr. These kinetic data are consistent with the notion that the proteins are first synthesized in the SON region and then transported intraaxonally to the posterior pituitary via axons traversing the median eminence. The time lag between injection in the SON and arrival of labeled proteins in the posterior pituitary of 60–120 min is comparable to values found by others in similar experiments with rats (Nordström and Sjöstrand, 1971*a,b;* Pickering and Jones, 1971), and consistent with the expected axonal transport rates of neurohypophyseal proteins (see Table 4 in Valtin *et al.,* 1974). Further evidence for the role of axonal transport in the emergence of labeled proteins in the median eminence and posterior pituitary comes from the colchicine studies shown in Fig. 8. While colchicine treatment did not appreciably effect incorporation of [^{35}S]cysteine into proteins in the SON (Fig. 8A, dashed line), the appearance of labeled proteins in the median eminence (Fig. 8B, dashed line) and posterior pituitary (Fig. 8C, dashed line) were dramatically reduced. Similar effects of colchicine on axonal transport of neurohypophyseal principles were demonstrated by Norström (1975) and Sachs *et al.* (1975).

5.4. Analysis of Proteins Transported to the Neurohypophysis

In an attempt to evaluate the diversity of proteins transported to the posterior pituitary following injection of [^{35}S]cysteine into the SON, we have separated the labeled proteins from the posterior pituitary (24 hr postinjection), using 4 different PAGE systems. These systems included a gel system designed to separate acidic proteins (comparable to the Ornstein-Davis system, running pH 9.5), an acid–urea gel system (Loh and Gainer, 1975*a;* running pH 2.7), isoelectric focusing (nominal pH range 3–10), and disc electrophoresis in SDS. The results of these separations are shown in Fig. 9.

Figure 9A shows that the labeling pattern of posterior pituitary proteins separated on pH 9.5 gels was relatively simple. One major peak of radioactivity was detected, which we assume to be neurophysin, for several reasons. First, neurophysin is known to be the major cysteine-rich, acidic protein transported to the posterior pituitary (Hope and Pickup, 1974). Second, this radioactive protein peak was highly soluble in alcohol, as has been reported for neurophysin elsewhere

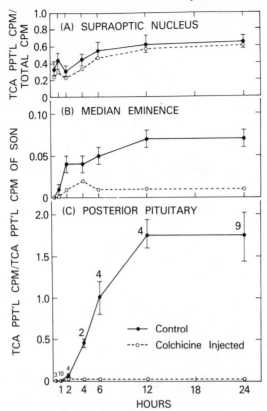

FIG. 8. Time-course of appearance of ^{35}S-labeled proteins in the supraoptic nucleus (A), median eminence (B), and posterior pituitary (C) following a bilateral microinjection of [35S]cysteine into the SON of female rats. Solid circles show data from control rats; open circles show data from rats that were intraventricularly injected with colchicine 12 hr before injection of the [35S]cysteine. Note that although there is considerable labeled protein in the SON after 30 min, labeled protein first appears in the median eminence only after 1 hr, and in the posterior pituitary after 2 hr. In addition, the appearance of axonally transported labeled protein is inhibited by colchicine (B, C), whereas the percentage of incorporation in the SON was unaffected (A) by colchicine. The incorporation in the SON (A) is expressed as the fraction of total counts per minute in the punched tissue that was TCA-precipitable. In (B) and (C), the TCA-precipitable counts per minute in the tissue was expressed as a fraction of the TCA-precipitable counts per minute in the punched SON sample of the same rat. Data (solid circles) expressed as means ± SDs. The number of experiments for each time point is shown in (C). From Gainer et al. (1977b).

(Albers and Brightman, 1959; MacArthur, 1931; Wuu and Saffran, 1969). Third, this peak corresponds in migration position on the gel to a major coomassie blue–stained band that is specifically decreased in intensity if the animals are osmotically stressed by 2% NaC1 in their drinking water (illustrated here only in reference to a similar peak after separation on acid–urea gels; i.e., compare Figs. 9B and 10). Finally, a similar labeling pattern of transported neurophysins has been detected by Norström et al. (1971), using a comparable gel system.

Initial reports of a single neurophysin in the rat (Norström et al., 1971; Coy

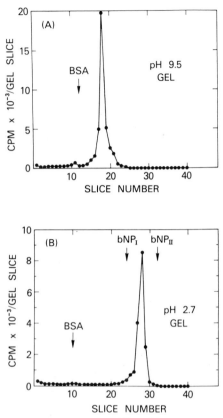

FIG. 9. Polyacrylamide-gel electrophoresis of ^{35}S-labeled proteins extracted from rat pituitaries 24 hr after injection of [^{35}S]cysteine into the SON. (A) Disc PAGE (7.5% acrylamide) in a pH 9.5 (running pH) system. Internal marker protein was bovine serum albumin (BSA). (B) Disc PAGE (7.5% of acrylamide) in a pH 2.7 system (acid–urea gel). Internal markers: BSA; bovine neurophysins I and II (bNP$_I$, bNP$_{II}$). (C) Isoelectric focus-

and Wuu, 1972) were followed by the reported detection of three neurophysins in the rat (Burford and Pickering, 1972). The latter authors found that by increasing the concentration of the bromphenol blue tracking dye in the upper reservoir buffer of the electrophoresis system, it was possible to reveal three distinct radioactive peaks (see also Pickering *et al.*, 1975; Sunde and Sokol, 1975) in the posterior pituitary following the intracisternal injection of [^{35}S]cysteine. One of these three peaks (the most rapidly migrating) is believed to be the vasopressin-associated neurophysin, since it was absent in the posterior pituitaries of rats from the vasopressin-deficient Brattleboro strain (Burford and Pickering, 1972; Pickering *et al.*, 1975). We have altered the bromphenol blue concentration in our gel runs in the manner suggested by Burford and Pickering (1972), but have consistently obtained the labeling pattern shown in Fig. 9A.

Separation of the labeled neurohypophyseal proteins on acid–urea gels also produced only one major radioactive peak, which migrated on the gel at a rate be-

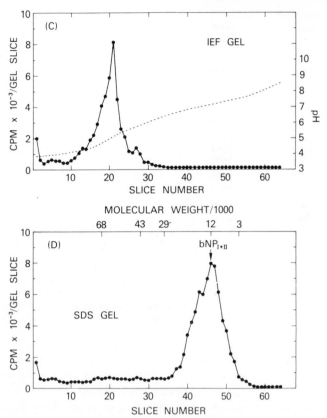

ing (7.5% acrylamide) using pH 3–10 ampholytes. The dashed line shows the actual pH gradient measured using a co-run gel. (D) Disc PAGE (11% acrylamide) in SDS. Upper abscissa shows slice positions of internal marker proteins of known molecular weight; arrow shows position of bNP$_I$ and bNP$_{II}$ in a co-run gel. From Gainer et al. (1977b).

tween the rates of bovine neurophysins I and II (Fig. 9B). The labeling pattern shown in Fig. 9B was the same regardless whether the experiments were done on normal or salt-treated (i.e., after 10 days of 2% NaCl in the drinking water) rats, the only difference being that the absolute counts per minute in the peak from salt-treated rats was about 4–5 times greater than from normal rats. The Coomassie blue–staining pattern of proteins separated by acid–urea PAGE is illustrated in Fig. 10. Approximately seven densely stained bands were detected using normal rats. However, only the most rapidly migrating band contained the ^{35}S-labeled protein. Consistent with this finding was the observation that this stained band selectively disappeared with salt treatment, although the same radioactive peak was still present. Since both vasopressin and oxytocin are depleted in the posterior pituitary of rats after prolonged osmotic stress (Jones and Pickering, 1969), it is possible that both the vasopressin- and oxytocin-associated rat neurophysins, if they are distinct proteins, comigrated in this gel system. In this regard, it is important to note that we have detected both ^{35}S-labeled arginine vasopressin and oxy-

Normal   Salt-Treated

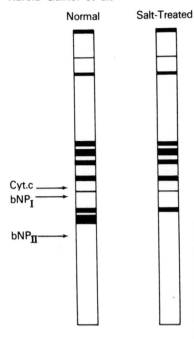

Cyt.c ⟶
bNP$_I$ ⟶

bNP$_{II}$ ⟶

FIG. 10. Diagram of Coomassie blue–stained bands of proteins extracted from posterior pituitaries of normal rats and rats that had been drinking 2% NaCl for 10 days (salt-treated), which were separated by electrophoresis on 15% acid–urea gels. Note disappearance of the most rapidly migrating band after salt treatment. Arrows denote migration positions of cytochrome c and bovine neurophysins I and II in co-run gels. From Gainer *et al.* (1977b).

tocin in the posterior pituitary 24 hr after injection of [^{35}S]cysteine in the SON. The ratio of labeled vasopressin to that of oxytocin in the neurohypophysis was about 1.8:1 in our experiments. This ratio was comparable to the value of 1.49:1 obtained by Burford *et al.* (1974) for the ratio of radioactivity in vasopressin-neurophysin to that in oxytocin-neurophysin after intracisternal injection of [^{35}S]cysteine.

In view of our inability to resolve more than one major radioactive peak using the gel systems described above, which separate proteins on the basis of both electrical charge and molecular size properties, we turned to isoelectric focusing separation procedures. Figure 9C shows the results of such experiments. The labeled neurohypophyseal proteins were electrofocused principally in the acidic region of the gel ranging in isoelectric points from pH 4.5 to 5.1 with the peak value at about pH 4.8. The width of this peak of radioactivity suggested that it was composed of several distinct zones of radioactivity, which were very close to one another, and were therefore obscured by the relatively low-resolution technique of gel-slicing and -counting. Subsequent electrofocusing experiments on slab gels with analysis by radioautography showed that this broad zone of radioactivity was composed of two major peaks of radioactivity (at pH 4.6 and 4.8), and a minor peak at about pH 5.1 (not illustrated). Whether these three peaks correspond to the three neurophysin peaks reported by Burford and Pickering (1972) remains to be determined.

Separation of the labeled neurohypophyseal proteins by PAGE in SDS showed that the transported proteins were of relatively low molecular weight (see Fig. 9D). Similar findings that the transported, labeled neurohypophyseal proteins showed a peak on SDS gels at around 12,000 daltons (ranging between 10,000 and 14,500 daltons) have been reported by others (Norström *et al.*, 1971; Norström, 1975; Burford *et al.*, 1971). It should be pointed out that although the peak in Fig. 9D comigrated with the 12,000-dalton marker protein in the gel, bovine neurophysins I and II (which are about 10,000 daltons) also comigrated with this marker protein (see arrow in Fig. 9D).

5.5. Biosynthetic Evidence for a Precursor

As was pointed in Section 3.1, the initial step in the identification of a precursor requires the demonstration, in biosynthetic studies, that a labeled peptide (or polypeptide) of higher molecular weight is first synthesized and then decreases in radioactivity with time as the labeled peptide product is formed. Since the biosynthetic process is located in the hypothalamic nuclei, we have analyzed the labeling profiles on acid–urea gels of proteins extracted from the SON at various times after injection of [^{35}S]cysteine. Figure 11 shows such data for four time points.

At 1 hr postinjection, the SON labeling profile is dominated by a very large peak of radioactivity with a much lower migration rate than the labeled neurophysin (Fig. 11A; Pit. Peak arrow indicates the mobility of labeled neurophysin). Thus, after 1 hr, very little label has appeared in the neurophysin peak in comparison to the radioactivity in the less mobile peak. As can be seen in Figs. 11B–D, this less mobile peak decreases in radioactivity with time as the labeled neurophysin peak is relatively increased. After 24 hr (Fig. 11D), there is virtually no radioactivity remaining in the less mobile peak, while the labeled neurophysin peak has become clearly dominant. Three different methods have been used to estimate the molecular weights of the less mobile peak (Fig. 11A) and the neurophysin peak (Figs. 11D and 12C). These methods included SDS gel electrophoresis, "Ferguson" plot analysis (see Chrambach and Rodbard, 1971; Rodbard and Chrambach, 1971), and gel chromatography using Sephadex G-75. All three methods were in close agreement. The molecular weights of the less mobile, putative precursor peak and neurophysin peak were 19–20,000 daltons and 10–12,000 daltons, respectively. Thus, we have observed in the SON a biosynthetic process consistent with the requirements of a precursor–product relationship. Because of these data, we tentatively propose that the dominant peak in Fig. 11A represents the precursor to neurophysin (and presumably the neurohypophyseal peptides).

As we have discussed earlier, data such as those discussed above represent necessary but not sufficient evidence to prove the identity of a precursor. In recent immunoprecipitation experiments, using three different antibodies to neurophysin, we have shown that both the labeled neurophysin peak (in Figs. 11D and 12C) and the labeled, putative precursor peak (in Fig. 11A) are immunoreactive to antibodies to neurophysin (Brownstein *et al.*, in preparation). This finding would indicate that there is significant amino acid sequence homology between the putative precursor and neurophysin, thereby strengthening the case for identification of the precursor. Further immunological studies as well as peptide mapping and amino acid sequencing studies are clearly necessary.

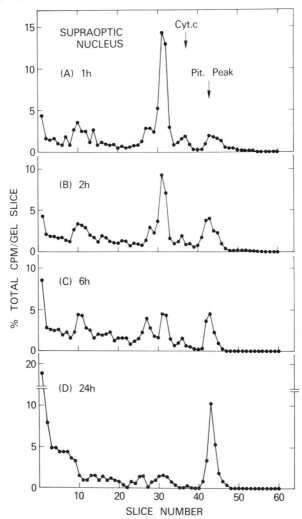

FIG. 11. Labeling patterns on 15% acid–urea gels of proteins extracted from punched SON regions at various times after injection of [35S]cysteine into the SON. Note that the major radioactive peak after 1 hr (A) is around slice 31, and that this peak decreases progressively with time relative to a radioactive peak (slice 43) that is dominant after 24 hr (B–D). The latter peak comigrates with the major radioactive peak found in the posterior pituitary after 24 hr (Pit. Peak; see Fig. 12C). Cytochrome c (arrow) was the internal protein marker in these gel runs. From Gainer, Sarne, and Brownstein (1977a).

5.6. Axonal Transport and Processing of the Precursor

In section 4.4, we presented a model of neuronal peptide biosynthesis and processing (see Fig. 5) based on data from *Aplysia* peptidergic neurons. This model suggested that the precursor (or an intermediate form) is packaged in a translocatable compartment (e.g., the neurosecretory granule), and hence may un-

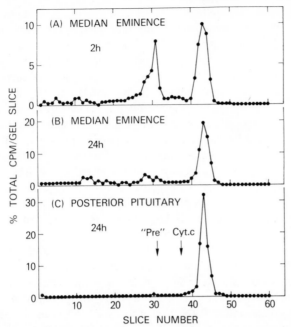

FIG. 12. Labeling patterns on 15% acid–urea gels of proteins transported to the median eminence at 2 hr (A) and 24 hr (B), and the posterior pituitary at 24 hr (C), after SON injection with [³⁵S]cysteine. Arrows designate the putative "precursor" peak (see Fig. 11A). Note that a major radioactive peak that comigrates with the "precursor" is found in the median eminence at 2 hr postinjection. From Gainer, Sarne, and Brownstein (1977a).

dergo posttranslational cleavage to the product intragranularly during axonal transport. The data illustrated in Fig. 12 provide further support for this model.

The model would predict that the labeling profile of proteins transported into the axon at various times postinjection of [³⁵S]cysteine into the SON would reflect the distribution of labeled proteins in the SON. That is, at earlier times (e.g., 1–2 hr postinjection), when the precursor is heavily labeled in the SON (Figs. 11A, B), the axonally transported proteins should contain significant labeled precursor. However, at later times (e.g., 24 hr postinjection), when virtually only labeled neurophysin is present in the SON, the axonally transported proteins should contain only labeled neurophysin. These predictions are borne out by the data shown in Fig. 12. At 2 hr postinjection, the labeling profile in the median eminence contains *two* peaks—the putative precursor peak and the neurophysin peak (Fig. 12A). If one considers the time delay between synthesis in the SON and transport to the axons in the median eminence, it is apparent that the labeled proteins arriving after 2 hr in the median eminence (Fig. 12A) must have been derived from the population of labeled proteins in the SON after more than 1 hr (Fig. 11A). Hence, the precursor-to-product conversion process resulting in the labeling pattern seen in Fig. 12A must have occurred intraaxonally during transport. At 24 hr postinjection, the median eminence labeling profile displays virtually *only* neurophysin (Fig. 12B), consistent with the pattern seen in the SON at that time (Fig. 11D).

In addition to the biosynthetic data presented above, several other lines of evidence suggest that the processing of the precursor is occurring intragranularly during axonal transport. That the ratio of vasopressin to oxytocin is greater in the hypothalamus than in the neurohypophysis (Vogt, 1953; Lederis and Jayasena, 1970), and that the vasopressin content of granules in the neurohypophysis is 5-fold greater than in the hypothalamus (Sachs, 1963), strongly suggests that the generation of these peptides from the precursor is principally occurring in a region between the perikaryon in the hypothalamus and the neural lobe (i.e., the axon). More recently, morphological evidence in support of intragranular processing has been reported. Using a triple aldehyde fixative, Morris and Cannata (1973) found that the dense core of neurosecretory granules in the neural lobe was best preserved at pH 5.0–6.0, but that a marked loss of density of the granule core occurred at pH 8.0 (only 5% of all granules in the neural lobe remained dense-cored at pH 8.0). In contrast, granules in the perikarya in the SON and PVN showed a high degree of preservation of the dense core at pH 8.0 (Cannata and Morris, 1973). These authors suggest that this difference in response to fixation at pH 8.0 between hypothalamus and neural lobe may reflect "maturation" changes in the granule during axonal transport. This "maturation" presumably reflects the precursor-to-product conversion process. Since preservation of the dense granule core at pH 8.0 was greater in the PVN than in the SON (Cannata and Morris, 1973), the authors further suggested that this difference may be related to the faster "maturation" of vasopressin-containing granules (concentrated in the SON) vs. oxytocin-containing granules (concentrated in the PVN) (Vogt, 1953). This latter explanation, however, is not consistent with the more recent data on the comparative vasopressin and oxytocin contents of the SON and PVN (Swaab et al., 1975; see also Chapters 4 and 6).

5.7. Summary and Conclusions

In the preceding sections, new biosynthetic evidence for a precursor of neurophysin and the neurohypophyseal peptides oxytocin and vasopressin has been presented in the context of the long history of support for such an idea. The protein that we have proposed as the candidate for the precursor is about 19,000 daltons, and appears to have an isoelectric point between that of neurophysin and arginine vasopressin (unpublished data). Assuming that the precursor, with a molecular weight of about 19,000 daltons, contains both the neurophysin (about 12,000 daltons) and the vasopressin or oxytocin (1100 daltons) amino acid sequences, then it is still not possible to account for the size of the entire precursor molecule.

Recently, we have examined the diversity of [^{35}S]cysteine-labeled peptides that are synthesized in the SON and transported to the posterior pituitary. Using two-dimensional thin-layer chromatography and high-voltage electrophoresis, we have detected, in addition to arginine vasopressin and oxytocin, at least four other highly labeled peptides that are transported to the neurohypophysis. These peptides could conceivably be the posttranslational cleavage products of the unaccounted-for part of the precursor. If this is the case, then one might expect that they would also reside in the secretory granule, and would be released together with the neurophysins and the octapeptides. Experiments are currently under way to test this possibility and to further characterize these new posterior pituitary peptides. If these peptides are indeed released, then a search for possible neurohormonal or other physiological functions for these peptides would be in order.

Although both vasopressin and oxytocin were synthesized in our experiments, we are unable at present to clearly resolve two distinct precursors or neurophysins in the rat using the normally high-resolution acid–urea PAGE system. Obviously, these proteins must be very similar structurally. Preliminary experiments using expanded isoelectric focusing gradients have been promising in this regard, and may yet allow for the resolution of separate oxytocin and vasopressin precursors. Another issue of concern is the interpretation of recent immunocytochemical and radioimmunoassay (RIA) studies in the SON and PVN of rats (see Chapters 4 and 6), utilizing antibodies to neurophysin. If the precursor is also immunoreactive, as appears to be the case, then the problems regarding heterogeneity and RIA discussed by Straus and Yalow (Chapter 3) must be considered in quantitative studies on these nuclei.

6. Regulation of Neuronal Peptide Biosynthesis

The synthesis of peptides by peptidergic neurons in the *Aplysia* abdominal ganglion and the rat hypothalamo-neurohypophyseal system appears to occur via ribosomal mechanisms, and hence would be restricted to neuronal perikarya. Therefore, any regulatory mechanisms of peptide biosynthesis must ultimately be located in the neuron soma. To date, very few experiments addressing such regulatory mechanisms have been attempted.

The regulation of biosynthesis of the 12,000-dalton precursor in the peptidergic neuron, R_{15}, in *Aplysia* has been demonstrated to be dependent on synaptic input (Gainer and Barker, 1974). The incorporation of [³H]leucine in the 12,000-dalton peak was *selectively* reduced by 30–50% when R_{15} was inhibited by synaptic input. Subsequent work showed that this inhibition of synthesis was not due to the various electrophysiological events that accompany synaptic inhibition (i.e., inhibition of action potentials, hyperpolarization, and increased membrane conductance to K^+), but was due to a direct effect of the putative neurotransmitter, dopamine (Gainer and Barker, 1975). Since R_{15} is a bursting pacemaker neuron, and presumably would release peptides from its axon terminals with each burst cycle, it is reasonable in a teleological sense that inhibition of this spike activity would reduce the demand for the released peptide, and therefore the biosynthesis of its precursor. However, the spike activity itself did not prove to be the relevant signal for the modulation of synthesis. The signal appeared to be the presence of the inhibitory transmitter. The events that coupled this signal to the selective inhibition of synthesis remain unclear.

Another form of regulation was observed in an identified peptidergic neuron in the nervous system of a different mollusk, *Otala lactea* (Gainer, 1972). This neuron, designated cell 11, appears to be homologous to R_{15} in *Aplysia* (Loh *et al.*, 1976). In active animals, cell 11 is a bursting pacemaker neuron, and synthesizes a group of specific low-molecular-weight proteins that are transported into the axon. However, when the animals are in a state of hibernation (aestivation), the same cell is electrically silent, and does not synthesize this transported group of low-molecular-weight proteins, although in all other respects, protein synthesis appears normal in this cell (Gainer, 1972; Loh *et al.*, 1976). This dramatic phenomenon of the regulation of synthesis of specific proteins in a peptidergic neuron was entirely under the control of the animal's behavioral mechanisms, and the relevant variables (signals) responsible for this transformation are unknown. Although we have been able to induce bursting pacemaker activity in a dormant cell 11 by a variety of means (Barker and Gainer, 1974, 1975), we have not yet succeeded in inducing the synthesis of the specific set of low-molecular-weight proteins.

As was pointed out earlier, when rats are dehydrated by water deprivation or by the addition of 2% NaC1 to their drinking water, most of their stores of neurohypophyseal hormones are released from the neural lobe (Jones and Pickering, 1969). After 4–5 days of dehydration, there is a 3–5 fold increase in synthesis of neurohypophyseal hormones and neurophysin (Takabatake and Sachs, 1964; see Table 7 in Valtin *et al.*, 1974; confirmed in our laboratory). Valtin *et al.* (1974) have discussed extensively the possible mechanisms that may underlie this increased rate of biosynthesis. It is evident, however, that such complex phenomena as dehydration in the rat and hibernation in the snail probably provide several ''signals'' to the responding neurons. Until the relevant ''signals'' are determined, no progress can be made in understanding the intracellular regulatory mechanisms that are involved.

7. Biological Significance of the Precursor Mode of Peptide Biosynthesis

A number of biological advantages are associated with the precursor mode of biosynthesis. Posttranslational cleavage mechanisms are not restricted to higher organisms. In a recent review, Hershko and Fry (1975) have pointed out that both bacteriophages and animal viruses utilize such mechanisms in the assembly of coat proteins. These authors have classified viral cleavage mechanisms into two categories called *formative* and *morphogenetic* cleavages. Formative cleavages are mechanisms by which certain RNA-containing animal viruses derive *all* functional proteins through the limited proteolysis of high-molecular-weight biosynthetic precursors. It has been suggested that this mechanism evolved because the protein-synthesis mechanisms of eukaryotes, in contrast to prokaryotes, are incapable of carrying out internal initiation in the translation of polycistronic viral messages. Thus, posttranslational cleavage, in this case, would replace translational punctuation signals. Such a mechanism would also effectively generate equimolar production of individual polypeptides. Morphogenetic cleavages, found in a wider variety of viruses, are the specific cleavages of structural protein during viral assembly. These cleavages occur at very specific steps in viral morphogenesis, and appear to play an important role in the sequential assembly of relatively complex viral structures (Hershko and Fry, 1975). Thus, several biological advantages of precursor biosynthesis are apparent from these studies: (1) the economical generation of equimolar polypeptides when necessary, and (2) the precise regulation of activation of a preformed molecule either for enzymatic purposes (as for the zymogens in eukaryotes; see Kassell and Kay, 1973) or for structural assembly purposes (e.g., in head maturation in T_4 bacteriophage; see Hershko and Fry, 1975).

The formation and secretion of collagen by fibroblasts is, in many regards, akin to the morphogenetic cleavage mechanisms described above. The biosynthetic precursor, procollagen, consists of three independent chains (two α_2 chains and one α_1 chain) with additional sequences at their N-termini (Bornstein, 1974). These additional sequences are linked by disulfide bonds, thus providing chain alignment so that helix formation is facilitated. Because procollagen (unlike collagen) is soluble under physiological conditions (Layman et al., 1971), it has been proposed that this is the translocatable form (i.e., for intracellular and transmembrane transport), and that after secretion into the extracellular space, specific proteolytic cleavages produce tropocollagen and ultimately collagen fibers (Layman et al., 1971; Goldberg et al., 1972; Goldberg and Sherr, 1973; Bornstein, 1974).

Another role for precursor formation may be to ensure the correct formation of the secondary or tertiary structure of a polypeptide following biosynthesis. Probably the best example of this role is the formation of insulin from proinsulin (Steiner et al., 1974). Insulin is composed of two different subunits (the A and B chains) connected by disulfide bonds. The invariant and correct assembly of this molecule would be improbable if the two chains had been synthesized as independent polypeptides. The synthesis of both chains as a single polypeptide, proinsulin, allows for proper folding, and hence correct disulfide bond formation. After correct assembly, the connecting peptide link between the A and B chains (i.e., the C-peptide) is removed by proteolytic action.

Blobel and Sabatini (1971) have proposed the "signal hypothesis" to explain why various secretory proteins are synthesized as precursors. This hypothesis states that proteins synthesized on rough endoplasmic reticulum (RER) are translated from mRNAs that contain "signal" codons following the initiation codon. Therefore, when the mRNA is translated, the protein chain then contains at its amino-terminus a "signal" sequence of amino acid residues (about 20–25 amino acids long) that directs the nascent polypeptide through the RER membrane into the cisternal space. The "signal" sequence is removed intracisternally by an endopeptidase (see Blobel and Dobberstein, 1975a,b), and the protein that is to be secreted is then translocated to the Golgi for packaging. This hypothesis would predict that considerable homology exists in the "signal" sequence of diverse secretory proteins. Extensive homology in the amino-terminal sequences of precursors to several pancreatic secretory proteins, synthesized in vitro from mRNA, has recently been reported (Devillers-Thiery et al., 1975).

The roles postulated above for a precursor–product relationship do not exhaust all the possibilities. Palade (1975) has outlined the sequence of events that occur between the biosynthesis and secretion of polypeptides. Two of these events, "segregation" and "concentration," may provide further rationales for precursor biosynthesis. "Segregation" refers to the intracisternal localization (in the RER) of the newly synthesized polypeptides. The RER membrane appears to be relatively poor in sphingomyelin and cholesterol (Meldolesi *et al.*, 1971), and is more permeable to small molecules (Palade, 1975; Tedesche *et al.*, 1963) than Golgi, secretory granule, and plasma membranes. Therefore, if the relevant peptide product of posttranslational cleavage is small enough to leak across the RER membrane, the peptide may be synthesized in a larger precursor form in order to avoid leakage before it is packaged in a granule membrane with permeability characteristics similar to the plasma membrane. The appropriate intracellular stage for posttranslational cleavage would then be after the protein has reached the Golgi organelle. The "concentration" step refers to the concentration of the secretory material ultimately in the secretory granule. This step appears to be energy-independent (Palade, 1975), and for exocrine proteins in the pancreas, appears to occur as a result of the formation of an insoluble complex of the basic peptide and a sulfated polyanionic peptglycan (Tartakoff *et al.*, 1974). The osmotic dead space created by this complex in the granule would shift the equilibrium between free and bound peptide, and hence would allow for energy-independent concentration. A similar phenomenon may exist for the neurosecretory granules in the hypothalamo-neurohypophyseal system. Since the peptides oxytocin and vasopressin form insoluble complexes with neurophysin (Breslow, 1974), and the granule core *in vivo* appears to be in an osmotically inactive form (Bargmann and Von Gaudecker, 1969; Holmes and Kiernan, 1964; Rodriguez, 1971; Livingstone and Lederis, 1971), it is possible that the peptide (e.g., vasopressin) and the carrier protein to which it complexes (i.e., neurophysin) are synthesized on a common precursor in order to ensure an equimolar content of these components in the granule. This would provide a highly efficient mechanism for the translational, packaging (in the Golgi), and concentration steps. In addition, more than one peptide would be coded for by a single mRNA, and efficiently packaged in the same granule, if the physiological effect was dependent on the simultaneous release of more than one peptide.

8. Conclusion

In this chapter, we have concentrated on the ribosomal-dependent precursor mode of peptide biosynthesis in neurons, in contrast to the obvious alternative, the enzymic (synthetase) mechanism. Although "synthetase" mechanisms do exist in the nervous system (see Section 1 and Chapter 7), we tentatively feel that such a mechanism cannot provide the reliability required for the synthesis of peptides (larger than tripeptides) that are to be used as intercellular messengers. Data have been presented here showing that certain peptidergic neurons, in mollusks as well as in mammals, utilize precursor and posttranslational cleavage mechanisms, and a hypothetical model of peptide biosynthesis in neurons has been described (see Fig. 5, Section 4.4). In addition to providing support for the precursor mode of biosynthesis, evidence has been presented indicating that the posttranslational cleavage events occur in a translocatable compartment (i.e., the secretory granule) in the axon during axonal transport. It will be very important, in future work, to identify the proteolytic enzymes involved (see Chapter 9), to determine whether they are "specific" enzymes, and to understand how they are incorporated into the granule.

The question of proteolytic enzyme specificity is of particular interest, since it is conceivable that different neurons could synthesize and subsequently release different peptides, in large part, because of the specific proteolytic enzyme packaged together with the precursor. For example, the same precursor could be synthesized and packaged into granules in two different neurons. However, if different proteolytic enzymes are copackaged in each neuron, then the resultant peptides would be different. Thus, there would be no necessity to provide extra information in the genome (i.e., new DNA sequences) for each new peptide, but only to use different but existent cell programs for the copackaging of different specific proteases. An example, of such a possibility comes from the work of Scott *et al.* (1973), who have proposed that the final peptide hormone products in the anterior pituitary, ACTH and β-lipotropin, are the precursors for α-MSH and β-MSH, respectively, in cells of the pars intermedia. In addition, endorphin, the opiate peptide, appears to be derived from β-lipotropin (see Chapters 9 and 13).

Finally, the precursor mode of biosynthesis suggests that the process of peptide biosynthesis in neurons is restricted to the neuron perikaryon. Therefore, the importance of retrograde axonal transport and synaptic input to the dendrites and soma are emphasized as possible sources of "signals" for regulatory mechanisms (see the discussion in Section 6). Studies on the regulation of neuronal peptide synthesis are only in their infancy, and model systems such as the *Aplysia* peptidergic neurons and magnocellular neurons in the SON and PVN will be of great value for these studies.

9. References

Ascher, R., 1974, Chemistry of the neurohypophysial hormones: An example of molecular evolution, in: *Handbook of Physiology, Section 7: Endocrinology,* Vol. IV (E. Knobel and W.H. Sawyer, eds.), pp. 119–130, American Physiology Society, Washington, D.C.

Albers, R.W., and Brightman, M.W., 1959, A major component of neurohypophysial tissue associated with antidiuretic activity, *J. Neurochem.* **3:**269–276.

Arch, S.W., 1972*a*, Polypeptide secretion from the isolated paravisceral ganglion of *Aplysia californica, J. Gen. Physiol.* **59:**47–59.

Arch, S.W., 1972*b*, Biosynthesis of the egg-laying hormone (ELH) in the bag cell neurons of *Aplysia californica, J. Gen. Physiol.* **60:**102–119.

Bargmann, W., 1968, Neurohypophysis: Structure and function, in: *Neurohypophysial Hormones and Similar Polypeptides* (B. Berde, ed.), pp. 1–39, Springer-Verlag, Berlin.

Bargmann, W., and Von Gaudecker, B., 1969, Über die Ultrastruktur neurosekretorischer Elementasgranula, *Z. Zellforsch. Mikrosk. Anat.* **96:**495–504.

Barker, J.L., and Gainer, H., 1974, Peptide regulation of bursting pacemaker activity in a molluscan neurosecretory cell, *Science* **184:**1371–1373.

Barker, J.L., and Gainer, H., 1975, Studies on bursting pacemaker potential activity in molluscan neurons. II. Regulation by divalent cations, *Brain Res.* **84:**479–500.

Blobel, G., and Dobberstein, B., 1975*a*, Transfer of proteins across membranes. I. Presence of proteolytically processed and unprocessed nascent immunglobulin light chains on membrane-bound ribosomes of murine myeloma, *J. Cell Biol.* **67:**835–851.

Blobel, G., and Dobberstein, B., 1975*b*, Transfer of proteins across membranes. II. Reconstitution of functional rough microsomes from heterologous components, *J. Cell Biol.* **67:**852–862.

Blobel, G., and Sabatini, D.D., 1971, Ribosome–membrane interaction in eukaryotic cells, in: *Biomembranes* (L.A. Manson, ed.), Vol. 2, pp. 193–195, Plenum Press, New York.

Bornstein, P., 1974, The biosynthesis of collagen, *Annu. Rev. Biochem.* **43:**567–604.

Breslow, E., 1974, The neurophysins, *Adv. Enzymol.* **40:**271–333.

Burford, G.D., and Pickering, B.T., 1972, The number of neurophysins in the rat. Influence of the concentration of bromphenol blue, used as a tracking dye, on the resolution of proteins by polyacrylamide gel eletrophoresis, *Biochem. J.* **128:**941–944.

Burford, G.D., Ginsburg, M., and Thomas, P.J., 1971, The effect of denaturants and Ca^{2+} on the molecular weight and polymerization of neurophysin, *Biochim. Biophys. Acta* **229:**730–738.

Burford, G.D., Dyball, R.E.J., Moss, R.L., and Pickering, B.T., 1974, Synthesis of both neurohypophysial hormones in both the paraventricular and supraoptic nuclei of the rat, *J. Anat.* **117:**261–269.

Cannata, M.A., and Morris, J.F., 1973, Changes in the appearance of hypothalamus–neurohypophysial neurosecretory granules associated with their maturation, *J. Endoctrinol.* **57:**531–538.

Chaiken, I.M., Randolph, R.E., and Taylor, H.C., 1975, Conformational effects associated with the interaction of polypeptide ligands with neurophysins, *Ann. N. Y. Acad. Sci.* **248:**442–450.

Chambach, A., and Rodbard, D., 1971, Polyacrylamide gel electrophoresis, *Science* **172:**440–451.

Coy, D.H., and Wuu, T.C., 1972, Purification and amino acid composition of constituents of rat neurophysin, *Biochim. Biophys. Acta* **263:**125–132.

Coggeshall, R.E., 1967, A light and electron microscope study of the abdominal ganglion of *Aplysia californica, J. Neurophysiol.* **30:**1263–1268.

Coggeshall, R.E., 1970, A cytologic analysis of the bag cell control of egg laying in *Aplysia, J. Morphol.* **132:**461–469.

Coggeshall, R.E., Kandell, E.R., Kupfermann, I., and Waziri, R.A., 1966, A morphological and functional study on a cluster of identifiable neurosecretory cells in the abdominal ganglion of *Aplysia californica, J. Cell Biol.* **31:**363–368.

Cross, B.A., Dyball, R.E., Dyer, R.G., Jones, C.W., Lincoln, D.W., Morris, J.F., and Pickering, B.T., 1975, Endocrine neurons, *Recent Prog. Horm. Res.* **31:**243–294.

Dahlstrom, A., 1968, Effect of colchicine on transport of amino storage granules in sympathetic nerves of rat, *Eur. J. Pharmacol.* **5:**111–113.

Dahlstrom, A., 1971, Effects of vinblastine and colchicine on monamine containing neurons of the rat, with special regard to the axoplasmic transport of amine granules, *Acta Neuropathol. Berlin* **5:**(Suppl.) 226–237.

Davidson, J.M., McEneany, L.S.G., and Bornstein, P., 1975, Intermediates in the limited proteolytic conversion of procollagen to collagen, *Biochemistry* **14:**5188–5194.

Dean, C.R., Hollenberg, M.D., and Hope, D.B., 1967, The relationship between neurophysin and the soluble proteins of pituitary neurosecretory granules, *Biochem. J.* **104:**8–10c.

Devillers-Thiery, A., Kindt, T., Scheele, G., and Blobel, G., 1975, Homology in amino-terminal sequence of precursors to pancreatic secretory proteins, *Proc. Nat. Acad. Sci. U.S.A.* **72:**5016–5020.

Edstrom, J.E., Eichner, D., and Schor, N., 1961, Quantitative ribonucleic acid measurements in functional studies of nucleus supraopticus, in: *Regional Neurochemistry* (S.S. Kety and J. Elkes, eds.), pp. 274–278, Pergamon Press, New York.

Eneström, S., and Hamberger, A., 1968, Respiration and mitochondrial content in single neurons of the supraoptic nucleus, *J. Cell Biol.* **38**:483–493.

Frazier, W.T., Kandel, E.R., Kupfermann, I., Waziri, R., and Coggeshall, R.E., 1967, Morphological and functional properties of identified neurons in the abdominal ganglion of *Aplysia californica, J. Neurophysiol.* **30**:1288–1351.

Gainer, H., 1972, Effects of experimentally induced diapause on the electrophysiology and protein synthesis of identified molluscan neurons, *Brain Res.* **39**:387–402.

Gainer, H., 1976, Peptides and neuronal function, *Adv. Biochem. Psychopharmacol.* **15**:193–210.

Gainer, H., and Barker, J.L., 1974, Synaptic regulation of specific protein synthesis in an identified neuron, *Brain Res.* **78**:314–319.

Gainer, H., and Barker, J.L., 1975, Selective modulation and turnover of proteins in identified neurons of *Aplysia, Comp. Biochem. Physiol.* **51B**:221–227.

Gainer, H., and Wollberg, Z., 1974, Specific protein metabolism in identifiable neurons of *Aplysia californica, J. Neurobiol.* **5**:243–261.

Gainer, H., Sarne, Y., and Brownstein, M.J., 1977*a,* Neurophysin biosynthesis: Conversion of a putative precursor during axonal transport, *Science,* in press.

Gainer, H., Sarne, Y., and Brownstein, M.J., 1977*b,* Biosynthesis and axonal transport of rat neurohypophysial proteins and peptides, *J. Cell Biol.,* in press.

Gewirtz, G., and Yalow, R.S., 1974, Ectopic ACTH production in carcinoma of the lung, *J. Clin. Invest.* **53**:1022, 1023.

Giller, E., and Schwartz, J.H., 1971, Choline acetyltransferase in identified neurons of abdominal ganglion of *Aplysia californica, J. Neurophysiol.* **34**:93–107.

Goldberg, G., and Sherr, C.J., 1973, Secretion and extracellular processing of procollagen by cultured human fibroblasts, *Proc. Nat. Acad. Sci. U.S.A.* **70**:361–365.

Goldberg, G., Epstein, E.H., and Scherr, 1972, Precursors of collagen secreted by cultured human fibroblasts, *Proc. Nat. Acad. Sci. U.S.A.* **69**:3655–3659.

Grollman, A.P., and Huang, M.T., 1973, Inhibitors of protein synthesis in eukaryotes: Tools in cell research, *Fed. Proc. Fed. Amer. Soc. Exp. Biol.* **32**:1673–1678.

Gross, G.W., 1973, The effect of temperature on the rapid axoplasmic transport in C-fibers, *Brain Res.* **56**:359–363.

Heller, H., 1974, History of neurohypophysial research, in: *Handbook of Physiology,* Section 7: *Endocrinology,* Vol. IV (E. Knobil and W.H. Sawyer, eds.), pp. 103–117, American Physiology Society, Washington, D.C.

Hershko, A., and Fry, M., 1975, Post-translational cleavage of polypeptide chains: Role in assembly, *Annu. Rev. Biochem.* **44**:775–797.

Hirata, Y., Yamamoto, H., Matsukura, S., and Imura, H., 1975, *In vitro* release and biosynthesis of tumor ACTH in ectopic ACTH producing tumors, *J. Clin. Endocrinol. Metab.* **41**:106–114.

Holmes, R.L., and Kiernan, J.A., 1964, The fine structure of the infundibular process of the hedgehog, *Z. Zellforsch. Mikrosk. Anat.* **61**:894–912.

Hope, D.B., and Pickup, J.C., 1974, Neurophysins, in: *Handbook of Physiology,* Section 7: *Endocrinology,* Vol. IV (E. Knobil and W.H. Sawyer, eds.), pp. 173–189, American Physiology Society, Washington, D.C.

Jacobs, J.W., Kemper, B., Niall, H.D., Habener, J.F., and Potts, J.T., 1974, Structural analysis of human proparathyroid hormone by a new microsequencing approach, *Nature London* **249**:155–157.

Jones, C.W., and Pickering, B.T., 1969, Comparison of the effects of water deprivation and sodium chloride inhibition on the hormone content of the neurohypophysis of the rat, *J. Physiol. London* **203**:499–458.

Judah, J.D., Gamble, M., and Steadman, J.H., 1973, Biosynthesis of serum albumin in rat liver. Evidence for the existence of "proalbumin," *Biochem. J., ,* **134**:1083–1091.

Kandel, E.R., and Kupfermann, I., 1970, The functional organization of invertebrate ganglia, *Annu. Rev. Physiol.* **32**:193–258.

Kandel, E.R., and Wachtel, H., 1968, The functional organization of neural aggregates in *Aplysia,* in:

Physiological and Biochemical Aspects of Nervous Integration (F.D. Carlson, ed.), pp. 17–65, Prentice-Hall, Englewood Cliffs, New Jersey.

Karlson, J.O., and Sjöstrand, J., 1969, The effect of colchicine on the axonal transport of protein in the optic nerve and tract of the rabbit, *Brain Res.* **13**:617–619.

Kassell, B., and Kay, J., 1973, Zymogens of proteolytic enzymes, *Science* **180**:1022–1027.

Kemmler, W., Steiner, D.F., and Borg, J., 1973, Studies on the conversion of proinsulin to insulin. III. Studies *in vitro* with a crude secretion granule fraction isolated from islets of Langerhans, *J. Biol. Chem.* **248**:4544–4551.

Kemper, B., Habener, J.F., Potts, J.T., and Rich, A., 1972, Proparathyroid hormone: Identification of a biosynthetic precursor to parathyroid hormone, *Proc. Nat. Acad. Sci. U.S.A.* **69**:643–647.

Kemper, B., Habener, J.F., Mulligan, R.C., Potts, J.T., and Rich, A., 1974, Preparathyroid hormone: A direct translation product of parathyroid messenger RNA, *Proc. Nat. Acad. Sci. U.S.A.* **71**:3731–3733.

Kreutzberg, G.W., 1969, Neuronal dynamics and axonal flow, IV. Blockage of intra-axonal enzyme transport by colchicine, *Proc. Nat. Acad. Sci. U.S.A.* **62**:723–728.

Kupfermann, I., 1967, Stimulation of egg-laying: Possible neuroendocrine function of bag cells of abdominal ganglion of *Aplysia californica, Nature London* **216**:814–815.

Kupfermann, I., 1970, Stimulation of egg laying by extracts of neuroendocrine cells (bag cells) of abdominal ganglion of *Aplysia, J. Neurophysiol.* **33**:877–881.

Kupfermann, I., 1972, Studies on the neurosecretory control of egg laying in *Aplysia, Amer. Zool.* **12**:513–516.

Kupfermann, I., and Weiss, K.R., 1976, Water regulation by a presumptive hormone contained in identified neurosecretory cell R_{15} of *Aplysia, J. Gen. Physiol.* **67**:113–123.

Kurahashi, K., 1974, Biosynthesis of small peptides, *Annu. Rev. Biochem.* **43**:445–459.

Laland, S.G., and Zimmer, T.L., 1973, The protein thiotemplate mechanism of synthesis for the peptide antibiotics produced by *Bacillus brevis, Essays Bichem.* **9**:31–57.

Layman, D.L., McGoodwin, and Martin, G.R., 1971, The nature of the collagen synthesized by cultured human fibroblasts, *Proc. Nat. Acad. Sci. U.S.A.* **68**:454–458.

Lederis, K., 1974, Neurosecretion and the functional structure of the neurohypophysis, in: *Handbook of Physiology*, Section 7: *Endocrinology*, Vol. IV (E. Knobil and W.H. Sawyer, eds.), pp. 81–102, American Physiology Society, Washington, D.C.

Lederis, K., and Jayasena, K., 1970, Storage of neurohypophysial hormones and the mechanism for their release, in: *International Encyclopedia of Pharmacology and Therapeutics*, Section 41, Vol. 1 (H. Heller and B.T. Pickering, eds.), pp. 111–154, Pergamon Press, Oxford.

Lipmann, F., 1971, Attempts to map a process evolution of peptide biosynthesis, *Science* **173**:875–884.

Livingstone, A., and Lederis, K., 1971, Functional ultrastructure of the neurohypophysis, *Mem. Soc. Endocrinol.* **19**:233–261.

Loh, Y.P., and Gainer, H., 1975*a*, Low molecular weight specific proteins in identified molluscan neurons. I. Synthesis and storage, *Brain Res.* **92**:181–192.

Loh, Y.P., and Gainer, H., 1975*b*, Low molecular weight specific proteins in identified molluscan neurons. II. Processing, turnover, and transport, *Brain Res.* **92**:193–205.

Loh, Y.P., and Peterson, R.P., 1974, Protein synthesis in phenotypically different, single neurons of *Aplysia, Brain Res.* **78**:83–98.

Loh, Y.P., Sarne, Y., and Gainer, H., 1975, Heterogeneity of proteins synthesized, stored and released by the bag cells of *Aplysia californica, J. Comp. Physiol.* **100**:283–295.

Loh, Y.P., Barker, J.L., and Gainer, H., 1976, Neurosecretory cell protein metabolism in the land snail, *Otala lactea, J. Neurochem.* **26**:25–30.

Loh, Y.P., Sarne, Y., Daniels, M.P., and Gainer, H., 1977, Subcellular fractionation studies related to the processing of neurosecretory proteins in *Aplysia* neurons, *J. Neurochem.*, in press.

Lowry, P.J., and Scott, A.P., 1975, The evolution of vertebrate corticotrophin and melanocyte stimulating hormone, *Gen. Comp. Endocrinol.* **26**:16–23.

MacArthur, C.G., 1931, A new pituitary preparation, *Science* **73**:448.

McCaman, R.E., and Dewhurst, S.A., 1971, Metabolism of putative transmitters in individual neurons of *Aplysia californica, J. Neurochem.* **18:**1329–1335.

McGregor, A.M., Komija, Y., Kidman, A.D., and Austin, L., 1973, The blockage of axoplasmic flow of proteins by colchicine and cytochalasins A and B, *J. Neurochem.* **21:**1059–1066.

McKelvy, J.L., 1975, Phosphorylation of neurosecretory granules by c-AMP-stimulated protein kinase and its implication for transport and release of neurophysin proteins, *Ann. N. Y. Acad. Sci.* **248:**80–91.

Meister, A., 1973, Glutathione: Metabolism and function via the γ-glutamyl cycle, *Life Sci.* **15:**177–190.

Meldolesi, J., Jamieson, J.D., and Palade, G.E., 1971, Composition of cellular membranes in the pancreas of the guinea pig. II. Lipids, *J. Cell Biol.* **49:**130–149.

Milstein, C., Brownlee, G.G., Harrison, T.M., and Mathews, M.B., 1972, A possible precursor of immunoglobulin light chains, *Nature London New Biol.* **239:**117–120.

Mitnick, M.A., and Reichlin, S., 1971, Thyrotropin-releasing hormone: Biosynthesis by rat hypothalamic fragments *in vitro, Science* **172:**1241–1243.

Mitnick, M.A., Valverde, R.C., and Reichlin, S., 1973, Enzymatic synthesis of prolactin-releasing factor (PRF) by rat hypothalamic incubates and by extracts of rat hypothalamic tissue: Evidence for "PRF" synthetase, *Proc. Soc. Exp. Biol. Med.* **143:**418–421.

Morris, J.F., and Cannata, M.A., 1973, Ultrastructural preservation of the dense core of posterior pituitary neurosecretory granules and its implications for hormone release, *J. Endocrinol.* **57:**517–529.

Moya, F., Nieto, A., R.-Candela, J.L., 1975, Calcitonin biosynthesis: Evidence for a precursor, *Eur. J. Biochem.* **55:**407–413.

Noe, B.D., and Bauer, G.E., 1971, Evidence for glucagon biosynthesis involving a protein intermediate in islets of the angler fish (*Lophius americanus*), *Endocrinology* **89:**642–651.

Norström, A., 1975, Axonal transport and turnover of neurohypophysial proteins in the rat, *Ann. N. Y. Acad. Sci.* **248:**46–63.

Norström, A., and Sjöstrand, J., 1971a, Axonal transport of proteins in the hypothalamo-neurohypophysial system of the rat, *J. Neurochem.* **18:**29–39.

Norström, A., and Sjöstrand, J., 1971b, Transport and turnover of neurohypophysial proteins of the rat, *J. Neurochem.* **18:**2007–2016.

Norström, A., Sjöstrand, J., Livett, B.G., Uttenthal, O., and Hope, D.B., 1971, Electrophoretic and immunological characterization of rat neurophysin, *Biochem. J.* **122:**671–676.

Osinchak, J., 1964, A fine structure and cytochemical study of neurosecretory cells in the rat, *Int. Congr. Cell Biol.* **77:**33–34.

Palade, G., 1975, Intracellular aspects of the process of protein synthesis, *Science* **189:**347–358.

Palkovitz, M., 1973, Isolated removal of hypothalamic or other brain nuclei of the rat, *Brain Res.* **59:**449, 450.

Peschle, C., and Condorelli, M., 1975, Biogenesis of erythropoietin: Evidence for pro-erythropoietin in a subcellular fraction of kidney, *Science* **190:**910–912.

Pickering, B.T., and Jones, C.W., 1971, The biosynthesis and intraneuronal transport of neurohypophysial hormones: Preliminary studies in the rat, in: *Subcellular Organization and Function in Endocrine Tissues* (H. Heller and K. Lederis, eds.), pp. 337–351, Cambridge Press, New York.

Pickering, B.T., Jones, C.W., Burford, G.D., McPherson, M., Swann, R.W., Heap, P.F., and Morris, J.F., 1975, The role of neurophysin proteins: Suggestions from the study of their transport and turnover, *Ann. N. Y. Acad. Sci.* **248:**15–35.

Quinn, P.S., Gamble, M., and Judah, J.D., 1975, Biosynthesis of serum albumin in rat liver. Isolation and probable structure of proalbumin from rat liver, *Biochem. J.* **146:**389–393.

Reichlin, S., and Mitnick, M.A., 1972, Enzymatic synthesis of GH–RF by rat incubates and by extracts of rat and porcine hypothalamic tissue, *Proc. Soc. Exp. Biol. Med.* **142:**497–501.

Reichlin, S., and Mitnick, M.A., 1973, Biosynthesis of hypothalamic hypophysiotropic factors, in: *Frontiers of Neuroendocrinology,* (W.F. Ganong and L. Martini, eds.), pp. 61–88, Oxford University Press, London.

Reichlin, S., Martin, J.B., Mitnick, M.A., Boshans, R.L., Grimm, Y., Bollinger, J., Gordon, J., and Malacaia, J., 1972, The hypothalamus in pituitary–thyroid regulation, *Recent Prog. Horm. Res.* **28:**229–286.

Reichlin, S., Saperstein, R., Jackson, I.M.D., Boyd, A.E., and Patel, Y., 1976, Hypothalamic hormones, *Annu. Rev. Physiol.* **38:**389–424.

Richards, F.M., and Vithayathil, P.J., 1959, The preparation of subtilisin modified ribonuclease and the separation of the peptide and protein components, *J. Biol. Chem.* **234:**1459–1465.

Rodbard, D., and Chrambach, A., 1971, Estimation of molecular radius, free mobility and valence using polyacrylamide gel electrophoresis, *Anal. Biochem.* **40:**95–134.

Rodriguez, E.M., 1971, The comparative morphology of the neural lobes of species which have different neurohypophysial hormones, *Mem. Soc. Endocrinol.* **19:**263–291.

Sachs, H., 1960, Vasopressin biosynthesis. I. *In vivo* studies, *J. Neurochem.* **5:**297–303.

Sachs, H., 1963, Studies on the intracellular distribution of vasopressin, *J. Neurochem.* **10:**289–297.

Sachs, H., and Takabatake, Y., 1964, Evidence for a precursor in vasopressin biosynthesis, *Endocrinology* **75:**943–948.

Sachs, H., Fawcett, P., Takabatake, Y., and Portanova, R., 1969, Biosynthesis and release of vasopressin and neurophysin, *Recent Prog. Horm. Res.* **25:**447–491.

Sachs, H., Saito, S., and Sunde, D., 1971, Biochemical studies on the neurosecretory and neuroglial cells of the hypothalamo-neurohypophysial complex, in: *Subcellular Organization and Function in Endocrine Tissues* (H. Heller and K. Lederis, eds.), pp. 325–336, Cambridge Press, New York.

Sachs, H., Pearson, D., and Nureddin, A., 1975, Guinea pig neurophysin: Isolation, developmental aspects, biosynthesis in organ culture, *Ann. N. Y. Acad. Sci.,* **248:**36–45.

Sano, I., 1970, Simple peptides in brain, *Int. Rev. Biol.* **12:**235–263.

Sawyer, W.H., 1971, Evolution of neurohypophysial peptides among the non-mammalian vertebrates, in: *Neurohypophysial Hormones* (G.E.W. Wolstenholme and J. Birch, eds.), pp. 5–13, Churchill Livingstone Co., London.

Schmeckpeper, B.J., Cory, J.S., and Adams, J.M., 1974, Translation of immunoglobulin mRNAs in a wheat germ cell free system, *Mol. Biol. Rep.* **1:**335–363.

Scott, A.P., Ratcliffe, J.G., Rees, L.H., Landon, J., Bennett, H.P.J., Lowry, P.J., and McMartin, C., 1973, Pituitary peptide, *Nature London New Biol.* **244:**65–67.

Sherwood, L.M., Rodman, J.S., and Lundberg, W.B., 1970, Evidence for a precursor to circulating parathyroid hormone, *Proc. Nat. Acad. Sci. U.S.A.* **67:**1631–1638.

Stachura, M.E., and Frohman, L.A., 1974, "Large" growth hormone: Ribonucleic acid–associated precursor of other growth hormone forms in rat pituitary, *Endocrinology* **94:**701–712.

Stachura, M.E., and Frohman, L.A., 1975, Growth hormone: Independent release of big and small forms from rat pituitary *in vitro, Science* **187:**447–449.

Steiner, D.F., and Oyer, P.E., 1967, The biosynthesis of insulin and a probable precursor of insulin by a human islet cell adenoma, *Proc. Nat. Acad. Sci. U.S.A.* **57:**473–480.

Steiner, D.F., Cunningham, D., Spigelman, L., and Aten, B., 1967, Insulin biosynthesis: Evidence for a precursor, *Science* **157:**697–700.

Steiner, D.F., Kemmler, W., Tager, H.S., and Peterson, J.D., 1974, Proteolytic processing in the biosynthesis of insulin and other proteins, *Fed. Proc. Fed. Amer. Soc. Exp. Biol.* **33:**2105–2115.

Strumwasser, F., Jacklet, J.W., Alvarez, R.B., 1969, A seasonal rhythm in the neural extract induction of behavioral egg-laying in *Aplysia, Comp. Biochem. Physiol.* **29:**197–206.

Sunde, D.A., and Sokol, H.W., 1975, Quantification of rat neurophysins by polyacrylamide gel electrophoresis (PAGE): Application to the rat with hereditary hypothalamic diabetes insipidus, *Ann. N. Y. Acad. Sci.* **248:**345–364.

Sussman, P.M., Tushinski, R.J., and Bancroft, F.C., 1976, Pregrowth hormone: Product of the translation *in vitro* of messenger RNA coding for growth hormone, *Proc. Nat. Acad. Sci. U.S.A.* **73:**29–33.

Swaab, D.F. Nijreldt, F., and Pool, C.W., 1975, Distribution of oxytocin and vasopressin in the rat supraoptic and paraventricular nucleus, *J. Endocrinol.* **67:**461, 462.

Tager, H.S., and Steiner, D.F., 1974, Peptide hormones, *Ann. Rev. Biochem.* **43:**509–538.

Takabatake, Y., and Sachs, H., 1964, Vasopressin biosynthesis. III. *In vitro* studies, *Endocrinology* **75:**934–942.

Tartakoff, A., Greene, L.J., and Palade, G.E., 1974, Studies on the guinea pig pancreas. Fractionation and partial characterization of exocrine proteins, *J. Biol. Chem.* **249:**7420–7431.

Tedeschi, H., James, J.M., and Anthony, W., 1963, Photometric evidence for the osmotic behavior of rat liver microsomes, *J. Cell Biol.* **18:**503–513.

Toevs, L., and Brackenbury, R., 1969, Bag cell–specific proteins and the humoral control of egg-laying in *Aplysia californica, Comp. Biochem. Physiol.* **29:**207–216.

Trakatellis, A.C., Tada, K., Yamaji, K., and Gasdiki-Kouidou, P., 1975, Isolation and partial characterization of angler fish proglucagon, *Biochemistry* **14:**1508–1512.

Urban, J., and Schreiber, G., 1975, Biological evidence for a precursor protein of serum albumin, *Biochem. Biophys. Res. Comm.* **64:**778–782.

Valtin, H., Stewart, J., and Sokol, H.W., 1974, Genetic control of the production of posterior pituitary principles, in: *Handbook of Physiology, Section 7: Endocrinology,* Vol. IV (E. Knobil and W.H. Sawyer, eds.), pp. 131–171, American Physiology Society, Washington, D.C.

Valtin, H., Sokol, H.W., and Sunde, D., 1975, Genetic approaches to the study of the regulation and actions of vasopressin, *Recent Prog. Horm. Res.* **31:**447–487.

Vogt, M., 1953, Vasopressor, antidiuretic, and oxytocic activities of extracts of the dog's hypothalamus, *Br. J. Pharmacol.* **8:**193–196.

Wilson, D.L., 1974, Protein synthesis and nerve cell specificity, *J. Neurochem.* **22:**464–467.

Wuu, T.C., and Saffran, M., 1969, Isolation and characterization of a hormone-binding polypeptide from pig posterior pituitary powder, *J. Biol. Chem.* **244:**482–490.

Yalow, R.S., and Berson, S.A., 1973, Characteristics of "big ACTH" in human plasma and pituitary extracts, *J. Clin. Endocrinol. Metab.* **36:**415–423.

Zambrano, D., and De Robertis, E., 1966, The secretory cycle of supraoptic neurons in the rat. A structural–functional correlation, *Z. Zellforsch. Mikrosk. Anat.* **73:**414–431.

9

Conversion and Inactivation of Neuropeptides

Neville Marks

1. Introduction

The potent actions and wide distribution within the CNS of active peptides has stimulated considerable interest in their formation and inactivation. Such processes involve participation of proteolytic enzymes in (1) a "constructive" role—formation of active peptides by breakdown of protein or polypeptide precursors; and (2) a "destructive" role—the inactivation of active peptides. If some neuropeptides act like conventional neurotransmitters, as has been proposed, then their inactivation at target sites by specific hydrolases could assume considerable significance. This chapter is focused, therefore, on the various brain proteolytic enzymes available for this purpose. As a general comment, it must be emphasized that current knowledge concerning the pathways involved in intracellular breakdown is still rudimentary, and for this reason only simple schemes can be advanced. We hope a better understanding of mechanisms regulating peptide turnover (synthesis and breakdown) with respect to brain proteolytic enzymes may help to fill some of the gaps and provide new insights.

1.1. Comment on Classification of Proteolytic Enzymes

As noted by the commission on nomenclature (*Enzyme Nomenclature,* 1972), enzymes comprising this group are large in number and are difficult to classify, since they frequently lack specificity. Enzymes fall into two subgroups, exopep-

Neville Marks · New York State Research Institute for Neurochemistry and Drug Addiction, Ward's Island, New York, New York.

TABLE 1. Tentative Classification of Brain Enzymes Available for Hormone Conversion or Inactivation [a]

Nomenclature (E.C. listing)	pH	Major substrate [b]	Hormone substrate	Comments
I. Exopeptidases (3.4)				
1. N-Terminal aminopeptidases (3.4.11.1–4)	7.6	Peptides with free α-N-terminal	MIF, somatostatin, kinins, angiotensins, Substance P	Largely cytoplasmic. No definitive separation of di- and tripeptidases in brain. Some peptidases can cleave acylated amides (arylamidases), and can be differentiated by puromycin inhibition. (Marks, 1970)
Dipeptidyl peptidase (3.4.14.1)	5.5	Dipeptidyl amides or polypeptides	Insulin, glucagon, ACTH	SH-groups at active center. Dependency on C1 can polymerize dipeptidyl amides. Can remove N-terminal dipeptide substrates formerly known as cathepsin C. (Marks and Lajtha, 1971)
Pyroglutamate peptidase (3.4.11.8)	7.6	Pyroglutamyl peptides	TRF	Partially characterized in pituitary. (Mudge and Fellows, 1973; Prasad and Peterkofsky, 1976)
2. C-Terminal				
Lysosomal carboxypeptidase A	5.5	Z–Glu–Tyr	Angiotensin	Inhibited by DFP, PCMB. Specificity similar to carboxypeptidase A. Formerly known as cathepsin A. (Grynbaum and Marks, 1976)
Lysosomal carboxypeptidase B	5.5	BAA	Kinins and other hormones	Inhibited by PMSF, TPCK. Broad specificity. Formerly known as cathepsin B$_2$ (Otto, 1976)
Carboxypeptidase A (3.4.12.2)	7.6	Z–Phe–Leu	Peptides, except those with C-terminal Arg, Lys, and Pro	Not available in purified form from CNS tissues. (Grynbaum and Marks, 1976; Marks, 1970)
Carboxypeptidase B (3.4.12.3)	7.6	Z–Ala–Arg	Peptides with C-terminal Lys or Arg, kinins	Not available in purified form from CNS tissues. (Grynbaum and Marks, 1976; Marks, 1970)
C-Terminal cleaving enzymes and deamidases[c]	7.6	Polypeptides with C-terminal amide	Oxytocin, LH-RH, Substance P (?) TRF	Not characterized in CNS tissue. May be more than one enzyme. One may show preference for C-terminal containing Pro-X or Pro-X–X. (Marks et al., 1973c; Walter, 1976; Bauer and Lipmann, 1976; Prasad and Peterkofsky, 1976)
Peptidyl dipeptidase (3.4.15.1)	7.6	Bradykinin, Hip–His–Leu	Proangiotensin (C1)	Was formerly kininase II, peptidase P, converting enzyme. Specificity dependent on presence or absence of C1. (Marks and Lajtha, 1971; Oliveira et al., 1976)

Enzyme	pH	Substrate	Products/Specificity	Comments
II. Serine proteinases (3.4.21)				
Kininogenin (3.4.21.8)	7.6	Kininogens	Precursors of kinin-9 and kinin-10	Was formerly kallikrein. Only indirect evidence for presence in brain. (Hori, 1968)
III. Thiol proteinases (3.4.22)				
Cathepsin B (3.4.22.1)	5.5	Proteins, BANA	—	Endopeptidase active at pH 5.5. Activity in brain lower than cathepsin D. Specificity toward hormones not determined. Inhibited by leupeptin. (Barrett, 1975)
IV. Acid proteinases (3.4.23)				
Cathepsin D (3.4.23.5)	3.0	Proteins, myelin basic protein	Insulin B chain, hexapeptides with –Phe–Phe groupings	No known cofactor requirements. Inhibited by pepstatin. Exists in isomeric forms. (Marks et al., 1973b)
V. Unclassified (3.4.99)				
Renin (3.4.99.19)	6.0	Proteins	Angiotensinogen	Presence in brain disputed. Inhibited by pepstatin. Renin substrate in C.S.F. (Day and Reid, 1976; Ganten et al., 1971a)
Cathepsin M (—)	7.6	Proteins	Myelin basic proteins, histones, LH-RH, somatostatin, Substance P	Not characterized, –SH at active center. Inhibited by PCMB, DFP. Also known as neutral proteinase or neutral endopeptidase. (Marks and Lajtha, 1971; Benuck and Marks, 1975, and unpublished observations)

[a] The classification is taken from *Enzyme Nomenclature* (1972) and recommendations made at the Second International Conference on Intracellular Catabolism (Turk and Marks, 1976). Such classifications contain other intracellular proteolytic enzymes, but evidence for their presence in brain is lacking. Only one or two references are cited; others can be found in the text. Abbreviations are: (Z) carbobenzoxy; (BAA) benzoyl arginyl amide; (BANA) benzoyl arginyl β-naphthylamide; (Hip) hippuryl.

[b] Only one or two examples are cited.

[c] Deamidases represent enzymes removing ammonia from C-terminal residues of polypeptides and can be differentiated from aminopeptidases (leucineaminopeptidase) hydrolyzing monoacylated amides.

tidases (3.4.11–15) and proteinases (3.4.21–24). Exopeptidases can act on peptides of various size range by removal of N- or C-terminal groups (single amino acids or dipeptides); they comprise two additional subgroups, amino- and carboxypeptidases (Table 1). *Proteinase* is the term selected for enzymes cleaving internal peptide bonds, although they can in addition cleave N- and C-terminal groups; this group of enzymes is divided into subgroups based on their catalytic mechanisms. In several instances, intracellular proteinases are not available in sufficient purity for definitive studies on specificity, and are assigned to an "unclassified" category (E.C.3.4.99). For purposes of description, it has been proposed that all intracellular proteinases be referred to as *cathepsins* and receive an alphabetical designation (see Barrett, 1975; Turk and Marks, 1976). It is of interest in terms of regulatory mechanisms that brain lysosomes contain exopeptidases in addition to proteinases, indicating that these organelles have the capacity to completely degrade selected proteins and polypeptides (Table 1) (Grynbaum and Marks, 1976; Marks *et al.*, 1976*a*).

Since many neuropeptides have unusual chemical groupings (N-terminal pyroGlu or acetyl, or C-terminal amide), many of the inactivating enzymes may prove to have novel specificities hitherto undescribed. The presence of N-terminal pyroGlu or acetyl and C-terminal amide (in some cases close to or adjacent to prolyl groups) prohibits the action of classic peptide hydrolases, and points to a role for a rate-limiting internal cleavage by neutral endopeptidases or similar enzymes (Marks and Stern, 1974*a*,*b*).

The practice of classifying an enzyme on the basis of one substrate has encountered problems. Enzymes believed to be highly specific for oxytocin, angiotensins, kinins, and insulin can often degrade others, and their designation may be misleading (Marks, 1970).

1.2. Methodological Considerations

Any account of the fate of neuropeptides must consider some of the problems associated with catabolic measurements. The level of peptides in tissues is governed by a number of factors, among which are the *relative* levels of synthesis vs. those of breakdown, binding and transport phenomena, compartmentation of enzyme and substrate, and role of inhibitor or accessory factors or both. Many of these considerations are identical to those involved in studies on turnover in brain, as described in detail elsewhere (Lajtha and Marks, 1971; Marks and Lajtha, 1971).

Information on breakdown can be gained from studies *in vivo* by noting the disappearance of peptide with time after systemic administration. Peptides are active down to 10^{-12} M, which can pose a formidable problem in terms of peptide detection. This problem can be partly overcome by the use of sensitive bioassay, radioimmunoassay (RIA) and isotope dilution methods, although these methods rarely provide sufficient information on the sites of inactivation or the biochemical specificity of the enzymes involved. Studies with labeled isotopes *in vivo* can also present problems, since a number of assumptions are involved that cannot al-

TABLE 2. Comparison of Exo- and Endopeptidases in Brain and Neurosecretory Areas [a]

Enzyme	Substrate	Enzyme activity (μmol/g fresh wt per hr)		
		Cortex	Hypothalamus	Pituitary
Neutral proteinase	Hb (pH 7.6)	9.3	9.1	11.5
Acid proteinase	Hb (pH 3.2)	15	17	45
Aminopeptidase	Leu–Gly–Gly	360	420	700
Arylamidase	Arg–βNA	140	109	455
Dipeptide hydrolase	Arg–Arg–βNA	27	18	29
—	MIF	3.2	3	4.1
—	TRF	0.6	2.3	0.5
—	LH–RH	1.0	4.6	3.9
—	Substance P	13.2	13.2	18
—	Somatostatin	4.3	—	—

[a] Enzymes were assayed by the procedures of Serra *et al.* (1972) and Marks and Stern (1974*a, b*). For hormone assays, 50 nmol substrate was incubated in a volume of 0.5 ml 10 mM Tris–HCl buffer containing 0.5 mM Cleland's reagent and 0.25 mg homogenate protein for 4 hr at 37°C, and then fixed with 3% (wt/vol) sulfosalicylic acid. Breakdown was measured by the appearance of free amino acids as detected in an analyzer, modified where necessary to detect glycinamide. See Figs. 5 and 6 and Table 3 for the structures of MIF, LH–RH, SRIF (somatostatin), and Substance P. The structure of MIF is Pro–Leu–Gly·NH₂.

together be resolved by current methodology (see Witter, 1976; Lajtha and Marks, 1971). Most peptides do not penetrate readily into brain in sufficient amounts for easy detection, and incorporation of label into tissue by itself is not sufficient proof of penetration, since it may represent radioactive breakdown products.

Studies *in vitro* are beset by other problems, among which are identification and purification of the hydrolase in question and optimal conditions for testing (extraction, stability, and effect of cofactors). Progress in characterization of intracellular proteolytic hydrolases has lagged seriously behind that of other enzymes, and in many instances, their catalytic mechanisms cannot be defined (Table 1). This lack of knowledge has hindered our understanding of mechanisms regulating peptide turnover. Since brain, like other tissues, contains a spectrum of proteolytic enzymes, more than one mechanism of peptide inactivation may exist (Marks *et al.,* 1974) (Table 2). Studies on the distribution of degradative hydrolases in brain regions may not provide the needed information on regulation; e.g., in regional studies on factors that regulate ACh levels, the synthetic enzyme choline acetyltransferase provides a better correlation than acetylcholinesterase. Studies on breakdown in crude tissues of proangiotensin, luteinizing hormone–releasing hormone (LH–RH), and Substance P show a ubiquitous distribution of enzyme, often sufficient to account for inactivation of physiological levels of hormone. There are several ways of expressing such data, e.g., in terms of milligrams fresh weight of protein or as total capacity of the tissue or region in question. In studies on Substance P, for example, the levels in neurosecretory areas were comparable to cortex, but if account is taken of area weight, the cortex has a capacity for exceeding that of any other region (Table 3). Clearly, the total capacity of a tissue as measured in cell free extracts may have little relationship to regulation, unless other factors such as those enumerated above are considered.

TABLE 3. Regional Distribution of Rat Brain Enzymes Degrading Luteinizing
Hormone–Releasing Hormone and Substance P [a]

| | Activity (pmol/mg fresh wt per min or brain region) | | | |
| | LH–RH | | Substance P | |
Brain region pmol	mg/fresh wt	total per region	mg/fresh wt	total per region
Cortex	12	21×10^3	220	40×10^3
Anterior pituitary	8	20	230	1250
Posterior pituitary	33	30	360	295
Hypothalamus	12	120	220	2210
Pineal gland	192	183	320	255

[a] Tissues from neurosecretory areas dissected from 10 adult male rats and then analyzed for LH–RH (based on Leu release) and Substance P (guinea pig ileum). Results are expressed per milligram weight, and also as the total inactivating present in that region. Other areas examined for Substance P and expressed as picomoles per milligram per minute were: pons–medulla, 185; cerebellum, 195; spinal cord, 170; optic chiasma, 330.

If studies on biochemical specificity are required, some information can be gained by measuring release of end-products such as amino acids by standard chromatographic techniques (Marks and Stern, 1974a,b). In recent developments, we have adapted a microdansylation method capable of detecting breakdown products in the range of 10^{-9} to 10^{-12} M that may be applicable to the study of hormonal degradation in the physiological range (Neidle and Marks, unpublished findings). For elucidation of absolute specificity, pure enzymes are essential, since contamination by peptidases and other enzymes can lead to secondary cleavage of peptide fragments. The problem can be overcome partly by the use of inhibitors or analogues to prevent adventitious reactions. One approach recently used with considerable success was to measure the release of amino acids or small peptides with time, using an autoanalyzer for detection, on the assumption that some points of cleavage are rate-limiting. Thus, for LH–RH, which has N- and C-terminal (blocked) groups, there was a preferential release of internal amino acids indicative of a neutral endopeptidase (see Section 3.1).

Studies involving detection of end-products by bioassay or RIA must be careful to guard against cross-reactivity of breakdown products or cleavage of the labeled antigen by peptidases in the assay mixture (see Chapter 3). If studies are broadened to include the role of enzymes for conversion, other criteria must be applied for identification of precursor and products, as outlined by Tager et al. (1975) and Hew and Yip (1976).

There has been interest in the direct correlation of breakdown with hormonal function (see Section 3.1.3). Studies involving isolated intact cells can provide information on this aspect, although these cells do not encompass possible breakdown of peptides in body fluids enroute to the target organ or the role of the blood–brain and other barriers. The following are among the critical questions in studies of hormonal interactions: Are some hormones or peptide factors protected during transport? What is the stability of the receptor–hormone complex? Does breakdown of the peptide when attached to the membrane receptor terminate the hormonal signal? What explanation is there for the longer in vivo action of peptides, as compared with their far shorter biological half-lives? In general, the rapid inactivation of peptides in both the gut and tissues has hindered their use in routine clinical studies. As will be discussed, an understanding of mechanisms involved in breakdown may be of assistance in the preparation of analogues with more potent and longer duration of action.

The discovery of new peptides in the CNS continues unabated, but until their structures are definitively established, studies on degradation can be misleading. For example, early studies on the effects of proteolytic enzymes on thyrotropin-releasing factor (TRF) were completely negative, leading some investigators to question its peptide nature. Among peptides that may have considerable importance in future studies are those involved in spermatogenesis (inhibin) and sleep, vasoactive gut peptides, and peptides affecting coronary circulation (Galoyan, 1973; Monnier *et al.*, 1975; Short, 1975; Pappenheimer *et al.*, 1975; Said and Rosenberg, 1976). Several peptides formerly associated only with gut cross-react immunologically with CNS components (gastrin and vasoactive gut peptide), and these data, coupled with the wide distribution of "CNS peptides" in peripheral tissues (Substance P, somatostatin), pose some fascinating questions regarding the functional roles of these peptides in different organs.

2. Conversion of Prohormones

Several examples exist in peripheral tissues for the formation of active peptides by breakdown of macromolecular precursor materials (synthesized on ribosomes). These examples provide ample precedent that such mechanisms constitute a major pathway for formation of several peptides in brain, but evidence to date is still only fragmentary. Macromolecules do not penetrate brain sufficiently well to account for their CNS actions, and formation within brain would be in keeping with many of its autonomous functions. In other cases, peripheral conversion may lead to formation of peptides with direct or indirect actions on the CNS, and this aspect represents an important area for investigation. With respect to formation by breakdown of prohormones, a number of critical questions remain unanswered: (1) the relative importance of peripheral vs. conversion within the organ in question; (2) identification of the precursor material and characterization of the degradative enzymes; (3) localization of the precursor material and questions related to its transport; and (4) the importance of the subsequent metabolic conversion of split peptides (formation of pyroGlu, acetylation, formation of C-terminal amides).

2.1. Corticotropin and Lipotropin

This group includes α- and β-melanocyte-stimulating hormone (MSH), β- (91 amino acids) and γ-lipotropin (58 amino acids) (LPH), and ACTH; all contain a common heptapeptide sequence, Met–Glu–His–Phe–Arg–Trp–Gly (Fig. 1). ACTH, γ-LPH are formed in the pars distalis; α-MSH and β-MSH are present in pars intermedia and may be formed at this site.

Precursors of corticotropin (big ACTH) that can release immunoassayable

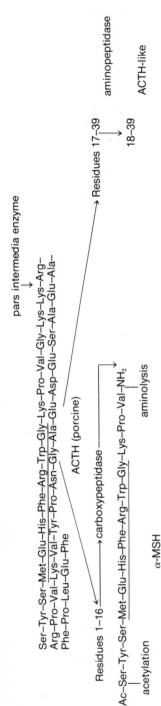

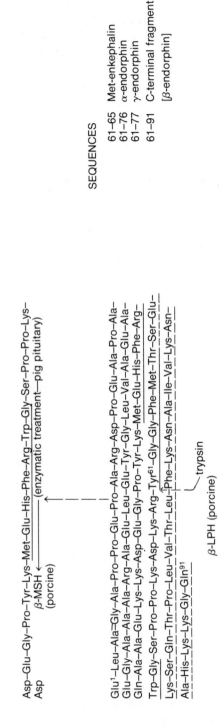

pars intermedia enzyme
↓

Ser–Tyr–Ser–Met–Glu–His–Phe–Arg–Trp–Gly–Lys–Pro–Val–Gly–Lys–Lys–Arg–
Arg–Pro–Val–Lys–Val–Tyr–Pro–Asn–Gly–Ala–Glu–Asp–Glu–Ser–Ala–Glu–Ala–
Phe–Pro–Leu–Glu–Phe

ACTH (porcine)

⟶ Residues 17–39   aminopeptidase

18–39 ⟶   ACTH-like

Residues 1–16 ⟶ carboxypeptidase

Ac–Ser–Tyr–Ser–Met–Glu–His–Phe–Arg–Trp–Gly–Lys–Pro–Val–NH₂

aminolysis

acetylation

α-MSH

Asp–Glu–Gly–Pro–Tyr–Lys–Met–Glu–His–Phe–Arg–Trp–Gly–Ser–Pro–Pro–Lys–
Asp

β-MSH ⟵ (enzymatic treatment—pig pituitary)
(porcine)

Glu^1–Leu–Ala=Gly–Ala–Pro–Ala–Pro–Glu–Pro–Ala–Arg–Asp–Pro–Glu–Ala–Pro–Ala–
Gly–Gly–Ala–Ala–Arg–Ala–Glu–Leu–Glu–Tyr–Gly–Leu–Val–Ala–Glu–Ala–
Gln–Ala–Ala–Glu–Lys–Lys–Asp–Glu–Gly–Pro–Tyr–Lys–Met–Glu–His–Phe–Arg–
Trp–Gly–Ser–Pro–Pro–Lys–Asp–Lys–Arg–Tyr^{61}–Gly–Gly–Phe–Met–Thr–Ser–Glu–
Lys–Ser–Gln–Thr–Pro–Leu–Val–Thr–Leu–Phe–Lys–Asn–Ala–Ile–Val–Lys–Asn–
Ala–His–Lys–Lys–Gly–Gln^{91}

trypsin

β-LPH (porcine)

SEQUENCES

61–65	Met-enkephalin
61–76	α-endorphin
61–77	γ-endorphin
61–91	C-terminal fragment [β-endorphin]

FIG. 1. Lipotropin and corticotropin conversion. The sequence is based on sequences supplied by Croft (1973) and Chretien (1973). ACTH can be cleaved by a trypticlike enzyme in pars intermedia to release the fragments, which can be converted by secondary metabolic reaction to α-MSH and an ACTH-like fragment. α-MSH and the 4–10 sequence (underlined in all peptides) can be degraded by brain enzymes to release all constituent amino acids (Marks and Kastin, unpublished findings). β-MSH can be derived from enzymatic treatment of pig pituitary granules (Bradbury et al., 1976a). β-Lipotropin (porcine) contains sequences identical to β-MSH (41–58) and to Met-enkephalin (61–65), α-endorphin, (61–76), and the C-fragment (61–91), several of which have morphinelike properties in vitro (Bradbury et al., 1976b; Guillemin et al., 1976; Li and Chung, 1976; Graf et al., 1976; Lazarus et al., 1976). The structure of porcine β-LPH is shown; camel β-LPH differs in its β-endorphin fragment at residue 83 in having Ile instead of Val. Other differences between porcine and camel β-LPH are noted by Li and Chung (1976).

ACTH or can be converted by controlled tryptic digestion are present in lung tumors (Yalow, 1976). In pig, Bradbury *et al.* (1976*a*) isolated and sequenced a 38-residue peptide believed to be the *N*-terminal of an ACTH prohormone and β-MSH (an 18 amino acid sequence in pig and ox), from secretory particles of pig pituitary following enzymatic treatment. It must be noted that in man, β-MSH has 22 residues of different sequence. ACTH-producing tumors contain β-MSH- and β-LPH-like peptides (Hirata *et al.*, 1976). The pituitary is the source of several peptides, Met– and Leu–enkephalins, and α- and β-endorphins, which have sequences identical to that present in porcine and camel B-LPH, and which have morphinelike properties in a number of different test systems (see Fig. 1) (Hughes *et al.*, 1975; Bradbury *et al.*, 1976*b*; Guillemin *et al.*, 1976; Li and Chung, 1976; Lazarus *et al.*, 1976). Treatment of β-LPH directly with trypsin can release the morphinelike *C*-fragment (residues 61–91). In a study on the binding of fragments derived from lipotropin to brain opiate receptors, Bradbury *et al.* (1976*b*) found the highest activity for *C*-fragments 61–91 and 61–89; lower activities for 61–87, 61–69, and 61–65 (methionine–enkephalin); and virtually none for β- and γ-LPH, an *N*-terminal fragment 1–38, and β-MSH (Fig. 1). Li and Chung (1976) also isolated a unitriakotapeptide from camel β-LPH differing only slightly in structure from porcine *C*-terminal fragments (Fig. 1) that had opiate activity *in vitro* and on intracerebral injection (Cox *et al.*, 1976). They concluded that this fragment was a degradation product of camel β-LPH. Graf *et al.* (1976) isolated a tryptic fragment of porcine β-LPH (residues 61–69) and a second peptide by endogenous hydrolysis using a pituitary homogenate (sequence 1–69), indicating hydrolysis of the $-Lys^{69}-Ser^{70}$ bond. Purified 1–69 peptide was not an agonist, but tryptic fragments at high concentrations were active on guinea pig gut; the activity of the 61–69 fragment was lower than that of Met–enkephalin. Graf and Kenessey (1976) also reported cleavage of bond 77–78 by an endopeptidase active at pH 6.5 followed by removal of Phe to yield α-endorphin. In a study of synthetic hexa- and heptapeptides corresponding to sequences 37–43 and 57–62 of β-LPH Bradbury *et al.* (1976*c*) inferred that the pituitary contains tryptic-like enzymes capable of releasing the *C*-fragment by cleavage of the $Arg^{60}-Tyr^{61}$ bond but that cleavage of the paired lysine residues close to the *N*-terminal of the β-MSH sequence was slow and possibly attributable to a different endopeptidase.

It might be noted that peptides structurally unrelated to enkephalins and endorphins have (low) opiate activity *in vitro*, including Leu–Trp–Met–Arg–Phe–Ala, a basic peptide, ACTH fragments, and casein hydrolysates (see Wajda *et al.*, 1976). The same group also showed morphinelike factors present in liver and kidney. There are reports that the pituitary and the pars intermedia of animals contain trypticlike enzymes, along with a carboxypeptidase capable of cleaving ACTH to form α-MSH and an ACTH-like peptide (residues 18–39 of ACTH) (see Fig. 1) (Scott *et al.*, 1973; Glas and Astrap, 1970). Pars intermedia is vestigial in adult man, but Smith *et al.* (1976) have shown that rudimentary pars intermedia in human fetus does contain peptides resembling α-MSH and ACTH-like peptide (CLIP), which may be the dominant hormones of fetal life, replaced by ACTH only before parturition. Conversion of split fragments of ACTH would necessitate a series of secondary metabolic reactions such as acetylation and aminolysis to yield N-terminal acetyl and C-terminal amide groups. β-LPH itself is claimed to be resistant to degradation during isolation; γ-LPH is present in pituitary, when mild conditions of extractions are employed (Chretien, 1973).

2.2. Pituitary–Hypothalamic Hormones

Several factors—TRF, LH–RH and prolactin release–inhibiting factor (PIF)—are claimed to be synthesized *de novo* from amino acids by a nonribosomal (soluble) system (Reichlin and Mitnick, 1973). As in the case of α-MSH, synthesis necessitates supplementary reactions such as aminolysis (C-terminal amide) and formation of N-terminal pyroGlu (although pyroGlu formation could occur nonenzymatically by cyclization of Gln·NH_2). Synthesis of factors has been demonstrated *in vitro* using intact cell preparations that presumably involve the classic pathways of protein synthesis (McElvey *et al.*, 1975). There is currently no good evidence for the formation by breakdown of prohormones, except possibly for melanocyte-inhibiting factor (MIF) from oxytocin (Walter *et al.*, 1973) and for LH–RH by tryptic digestion of lyophilized sheep hypothalamus, as described in a brief report by Miller *et al.* (1975). There is some evidence in pituitary for possible macromolecular percursors for growth hormone (GH) (Stachura and Frohman, 1975) and for prolactin (Friesen *et al.*, 1970), based on immunoprecipitation and other methods.

An important question in studies on conversion is the nature or minimal size of the active peptide sequence, since many native hormones can be split further to yield shorter active fragments (Fujita *et al.*, 1975). Human growth hormone (HGH), for example, can be digested with plasmin or other proteinases to yield an active hormone fragment (Reagan *et al.*, 1975; Lewis *et al.*, 1975). There is a considerable species difference, since fragments of bGH are inactive in man.

2.3. The Angiotensin–Renin System

Formation of octa-, deca-, and possibly heptapeptides (angiotensins I, II, and III) is a good illustration of the potential significance of proteolytic enzymes in peptide function (Fig. 2). There is a large body of literature on this topic (see Marks and Lajtha, 1971; Severs and Severs, 1973; Edros, 1975; Davis and Freeman, 1976), and this review will be limited only to evidence for the presence of the "angiotensin–renin system" in nerve tissue (Fig. 2).

2.3.1. Renin

Renin (E.C.3.4.99.19) acts on protein precursors present in plasma (angiotensinogens) to yield the decapeptide (angiotensin I). The biochemical and functional significance of brain renin is unclear, although renin substrates are present in high concentration in the CSF, and reninlike activity is reported present in extracts of rat, dog, and human brain, and is present in neurosecretory regions such as the pituitary and pineal glands (Fisher-Ferraro *et al.*, 1971; Ganten *et al.*, 1971a,b, 1975; Poth *et al.*, 1975; Daul *et al.*, 1975; Reid and Ramsey, 1975). Renin does not appear to cross the blood–CSF barriers, but intraventricular administration can stimulate dipsogenic and blood pressure responses (Reid and Ramsey, 1975). Brain areas mediating the central actions of angiotensin are the subfornical organ, area postrema, and subnucleus medalis (see Severs and Severs, 1973; Severs and Summy-Long, 1976), several of which lie outside the major anatomical barriers associated with the blood–CSF systems (Simpson and Routtenberg, 1973). This may partly account for actions of systemic angiotensins that fail to penetrate the CNS. Reninlike activity is reported present in subcellular and soluble fractions of human brain, with the highest concentration in mitochondrial particulates; activity has been found in all regions examined, with the highest concentrations in pituitary and cerebellum, and lower concentrations in hypothalamus, cerebral cortex, and brainstem areas (Daul *et al.*, 1975; Day and Reid, 1976). In recent attempts to characterize brain renin, Day and Reid (1976) found that the properties of a 100-fold purified enzyme (assayed with a plasma or CSF substrate or both) were identical in most respects to brain cathepsin D. These data lend doubt to the physiological significance of brain renin, since no activity was detectable with purified preparations above pH 6.0. Reninlike activity in crude extracts at pH 4–5 might be attributable to lysosomal endopeptidases with pepsinlike specificity, as implied by studies on angiotensin formation by action of pepsin and ox plasma (Fernandez *et al.*, 1965). The purified "reninlike" enzyme of dog brain showed a regional distribution identical to that of cathepsin D and an angiotensinase (measured with angiotensin I substrate at pH 5.5). The pH optimum with renin substrate was 4.5, which is higher than cathepsin D, but pH optima of proteolytic enzymes are known to vary considerably with the type of substrate employed. Brain reninlike activity is inhibited by pepstatin, and therein shows similarities to plasma renin and to cathepsin D (Day and Reid, 1976; Marks *et al.*, 1973b).

2.3.2. Converting Enzyme

Converting enzyme (E.C.3.4.15.1) is a dipeptide hydrolase that converts proangiotensin (angiotensin I) into the active octapeptide (angiotensin II) by removal of C-terminal His–Leu (Fig. 2). Although it represents an excellent example of a proteolytic enzyme involved in regulating active peptide formation, studies on purified enzyme from nerve tissues are not available. Angiotensins are present in brain, with concentrations of the decapeptide precursor (angiotensin I) higher than the octapeptide (Ganten *et al.*, 1971b; Fisher-Ferraro *et al.*, 1971). This finding points to a possible regulatory role for the brain dipeptide hydrolase. In studies on rat brain homogenate, Cushman and Cheung (1971), using Hip–His–Leu as a substrate, reported specific activity in brain lower than that of lung, but higher than that of spleen, kidney, and liver. Studies with synthetic substrate show activity in all regions examined, and especially high in caudate nucleus, pituitary,

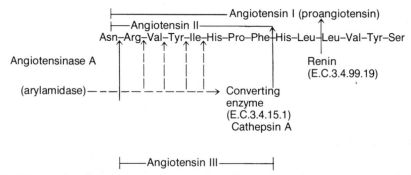

FIG. 2. The renin–angiotensin system, showing metabolic conversions and inactivation of angiotensins in brain. Reninlike substrate is present in brain and CSF, but evidence for reninlike enzymes is less certain (see the text). Arylamidase specific for acidic arylamides (arylamidase A) would lead to formation of angiotensin III. Arylamidases and cathepsin A–inactivating angiotensin have been purified from rat brain. Other potential inactivating enzymes present in brain are cathepsin C and carboxypeptidases active at physiological pH.

striatum, and cerebellum (Yang and Neff, 1972; Poth *et al.*, 1975); in Huntington's chorea, levels are reported to decrease 60% in globus pallidus (Arregu *et al.*, 1976). The enzyme is present in all subcellular fractions, highest in crude mitochondria but also associated with soluble fractions (Yang and Neff, 1972). Properties of the crude enzyme were similar to that purified from lung, and showed a dependence on Cl^-; among other substrates hydrolyzed were Z–Gly–Gly–Val and Hip–Gly–Gly, with release of C-terminal dipeptides. The enzyme was inhibited by EDTA, *o*-phenanthroline, the bradykinin-potentiating factor SQ 20881 (pyroGlu–Trp–Pro–Arg–Pro–Gly–Ile–Pro–Pro), and an angiotensin analogue (Phe^4–Tyr^8–angiotensin I).

Recently, several groups have postulated the existence of des-Asp–Angiotensin II (angiotensin III) formed by the action of an aminopeptidase (Chiu and Peach, 1974; Chiu *et al.*, 1976) (Fig. 2). Since brain contains aminopeptidase-removing N-terminal acidic groups, the role of the heptapeptide as an agonist may be of significance (Marks, 1970; Goodfriend and Peach, 1975).

2.4. The Kinin System

A series of related hypotensive peptides can be isolated by treatment of plasma α-globulin (kininogens) with proteolytic enzymes isolated from blood or different tissues (Prado, 1970; Rocha e Silva, 1975). These peptides consist of kinin-9 and larger peptides (pachykinins), kinin-10 (kallidin), kinin-11 (Met–Lys–bradykinin), and GAML–bradykinin (Gly–Arg–Met–Lys–bradykinin) (Fig. 3). Kinin-9 is formed by plasma kallikrein, kinin-10 by glandular enzyme of pancreas and salivary gland, kinin-11 by trypsin or plasmin, and GAML–bradykinins by proteinases with specificity similar to pepsin (Prado, 1970). The pathophysiological implications of kinin research are of considerable interest and are summarized in reviews (see Erdos, 1971; Rocha e Silva, 1975). Very few of these studies have

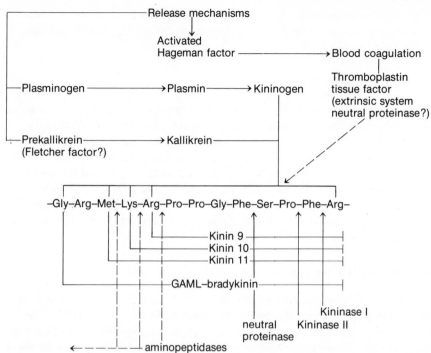

FIG. 3. The kinin system, showing enzymes involved in formation and inactivation of kinins. The scheme is based on that of Rocha e Silva (1975) and Marks and Lajtha (1971). The peptide portion of a kininogen that might generate kinins 9–11 or larger pachykinins is indicated. This portion is acted on by kallikrein (kininogenin E. C. 4.3.21.8) to yield kinins that in turn can be interconverted by aminopeptidases or arylamidases. Among enzymes inactivating bradykinin in brain are a neutral proteinase (cathepsin M) and amino- and carboxypeptidases. Kininases I and II as such have not been studied in detail in nerve tissue. The Fletcher factor is important for activation of kallikrein and the thromboplastin factor.

been concerned with brain kininogens and converting enzymes (Marks and Lajtha, 1971).

2.4.1. Kallikrein

Only indirect evidence exists for the presence of kininogens and kinin-releasing enzymes in brain (Werle and Vogel, 1961; Inouye et al., 1961; Hori, 1968; Werle and Zach, 1970; Webster, 1970). Based on inhibition studies with trasylol, Hori (1968) reported levels of kallikrein (E.C.3.4.21.8) in brain higher than in spleen and liver, but lower than in other organs. It has been established for other tissues that kallikrein is activated from a zymogen form by a diverse series of interactions involving the Hageman and Fletcher factors, proteolytic enzymes, alterations in pH, effects of heat, and blood coagulation (see Marks and Lajtha, 1971; Rocha e Silva 1975). There are also indications of a control mechanism exerted by specific inhibitors that appear to increase in shock, pregnancy, and cerebral apoplexy; in pathological conditions, this inhibition may represent a delicate defense mechanism against some inflammatory disorders (Marks and Lajtha, 1971; Yasuo et al., 1975).

2.4.2. Aminopeptidases

Interconversion of kinins by peptide hydrolases represents a further example of peptide formation by specific cleavage of polypeptides (Fig. 3). Camargo et al. (1972) purified a brain aminopeptidase active at pH 7.6 capable of forming kinin-9 from kinin-10 and kinin-11. This enzyme was identical in its properties to brain arylamidases that are known to be inhibited by puromycin and that do not appear to attack kinin-9 itself (Marks, 1970; Marks and Pirotta, 1971) (see Section 3.10). It is known that conversion of kinins can be mediated by plasma aminopeptidases with properties that are also akin to those of arylamidases (Hopsu-Harvu et al., 1966; Guimaraes et al., 1973).

2.5. Non-CNS Peptides

Insulin is formed from proinsulin in β-cells of pancreas, with the Golgi apparatus playing a role in progressive conversion (see Steiner, 1974, 1975). The enzymes most studied are an endopeptidase with trypticlike specificity and a carboxypeptidase β–like enzyme. Studies with rat proinsulin indicate a role for a chymotrypticlike enzyme acting within the C-peptide, but its participation is not considered essential for overall conversion (Fig. 4). Studies on proinsulin transformation provide a prototype for defining the necessary criteria needed to isolate and identify precursor materials (Tager et al., 1975). It has been postulated that a change in the conversion process might account for pancreatic dysfunction in diabetes (Kemmler, 1975). The half-life of proinsulin and the C-peptide based in immunological studies is relatively short, in the range of 8–10 min (Oyama et al., 1975). Less is known about glucagon formation, although there is evidence for the presence in tissues of macromolecular precursors in fish (Noe and Bauer, 1973) and in animal pancreas (O'Conner and Lazarus, 1976).

Another polypeptide of interest is gastrin that can be formed from "big" to "very big" precursors (17–34 residues or larger); the C-terminal pentapeptide retains biological activity (Gregory and Tray, 1972; see also Chapter 3). It is of interest that peptides cross-reacting with gastrin antisera have been reported as present in the CNS (Vanderhaeghen et al., 1975). Parathyroid hormone (PTH) is also formed from a precursor (pro-PTH) (Habener and Potts, 1975). The precursor is a single peptide chain with an N-terminal basic hexapeptide (Lys–Ser–Val–Lys–Lys–Arg) and possibly some additional C-terminal sequences in human and bovine prohormone (Chu et al., 1975b). Activation involves participation by endopeptidases, which have not been characterized.

Synthetic N-terminal sequences of PTH are more stable, and this stability may account for their longer retention in vivo (Fujita et al., 1975). PTH is degraded by plasma membranes of renal cortex and other organs (Chu et al., 1975a).

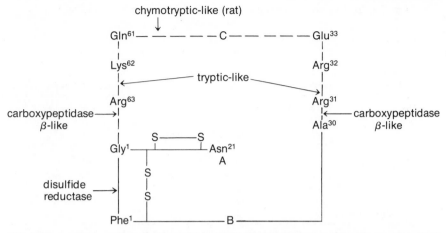

FIG. 4. Conversion and inactivation of insulin. Points of cleavage of porcine proinsulin A chain 1–21, B chain 1–30, and C chain 31–63 (dashed lines) by pancreatic enzymes involved in formation of insulin (solid lines). The chymotryptic cleavage is not regarded as essential for overall conversion, but would result in a shortened *C*-peptide. Insulin can be split into A and B chains by the action of glutathione disulfide reductases. The oxidized linear B chain is a substrate for brain proteolytic enzymes (Lajtha and Marks, 1966, 1969).

3. Inactivation of Active Peptides

3.1. Luteinizing Hormone–Releasing Hormone

This decapeptide has been extensively studied in terms of degradation by brain and in neurosecretory extracts (Marks and Stern, 1974*a,b;* Koch *et al.,* 1974; Kochman *et al.,* 1975; Griffiths *et al.,* 1975*a,b*). The presence of blocked *N*- and *C*-terminal groups implies action of novel enzymes, since these groups cannot be removed by conventional peptide hydrolases (Table 4). That this is so has been confirmed by studies on the timed release of amino acid components using brain extracts that indicate two basic mechanisms: (1) internal cleavage by a neutral endopeptidase, followed by the secondary action of peptide hydrolases; and (2) a slower removal of *C*-terminal glycinamide by a second enzyme acting on the *C*-terminus (Marks and Stern, 1974*a,b*). Evidence for endopeptidase action was obtained by Koch *et al.* (1974), with isolation of the intermediate pyro-Glu^1---Gly^6 pointing to –Gly^6–Leu^7 as one of the primary cleavage points. The same group reported rates of breakdown by hypothalamic homogenates using RIA that were comparable to those obtained by biochemical procedures (see Table 3) (Marks and Stern, 1974*b*). Information on the internal and other sites of cleavage was provided by the use of analogues. With D-Ala^6–LH–RH, there is a considerable reduction in release of residues adjacent to the substituted group without change in Gly·NH$_2$ release. That this can be attributed to a second (*C*-terminal-cleaving) enzyme was confirmed by the use of D-Ala^6–LH–RH–ethylamide, which blocked Gly·NH$_2$ release (Table 4). The reduced rate of degradation for both analogues was accompanied by a significant increase in the biological activity *in*

TABLE 4. Breakdown of Luteinizing Hormone–Releasing Hormone by Rat Brain Homogenate [a]

Residue released	Position	Breakdown (percent amino acid released)		
		LH–RH[b]	D-Ala[6]	D-Ala[6] ethylamide
His	2	20	5	5
Ser	4	30	trace	trace
Tyr	5	45	trace	trace
Gly	6	46	(−) [c]	(−) [c]
Leu	7	50	0	0
Arg	8	50	18	trace
Pro	9	30	15	trace
Gly·NH$_2$	10	15	15	(−) [c]
Ovulatory activity		1	7	80

[a] Breakdown was based on the release of amino acids following incubation for 4 hr at 37°C, as detected by an autoanalyzer (Marks and Stern, 1974a,b). Trp and pyroGlu were not measured. Biological activity is based on the data of Fujino et al. (1974).
[b] LH–RH: pyroGlu–His–Trp–Ser–Tyr–Gly–Leu–Arg–Pro–Gly·NH$_2$.
[c] D-Ala not detected as a breakdown product.

vivo (Fujino et al., 1974). In general, analogues substituted in position 6 with D-amino acids (Ala, Glu, Leu, Phe, Trp) have biological activities in vivo 2–20 times that of LH-RH as compared with 80-fold for D-Ala[6]–LH–RH–C$_2$H$_5$NH$_2$, reinforcing the view that there may be a correlation to rates of biodegradation (Fujino et al., 1974; Coy et al., 1975, 1976; Ling et al., 1976). The ideal analogue would be one that retains its affinity for the relevant receptor and has at the same time a decreased rate of degradation. Also of particular interest in relation to breakdown are analogues that act as antagonists; two of these analogues contain D-substituted amino acids, D-Phe^2–D-Leu^6–, D-Phe^2, D-Ala^6, and D-Phe^2–Phe^3–D-Phe^6–LH–RH (Cruz et al., 1976).

3.1.1. Distribution of Inactivating Enzyme

Enzymes inactivating LH–RH as measured by RIA and biochemical procedures are ubiquitously distributed in all brain areas studied to date, including cerebral cortex, cerebellum, different anatomical areas of the hypothalamus, thalamus, and pineal gland (Griffiths et al., 1973, 1974; Koch et al., 1974; Marks and Stern, 1974b; Kochman et al., 1975; Marks, 1976; see also Table 3). Over 90% was associated with the cytosol and less than 5% with synaptosomes and mitochondrial and nuclear fractions (Benuck and Marks, unpublished findings). The low levels in particulates may be significant, since LH–RH is reported to be present in specific organelles and axon terminals (Taber and Karavolas, 1975; Zimmerman, 1976).

Studies on inactivation by other tissues are not available, but inactivating enzyme is present in serum, higher in rat as compared with man (Table 5), and the presence of this enzyme may be an important factor in its short half-life (Redding *et al.*, 1973; Marks, 1976; Benuck and Marks, 1976). Studies with analogues using serum indicate a similar mechanism for inactivation as proposed for brain; breakdown was still apparent with LH–RH-ethylamide (action of an endopeptidase followed by peptide hydrolases), but virtually undetectable with D-Ala^6–LH–RH-ethylamide (blockage of endopeptidase action) (Table 5). Redding *et al.* (1973), using tritium-labeled LH–RH given intravenously, found that the major excretion products in urine were pyroGlu and pyroGlu–His.

3.1.2. Enzyme Purification

Attempts to purify enzyme have been handicapped by enzyme lability and absence of a rapid (biochemical) method for detecting LH–RH breakdown. Enzyme activity monitored by a simple endopeptidase assay (the use of hemoglobin, casein, and myelin basic protein) and retested later by radioimmunoassay was purified 20–30 fold from rat brain homogenates (Benuck and Marks, unpublished findings). The enzyme was stable only 1–2 weeks when frozen, and showed a broad pH optimum of 5.5–8, depending on the substrate used. Further characterization is required to ascertain whether this is a specific LH–RH-degrading enzyme and can be differentiated from brain neutral proteinase and other hormonal degrading systems (see Table 1). Inhibition of LH–RH-inactivating enzyme is of clinical inter-

TABLE 5. Degradation of Neuropeptides by Rat and Human Serum [a]

		Breakdown in 4 hr (%)	
Substrate	End-product measured	Rat	Man
MIF	Leu	100	trace
	Gly·NH$_2$	100	trace
LH–RH	Leu	52	14
	Gly·NH$_2$	68	0
des-Gly–^{10}LH–RH–C$_2$H$_5$·NH$_2$	Leu	24	34
D-Ala–^{6}LH–RH–C$_2$H$_5$·NH$_2$	Leu	0(6)	0(12)
H$_2$-Somatostatin	Lys	63	58
	Phe	50	31
des-Ala–Gly–	Lys	25	14
N-Ac-Cys–somatostatin	Phe		
D-Trp^8–somatostatin	Lys	trace	0
	Phe	trace	0
TRF	His	25(80)	5(30)
3-Me–His^2–TRF	3-Me–His	90	80
Met–Enkephalin amide	Tyr	30	25
α-endorphin	Tyr	20	5
α-MSH	Tyr	23	8
Oxytocin	Gly·NH$_2$	56	25

[a] The incubation mixture consisted of 50 nmol peptide in 0.2 ml serum incubated for 4 hr at 37°C, and the end-products were determined by autoanalysis (Benuck and Marks, 1976). In selected cases, the results for prolonged incubation (24 hr) are indicated in parentheses. Controls were run in parallel for equal periods of time and background subtracted to obtain net breakdown. Results are the means of 3–6 determinations and agree with 5%.

est, but studies using specific inhibitors of trypsin are inconclusive (Koch et al., 1974; Kochman et al., 1975). In our own work, partially purified enzyme assayed with proteins was inhibited by DFP, and PMSF. In most of its properties, it resembled a neutral proteinase (Cathepsin M, Table 1) (Marks and Lajtha, 1965).

3.1.3. Hormonal Interactions

There are several studies suggesting steroidal interactions with LH–RH-inactivating enzymes, carrying the implication that a "feedback" mechanism might affect gonadotropic secretion. Enzymes inactivating LH–RH were higher in soluble hypothalmic extracts of male as compared with female rats, and varied in concentration in different anatomical regions, with the highest levels reported for hypophysiotropic and median eminence regions (Griffiths et al., 1974; Griffiths and Hooper, 1974). Activity fell on gonadectomy, but could be reversed by administration of estradiol to females and testosterone to males (Griffiths et al., 1975a). These data are consistent with those of Shin et al. (1974), who found a paradoxical rise in LH–RH in hypothalami of castrated rats that was reversible by steroids.

Kuhl and Taubert (1975a,b) have observed an altered level of a "cystine arylamidase" on treatment with LH–RH in vivo, and inhibition of this enzyme, on addition of other peptidyl amides such as LH–RH itself, TRH, oxytocin, and lysine vasopressin, in crude hypothalamic supernatant. A similar competitive inhibition was noted by Griffiths and Hooper (1974) between oxytocin and LH–RH for a crude hypothalamic enzyme in vitro, indicating that this might represent a regulatory mechanism; characterization of the inactivating enzymes involved, however, is essential to clarify a possible role in gonadotropic function. Caution is required in any interpretation of hormonal effects in peptide metabolism, owing to the large number of secondary effects involved (Marks and Lajtha, 1971). Gonadectomy and hormone treatment can affect a large variety of cellular systems involved in electrolyte balance, glucose and biogenic amine metabolism, and a second messenger (cAMP).

3.2. Somatostatin

The cyclic tetradecapeptide present in the hypothalamus inhibits the release of GH from the pituitary, and also of glucagon and insulin from the pancreas of fasted animals. Its very short biological half-life, however, limits its clinical usefulness. In studies on inactivation in serum and brain, we have used the same strategy as adopted for breakdown of LH–RH, namely, the timed appearance of amino acids. Results showed a preferential release of internal amino acids with a rate-limiting cleavage at the Trp^8–Lys^9 linkage and the bonds adjacent to the Phe residues (Fig. 5) (Marks and Stern, 1975; Marks et al., 1976b). The slower release of Ala and Cys largely excludes the action of an aminopeptidase and a carboxypeptidase as the primary mechanism for inactivation; also, it is known that

1. (somatostatin)

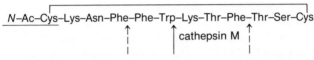

2. D-Trp⁸–analogue

Ala–Gly–Cys–Lys–Asn–Phe–Phe–D-TRP–Lys–Thr–Phe–Thr–Ser–Cys

aminopeptidase

3. Des Ala–Gly–(N–Ac–Cys³) analogue

N–Ac–Cys–Lys–Asn–Phe–Phe–Trp–Lys–Thr–Phe–Thr–Ser–Cys

cathepsin M

FIG. 5. Sites of cleavage of somatostatin by brain enzymes and serum. Established (solid arrows) and postulated (dashed arrows) points of cleavage by rat brain enzymes. See Fig. 6 for other details. From Marks *et al.* (1976*b*).

the first two residues, Ala and Gly, are not essential for biological activity (Brown *et al.*, 1975). Further evidence that internal cleavage is responsible for inactivation was provided by studies with analogues: incubation of the des-Ala–Gly (Ac-Cys³) somatostatin led to the release of internal amino acids, showing that N-blocked somatostatin was still cleaved. Additional confirmation for an endopeptidase was provided by the failure to detect Lys using D-Trp⁸–somatostatin (Marks *et al.*, 1976*b*) (Fig. 5). Rivier *et al.* (1975) showed that the D-Trp⁸ analogue was 8 times more potent than the native analogue, but was not longer-acting *in vivo*. The latter result is understandable, since the –Trp⁸–Lys⁹ bond is only one of three potential bonds cleaved by cathepsin M. In a systematic study on substitution of each individual residue with Ala, no major increase in activity was found, except in the case of those with L-Ala at positions 2 and 8 (Rivier *et al.*, 1975). These results can be attributed to changes in receptor occupancy rather than biodegradation, since L-Ala amino acids probably do not affect breakdown to the same extent as the D-isomer. In studies on the proposed tertiary structure of somatostatin, it was recently shown that there is a β-structure stabilized by stacking of the three aromatic rings, Phe⁷, Trp⁸, and Phe¹¹. It is of interest that these three residues appear to be the ones most involved in the action of the central endopeptidase and may partially account for the increased potency of the D-Trp⁸ analogue. The enzyme inactivating somatostatin is present largely in brain supernatant fractions. It can be purified about 10-fold using the identical procedures for purification of other hormonal inactivating enzymes, and further studies are required to differentiate this enzyme from those inactivating LH–RH, Substance P, and bradykinin.

Studies on inactivation in serum show higher rates of degradation in rat as compared with man, with a significant reduction using the D-Trp[8] analogue (see Table 5) (Benuck and Marks, unpublished observations).

3.3. Thyrotropin-Releasing Factor

TRF was the first hypothalamic factor to be characterized, and although it is only a tripeptide, the mechanisms for its inactivation are unclear. Several groups have shown destruction by extracts of brain or neurosecretory regions, with activity present in soluble and particulate fractions (Mudge and Fellows, 1973; Bassiri and Utiger, 1974; Marks and Stern, 1974a; McKelvey et al., 1975; Griffiths et al., 1975c, 1976). In a recent report, Grim-Jorgensen et al. (1976) observed a considerably lower degradation in extracts of a clonal cell line derived from a CNS rat tumor. The level of degradative enzymes in serum may be dependent on thyroid function (Bauer, 1976).

In an attempt to clarify mechanisms of inactivation, we have studied the release of different amino acid components. Incubation with rat brain extracts led to the release of His, pointing to cleavage of the pyroGlu–His and His–Pro·NH$_2$ bonds (Table 2 and Fig. 6) Marks and Stern, 1974a,b). Dansylation on polyamide plates revealed that the deamido form and NH$_3$ were products at short incubation periods. Human serum differed from rat brain, since one of the end-products was Pro, indicating that deamidation had occurred. On a comparative basis, the inactivation in rat serum exceeded that of man; also, rates of breakdown were lower

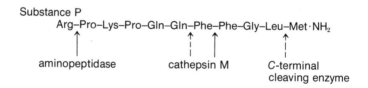

Substance P
Arg–Pro–Lys–Pro–Gln–Gln–Phe–Phe–Gly–Leu–Met·NH$_2$

aminopeptidase   cathepsin M   C-terminal cleaving enzyme

Neurotensin

pyroGlu–Leu–Tyr–Glu–Asn–Lys–Pro–Arg–Arg–Pro–Tyr–Ile–Leu

carboxypeptidase (cathepsin A)

TRF
pyroGlu–His–Pro·NH$_2$

pyroglutamyl peptidase   deamidation
amino–or carboxypeptidases (?)

FIG. 6. Sites of cleavage for Substance P, neurotensin, and TRF by brain enzymes. Established (solid arrows) and postulated (dashed arrows) points of cleavage. The term *cathepsin M* is used for brain neutral proteinases (for literature citations, see the text and Table 1).

for TRF than for LH–RH (see Table 5). The more active analogue 3-Me–His^2–TRF was degraded by brain and serum, with release of 3-Me–His at rates higher than that of the native hormone (Benuck and Marks, 1976).

Among enzymes potentially capable of inactivating TRF is a pyroglutamyl peptidase purified from bovine pituitaries by Mudge and Fellows (1973), and also observed in hamster hypothalamus by Prasad and Peterkofsky (1976). This enzyme cleaved TRF with release of His–Pro·NH$_2$, pointing to cleavage at the pyroGlu–His band. Pituitary enzyme was unstable, and showed similarities to a bacterial enzyme (E.C.3.4.11.8), since it could be stabilized by addition of 2-pyrrolidone. In addition to TRF, it also cleaved pyroGlu–Ala (K_m2.3 × 10^{-4}M), the deamido form of TRF, and its methylester. In hypothalamus, enzyme was soluble and separable in two fractions, one of which displayed a pyroglutamyl peptidase activity and the other a separate amidase; among products found were Pro, the deamido form of TRF, His–Pro, and His–Pro·NH$_2$ (Prasad and Peterkofsky, 1976). Inhibition of inactivating enzymes was reported on addition of high concentrations of other peptidyl amides (LH–RH, Substance P, tetragastrin), indicating that the enzymes involved in TRF breakdown are probably highly specific; tryptic inhibitors from soybean inhibited amidase activity, whereas Tos–Phe–CH$_2$Cl, benzamidine, and COCl$_2$ preferentially inhibited the pyroglutamyl peptidase. We have attempted to purify this enzyme further, using pyroGlu–βNA as the substrate. Activity in brain is very low, indicating that brain enzyme may differ from the bacterial enzyme in terms of its substrate requirements. Experience with LH–RH (N-terminal pyroGlu) shows that His adjacent to pyroGlu and Trp is not rapidly liberated, but pyroGlu–His as a dipeptide is rapidly cleaved (Marks and Stern, 1974b).

Another potential mechanism for inactivation is deamidation, but the data on this aspect are conflicting. Bauer et al. (1973) and Bauer and Lipmann (1976) observed formation of deamido TRF along with Pro·NH$_2$ in extracts of lyophilized hypothalami and serum; McKelvey et al. (1975) found in organ cultures of median eminence the deamido form along with Pro as the principal end-products. Nair et al. (1971) and Redding and Schally (1972) found in blood the deamido form, H–Glu–His–Pro–OH, and His–Pro·NH$_2$ following a systemic injection of labeled TRF. Knigge and Schock (1975) found pyroGlu–His and Pro as the major end-products in rat serum. Prasad and Peterkofsky (1976) have separated an enzyme from hypothalamus but absent in pituitary that is capable of deamidating added TRF. Visser et al. (1975), however, were unable to detect in blood the deamido form as an end-product by RIA using an antiserum directed against this form. Inactivating enzyme is reported to increase in plasma of rats with increase of age (Neary et al. 1976).

Comment on the pyroGlu Moiety. It has not been established whether pyroGlu peptides are of natural occurrence or are formed as an artifact by cyclization of glutamine. It may be of significance that free pyroGlu, along with a pyroglutamyl peptidase, is reported as present in mammalian tissues (Szewezuk and Kwiatkowski, 1970; Mudge and Fellows, 1973). In brain, the level of pyroGlu is about 0.13 μmol/g weight, and is higher compared with all other organs except kidney. Thus, the presence of a pyroGlu in tissues is of interest in relation to the γ-glutamyl cycle and transport phenomena (Orlowski *et al.*, 1974).

3.4. Substance P and Neurotensin

The potent pharmacolgocial properties of Substance P coupled to its wide distribution in nerve tissue point to a role in nerve function (see Chapter 5). Studies with synthetic unidecapeptide show that inactivating enzyme is associated with all regions of rat brain, being highest in pineal gland, followed by posterior pituitary, hypothalamus, pons–medulla, and cortex, and lowest in spinal cord (see Table 3). In subcellular fractions, the mitochondrial–synaptosomal particulates were characterized by low activity as compared with the cytosol (Benuck and Marks, unpublished findings).

An enzyme that cleaves Substance P has been purified partially using simple extraction procedures and column chromatography (Benuck and Marks, 1975). The major peak of activity was coincident with an enzyme that degrades hemoglobin and histones, and exhibited the properties associated with a brain neutral proteinase (cathepsin M). Evidence that internal cleavage preceded release of free amino acids was provided by bioassay utilizing contraction of guinea pig ileum. At longer periods of incubation, we observed free Met, which was shown to be derived from released methioninamide.

Studies on Substance P breakdown using bioassay and RIA must take into account the possible reactivity of breakdown products, which can elicit contraction of guinea pig gut or depolarize frog spinal motoneurons (see Otsuka *et al.*, 1975; Niedrich *et al.*, 1975). This is illustrated by the work of Yajima *et al.* (1973), who found gut contraction by N-blocked Lys^3-, Gln^5-, and Gln^6- – –Met-NH_2^{11} peptides equal to or exceeding Substance P, and by Bergmann *et al.* (1975), who observed good activities for a series of synthetic C-terminal peptides. Niedrich *et al.* (1975), using a large series of synthetic peptides, found that C-terminal pentapeptides of Substance P, eledoisins, and physalaemins had potent contracting ability, equal to that found for acylated hexa- and heptapeptides. Similar observations for C-terminal penta- and heptapeptides were found by Bury and Mushford (1976). These studies emphasize the possible importance of N-terminal peptidases in metabolic studies on the fate of Substance P. An unresolved question is the possible heterogeneity of Substance P observed in older studies (see Zetler, 1970); a partial explanation may be forthcoming in terms of breakdown products of Substance P or its precursors.

Neurotensin is the name given to a vasoactive peptide isolated from bovine hypothalamus that has kininlike properties (see Chapter 5). Studies on its breakdown in rat brain homogenates show a C-terminal release of aminoacids, which suggests that carboxypeptidases are the major route for inactivation. The presence of pyroGlu, as in the case of LH–RH, appears to prevent action by peptide hydrolases. Failure to detect stoichiometric amounts of Leu and Tyr indicates that pyroglutamyl peptidases do not play a role under the conditions used for testing brain extracts *in vitro* (see Fig. 6).

3.5. Oxytocin and Vasopressin

These two major neurohypophyseal hormones are cyclic nonapeptides, but other components of biological interest are present in tissue extracts, depending on the species examined (vasotocins, mesotocin, isotocin, and glumitocin; see Ressler and Popenoe, 1973). Vasotocins are present in the pineal gland, and together with its breakdown products may play a role in regulating gonadotropic function. Among enzymes known to inactive nonapeptides are chymotrypsins A and C, which cleave the Leu^8–Gly^9–NH_2 bond of oxytocin, and trypsin, which cleaves $-Arg^8$–Gly^9–NH_2 of Arg–vasopressin to yield glycinamide (Tuppy, 1968). Intracellular aminopeptidases act only slowly; carboxypeptidases and pepsin have no action; papain acts slowly with release of Asn and Glu. Brain extracts act on cyclic nonapeptides, if results are based on the release of glycinamide and the C-terminal dipeptide (Marks et al., 1973a). Product-precursor studies show that dipeptide formation precedes release of glycinamide, probably indicating secondary cleavage by an aminopeptidase (Fig. 7). The brain enzyme cleaving the $-Pro^7$–Leu^8–bond can be purified, albeit with large loss of activity (Marks and Stern, 1974a). There is no conclusive evidence in brain that glycinamide is released by a separate (chymotrypticlike) enzyme, owing to the presence of contaminating arylamidases in purified preparations.

In studies on "postproline" cleaving enzymes in purified preparations from

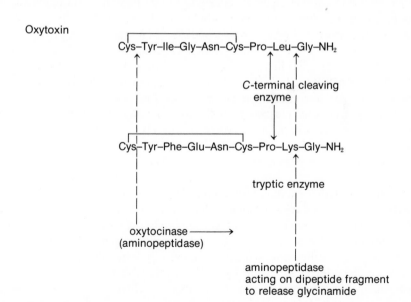

FIG. 7. Oxytoxin and vassopressin inactivation. Inactivation can proceed via two mechanisms: (1) reductive scission of the disulfide ring, followed by aminopeptidase cleavage; (2) removal of C-terminal dipeptides by C-terminal carboxypeptidases with specificity directed toward peptidyl amides. Aminopeptidases in brain have only slow actions as compared with plasma enzymes. The C-terminal-cleaving enzyme has been partially purified (Marks and Stern, 1974a).

kidney, there is some evidence for enzymes with specificity for –Pro–X or –Pro–X–X releasing Gly·NH$_2$ (LH–RH), Leu–Gly·NH$_2$ (oxytocin), and Lys–Gly–NH$_2$ (vasopressin) (Walter, 1976). It may be of interest to determine whether this enzyme has properties akin to a prolylcarboxypeptidase that is active at pH 5–7 described by Erdos (1971).

The role of elevated plasma oxytocinases during pregnancy is unclear. Oxytocin consists of more than one enzyme, and results in N-terminal cleavage, yielding Cys–Pro–Leu–Gly·NH$_2$ (Molnarova and Sova, 1975; Ferrier et al., 1974). Its action in plasma could be preceded by a disulfide transdehydrogenase yielding linear peptides that are good substrates for aminopeptidases (Marks et al., 1973a). There is considerable scope for examining the clinical use of N-substituted oxytocins that resist digestion by the action of aminopeptidases or other enzymes, or the use of deaminocarbo or dicarbaoxytocins that resist nonenzymatic interchanges, or enzymatic cleavage of the disulfide bonds (Cort et al., 1975). There is considerable interest in the role that conformation plays in breakdown. Studies in solution show that peptide bonds outside the 20-membered ring might be more susceptible; these bonds include the C-terminal dipeptide and the adjacent bond in positions 7 and 8 (Walter, 1976).

3.6. Melanocyte-Inhibiting Factor

The structure of this tripeptide is identical to the C-terminus of oxytocin, and may be present as a breakdown product in brain extracts (see Section 2.2). MIF is degraded by purified and crude aminopeptidases of rat brain, leading to the liberation of all three residues (Marks and Walter, 1972; Simmons and Brecher, 1973). MIF is degraded by sera, higher in rat as compared with man (see Table 5). The possibility exists that the Leu–Glu·NH$_2$ bond is split prior to Pro–Leu, but present data do not permit any conclusion concerning the first point of cleavage. A C-terminal-cleaving enzyme purified partially from rat kidney capable of cleaving –Pro^7–Leu^8– of oxytocin failed to hydrolyze MIF (Walter, 1976).

3.7. Insulin

Extensive studies exist on insulin breakdown in relation to (1) the role of the disulfide bond, (2) the role of binding phenomena to breakdown, and (3) the enzymatic mechanism related to formation from proinsulin (see Section 2.5 and Fig. 4). Insulin does not penetrate the CNS, and its rapid disappearance from blood is attributed to uptake by various organs, followed by intracellular breakdown. Alteration in blood levels of insulin indirectly affects CNS function, owing to profound changes in the levels of glucose and its metabolites.

Insulin is degraded in vitro by fat, liver, muscle, and blood cells (Mirsky and Broh-Kahn, 1949; Brush, 1971; Ansorge, 1971; Tscheshe et al., 1974; Duckworth et al., 1975; Izzo, 1975). Since the intact insulin polypeptide is required for biological activity, cleavage of the disulfide bonds might account for inactivation. Most tissues contain glutathione disulfide oxidoreductases (E.C.1.8.4.2) available for the purpose (see Varandani, 1972; Ansorge *et al.*, 1973), but this enzyme has not been studied in brain, probably because it is not a target organ for insulin. The products, linear A and B chains, are potential substrates for a large number of lysosomal and cytoplasmic enzymes active at acidic and physiological pH (Lajtha and Marks, 1966; Izzo, 1975). Brain cathepsin D, like that of other tissues, cleaved the B chain with a specificity similar to that of pepsin (Marks and Lajtha, 1971). Varandani (Varandani *et al.*, 1972; Varandani, 1974), has proposed a stepwise degradation of insulin preceded by reductive scission of the disulfide bonds, but participation of intracellular enzymes would necessitate transport of insulin across the plasma membrane by pinocytotic or other mechanisms. Since many peptides have disulfide bridges, the reductive scission of this ring may represent a regulatory mechanism; however, the physiological importance of the reductive enzyme has been questioned because of its low affinity for insulin (K_m 21 μM, as compared with 0.1 for "insulinase"; see Thomas, 1973). Both enzymes are present in cytosol and in particulates such as purified rat liver lysosomes, indicating that insulin degradation can occur at more than one cellular site (Grisolia and Wallace, 1976). The catalytic mechanisms of proteinases specifically recognizing intact insulins (referred to as "insulinase," 3.4.99.10) are unknown.

Studies on insulin degradation in perfused organs indicate that insulin is degraded intracellularly (Berson *et al.*, 1956; Mortimore and Tietze, 1959). Attachment to a membrane may be a prelude to degradation, as indicated by a diminished breakdown in the intact fat cell membranes, in hepatocytes treated with trypsin so as to reduce insulin binding and human peripheral granulocytes (Crofford *et al.*, 1972; Kono, 1975; Hammond and Jarett, 1975; Cuatrecasas, 1974; Fussganger *et al.*, 1976). The failure to depress degradation in hepatocytes at very high levels of insulin indicates that the degradative capacity may exceed that of binding (Terris and Steiner, 1975). The same group found that insulin analogues reduced degradation in liver cells *in vitro,* but insulin A and B chains, neurohypophyseal peptides, nerve growth factor, and somatostatin were ineffective. There is considerable scope for examining the role that binding plays in the degradation of insulins and other peptides. (Zahn *et al.*, 1972; Phol and Crofford, 1975; Pullen *et al.*, 1976). Insulin is a well-studied peptide, and varies only slightly in different species. Most insulins have 49 peptide bonds, and if account is taken of residues such as Lys, Glu, His, and Arg, they have three free NH_2 groups, and six COOH groups, two imidazole, and one guanidium group. Removal of the *N*-terminal Phe of the B chain does not affect activity significantly, but removal of the *N*-Gly of the A chain reduces activity by 90%. Studies have shown that the *C*-tripeptide of the B chain is not essential for biological activity.

In summary, there appear to be several distinct systems for inactivation of insulin in tissues, including reductive scission of the disulfide bond and degradation by cytoplasmic and lysosomal enzymes. Breakdown may be sequential, since the linear A and B chains are good substrates for tissue proteolytic enzymes (Varandani *et al.*, 1972; Marks and Lajtha, 1971; Kohnert *et al.*, 1976). The role of binding phenomena as a regulatory mechanism in fat and liver cells may be an important homeostatic mechanism. None of the proteolytic enzymes active at physiological pH catabolizing intact insulin has been purified sufficiently to permit explanation of their catalytic action (Baskin *et al.*, 1975; Hausmann *et al.*, 1975).

3.8. Angiotensin

The native Asn–octapeptide (referred to as angiotensin amide) is a substrate for a large number of intra- and extracellular enzymes, indicating that more than one mechanism is available for inactivation (see Fig. 2). Matsunga *et al.* (1968) observed inactivation of synthetic Asp–angiotensin by crude brain extracts at pH 5.0 (inhibited by DFP) and pH 7.6. Incubation of Asp–angiotensin with a purified brain aminopeptidase was shown to lead to the rapid release of the remaining three C-terminal ones (Abrash *et al.*, 1971). An enzyme extracted from rat and dog brain was purified partially by Goldstein *et al.* (1972), and found to inactivate angiotensin–amide at pH 5.8, but not bradykinin or vasopressin, using bioassay procedures. Recently, a carboxypeptidase purified 300-fold from rat brain active at pH 5–6 was shown to inactivate angiotensins by C-terminal release of Phe (Grynbaum and Marks, 1976; Marks *et al.*, 1976*a*). This enzyme was lysosomal, and showed properties akin to that of cathepsin A, based on cleavage of model N-protected dipeptides and by the effects of inhibitors, and can be differentiated from a "prolyl carboxypeptidase" (angiotensinase C) present in peripheral tissues (see Erdos, 1971). The presence of aminopeptidase capable of cleaving acidic arylamidases (α-Asp-βNA, Asn–βNA, Glu–βNA) raises some interesting questions as to their role in the formation of des-Asn analogues (angiotensin III) (see Section 2.3). Such enzymes, referred to as arylamidase A, are present in brain particulates and in plasma, and show dependence for Ca^{2+} (Marks, 1970; Erdos, 1971).

All together, no fewer than five tissue enzymes have been described that specifically degrade angiotensin (angiotensinases A, A_2, B, and G). Angiotensins A and A_2 are plasma aminopeptidases, and B is an endopeptidase with chymotrypticlike specificity. Angiotensin C is specific for Z–Pro–Phe, and is present in kidney cortex and human urine (see Erdos, 1971). In addition, tissues such as brain contain cathepsin C (dipeptidyl peptidase E.C.3.4.14.1), a potential candidate for inactivation of angiotensins by removal of the N-terminal dipeptide sequences. In a recent study on binding of labeled angiotensin II to brain membranes, Bennett and Snyder (1976) found that degradation was prevented by addition of glucagon and PMSF to the medium. This finding implies that if more than one enzyme is responsible for degradation by membranes, they can be inhibited by a competitive substrate or a chymotrypticlike inhibitor or both.

3.9. Kinins

There is only indirect evidence for the presence of the kinin system in brain (see Section 2.4). There are a number of reports that kinins are inactivated on incubation with crude brain extracts (see Marks and Lajtha, 1971). The role of kinins in nerve function and as a potential neurotransmitter has received considerable attention (see Rocha e Silva, 1975). Among its actions are alteration if the threshold to convulsions induced by analeptics, and altered behavior in mice, and there are reports of kinin release on stimulation of the vagus nerve (see Shikimi *et al.*, 1973). Enzymes inactivating kinins have been purified partially, and shown to have properties of both peptide hydrolases and endopeptidases (Shikimi *et al.*, 1970, 1973; Shikimi and Iwata, 1970; Marks and Pirotta, 1971; Camargo *et al.*, 1973). The enzyme purified 23-fold by Shikimi and his group released Arg and Phe end groups, and was inhibited by metal ions, GSH, mercaptoethanol, and trasylol. Studies on its distribution indicate activity in cell regions, highest in the cerebral cortex cerebellum (Shikimi *et al.*, 1973). In a study on subcellular fractions, Hori (1968) observed activity in all fractions, highest in the supernatant. We showed that a soluble enzyme from brain could inactivate kinins 9–11 by preferential release of internal amino acids, and postulated that the primary point of cleavage occurred at the $-Phe^5-Ser^6$ bond. This was confirmed by Carmago *et al.* (1973) with the isolation of two polypeptide intermediates, $Arg^1---Phe^5$ and $Ser^6---Arg^9$. In both studies, the enzyme had properties akin to a neutral endopeptidase, but differing slightly in specificity as compared with cathepsin M (see Table 1 and Fig. 3). The release of *N*- and *C*-terminal groups by supernatant peptidases also indicates that more than one mechanism exists for inactivation; Arg and Phe were rapidly released, but the Pro–Pro–Gly was only slowly hydrolyzed.

A much-studied enzyme present in lung and kidney but not studied in brain is a peptidyl dipeptidase (E.C.3.4.15.1) (see Section 2.3) that is also known as angiotensin-converting enzyme, but that also hydrolyzes bradykinin *in vitro* at the $-Pro^7-Phe^8$ bond, with release of Phe^8-Arg^9 (Erdos, 1971; Overturf *et al.*, 1975). The substrate specificity of this enzyme, purified from lung, indicates a requirement for a single free carboxyl group at the *C*-terminal end of the peptide; *C*-terminal esters, or amides, and *C*-terminal peptides with acidic residues are not hydrolyzed (Bakhle, 1974). There is evidence for an optimal chain length; peptides with 13 or more residues are hydrolyzed more slowly than nonapeptides such as kinin 9. The degree of Cl^- dependence varies with the substrate; kinin 9 is hydrolyzed in its absence, and only at low rates in its presence. A number of nona- and pentapeptides from *Bothrops jaraca* venom can activate kininase II, but inhibit the angiotensin I conversion (Erdos, 1976). In recent studies, Oliveira *et al.* (1976) separated two thiol peptidases from supernatant of rat brain homogenate, one of which (termed kininase A by the authors) cleaved the $-Phe-Ser-$ bond and the second (kininase B), the Pro–Phe linkage. The first enzyme is an endopeptidase with properties differing from that described by Mark and Pirotta (1971) in terms of protein substrates, but having the identical specificity with respect to kinins 9–11. Kininase B (mol. wt. approximately 68,000) appears to differ in some of its properties from other peptidyl dipeptidases known to degrade kinins and angiotensin I. Neither enzyme hydrolyzed kininlike sequences contained within the precursor kininogen protein, but both hydrolyze an octadecapeptide containing kinin-9.

Another enzyme described in the literature but not studied in detail in brain is a carboxypeptidase N (kininase I) that removes C-terminal Arg at pH 7.6. Since brain contains carboxypeptidases active at this pH, this would provide a potential mechanism for inactivation. Other candidates for inactivation include intracellular cathepsins, notably cathepsin B_2 and possibly cathepsin C. In studies with purified enzymes, we found no evidence for inactivation by brain cathepsin D and a highly purified arylamidase (Marks and Pirotta, 1971).

Tissue extracts also contain a "tissue factor" (thromboplastin) that can extensively hydrolyze kinins, Substance P, angiotensins I and II, bradykinin, LH–RH, oxytocin, and bradykinin-potentiating pentapeptides, in addition to human fibrinopeptide (Glas and Astrap, 1970; Pitlick et al., 1971; Ryan, 1974; Zeldis et al., 1972; Simmons et al., 1976). Tissue factor can promote blood coagulation via the extrinsic system; such factors have been solubilized from bovine brain and lung, and may have properties akin to neutral endopeptidases (Nemerson and Pitlick, 1970). There is some information that the Fletcher factor involved in kallikrein activation also affects activation of prothromboplastin, thus linking two potential systems for kinin turnover (see Rocha e Silva, 1975; see also Fig. 3).

3.10. Lipotropic-Related Peptides

Studies on hypothalamic factors stimulating ACTH go back several decades, but isolation of the corticotropin-releasing factor (CRF) has not been achieved. Studies with proteolytic enzymes indicate that it probably is a peptide. There is renewed interest in ACTH and derived peptides based on their potent behavioral effects (de Wied et al., 1975; Strand and Cayer, 1975; Kastin et al., 1975; Gispen et al., 1976). Many ACTH analogues have been prepared, and their varying potency has been attributed to relative rates of breakdown (Tanaka, 1971; Hofmann et al., 1974; Geiger and Schroeder, 1972; Saez et al., 1975; Draper et al., 1975). As noted, ACTH is cleaved by a trypticlike enzyme in pars intermedia to form two fragments, one of which can be converted to α-MSH and the other into an ACTH-like peptide (Scott et al., 1973) (see Section 2.1 and Fig. 1). Unlike insulin, breakdown of ACTH by adrenal membranes does not appear to be correlated with binding phenomena (Saez et al., 1975). Among peptides of interest to neurobiology are α- and β-MSH, ACTH sequences 1–10 and 4–10, and peptides prepared from brain or pituitary that have sequences in common with β-LPH (enkephalins, α-endorphin, and C-fragments; see Fig. 1 and Section 2.1). In preliminary experiments, we observed degradation by brain extracts and serum for selected peptides when incubated for periods up to 4 hr; with Met– and Leu-enkephalins, there was a rapid appearance of N-terminal Tyr within 1 min when tested under conditions selected for assay (50 nmol substrate with 0.2 ml 10% brain extract or 0.2 ml serum), with degradation exceeding 70% by 10 min. In contrast, no breakdown was detected for α-endorphin at short incubation periods, although substantial degradation occurred at 1–4 hr (Table 6). The rapid degradation of enkephalins may explain their transient central effects (Jacquet et al., 1976). The release of split products indicate participation by brain amino- and carboxy-

TABLE 6. Breakdown of Peptides Related to ACTH and β-Lipotropin by Rat Brain Homogenate [a]

Substrate	Breakdown (percent amino acid or peptide released)
Met–Enkephalin	Gly (75), Tyr (78), Phe (95), Met (80)
Leu–Enkephalin	Gly (72), Tyr (90), Phe (95), Leu (95)
Met–Enkephalin amide	Gly–Gly (40), Tyr (40), Phe (40), Met (trace)
α-Endorphin	Tyr (60), Gly–Gly (60), Phe (36), Met (48), Thr (60), Ser (40), Lys (46), Val (50)
α-MSH	Tyr (42), Ser (60), Met (60), His (46), Phe (80), Arg (20), Gly (34), Lys (24), Pro (trace)
ACTH$_{4-10}$	Met (48), His (76), Phe (51), Arg (85), Gly (22)

[a] The reaction mixture contained 50 nmol peptide incubated with 0.2 ml 10% rat brain homogenate for 4 hr at 37°C. Amino acids and peptides were determined as described in Marks and Stern (1974a). In studies on breakdown with time, Met– and Leu–enkephalin were the most rapidly degraded, with appearance of Tyr and Gly–Gly within 1–5 min of incubation (see the text).

peptidases with release of Gly–Gly as an intermediate product. The structure of the pentapeptide indicates that it is a substrate for known brain aminopeptidases (Marks, 1970); Gly–Gly, however, is degraded by a specific Co^{2+}-requiring dipeptidase that can be purified from brain extracts (Marks and Stern, unpublished findings). It can be predicated that blocking *N*- and *C*-terminal groups would increase stability and analogues already prepared show enhanced *in vivo* (Pert *et al.*, 1976) and *in vitro* actions (Lazarus *et al.*, 1976). Hambrook *et al.* (1976) has demonstrated that *N*-terminal cleavage can deactivate enkephalin and presumably the same applies to *C*-terminal removal. Despite an improved action of enkephalin none of these compare to β-endorphin which is 50-fold more active than morphine when applied to a known central opiod site (periaqueductal gray) (Jacquet *et al.*, 1976). In addition to analgesia, the acute changes in behavior (sedation, catatonia) suggest promising applications for treatment of psychosis (Jacquet and Marks, 1976). Since β-endorphin is a substrate for brain enzymes (breakdown was slower than enkephalins) it could act as a precursor for the production of shorter active fragments (Marks, unpublished findings).

Metabolism of MSH/ACTH peptides is of interest in view of their structural similarities to lipotropic peptides. Peptides with the active (ACTH$_{4-10}$) sequence show opiod binding properties and can induce behavioral changes (Terrenius *et al.*, 1975). A hexapeptidyl analogue Met (O)–Glu–His–Phe–D–Lys–Phe is slowly degraded *in vivo* and *in vitro* (Witter *et al.*, 1975; Vorhoef and Witter, 1976; Marks *et al.*, 1976c) with cleavage of the –His–Phe bond followed by a slow release of Phe from one of the degradation products. This hexapeptide is 1000-fold more active than α-MSH in typical behavioral paradigms (Witter *et al.*, 1975). Studies with α-MSH and ACTH$_{4-10}$ show rapid cleavage by brain extracts, and by sera of various sources (Table 6), pointing towards endopeptidase cleavage as the initial event at the Phe–Arg bond and thus similar to the inactivation reported by intestinal enzymes by Lowery and McMartin (1974).

ACKNOWLEDGMENTS

I would like to accord my gratitude to the following persons for provision of peptide substrates and for many helpful discussions: Wilfred F. White, Abbott Laboratories, Chicago, Illinois (LH–RH and analogues); Roger Guillemin and Jean Rivier, The Salk Institute, San Diego, California, (somatostatin and α-endorphine); Abba Kastin, V.A. Hospital, New Orleans, Louisiana (α-MSH and $ACTH_{4-10}$); and also Yasuko Jacquet of this department. The investigations in this review were supported in part by USPHS Grants NB-03226 and NS-12578.

4. References

Abrash, L., Walker, R., and Marks, N., 1971, Inactivation studies of Angiotensin II by purified brain enzymes, *Experientia* **27**:1352, 1353.

Arregu, A., Bennett, J.P., and Synder, S.H., 1976, Regional distribution of angiotensin converting enzyme in calf brain and its activity in basal ganglia of Huntington's chorea, *Neurology* **4**:367.

Ansorge, S., Bohley, P., Kirschke, H., Langner, J., and Hanson, H., 1971, Metabolism of insulin and glucagon. Breakdown of radioiodinated insulin and glucagon in rat liver cell fractions, *Eur. J. Biochem.* **19**:283–288.

Ansorge, S., Bohley, P., Kirschke, H., Langner, J., Wiederanders, B., and Hanson, H., 1973, Glutathione-insulin transdehydrogenase from microsomes of rat liver, *Eur. J. Biochem.* **32**:27–35.

Bakhle, Y.S., 1974, Converting enzyme *in vitro* measurements and properties, in: *Handbook of Experimental Pharmacology,* Vol. 37, pp. 41–80, Springer-Verlag, New York.

Barrett, A.J., 1975, Lysosomal and related proteinases, *Cold Spring Harbor Conf. Cell Proliferation* **2**:467–482.

Baskin, F.C., Duckworth, W.C., and Kitabchi, E., 1975, Sites of cleavage of glucagon by insulin–glucagon protease, *Biochem. Biophys. Res. Commun.* **67**:163–168.

Bassiri, R.M., and Utiger, R.D., 1974, TRF in the hypothalamus of the rat, *Endocrinology* **94**:188–197.

Bauer, K., Jose, S., and Lipmann, F., 1973, Degradation of TRH by extracts of hypothalmus, *Fed. Proc. Fed. Amer. Soc. Exp. Biol.* **32**:489.

Bauer, K., 1976, Regulation of degradation of thyrotropin-releasing hormone by throid hormones, *Nature* **259**:591–593.

Bauer, K., and Lipmann, F., 1976, Attempts toward biosynthesis of the thyrotropin-releasing hormone and studies on its breakdown in hypothalamic tissue preparations, *Endocrinology* **90**:230–242.

Bennett, J.P., and Snyder, S.A., 1976, Angiotensin II binding to mammalian membranes, *J. Biol. Chem.* in press.

Bennett, G.W., and Edwardson, J.A., 1975, Release of corticotrophin releasing factor and other hypophysiotropic substances from isolated nerve endings (synaptosomes), *J. Endocrinol.* **65**:35–44.

Benuck, M., and Marks, N., 1975, Enzymatic inactivation of substance P by a partially purified enzyme from rat brain, *Biochem. Biophys. Res. Commun.* **65**:153–160.

Benuck, M., and Marks, N., 1976, Differences in the degradation of hypothalamic releasing factors by rat and human serum, *Life Sciences* **19**:1271–1276.

Bergmann, J., Oehme, P., Bienert, M., and Niedrich, H., 1975, Active mechanism of peptides attaching smooth muscles II. Differentiation of the biological activity of tachykinins into affinity and intrinsic efficiency, *Acta Biol. Med. Ger.* **34**:475–481.

Berson, S.A., Yalow, R.S., Bauman, A., Rothschild, M.A., and Neverly, K., 1956, Insulin-^{131}I metabolism in human subjects. Demonstration of insulin binding globulin in the circulation of insulin treated subjects, *J. Clin. Invest.* **35**:170–185.

Bradbury, A.F., Smyth, D.G., and Scott, C.R., 1976a, The peptide hormones: Molecular and cellular aspects, *Ciba Found. Symp.* **41**:61–75, Churchill, London.

Bradbury, A.F., Smyth, D.G., Snell, C.R., Birdsall, N.J.M., and Hulme, E.C., 1976b, C-fragment of lipotrophin has a high affinity for brain opiate receptors, *Nature London* **260**:793–795.

Bradbury, A.F., Smyth, D.G., and Snell, C.R., 1976c, Lipotropin: precursor to two biologically active peptides, *Biochem. Biophys. Res. Commun.* **69**:950–956.

Brown, M., Rivier, T., Vale, W., and Guillemin, R., 1975, Variability of the duration of inhibition of growth hormone release by N-Acylated-de(Ala'Gly^2)–somatostatin analogs, *Biochem. Biophys. Res. Commun.* **65**:752–756.

Brush, J.S., 1971, Purifications and characterisation of a protease with specificity for insulin from rat muscle, *Diabetes* **20**:140–155.

Brownstein, M., Arimura, A., Sato, H., Schally, A.V., and Kizer, J.S., 1975, The regional distribution of somatostatin in the rat brain, *Endocrinology* **96**:1456–1461.

Bury, W.R., and Mashford, M.L., 1976, Biological activity of C-terminal partial sequences of substance P, *J. Med. Chem.* **19**:854–856.

Camargo, A.C.M., Ramalho-Pinto, F.J., and Greene, L.T., 1972, Brain peptidases: Conversion and inactivation of brain hormones, *J. Neurochem.* **19**:37–49.

Camargo, A.C.M., Shapanka, R., and Greene, L.T., 1973, Preparation, assay, and partial characterization of a neutral endopeptidase from rabbit brain, *Biochemistry* **12**:1838–1844.

Chiu, A.T., and Peach, M.J., 1974, Inhibition of induced aldersterone biosynthesis with a specific antagonist, *Proc. Nat. Acad. Sci. U.S.A.* **71**:341–344.

Chiu, A.T., Ryan, J.W., Stewart, J.M., and Dover, F.E., 1976, Formation of angiotensin-III by angiotensin converting enzyme, *Biochem. J.* **155**:189–192.

Chretien, M., 1973, Lipotrophins LPH, in: *Methods in Investigative and Diagnostic Endocrinology* (S.A. Berson and R.S. Yalow, eds.), Vol. 2A, pp. 617–632, North Holland, New York.

Chu, L.L.H., Forte, L.R., Anast, C.S., and Cohn, D.W., 1975a, Interaction of PTH with membranes of kidney cortex, *Endocrinology* **97**:1014–1023.

Chu, L.L.H., Huang, D.W.Y., Littledike, E.T., Hamilton, J.W., and Cohn, D.V., 1975b, Porcine proparathyroid hormone. Identification, biosynthesis and partial amino acid sequence, *Biochemistry* **14**:3631–3635.

Cort, J.H., Schuck, O., Stribna, J., Skophova, J., Jost, K., and Mulder, J.F., 1975, Role of the disulfide bridge and the C-terminal bipeptide in the antidiuretic action of vasopressin in man and the rat, *Kidney Int.* **8**:292–302.

Coy, D.H., Labrie, F., Savary, M., Coy, E.J., and Schally, A.V., 1975, LH-releasing activity of potent LH–RH analogs *in vitro*, *Biochem. Biophys. Res. Commun.* **67**:576–581.

Coy, D.H., Vilchez-Martinez, J.A., Coy, E.J., and Schally, A.V., 1976, Analogs of LH–RH with increased biological activity produced by D-amino acid substitutions in position 6, *J. Med. Chem.* **19**:199.

Cox, B.M., Goldstein, A., and Li, C.H., 1976, Opioid activity of a peptide β-lipotropin (61–91) derived from β-lipotropin, *Proc. Nat. Acad. Sci. U.S.A.* **73**:1821–1823.

Crofford, O.B., Rogers, N.L., and Russell, W.G., 1972, The effect of insulin on fat cells. An insulin degrading system contracted from plasma membrane of insulin responsive cells, *Diabetes* **21**(Suppl.2):403–413.

Croft, L.R., 1973, *Handbook of Protein Sequences,* Joynson-Bruvvers, Oxford, England.

Cruz, A. de la, Coy, D.H., Vilchez-Martinez, J.A., Arimura, A., and Schally, A.V., 1976, Blockade of ovulation in rats by inhibitory analogs of LH–RH, *Science* **191**:195, 196.

Cuatrecasas, P., 1974, Commentary: Insulin receptors, cell membranes and hormone action, *Biochem. Pharmacol.* **23**:2353–2361.

Cushman, D.W., and Cheung, H.S., 1971, Concentrations of angiotensin converting enzyme in tissues of the rat, *Biochim. Biophys. Acta* **250**:261–265.

Daul, C.B., Heuter, R.G., and Garey, R.E., 1975, Angiotensin-forming enzyme in human brain, *Neuropharmacology* **14**:75–80.

Davis, J.O., and Freeman, R.H., 1976, Mechanisms regulating renin release, *Physiol. Rev.* **56**:1–36.

Day, R.P., and Reid, F.A., 1976, Renin activity in dog brain: Enzymological similarity to cathepsin D, *Endocrinology* **99**: 93–100.

de Wied, D., Witter, A., and Grevers, H.M., 1975, Commentary: Behaviorally active ACTH analogues, *Biochem. Pharmacol.* **24**:1463–1468.

Draper, M.W., Rizach, M.A., and Merrifield, R.B., 1975, Synthetic position 5 analogs of adrenocorticotropin fragments and their *in vitro* lipolytic activity, *Biochemistry* **14**:2933–2938.

Duckworth, W.C., Heniemann, M., and Kitabchi, A.E., 1975, Proteolytic degradation of insulin and glucagon, *Biochim. Biophys. Acta* **377**:421–430.

Enzyme Nomenclature, 1972, Elsevier, New York.

Erdos, E.G., 1971, Enzymes that inactivate vasoactive peptides, in: *Handbook of Experimental Pharmacology,* Vol. 28, pp. 620–653, Springer-Verlag, Berlin.

Erdos, E.G., 1976, Commentary: the kinins, a status report, *Biochem. Pharmacol.* **24**:1563–1569.

Erdos, E.G., 1975, Angiotensin I converting enzyme, *Circ. Res.* **36**:247–255.

Fawcett, C.P., Powell, A.E., and Sachs, H., 1968, Biosynthesis and release of neurophysins, *Endocrinology* **83**:1299–1310.

Fernandez, M.T.F., Paladini, A.C., and Delius, A.E., 1965, Isolation and identification of a pepsitensin, *Biochem. J.* **97**:540–546.

Ferrier, B.M., Hendrie, J.M., and Branda, L.A., 1974, Plasma oxytocinase: Synthesis and biological properties of the first product of the degradation of oxytocin by this enzyme, *Can. J. Biochem.* **52**:60–66.

Fisher-Ferraro, C., Nahmud, V.E., Goldstein, D.J., and Finkelman, S., 1971, Angiotensin and renin in rat and dog brain, *J. Exp. Med.* **133**:353–361.

Friesen, H., Guyda, J., and Hardy, J., 1970, Biosynthesis of human growth hormone and prolactin, *J. Clin. Endocrinol.* **31**:611–624.

Fujino, M., Yamazak, I., Kobayashi, S., Fukuda, T., Shinagawa, S., Nakayama, R., White, W.F., and Rippel, R.H., 1974, Some analogs of LH–RF having intense ovulation-inducing activity, *Biochem. Biophys. Res. Commun.* **57**:1248–1256.

Fujita, T., Okatu, M., Okano, K., and Yoshikawa, M., 1975, Differential hydrolysis of bovine PTH and its *N*-terminal peptide by rat kidney, *Endocrinol. Jpn.* **22**:39–42.

Fussganger, R.D., Kahn, C.R., Roth, J., and Meyts, P., 1976, Binding and degradation of insulin by human peripheral granulocytes, *J. Biol. Chem.* **251**:2761–2769.

Galoyan, A.A., 1973, New hormones of the hypothalamo-neurohypophyseal system, *Vopr. Biokhim. Mozga* **VIII**:107–125.

Ganten, D., Marguez-Julio, A., Granger, P., Hayduk, K., Karsunky, K.P., Boucher, R., and Genest, J., 1971a, Renin in brain, *Amer. J. Physiol.* **221**:1723–1737.

Ganten, D., Minnich, J.L., Granger, P., Hayduk, K., Brecht, H.M., Barbeau, A., Boucher, R., and Genest, J., 1971b, Angiotensin-forming enzyme in brain tissue, *Science* **173**:64, 65.

Ganten, D., Ganten, U., Schelling, P., Boucher, R., and Genest, J., 1975, The renin and iso-renin–angiotensin systems in rats with experimental pituitary tumors, *Proc. Soc. Expr. Biol. Med.* **148**:568–572.

Geiger, R., and Schroeder, H.-G., 1972, New short chain synthetic corticotrophin analogues with high corticotrophic activity, in: *Progress in Peptide Research,* Vol. II (S. Lande, ed.), pp. 273–279, Gordon and Breach, New York.

Gispen, W.H., Wiegant, V.M., Greven, H.M., and de Wied, D., 1976, The induction of excessive grooming in the rat by intraventricular application of peptides derived from ACTH: Structure–activity studies, *Life Sci.* **17**:645–652.

Glas, P., and Astrap, T., 1970, Thromboplastin and plasminogen activator in tissues of the rabbit, *Amer. J. Physiol.* **219**:1140–1146.

Goldstein, D.J., Diaz, A., Finkielman, S., Nahmod, V.E., and Fischer-Ferraro, C., 1972, Angiotensinase activity in rat and dog brain, *J. Neurochem.* **19**:2451–2452.

Goodfried, T.L., and Peach, M.J., 1975, Angiotensin-III: (des-Asp-acid)–angiotensin II, *Circ. Res. Suppl.* 1, **36**, **37**:1–38.

Graf, L., and Kenessey, A., 1976, Specific cleavage of a single peptide bond (residues 77–78) in β-lipotropin by a pituitary endopeptidase, *FEBS Lett.* **69**:255–260.

Graf, L., Ronai, A.Z., Bajusz, S., Cseh, G., and Szekely, J.I., 1976, Opioid agonist of β-LPH fragments: A possibly biological source of morphine-like substances in the pituitary, *FEBS Lett.* **64**:181–184.

Gregory, R.A., and Tray, H.J., 1972, Isolation of two "big gastrins" from Zollinger-Ellison tumour tissue, *Lancet* **2**:797–799.

Griffiths, E.C., and Hooper, K.C., 1974, Peptidase activity in different areas of the rat hypothalamus, *Acta Endocrinol. Copenhagen* **77**:10–18.

Griffiths, E.C., Hooper, K.C., and Hopkinson, C.R.N., 1973, Evidence for an enzymic component in the rat hypothalamus inactivating LH–RF, *Acta Endocrinol. Copenhagen* **74**:49–55.

Griffiths, E.C., Hooper, K.C., Jeffcoate, S.L., and Holland, D.T., 1974, Presence of peptidases in the rat hypothalamus inactivating luteinizing hormone releasing hormone (LH–RH), *Acta Endocrinol. Copenhagen* **77**(3):435–442.

Griffiths, E.C., Hooper, K.C., Jeffcoate, S.L., and Holland, D.T., 1975*a*, Peptidases in different areas of the rat brain inactivating LH–RH, *Brain Res.* **85**:161–164.

Griffiths, E.C., Hooper, K.C., Jeffcoate, S.L., and Holland, D.T., 1975*b*, The effects of gonadectomy and gonal steroids on the activity of hypothalamic peptidases inactivating LH–RH, *Brain Res.* **88**:384–388.

Griffiths, E.C., Hooper, K.C., Jeffcoate, S.L., and White, N., 1975*c*, Peptidases in the rat hypothalamus inactivating TRF, *Acta Endocrinol. Copenhagen* **79**:209–216.

Griffiths, E.C., Hooper, K.C., Jeffcoate, S.L., and White, N., 1976, Inactivation of TRF by peptidases in different areas of rat brain, *Brain Res.* **105**:376–380.

Grim-Jorgensen, Y., Pfeiffer, S.E., and McKelvy, J.F., 1976, Metabolism of thyrotropin releasing factor in two clonal cell lines of nervous system origin, *Biochem. Biophys, Res. Commun.* **70**:167–173.

Grisolia, S., and Wallace, R., 1976, Insulin degradation by lysosomal extracts from rat liver: Model for a role of lysosomes in hormone degradation, *Biochem. Biophys. Res. Commun.* **70**:22–27.

Grynbaum, A., and Marks, N., 1976, Characterisation of rat brain catheptic carboxypeptidase (cathepsin A) inactivating angiostein II, *J. Neurochem.* **26**:313–318.

Guillemin, R., Ling, N., and Burgus, R., 1976, Endorphins: Hypothalamic and neurohypophyseal peptides with morphinomimetic activity. Isolation and primary structure of α-endorphin, *C. R. Acad. Sci. Paris Ser. D* **282**:783–789.

Guimaraes, J.A., Borges, D.R., Prado, E.S., and Prado, J.L., 1973, Kinin-converting aminopeptidase from human serum, *Biochem. Pharmacol.* **22**:3157–3171.

Habener, J.F., and Potts, J.T., 1975, Techniques for the identification of a biosynthetic precursor of PTH, *Methods Enzymol.* **37**:345–359.

Hambrook, J.M., Morgan, B.A., Rance, M.J., and Smith, C.F.C., 1976, Mode of deactivation of the enkephalins by rat and human plasma and rat brain homogenates, *Nature* **262**:782–783.

Hammond, J.M., and Jarett, L., 1975, Insulin degradation by isolated fat cells and their subcellular fractions, *Diabetes* **24**:1011–1019.

Hausmann, L., Klissek, J., and Kamps, K., 1975, Radioimmunological proinsulin determination in serum with an "insulin degrading protease" from rat liver, *Endocrinologie* **66**:56–66.

Henry, J., Krnjević, K., and Morris, M.E., 1975, Substance P and spinal neurons, *Can. J. Physiol. Pharmacol.* **53**:423–432.

Hew, C.-L., and Yip, C.C., 1976, Biosynthesis of polypeptide hormones, *Canadian J. Biochem.* **54**:591–599.

Hirata, Y., Matsukura, S., Imura, H., Nakamura, M., and Tanaka, A., 1976, Size heterogeneity of β-MSH in ectopic ACTH-producing tumors: Presence of β-LPH–like peptide, *J. Clin. Endocrinol. Metab.* **42**:33–38.

Hofman, K., Montibeller, J.A., and Finn, F.M., 1974, ACTH antagonists, *Proc. Nat. Acad. Sci. U.S.A.* **71**:80–83.

Hopsu-Harvu, V.K., Makinen, K.K., and Glenner, G.G., 1966, Formation of bradykinin from kalliden-10 by aminopeptidase B, *Nature London* **212**:1271–1272.

Hori, S., 1968, The presence of bradykinin-like materials, kinin releasing and destroying action in brain, *Jpn. J. Physiol.* **18**:772–787.

Hughes, J., Smith, T.W., Kosterlitz, H.W., Fothergill, L.A., Morgan, B.A., and Morris, H.R., 1975, Identification of two related pentapeptides from the brain with potent opiate agonist activity, *Nature London* **258**:577–579.

Inouye, A., Kataoka, K., and Tsujioka, T., 1961, On a kinin-like substance in the nervous tissue elements treated with trypsin, *Jpn. J. Physiol.* **11**:319–334.

Izzo, J.T., 1975, Degradation of insulin, in: *Handbook of Experimental Pharmacology,* Vol. 32, pp. 229–248, Springer-Verlag, Berlin.

Jacquet, Y., Marks, N., and Li, C. -H. 1976, Behavioral and biochemical properties of "opiod" peptides, in: *Opiates and Endogenous Opiod Peptides* (H. W. Kosterlitz, ed.), pp. 411–414, Elsevier, Amsterdam.

Jacquet, Y., and Marks, N., 1976, The C-fragment of β-lipotropin: an endogenous neuroleptic or antipyschotogen, *Science* **194**:632–635.

Kastin, A.J., Sandman, C.A., Stratton, L.O., Schally, A.V., and Miller, L.H., 1975, Behavioral and electrographic changes in rat and man after MSH, *Prog. Brain Res.* **42**:143–150.

Kemmler, W., 1975, Conversion of pro insulin, in: *Handbook of Experimental Pharmacology,* Vol. 32, pp. 17–56, Springer-Verlag, Berlin.

Knigge, K.M., and Schock, D., 1975, Characteristics of the plasma TRH degrading enzyme, *Neuroendocrinology* **19**:277–287.

Koch, Y., Baram, T., Chobsieng, P., and Fridkin, M., 1974, Enzymic degradation of LH–RH by hypothalamic tissue, *Biochem. Biophys. Res. Commun.* **61**:95–103.

Kochman, K., Kerdelhue, B., Zor, U., and Jutisz, M., 1975, Studies of enzymatic degradation of luteinizing hormone-releasing hormone by different tissues, *FEBS Lett.* **50**:190–194.

Kohnert, K.D., Jahr, H., Schmidt, S., Hahn, J.J., and Zuhlke, H., 1976, Demonstration of insulin degradation by thiol-protein disulfide oxitoreductase and proteinases of pancreatic islets, *Biochim. Biophys. Acta* **422**:254–259.

Kono, I., 1975, Proteolytic Modification of the insulin receptor of adipose tissue cells, *Methods Enzymol.* **37**:211–213.

Kuhl, H., and Taubert, H.D., 1975*a*, Short-loop feedback mechanism of luteinizing hormone, *Acta Endocrinol. Copenhagen* **78**:649–663.

Kuhl, H., and Taubert, H.D., 1975*b*, Inactivation of LH–RF by rat hypothalamic L-crystine arylamidase., *Acta Endocrinol. Copenhagen* **78**:634–648.

Lajtha, A., and Marks, N., 1966, Cerebral protein breakdown, in: *Protides of the Biological Fluids* (H. Peeters, ed.), p. 103, Elsevier Publishing Co., Amsterdam.

Lajtha, A., and Marks, N., 1969, Metabolism in the nervous system (ed. by S. Bogoch), *Future of Brain Sciences,* pp. 181–196, Plenum Press, New York.

Lajtha, A., and Marks, N., 1971, Protein turnover, in: *Handbook of Neurochemistry* (A. Lajtha, ed.), Vol. 5, Part B, pp. 551–629, Plenum Press, N.Y.

Lazarus, L.H., Ling, N., and Guillemin, R., 1976, β-LPH as a prohormone for the morphinomimetric peptides endorphins and enkephalins, *Proc. Nat. Acad. Sci. U.S.A.* **73**:2156–2159.

Lewis, U.J., Pence, S.J., Smyth, R.N.P., and Vanderlaan, W.P., 1975, Enhancement of the growth activity of human growth hormone, *Biochem. Biophys. Res. Commun.* **67**:617–624.

Li, C.H., and Chung, D., 1976, Isolation and structure of an untriakontapeptide with opiate activity from camel pituitary glands, *Proc. Nat. Acad. Sci. U.S.A.* **73**:1145–1148.

Ling, N.C., Rivier, J.E., Monahan, M.W., and Vale, W.W., 1976, Analogues of luteinizing hormone-releasing factor modified at positions 2, 6, 10, *J. Med. Chem.* **19**:937–941.

Lowery, P.J., and McMartin, C., 1974, Metabolism of two adrenocorticotrophin analogues in the intestine of the rat, *Biochem. J.* **133**:87–95.

Marks, N., 1970, Peptide hydrolases, in: *Handbook of Neurochemistry* (A. Lajtha, ed.), Vol. 3, Chapt. 5, pp. 133–171, Plenum Press, New York.

Marks, N., 1976, Biodegradation of hormonally active peptides in the CNS, in: *Subcellular Mechanisms in Reproductive Neuroendocrinology* (F. Naftolin, ed.), pp. 129–147, ASP-Elsevier, Amsterdam.

Marks, N., and Lajtha, A., 1965, Separation of acid and neutral proteinases of brain, *Biochem. J.* **97**:74–83.

Marks, N., and Lajtha, A., 1971, Protein and polypeptide breakdown, in: *Handbook of Neurochemistry* (A. Lajtha, ed.), Vol. 5, Part A, pp. 49–139, Plenum Press, New York.

Marks, N., and Pirotta, M., 1971, Breakdown of bradykinin and its analogs by rat brain neutral proteinase, *Brain Res.* **33**:565–567.

Marks, N., and Stern, F., 1974a, Enzymatic mechanisms for the inactivation of luteinizing hormone–releasing hormone (LH–RH), *Biochem. Biophys. Res. Commun.* **61**:1458–1463.

Marks, N., and Stern, F., 1974b, Novel enzymes involved in the inactivation of hypothalamo–hypophyseal hormones, in: *Psychoneuroendocrinology* (N. Hatotani, ed.), pp. 153–162, S. Karger, Basel.

Marks, N., and Stern, F., 1975, Inactivation of somatostatin (GH–RIH) and its analogs by crude and partially purified rat brain extracts, *FEBS Lett.* **55**:220–224.

Marks, N., and Walter, R., 1972, MSH-Release inhibiting factor: Inactivation by proteolytic enzyme, *Proc. Soc. Exp. Biol. Med.* **140**:673–676.

Marks, N., Abrash, L., and Walter, R., 1973a, Degradation of neurohypophyseal hormones by brain extracts and purified brain enzymes, *Proc. Soc. Exp. Biol. Med.* **142**:455–460.

Marks, N., Grynbaum, A., and Lajtha, A., 1973b, Pentapeptide (pepstatin) inhibition of brain acid proteinase, *Science* **181**:949–951.

Marks, N., Galayon, A., Grynbaum, A., and Lajtha, A., 1974, Protein and peptide hydrolases of the rat hypothalamus and pituitary, *J. Neurochem.* **22**:735–739.

Marks, N., Grynbaum, A., and Benuck, M., 1976a, On the sequential degradation of myelin basic protein by cathepsins A and D, *J. Neurochem.* **27**:765–768.

Marks, N., Stern, F., and Benuck, M., 1976b, Correlation between biological potency and biodegradation of a somatostatin analogue, *Nature London* **261**:511–512.

Marks, N., Stern, F., and Kastin, A.J., 1976c, Breakdown of α-MSH and derived peptides by rat brain extracts, rat and human serum, *Brain Res. Bull.* **1**: in press.

Matsunga, M., Saito, N., Kira, J., Ogino, K., and Takagasu, M., 1968, Acid angiotensinase as a lysosomal enzyme, *Jpn. Cir. J.* **32**:137–150.

McElvey, J.F., Sheridan, M., Joseph, S., Phillips, C.H., and Perrie, S., 1975, Biosynthesis of TRF in organ culture of the guinea pig median eminence, *Endocrinology,* **97**:908–918.

Miller, R., Aehnelt, C., Rossier, G., and Hendricks, S., 1975, Evidence for the existence of higher-molecular-weight precursor of LH–RF, *IRCS-Med. Sci.* **3**:603.

Mirsky, I.A., and Broh-Kahn, R.H., 1949, The inactivation of insulin by tissue extracts, 1. The distribution and properties of insulin inactivation (insulinase), *Arch. Biochem.* **20**:1–25.

Molnarova, B., and Sova, O., 1975, The influence of pH in the medium on degradation of neurohormones by pregnancy serum, *Physiol. Bohemoslov.* **24**:199–208.

Monnier, M., Schoenenberger, G.A., Dudler, L., and Herkert, B., 1975, Production, isolation, and further characterisation of the sleep peptide delta, *Proc. Eur. Congr. Sleep Res.* **2**:41–46, 65–70.

Mortimore, G.E., and Tietze, F., 1959, Studies on the mechanism of capture and degradation of insulin-I^{131} by the cyclically perfused rat liver, *Ann. N. Y. Acad. Sci.* **82**:329–337.

Moss, R.L., and McCann, S.M., 1973, Induction of mating behavior in rats by luteinizing hormone releasing factor, *Science* **181**:177–179.

Mudge, A.W., and Fellows, R.E., 1973, Bovine pituitary pyrrolidone–carboxyl peptidase, *Endocrinology* **93**:1428–1434.

Nair, R.M.G., Redding, T.W., and Schally, A.V., 1971, Site of inactivation of TRH by human plasma, *Biochemistry* **10**:3621–3624.

Neary, J.T., Kieffer, D.J., Federico, P., Mover, H., Maloof, F., and Soodak, M., 1976, Thyrotropin releasing hormone: development of inactivation system during maturation of the rat, *Science* **193**:403–405.

Nemerson, Y., and Pitlick, F.A., 1970, Purification and characterisation of the protein component of tissue factor, *Biochemistry* **9**:5100–5113.

Niedrich, H., Bienert, M., Mehlis, B., Bergmann, J., and Oehme, P., 1975, Studies on the action mechanisms of peptides attacking smooth muscle. III. Effect of *N*-acylation upon effectiveness of *C*-terminal partial sequences of eledosin, physalemin, and substance P on guinea pig ileum, *Acta. Biol. Med. Ger.* **34**:483–489.

Noe, B.D., and Bauer, G.E., 1971, Evidence for glucagon biosynthesis involving a protein intermediate of angler fish, *Endocrinology* **89**:642–651.

O'Conner, K.J., and Lazarus, N.R., 1976, The purification and biological properties of pancreatic big glucagon, *Biochem. J.* **156**:265–277.

Oliveira, E.B., Martins, A.R., and Camargo, C.M., 1976, Isolation of brain endopeptidases: Influence of size and sequence of substrates structurally related to bradykinin, *Biochemistry* **15**:1967–1974.

Orlowski, M., Sessa, G., and Green, J.P., 1974, γ-Glutamyl transpeptidase in brain capillaries: Possible site of a blood brain barrier for amino acids, *Science* **184**:66–68.

Otsuka, M., Konishi, S., and Takahashi, T., 1975, Hypothalamic Substance P as a candidate for transmitter of primary afferent neurons, *Fed. Proc. Fed. Amer. Soc. Exp. Biol.* **34**:1922–1928.

Otto, K., 1976, On the specificity of cathepsin B_2, *2nd Symposium on Intracellular Protein Catabolism.* (V. Turk and N. Marks, eds.), Slovenji Press, Ljubljana.

Overturf, M., Wyatt, S., and Fitz, A., 1975, Angiotensin I ($Phe^8–His^8$) hydrolase and bradykininase from human lung, *Life Sci.* **16**:1669–1682.

Oyama, H., Horino, M., Matsumura, S., Kobayashi, K., Suetsugu, N., 1975, Immunological half-life of porcine proinsulin, *Horm. Metab. Res.* **7**:520, 521.

Pappenheimer, J.R., Koshi, G., Fenci, V., Karnovsky, M.L., and Krueger, J., 1975, Extraction of sleep-promoting factor S from CSF and from brains of sleep-deprived animals, *J. Neurophysiol.* **38**:1299–1311.

Pert, C.B., Pert, A., Chang, J. -K., and Fong, B.T.W., 1976, (D–Ala²)–Met–Enkephalinamide: A potent, long lasting synthetic pentapeptide analgesic, *Science* **194**:330–332.

Pitlick, F.A., Nemerson, Y., Gottlieb, A.J., Gordon, R.G., and Williams, W.J., 1971, Peptidase activity associated with the tissue factor of blood coagulation, *Biochemistry* **10**:2650–2657.

Pohl, S.L. and Crofford, O.B., 1975, Techniques for the study of polypeptide hormone inactivation at receptor sites, *Methods Enzymol.* **37**:211–213.

Poth, M.M., Heath, R.G., and Ward, M., 1975, Angiotensin-converting enzyme in human brain, *J. Neurochem.* **25**:83–86.

Prado, J.L., 1970, Proteolytic enzymes as kininogenases, *Handbook of Experimental Pharmacology,* Vol. 35, pp. 156–192, Springer-Verlag, Berlin.

Prasad, C., and Peterkofsky, A., 1976, Demonstration of pyroglutamyl peptidase and amidase activities toward thyrotropin-releasing hormone in hamster hypothalamus extracts, *J. Biol. Chem.* **251**:3229–3234.

Pullen, R.A., Lindsay, D.G., Wood, S.P., Tickle, I.J., Blundell, T.L., Wollmer, A., Krail, G., Brandenbury, D., Zahn, H., Gliemann, J., and Gammelfoft, S., 1976, Receptor-binding regions of insulin, *Nature London* **259**:369–373.

Reagan, C.R., Mills, J.B., Kostyo, J.L., and Wilhelmi, A.E., 1975, Isolation and biological characterisation of fragments of hGH produced by digestion with plasmin, *Endocrinology* **96**:625–636.

Redding, T.W., and Schally, A.V., 1972, On the half-life of TRF in rats, *Neuro-endocrinology,* **9**:250–256.

Redding, T.W., Kastin, A.J., Gonzalez-Barcena, D., Coy, D.H., Coy, E.J., Schalch, D.S., and Shally, A.V., 1973, The half-life, metabolism and excretion of tritiated LH–RH in man, *J. Clin. Endocrinol. Metab.* **37**:626–663.

Reichlin, S., and Mitnick, M., 1973, Biosynthesis of hypothalamic hypophysiotrophic factors, in: *Frontiers in Neuroendocrinology* (F. Ganong and L. Martini, eds.), pp. 61–88, Oxford Press, New York.

Reid, I.A., and Ramsey, D.J., 1975, The effects of intracerebroventricular administration of renin on drinking and blood pressure, *Endocrinology* **97**:536–542.

Ressler, C., and Popenoe, E.A., 1973, Oxytocin, in: *Methods of Investigative and Diagnostic Endocrinology* (S.A. Berson and R.S. Yalow, eds.), Vol. 2A, pp. 681–691, North-Holland, New York.

Rivier, J., Brown, M., and Vale, W., 1975, D-Trp⁸-Somatostatin: An analog of somatostatin more potent than the native molecule, *Biochem. Biophys. Res. Commun.* **65**:746–751.

Rocha e Silva, M., 1975, Present trends in kinin research, *Life Sci.* **15**:7–22.

Ryan, J.W., 1974, Angiotensin, in: *Handbook of Experimental Pharmacology,* Vol. 37, pp. 81–110, Springer-Verlag, New York.

Saez, J.M., Dazord, A., Morera, A.M., and Bataille, P., 1975, Interactions of adrenocorticotropic hormone with its adrenal receptors, *J. Biol. Chem.* **250:**1683–1689.

Said, S.I., and Rosenberg, R.N., 1976, Vasoactive intestinal polypeptide: Abundant immunoreactivity in neural cell lines and normal nervous tissue, *Science* **192:**907–909.

Scott, A.P., Ratcliffe, J.G., Rees, L.H., Landon, J., Bennett, H.P.J., and Lowery, P.J., 1973, Pituitary peptide, *Nature London* **244:**65–67.

Serra, S., Grynbaum, A., Lajtha, A., and Marks, N., 1972, Peptide hydrolases in spinal cord and brain of the rabbit, *Brain Res.* **44:**579–592.

Severs, W.B., and Severs, A.E., 1973, Effects of angiotensin on the CNS, *Pharm. Rev.* **25:**415–449.

Severs, W.B., and Summy-Long, J., 1976, The role of angiotensin in thirst, *Life Sci.* **17:**1513–1526.

Shikimi, T., and Iwata, H., 1970, Pharmacological significance of peptidase and proteinase in the brain. II. Purification and properties of a bradykinin inactivating enzyme from rat brain, *Biochem. Pharmacol.* **19:**1399–1407.

Shikimi, T., Shigemitsu, H., and Heitanoh, I., 1970, Substrate specificity and amino acid composition of partially purified enzyme inactivating bradykinin in brain, *Jpn. J. Pharmacol.* **20:**169–170.

Shikimi, T., Kema, R., Matsumolo, M., Yamahata, Y., and Miyata, S., 1973, Studies on kinin-like substances in brain, *Biochem. Pharmacol.* **22:**567–573.

Shin, S.H., Howitt, C., and Milligan, J.V., 1974, A paradoxical castration effect on LH–RH levels in male rat hypothalamus and serum, *Life Sci.* **14:**2491–2496.

Short, R.V., 1975, Commentary: Hormonal control of spermatogenesis, *Nature London* **254:**144.

Simmons, W.H., and Brecher, A.S., 1973, Inactivation of MIF by a Mn^{2+}-stimulated bovine brain aminopeptidase, *J. Biol. Chem.* **248:**5780–5784.

Simmons, W.H., Burkholder, D.E., and Brecher, A.S., 1976, The hydrolysis of biologically active peptides by bovine tissue factor (thromboplastin), *Soc. Exp. Biol. Med.* in press.

Simpson, J.B., and Routtenberg, A., 1973, Subfornical organ: Site of drinking elicitation by angiotensin II, *Science* **181:**1172–1175.

Smith, I., Mullen, P.E., Silman, R.E., Snedden, W., and Wilson, R.W., 1976, Human foetal pituitary peptides and parturition, *Nature London* **260:**716–718.

Stachura, M.E., and Frohman, L.A., 1975, Growth hormone: Independent release of big and small forms from rat pituitary *in vitro, Science* **187:**477–449.

Steiner, D.F., Kemmler, W., Tager, H.S., and Rubenstein, A.H., 1974, Molecular events taking place during intracellular transport of coportable proteins. The conversion of peptide hormone precursors., *Adv. Cytopharmacol.* **2:**195–205.

Steiner, D.F., Kemmler, W., Tager, H.S., Rubenstein, A.H., Lernmark, A., and Zuehlke, H., 1975, Proteolytic mechanisms in the biosynthesis of polypeptide hormones, *Cold Spring Harbor Conf. Cell Proliferation* **3:**250–255.

Strand, F.L., and Cayer, F.I., 1975, Modulatory effects of polypeptides on peripheral nerve and muscle, *Prog. Brain. Res.* **42:**187–194.

Szewezuk, A., and Kwiatkowska, J., 1970, Pyrrolidonyl peptidases in animal, plant, and human tissues. Occurrence and some properties of the enzyme, *Eur. J. Biochem.* **15:**92–96.

Taber, C.A., and Karavolas, H.J., 1975, Subcellular localization of LH releasing activity in the rat hypothalamus, *Endocrinology* **96:**446–452.

Tager, H.S., Rubenstein, A.H., and Steiner, D.F., 1975, Methods for the assessment of peptide precursors: Studies on insulin biosynthesis, *Methods Enzymol.* **37:**326–344.

Tanaka, A., 1971, Structure–activity relationships of synthetic adrenocorticotropic octadecapeptides with special reference to biological half-life, *Endocrino. Jpn.* **18:**155–168.

Terrenius, L., Gispen, H.W., and de Wied, D., 1975, ATCH-like peptides and opiate receptors in the rat brain: Structure–activity studies, *Eur. J. Pharmacol.* **33:**395–399.

Terris, S., and Steiner, D.F., 1975, Binding and degradation of ^{125}I-insulin by rat hepatocytes, *J. Biol. Chem.* **250:**8389–8398.

Thomas, J.H., 1973, The role of "insulinase" in the degradation of insulin, *Postgrad. Med. J. Suppl.* **49**:940–944.

Tschesche, H., Dietl, T., Kolb, H.J., and Standl, E., 1974, An insulin degrading proteinase from human erythrocytes and its inhibition by proteinase inhibitors, in: *Proteinase Inhibitors, Bayer-Symp.* **5**:586–593.

Tuppy, H., 1968, The influence of enzymes on neurophysiological hormones and similar peptides, *Handbook of Experimental Pharmacology*, Vol. 23, pp. 67–130, Springer-Verlag, New York.

Turk, V., and Marks, N. (eds.), 1976, *2nd International Conference on Protein Catabolism*, Slovenji Press, Ljubljana.

Vanderhaeghen, J.J., Signeau, J.C., and Gepts, W., 1975, New peptide in the vertebrate CNS reacting with antigastrin antibodies, *Nature London* **257**:604, 605.

Varandani, P.T., 1974, Feedback control of insulin by liver glutathione–insulin transdehydrogenase, *Diabetes*, **25**:117–125.

Varandani, P.T., Shroger, L.A., and Naez, M.A., 1972, Sequential degradation of insulin by rat liver homogenates, *Proc. Nat. Acad. Sci. U.S.A.* **69**:1681–1684.

Verhoef, J., and Witter, A., 1976, In vivo fate of a behaviorally active $ACTH_{4-9}$ analog in rats after systemic administration, *Pharmacol. Biochem. Behav.* **4**:583–590.

Visser, J.T., Kluotwijk, W., Docter, R., and Hennemann, G., 1975, RIA for the measurement of pyroGlu–His–Pro, a proposed TRF metabolite, *J. Clin Endocrinol. Metab.* **40**:742–745.

Wajda, I.J., Neidle, A., Ehrenpreis, S., and Manigault, I., 1976, Properties and distribution of morphine-like substances, in: *Opiates and Endogenous Opiod Peptides* (R.W. Kosterlitz, ed.), Elsevier-North Holland, Amsterdam.

Walter, R., 1976, Partial purification of post-proline cleaving enzyme, *Biochim. Biophys. Acta* **422**:138–158.

Walter, R., Griffiths, E.C., and Hooper, K.C., 1973, Production of MIF by a particulate preparation of hypothalami: Mechanisms of oxytocin inactivations, *Brain Res.* **60**:449–508.

Webster, M.E., 1970, Kallikreins in glandular tissue, in: *Handbook of Experimental Pharmacology*, Vol. 35, pp. 131–155, Springer-Verlag, New York.

Werle, E., and Vogel, R., 1961, Über die Freisetzung einer Kallikreinartigen Substanz aus Extracten verschiedener Organe, *Arch. Pharmacodyn. Ther.* **131**:257–261.

Werle, E., and Zach, P., 1970, Verteilung von Kinogen in Serum und Geweben bei Ratten und anderen Saugetieren, *Z. Klin. Chem. Klin. Biochem.* **8**:186–189.

Witter, A., 1976, Commentary: The in vivo fate of brain oligopeptides, *Biochem. Pharmacol.* **24**:2025–2030.

Witter, A., Greven, H.W., and de Wied, D., 1975, Correlation between structure, behavioral activity and rate of biotransformation of some $ACTH_{4-9}$ analogs, *J. Pharmacol. Exper. Ther.* **193**:853–860.

Yajima, H., Kitagawa, K., and Segawa, T., 1973, Studies on peptides XXXVIII. Structure–activity correlations of substance P, *Chem. Pharm. Bull.* **21**:2500–2506.

Yalow, R., 1976, Polypeptide hormones—Molecular and cellular aspects. Multiple forms of ACTH and their significance, *Ciba Found. Symp.* **41**:159–172.

Yang, H.Y.T., and Neff, N.H., 1972, Distribution and properties of angiotensin converting enzyme of rat brain, *J. Neurochem.* **19**:2443–2450.

Yasuo, I., 1975, Treatment of arteriosclerosis with kallikrein. Therapeutic effect and stimulating effects on platelet aggregation, *Yakuri To Chiryo* **3**:755–770; also *Chem. Abstr.* **84**:54276x.

Zahn, H., Brandenburg, D., and Gattner, H.-G., 1972, Molecular basis of insulin action: Contribution of chemical modification and synthetic approaches, *Diabetes*, **21**(Suppl.2):468–475.

Zeldis, S.M., Nemerson, Y., Pitlick, F.A., and Leutz, T.L., 1972, Tissue factor (thromboplastin): Localisation to plasma membranes by peroxidase-conjugated antibodies, *Science* **175**:766–768.

Zetler, G., 1970, Biologically active peptides (substance P), in: *Handbook of Neurochemistry* (A. Lajtha, ed.), Vol. 4, pp. 135–146, Plenum Press, New York.

Zimmerman, E.A., 1976, Localization of hypothalamic hormones by immunocytochemical techniques, *Front. Neuroendocrinol.* **4**:25–68.

10

Peptides in Invertebrate Nervous Systems

Nora Frontali and Harold Gainer

1. Introduction

Throughout the Metazoa, nervous and endocrine systems are involved in the coordination of the activities of various organs and tissues. However, in the less developed invertebrate phyla, conventional epithelial endocrine glands appear to be absent (Highnam and Hill, 1969), and organismic integration is primarily under the influence of neurosecretory cells. Since the pioneering studies of the Scharrers (Scharrer and Scharrer, 1937), a vast number of investigations into invertebrate neuroendocrine systems have been made. Several reviews on this subject have recently been published (Gersch, 1975; Golding, 1974; Goldsworthy and Mordue, 1974; Truman and Riddiford, 1974a).

In many of these studies on invertebrate neurosecretory cells, the biologically active secreted material has been shown to be peptidic. The purpose of this chapter is twofold: first, to provide an exposition of the present state of knowledge of peptides in invertebrate nervous systems; second, to alert the neurobiologist who is unaware of invetebrate systems to the potentials of these systems as models for vertebrate studies (for examples of invertebrate neurons being used in this fashion, see Chapters 8, 11, and 12). In this chapter, we have used as the minimum criterion for the characterization of the active factor (or hormone) as a peptide that the biological activity of the factor is destroyed by proteases. In many cases, considerably more sophisticated peptide chemistry has been done. The information about peptides in this chapter is presented phylogenetically. Table 1, however, presents the peptides according to their functional roles; the last column lists the chapter sections in which they are discussed.

Nora Frontali · Istituto Superiore di Sanitá, Viale Regina Elena, 229, Rome, Italy. **Harold Gainer** · Section on Functional Neurochemistry, Behavioral Biology Branch, National Institute of Child Health and Human Development, National Institutes of Health, Bethesda, Maryland.

TABLE 1. Biologically Active Peptides in Invertebrate Nervous Systems

Function of peptide	Phylum	Molecular weight (daltons)	Source	References[a]	Section
Chromatotropins					
Red pigment–concentrating hormone	Arthropoda (Crustacea)	1,030[b]	Sinus gland (eyestalk)	Fernlund and Josefsson (1968, 1972)	3.1
White pigment–concentrating hormone	Arthropoda (Crustacea)	<5,000	Sinus gland (eyestalk)	Skorkowski and Kleinholz (1973)	3.1
Black pigment–dispersing hormone (several forms)	Arthropoda (Crustacea)	~3,500	Sinus gland (eyestalk)	Hallahan and Powell (1971), Kleinholz (1972)	3.1
Retinal pigment light-adapting hormone	Arthropoda (Crustacea)	~2,000	Sinus gland (eyestalk)	Fernlund (1971)	3.2
Retinal pigment dark-adapting hormone	Arthropoda (Crustacea)	—	Sinus gland (eyestalk)	Fingerman and Mobberly (1960)	3.2
Hyperglycemic factors					
Hyperglycemic hormone	Arthropoda (Crustacea)	~6,300	Sinus gland (eyestalk)	Kleinholz and Keller (1973), Kleinholz (1975)	3.4
Hyperglycemic hormone (two factors)	Arthropoda (Insecta)	<1,500[c]	Corpora cardiaca	Natalizi and Frontali (1966), Natalizi et al. (1970), Traina et al. (1976)	4.8
Hyperglycemic hormone	Arthropoda (Crustacea)	~2,000	Corpora cardiaca	Stürzebecher (1969)	—
Adipokinetic hormone	Arthropoda (Crustacea)	<1,500	Corpora cardiaca	Mayer and Candy (1969)	4.9
Cardioactive peptides					
Cardioexciter substances (two)	Arthropoda (Crustacea)	<5,000	Pericardial organs	Belamarich (1963), Belamarich and Terwilliger (1966), Berlind and Cooke (1970)	3.3
Neurohormone C	Arthropoda (Insecta)	—	CNS, corpora cardiaca	Gersch et al. (1963)	4.1
Neurohormone D	Arthropoda (Insecta)	~2,000	CNS, corpora cardiaca	Gersch and Stürzebecher (1971)	4.1

Factor 2a	Arthropoda (Insecta)	~1,800	Brain, corpora cardiaca	Traina et al. (1976)	4.1
Factor 2b	Arthropoda (Insecta)	~1,300	Brain, corpora cardiaca	Traina et al. (1976)	4.1
Heart activator	Mollusca	~1,300	Neurohemal organ venae cavae	Blanchi (1973)	5.1
Substance "X"				Frontali et al. (1967), Welsh (1971)	5.1
A	Mollusca	>1,500	Ganglia and heart		
B	Mollusca	<1,500	Ganglia and heart		
C	Mollusca	<1,500[c]	Ganglia and heart		
Gut activation peptides P_1 and P_2	Arthropoda (Insecta)	—	Corpora cardiaca	Brown (1965)	4.2
Proctolin (neurotransmitter)	Arthropoda (Insecta)	648[b]	Hindgut, whole insects	Brown and Starratt (1975)	4.2
Salt and water regulation					
Diuretic hormone	Arthropoda (Insecta)	>30,000	6th abdominal ganglion (*Periplaneta*)	Goldbard et al. (1970)	4.3, 4.7
Antidiuretic hormone	Arthropoda (Insecta)	~8,000	6th abdominal ganglion (*Periplaneta*)	Goldbard et al. (1970)	4.3
Diuretic hormone (two forms)	Arthropoda (Insecta)	>60,000	Mesothoracic ganglion (*Rhodnius*)	Aston and White (1974)	4.3
Antidiuretic peptide	Mollusca	<2,000 / <10,000	Cell R_{15} in abdominal ganglion	Kupfermann and Weiss (1976)	5.3
Neuron spike activation					
Neurohormone D	Arthropoda (Insecta)	~2,000	Corpora cardiaca	Gersch and Stürzebecher (1971)	4.6
Spontaneous firing–increasing (SFI) factors	Arthropoda (Insecta)	<1,500[c]	Corpora cardiaca	Natalizi et al. (1970)	4.6
Circadian modulator	Arthropoda (Crustacea)	—	Eyestalk	Aréchiga et al. (1974)	3.6
BPP activator	Mollusca	1,100	Ganglia	Ifshin et al. (1975), Mayeri and Simon (1975)	5.2

(continued)

TABLE 1. Biologically Active Peptides in Invertebrate Nervous Systems (cont.)

Function of peptide	Phylum	Molecular weight (daltons)	Source	References[a]	Section
Reproductive processes					
Monogamic factor	Arthropoda (Insecta)	5,400	Male paragonial glands (*Drosophila*)	Bauman (1974)	4.10
PS1 (matrone)	Arthropoda (Insecta)	(2,700 subunit)	Male paragonial glands (*Drosophila*)	Bauman (1974)	4.10
PS2		2,000			4.10
Matrone	Arthropoda (Insecta)	~90,000	Whole males (*Aedes*)	Fuchs and Hiss (1970)	4.10
Egg-laying hormone	Mollusca	6,000	Bag cells (*Aplysia*)	Toevs and Brackenbury (1969), Arch *et al.* (1976*a,b*)	5.4
Egg capsule–laying hormone	Mollusca	—	Parietal ganglion (*Busycon*)	Ram (1975)	5.4
Radial nerve factor	Echinodermata	4,800	Radial nerve	Chaet (1967)	6.1
Radial nerve factor	Echinodermata	2,100	Radial nerve	Kanatani *et al.* (1971)	6.1
Growth and development					
Head-activator peptide	Coelenterata	1,000	*Hydra* hypostome Rat hypothalamus	Schaller (1973) Schaller (1975)	2.1 2.1
Neck-inducing factor	Coelenterata	1,650 (subunit)	*Chrysaora*	Loeb (1974*b*)	2.2
Molt-inhibition	Arthropoda (Crustacea)	—	Eyestalks	Rao (1965)	3.5
Prothoracicotropic hormone	Arthropoda (Insecta)			Ishizaki and Ichikawa (1967)	4.4
BI		~31,000	Brain (*Bombyx*)		
BII		~12,000	Brain (*Bombyx*)		
BIII		~9,000	Brain (*Bombyx*))		
Prothoracicotropic hormone	Arthropoda (Insecta)	~20,000	Brain (*Bombyx*)	Yamazaki and Kobayashi (1969)	4.4

Activation factors	Arthropoda (Insecta)			Gersch and Stürzebecher (1970)	4.4
I		~50,000	Brain		
II		10–20,000	Brain		
Embryonic hormones (two factors)	Arthropoda (Insecta)	5–10,000	Head (*Bombyx*)	Sonobe and Ohnishi (1971)	4.5
Embryonic hormone (one of two factors)	Arthropoda (Insecta)	2–4,000	Head (*Bombyx*)	Isobe et al. (1973)	4.5
Bursicon	Arthropoda (Insecta)	~40,000	Brain, thoracic ganglia	Mills and Nielsen (1967)	4.7
				Fraenkel et al. (1966)	
Puparium tanning factor (PTF) (two forms)	Arthropoda (Insecta)	~26,000 (subunits) ~80,000 (subunits)	Nervous system, hemolymph (*Sarcophaga*)	Sivasubramanian et al. (1974)	4.2
Puparium retraction factor (PRF)	Arthropoda (Insecta)	90,000 (subunits)	Hemolymph (*Sarcophaga*)	Sivasubramanian et al. (1974)	4.7

[a] A more detailed reference list is presented in the text.
[b] Amino acid sequence known.
[c] Chromatographic effect could cause this figure to be underestimated.

2. Coelenterata

Probably the simplest nervous system in the animal kingdom exists in the coelenterates as a primitive nerve net concentrated in the hypostome and at the base of the tentacles. Typical neurosecretory cells have been identified in these organisms, and appear to be involved in the control of growth (Burnett et al., 1964; Lentz, 1965).

2.1. Head-Activator Peptide

A growth-activating substance that induced head and bud formation and gonadal regression was found in extracts and subcellular fractions from the hypostome of the freshwater cnidarian *Hydra* (Burnett and Diehl, 1964; Lentz, 1965; Schaller and Gierer, 1973). This substance has been isolated, and is a peptide of about 1000 daltons, which is active at concentrations less than 10^{-10} M (Schaller, 1973). The peptide has a remarkable affinity to Sephadex and migrates 5 times slower than NaCl on G-10 Sephadex, thereby allowing a 10,000-fold enrichment of the peptide from crude extracts in a single separation step.

Recent studies by Schaller (1975) have shown that a relative of *Hydra* (*Actinia equina* L.) and rat brain contain a peptide very similar or identical to the "head-activator" of *Hydra*. While extracts from rat lung had no activity, similar extracts from rat brain contained activity equivalent to at least 20,000 *Hydra*. This peptide was principally concentrated in the hypothalamus. While this "head-activator" peptide from rat hypothalamus stimulated head or bud formation in regenerating *Hydra* at less than 10^{-10} M concentrations, application of thyrotropin-releasing hormone (TRH), luteinizing hormone–releasing hormone (LH–RH), somatostatin, or neurotensin at 10^{-8} M was ineffective. The function of this unique peptide in mammalian hypothalamus is unknown at present.

2.2. Neck-Inducing Factor

In the scyphozoan jellyfish *Chrysaora quinquecirra,* the metamorphosis from sessile polyp to free-swimming medusa is referred to as the process of *strobilation*. This process is regulated by the environmental temperature (Loeb, 1972) and is, in part, dependent on the presence of a "neck-inducing" factor (peptide) in the organism (Loeb, 1974a,b; 1975). This peptide is dissociated into subunits of about 1650 daltons in 7 M guanidinium hydrochloride, but in the absence of this dissociating agent can recombine in aggregates greater than 200,000 daltons (Loeb, 1974b). Loeb speculates that the unique aggregating property of this peptide would be advantageous both for affecting the animals that release it and for communication among animals in a colony in the sea.

3. Arthropoda (Crustacea)

3.1. Peptides That Act on Tegumentary Chromatophores

Many crustaceans have the ability to change their body color in order to match their background. These color changes are brought about by hypodermal chromatophores, i.e., specialized cells containing movable pigment granules and having richly ramified cell processes. These cells may be distinguished according to color in erythrophores, melanophores, leukophores, and xantophores; the color display is obtained when the pigment granules disperse throughout the cell, the opposite when they concentrate in a small central spot. These movements are controlled through neurohormones that are released from nerve endings in the sinus gland, a neurohemal organ that in most decapod Crustacea is located in the eyestalks, while the neuronal cell bodies are located in the X organ and in the CNS.

Several neurohormones that are effective in dispersion or concentration of pigment within chromatophores have been separated from extracts of crustacean eyestalks, and some have been shown to be peptides. A "red pigment–concentrating hormone" was separated from eyestalks of prawns by Edman *et al.* (1958), then by Josefsson and Kleinholz (1964), Kleinholz and Kimball (1965), and Josefsson (1967). The latter author showed that it is quickly inactivated following incubation with pepsin or chymotrypsin, but it resists treatment with trypsin. Fernlund and Josefsson (1968, 1972) further purified the hormone by means of gel filtration on columns of Sephadex G-25 and, using different mixtures of water and butanol as solvents, on Sephadex LH-20. The purified peptide had an ultraviolet (UV) absorption curve identical to tryptophan; quantitative amino acid analysis, in addition to UV spectrometry, allowed its composition to be established, but sequencing was made difficult by a blocked NH_2-terminal. Digestion of the hormone with thermolysin and fractionation of the digest on a column of Sephadex G-25 allowed the separation of two peptide fragments. The smaller, with a blocked NH_2-terminal, was analyzed by means of mass spectrometry; the larger was sequenced with the Edman-dansyl method. The result was the following sequence: pGlu–Leu–Asn–Phe–Ser–Pro–Gly–Trp–NH_2 (the presence of the aspartic residue as amide was inferred from the electrophoretic immobility of the hormone). This structure was confirmed by chemical synthesis and biological testing; the purified and the synthetic hormones were active in picogram amounts in concentrating the pigment of erythrophores (see also Fernlund 1974). The synthetic hormone was shown by Fingerman (1973) to be completely inactive on leukophores and on melanophores. Sodium ions are necessary for maximal response; ouabain inhibits the response, whereas tetrodotoxin enhances it (Fingerman and Connell, 1968). As the red pigment becomes concentrated in the presence of the hormone, the membrane potential of the red chromatophore becomes hyperpolarized (Freeman *et al.,* 1968).

F.A. Brown (1935) reported that the eyestalks of the prawn *Palaemonetes vulgaris* contain a substance that causes pigment dispersion in leukophores.

Fingerman and Rao (1969) and Fingerman (1970) showed that the movement of pigment in leukophores is controlled by two substances, a dispersing and a concentrating one. The white pigment–concentrating hormone has been isolated by Skorkowski (1971, 1972) from extracts of *Crangon crangon* and of *Rhithropanopeus harrisi* eyestalks by means of gel filtration on columns of Sephadex G-25. Further purification was achieved by Skorkowski and Kleinholz (1973) by means of partition between butanol and water. The purified hormone was quickly inactivated by incubation with trypsin and with pronase; the active peaks from two species were eluted from the column in the same position.

The presence of a black pigment–dispersing activity has been demonstrated by several authors in eyestalk extracts of various Crustacea. Perez-Gonzalez (1957) demonstrated that this activity of *Uca pugilator* extracts did not resist treatment with chymotrypsin. Stephens *et al.* (1956) separated electrophoretically three areas of melanin-dispersing activity from extracts of *Uca* sinus glands. Hallahan and Powell (1971) subjected heated aqueous extracts of *Uca pugilator* sinus glands to gel filtration on Sephadex G-25, and obtained a single peak of activity with a molecular weight of 3500, as estimated from calibration of the column with compounds of known molecular weight. Kleinholz (1972) subjected the material similarly purified from eyestalk extracts from different crustacean species (Sephadex G-25) to ion-exchange chromatography on DEAE cellulose and separated three peaks of melanin-dispersing activity, which he called *alpha, beta,* and *gamma.* Each of these peptide fractions still resolved into subfractions when subjected to ion-exchange chromatography on CM cellulose.

3.2. Peptides That Act on Retinic Pigments

The ommatidia of higher crustacean compound eyes have three sets of pigments distinguished as "distal," "proximal," and "reflecting." All three show photomechanical movements, but the best studied refer to the distal pigments. The physiological value of these movements is to screen the photosensitive component of the ommatidium (the rhabdom) in bright light and to uncover it in the dark or in light of low intensity (Kleinholz, 1961). These migrations are under the control of hormonal factors, again elaborated by neurosecretory cells and released by nerve endings in the sinus gland, situated in the eyestalks. Kleinholz (1934, 1936) found that extracts prepared from eyestalks of various crustacean species, injected into dark-adapted *Palaemonetes* prawns, caused migration of the distal pigment to the fully light-adapted position. Fingerman *et al.* (1959) and Fingerman and Mobberly (1960) (also working with *Palaemonetes*) injected eyestalk extracts into prawns adapted to an intermediate condition of light. During the first 2 hr, the distal pigment migrated to the light-adapted position, and in time to the dark-adapted position. Two hormonal responses were evidently taking place in temporal succession. Both hormones, which these authors called *LAH* (light-adapting) and *DAH* (dark-adapting), were heat-stable, could be inactivated by trypsin, and migrated in an electric field. They could be separated from each other by means of electrophoresis at pH 9.0 or (Fingerman *et al.,* 1971) by means of gel filtration on

Bio-gel P-6. As regards the LAH, the more studied of the two, it was separated from the red pigment–concentrating hormone by means of gel filtration on Sephadex G-25 (Kleinholz and Kimball, 1965) or ion-exchange chromatography on Dowex 50 (Josefsson, 1967). Fernlund (1971) subjected eyestalk extracts of *Pandalus borealis* to partition between an organic and an aqueous phase, and then to gel filtration on Sephadex G-25. When the active peak was chromatographed on a column of CM-Sephadex, it separated into four different active components. The main component was further purified and subjected to amino acid analysis. It turned out to be an octadecapeptide with a molecular weight around 2000, containing 12 different amino acid residues, i.e., Arg, Asp, Thr, Ser, Glu, Pro, Gly, Ala, Val, Met, IsoLeu, and Leu. The *N*-terminal amino acid was blocked.

3.3. Cardioactive Peptides

Alexandrowicz and Carlisle (1953) demonstrated that an aqueous extract of pericardial organs (neurosecretory structures located in the pericardial cavity of decapod and stomatopod Crustacea) acted as a powerful cardioexciter. The extracts increased the amplitude and the frequency of the heartbeat in all crustacean species studied, except for *Maja squinado,* in which only the amplitude was increased. Florey and Florey (1954) extracted from the central ganglia and peripheral nerves of Crustacea a substance that accelerated the crustacean heart and that was tentatively identified as 5-HT. They suggested that this could be the active substance present in the pericardial organs. Maynard and Welsh (1959) found 5-HT to be present in pericardial organ extracts, but at concentrations too low to account for the whole activity of the extract. Moreover, the active principle, which could be separated from 5-HT by means of paper chromatography, was destroyed by treatment with trypsin or with chymotrypsin. It was dialyzable, soluble in ethanol and in water but sparingly in acetone, and heat-stable in neutral and acid milieu. It was therefore identified as a peptide. Belamarich (1963) and Belamarich and Terwilliger (1966) separated two cardioexciter and peptidic factors by means of paper chromatography from crab pericardial organs. A preliminary analysis of their amino acid composition demonstrated the presence of Lys, Glu, Asp, Ser, Gly, Ala and Val, and the absence of Trp, His, Arg, and sugar residues. Berlind and Cooke (1970) also found the two peptidic cardioexciter substances, which they separated by means of gel filtration on Sephadex G-25, in the perfusion fluid of electrically stimulated pericardial organs. Terwilliger *et al.* (1970) subjected lobster pericardial organ sucrose homogenates to density-gradient centrifugation, and found that the cardioexciter peptidic material was associated with a subcellular fraction rich in electron-dense granules with a diameter of approximately 1500 Å. Cooke and Hartline (1975) assayed a pericardial organ extract on electrophysiological preparations of isolated lobster cardiac ganglia, and found that all cells reacted with an increase of the average firing frequency, but only if the application was made in the region between the soma and the proximal site of impulse initiation.

3.4. Hyperglycemic Factors

Abramowitz *et al.* (1944) demonstrated that injection of crab eyestalk extracts into crabs produced a rapid and large increase of hemolymph glucose levels lasting several hours. Kleinholz *et al.* (1967) found that this activity of eyestalk extracts was heat-labile and nondialyzable, and was abolished by treatment with chymotrypsin, pepsin, and pronase, but not by papain or by trypsin. They concluded that the factor was probably a protein. Gel filtration on Sephadex G-100 gave a peak of activity in an intermediate position between V_0 and V_t. This peak, subjected to several separation procedures, turned out to be not homogeneous, but made up of several active components. Crustacean hyperglycemic hormones exhibit a certain species-specificity. Keller (1969) and Kleinholz and Keller (1973) demonstrated that the maximal hyperglycemic response is obtained with eyestalk extracts of the same or of closely related species. The latter authors estimated (from gel filtration on Sephadex G-100) the molecular weights of the hyperglycemic hormones extracted from *Cancer magister, Pandalus jordani,* and *Orconectes limosus* to be all around 6300 daltons, but on chromatography on DEAE-Sephadex, their elution volumes were different. Acrylamide-gel electrophoresis revealed that each contained more than one active component. The amino acid composition of a hyperglycemic hormone from eyestalks of *Cancer magister* has recently been reported (Kleinholz, 1975).

As regards the mechanism of action of these hormones, Bauchau *et al.* (1968) have shown that they activate crustacean muscle phosphorylase, probably through stimulation of cyclic (c) AMP synthesis. Keller and Andrew (1973) observed that the hyperglycemic response is associated with a marked decrease in glycogen content of various organs, especially muscle and tegumentary tissue, and with decreased *in vivo* incorporation of C^{14}-labeled glucose in these tissues.

3.5. Molt-Inhibiting Factor

The earliest observations of accelerated molting in Crustacea as a consequence of the ablation of both eyestalks were made by Zeleny (1905). It was later found that this effect is due to the ablation of the X organ and the sinus gland, which synthesize and secrete, respectively, a molt-inhibition hormone (Passano, 1960). Rao (1965) purified the hormone extracted from eyestalks of the crab *Ocypode macrocera,* and reported it to be soluble in ethanol, methanol, and phenol, and to be inactivated by incubation with trypsin.

3.6. Circadian Modulator

The crab *Carcinus maenas* has been shown by Aréchiga *et al.* (1974) to exhibit circadian variations of rhythmic activities in the nervous system (responsiveness to light, level of excitement of efferent neurons, and reflex activity) that are modulated by humoral factors present in eyestalk extracts. The active material is dialyzable and heat-stable, and is inactivated by treatment with pronase.

4. Arthropoda (Insecta)

4.1. Peptides That Act on the Heartbeat Rate

Cameron (1953) demonstrated that an aqueous extract of insect corpora car-diaca (CC) caused an increase in the heartbeat frequency of a semiisolated heart preparation of the same species. Unger (1956, 1957) separated two cardio acceler-ating factors (which he called neurohormones C and D) from aqueous extracts of cockroach brains, subesophageal ganglia, or ventral cords. By means of differen-tial extraction with organic solvents, Gersch et al. (1960) separated four car-dioaccelerating factors from cockroach CC, Ralph (1962) as many as six. Davey (1961a,b) demonstrated that the heart-accelerating activity of cockroach CC ex-tracts resisted heating at 100°C, and was abolished by treatment with trypsin or pepsin. He concluded that the active factor appeared to be a peptide. Gersch et al. (1963) treated neurohormones C and D separately with trypsin, and found that both were inactivated. Gersch and Stürzebecher (1967) estimated the molecular weight of neurohormone D to be around 2000. In addition, B.E. Brown (1965) separated two heart-accelerating factors (which he called P_1 and P_3) from extracts of cockroach CC by means of paper chromatography. Both factors resisted heating at 100°C, and were inactivated by chymotrypsin; P_1 was less dialyzable than P_3. By combining ion-exchange chromatography and gel filtration, Natalizi et al. (1970) separated from watery extracts of cockroach CC heated at 100°C, four heart-accelerating factors (peaks 1, 2, 5, and 6) that were all inactivated by proteolytic enzymes. Peak 1 was found by Traina et al. (1974) to be relatively heat-labile, as is neurohormone D (Baumann and Gersch, 1973), and was tentatively identified with the latter on the basis of its chromatographic behavior as well, whereas peak 2 is better preserved if the extract is heated. Both neurohormone D (Richter, 1967) and peak 1 (Natalizi et al., 1970) characteristically increase the steepness of the cockroach electrocardiographically recorded waves. Improving their separation methods, Traina et al. (1976) later subdivided peaks 1 and 2 into two components each. Factors 1a and 1b were inactivated by trypsin, whereas factors 2a and 2b were inactivated by chymotrypsin, but not by trypsin. Purified preparations of fac-tors 2a and 2b were subjected to amino acid analysis. The data fit with the hypothesis of two peptides (or groups of similar peptides), both containing tryp-tophan, lacking sulfur-containing amino acids, and having molecular weights of approximately 1800 (2a) and 1300 (2b). Factors 1a, 1b, 2a, and 2b were also found to be present in cockroach brain extracts.

Besides brain and CC, the heart-accelerating activity has also been found in extracts of *Periplaneta americana* "perisympathetic organs" (Raabe et al., 1966), cardiac nerves (Johnson and Bowers, 1963), and heart (Natalizi et al., 1970), and in extracts of nervous or neurosecretory tissues of many other insects, includ-ing locusts (Cazal, 1967; Mordue and Goldsworthy, 1969), moths (Gersch, 1958), and honeybees (Natalizi and Frontali, 1966), but not in extracts of housefly heads (Normann, 1972). In locust CC, where a glandular lobe may be separated from a storage lobe, Mordue and Goldsworthy (1969) have found by means of paper chromatography that the storage lobe contains mainly a peptide with characteristics similar to Gersch's neurohormone D, whereas the glandular lobe contains mainly a peptide with the characteristics of neurohormone C.

Fractionating cockroach CC sucrose homogenates into subcellular components, Evans (1962) found that the heart-accelerating activity was associated with a particulate fraction that, at the electron-microscopy (EM) level, was shown to be largely made up of electron-dense, membrane-bound neurosecretory granules, similar to those seen in EM preparations of intact CC. Such granules are also visible in EM pictures of the nerves emerging from the CC and lining the heart (Johnson and Bowers, 1963), and in nerve endings making synapses on the heart muscle fibers (Normann, 1965; Johnson, 1966). These findings support the hypothesis of a direct neuronal transport of the neurohormones to the target organs. The hypothesis by Davey (1961b) of an indirect action of the cardioactive peptides through a hypothetical factor released from the pericardial cells has lost favor.

Several experiments on the release of heart-accelerating factors into the perfusion medium as a consequence of electrical stimulation have been performed on *in vitro* preparations of *Periplaneta americana:* release from isolated heads following stimulation of the brain (Kater, 1968), from 6th abdominal ganglia following stimulation of the same (Gersch, 1974a), and from isolated CC following stimulation of the cardiac nerves. For this release, the presence of calcium ions was found to be essential, and the released heart-accelerating factor was identified as neurohormone D (Gersch et al., 1970). This electrically stimulated release of neurohormone D was found by Gersch (1972) to be increased by eserine and inhibited by atropine; he therefore concluded that a cholinergic mechanism may be involved. Gersch et al. (1969, 1971) have assayed neurohormone D on an *in vitro* preparation of isolated toad bladder (used as a model of an excitable membrane), and found that it was active in increasing the permeability of the membrane to sodium ions.

4.2. Peptides That Act on the Gut

Cameron (1953) observed that aqueous extracts of *Periplaneta americana* CC stimulate the peristaltic activity of cockroach and locust hindgut *in vitro*. Davey (1962) showed that this effect on the hindgut consists in an increase in tonus and in frequency, amplitude, and coordination of the contractions. By means of paper chromatography, B.E. Brown (1965) separated from extracts of *Periplaneta americana* CC two factors that act on the hindgut of the same species, which he called factors P_1 and P_2. Factor P_1 coincided with one of the two factors that act on the heartbeat frequency (see section 4.1); both were heat-stable, dialyzable, and inactivated by chymotrypsin.

Later, Brown (1967) separated a heat-stable and dialyzable factor from foreguts and hindguts of American cockroaches by dialyzing their homogenates against distilled water. This "gut factor" (which was later called "proctolin" by Brown and Starratt, 1975) caused a slow, graded contraction of the longitudinal muscles of the hindgut (proctodeum), a response similar to that obtained through electrical stimulation of the nerves innervating the hindgut. This factor differed from P_1 and P_2 in its action on the hindgut and in not being inactivated by chymotrypsin. Surgical section of the proctodeal nerves was followed after a few days by depletion of the "gut factor" from the hindgut. Holman and Cook (1972) separated a similar factor by means of ion-exchange chromatography starting from ethanolic extracts of Madeira cockroach hindguts. The biological activity of this factor was not abolished by treatment with chymotrypsin, but was completely destroyed by treatment with pronase. On the basis of this study and of gel-filtration experiments, this material was estimated to be a peptide with a molecular weight between 400 and 600. The characteristic response of the isolated hindgut to this peptide was a sustained increase in tonus, frequency, and amplitude of the contractions, a response that was not affected by the addition of tetrodotoxin to the medium, although tetrodotoxin blocked neurally evoked contractions of the visceral muscle. The peptide could be extracted from hindguts, 6th abdominal ganglia, proctodeal nerves, and heads of American and Madeira cockroaches and of a locust, and was also found to be active in increasing the beat frequency of cockroach heart preparations. Sucrose homogenates of heads or hindguts were subjected to differential centrifugation, and the biological activity on the hindgut was found to be associated with a subcellular fraction rich in neurosecretory granules. Neurosecretory axons emerging from the 6th abdominal ganglion could be seen histochemically (with Victoria blue) and ultrastructurally to innervate the proctodeal longitudinal muscle layer. The association of the gut-stimulating activity of cockroach CC homogenates with particulate material was confirmed by Sowa and Borg (1975), who further fractionated the sediment by means of density-gradient centrifugation.

The hindgut-stimulating hormone has also been shown to be synthesized by the neurosecretory cells of cockroach brains *in vitro,* to pass through nerves to the CC, and to be released into the medium (Holman and Marks, 1974; Marks and Holman, 1974).

Cook and Holman (1975) studied the mechanical and electrical events following application of the purified hindgut-stimulating hormone to cockroach hindgut segments *in vitro,* and found that the presence of calcium was essential. They suggested that the hormone may act at one or more of three sites, certainly including the third one: excitation–secretion coupling at the nerve terminal, synaptic transmission, and direct action potential excitation in the muscle fiber. Cook *et al.* (1975) also suggested, on the basis of experiments with aminophyllin and AMP (which mimic the effects of the hormone on the hindgut), that AMP might serve as a mediator of the neurohormone action, by increasing calcium transport across the membrane of the muscle fibers.

Brown and Starratt (1975) purified about 180 μg of the hindgut-stimulating hormone, which, as noted above, they called proctolin, from 125 kg of whole *Periplaneta americana* adults by means of a complex procedure involving extraction in perchloric acid, several steps of ion-exchange chromatography (on Dowex 50W and on Rexin 101), alumina-adsorption chromatography, countercurrent partition, paper chromatography, paper high-voltage electrophoresis, and gel filtration on Sephadex G-15. The isolated proctolin gave a single ninhydrin-positive spot on paper and thin-layer chromatography and on high-voltage paper electrophoresis. Starratt and Brown (1975) subjected the hydrolyzate to amino acid analysis by means of two-dimensional thin-layer chromatography after dansylation, revealing the presence of Arg, Leu, Pro, Thr, and Tyr in approximately equimolar amounts. Sequence analysis by means of the Edman dansyl method indicated the following structure: Arg–Tyr–Leu–Pro–Thr (mol. wt. 648). The structure was confirmed by synthesis of the peptide, cochromatography on paper and thin-layer plates, and coelectrophoresis. Pharmacological assays demonstrated that the specific activity of the synthetic product on a hindgut preparation *in vitro* was identical to that of the natural peptide, that the dose–response curves were parallel, and that both were inhibited by tyramine. Removal or replacement of each of the terminal amino acids led to losses of 99% of the activity. Proctolin was shown to be present in several insect species, representatives of six different orders. Brown (1975) studied the effects of proctolin on *in vitro* preparations of cockroach hindgut with intact nerve supply (but without the 6th abdominal ganglion, in which the motor cell bodies are located): the graded contractions of the muscles (which are slow and striated) evoked by repetitive nerve stimulation were simulated by proctolin 10^{-9} M. Proctolin was fully active on tetrodotoxin-treated or surgically denervated mucles, indicating that the receptors are located on the muscle fiber membrane. Tyramine, at threshold levels of 5×10^{-8} M, reversibly antagonized the responses evoked by nerve stimulation and by proctolin. On the basis of this and of other experimental evidence, the author suggests that the peptide may function as an excitatory neurotransmitter at the level of the rectal neuromuscular junctions. These junctions contain neurosecretory granules 1000–2000 Å in diameter, in addition to electron-lucent vesicles.

4.3. Peptides That Act on Diuresis and on Malpighian Tubule Movements

Extracts of nervous or neurosecretory tissues of various insects have been shown by several authors to affect diuresis. The physiological assays have been performed on widely different insect preparations. Clearance of dyes from the hemolymph (e.g., in *Periplaneta americana:* Wall and Ralph, 1964; in *Carausius morosus:* Unger, 1965; in locusts: Mordue and Goldsworthy, 1969), water excretion from completely isolated Malpighian tubules (in *Rhodnius prolixus* and *Carausius morosus:* Maddrell, 1963; Maddrell *et al.,* 1969; Pilcher, 1970) or from Malpighian tubules isolated together with part of the gut (in *Periplaneta americana:* Mills, 1967; in locusts: Cazal and Girardie, 1968; in various insects: De Besse and Cazal, 1968), and water absorption from the rectum (in *Periplaneta americana:* Wall, 1967; in locusts: Mordue, 1969) were used as bioassays. Factors active on diuresis were found in extracts of the terminal abdominal ganglion of cockroaches; the mesothoracic ganglionic mass of assassin bugs; the CC, brain, and subesophageal ganglion of the stick insect; the CC, brain, subesophageal ganglion, and abdominal nerve chain of locusts; and the "perisympathetic" organs of various insects.

Different species have obviously different requirements as regards the maintenance of water balance; e.g., in *Rhodnius prolixus,* secretion by the Malpighian tubules is very intense immediately after a blood meal. Maddrell (1966) has shown that in this insect, more of the diuretic hormone is in the posterior portion of the mesothoracic ganglionic mass, and is released into the hemolymph from a series of swollen axons and nerve endings containing electron-dense granules that may be seen at EM to lie immediately beneath the fibrous nerve sheath of the abdominal nerves, shortly after their emergence from the ganglionic mass. Release from this true neurohemal system is probably triggered by the arrival of sensory inputs, originating from abdominal distension as a consequence of the blood meal, through the abdominal nerves (sectioning of these nerves interrupts this mechanism). It is not surprising that a neurohormone with the Malpighian tubules as target should be released into the hemolymph, since the tubules are immersed in it and do not seem to be innervated.

The chemical nature of the diuretic or antidiuretic factors demonstrated in the papers mentioned above has remained unclear until recently, when a few contributions indicated that some of these factors are peptides. With the help of a sensitive isotope-dilution technique applied to an *in vitro* preparation of the rectum, Goldbard *et al.* (1970) separated from watery extracts of 6th abdominal ganglia of *Periplaneta americana* two factors that were thought to be peptides on the basis of their molecular weight. Estimated from gel filtration, a diuretic hormone with a molecular weight above 30,000 and an antidiuretic hormone with a molecular weight around 8000 were found. Aston and White (1974) found that the diuretic activity on isolated Malpighian tubules of watery extracts of *Rhodnius prolixus* mesothoracic ganglionic masses or of freeze-dried heads and thoraxes was heat-labile, nondialyzable, and abolished by trypsin, chymotrypsin, and pronase. Of this activity, 80% was shown to be bound to particulate material, since it could be sedimented by centrifugation. Gel filtration of the supernatant was found to separate the diuretic activity into two distinct zones, one with high molecular weight (more than 60,000) and one with low molecular weight (less than 2000). When samples of hemolymph from freshly fed bugs were chromatographed under similar conditions, only the low-molecular active zone was found. Isolated mesothoracic ganglionic masses incubated *in vitro* in Ringer's solution with a high potassium content were found to release diuretic hormone in the medium. This activity has been shown to en.erge from the column together with the low-molecular-weight material. According to the authors, this material (which is very labile in solution at room temperature and has not yet been assayed for inactivation by proteolytic enzymes) qualifies as the physiologically active diuretic hormone. 5-HT, which is also active in the biological test, does not cochromatograph with the active zone.

Aston (1975) studied the relationship between the diuretic hormone released by *Rhodnius prolixus* mesothoracic ganglionic masses *in vitro* and cAMP. Direct measurements of cAMP levels during stimulation of the tubules by this preparation of diuretic hormone support the view that cAMP acts as a "second messenger" in this system. This had already been suggested by Maddrell and Gee (1974) on the basis of the observation that cAMP effectively mimicked the hormone's diuretic activity on the Malpighian tubules of *Rhodnius* and *Carausius.*

Not only diuresis, but also movements of the Malpighian tubules, are apparently under hormonal control. Pilcher (1971) found that the writhing movements of the Malpighian tubules of *Carausius* (produced by a few muscular fibers running along the tubules—fibers that do not seem to be innervated) are stimulated by 5-HT, 10^{-10} M, and also by extracts of brain, CC, subesophageal ganglion, and first thoracic and last abdominal ganglion. The factor is stable at 100°C at neutral or acid pH, but is inactivated by treatment with chymotryspin or pepsin. It therefore appears to be a peptide. Flattum *et al.* (1973) have observed a stimulatory effect on the frequency of the coiling movements of the Malpighian tubules of cockroaches and locusts *in vitro* resulting from addition to the medium of hemolymph from DDT-intoxicated animals. The authors suggest that the "autoneurotoxin" that was originally proposed by Beament (1958) and by Sternburg (1960) to be released from insect nervous tissue into the hemolymph under different stress conditions, including DDT intoxication, may consist of abnormally increased amounts of this hormone and perhaps other neurohormones.

4.4. Peptides (Proteins) That Act on the Prothoracic Gland (Prothoracicotropic Hormones)

The discovery by Wigglesworth (1940) that the molting of the bug *Rhodnius prolixus* was initiated by a hormonal factor secreted by a cluster of nerve cells located in the brain is considered one of the first demonstrations of a neurosecretory factor. Williams (1947) demonstrated that this factor, which he called "brain hormone," was active in promoting the secretion of the molting hormone (the steroid compound also called ecdysone) by the prothoracic glands. The protein nature of the brain hormone was first demonstrated by Ichikawa and Ishizaki (1963), who found the activity of methanolic extracts of *Bombyx mori* brains in promoting adult differentiation in brainless (permanent) pupae of *Phylosamia cynthia ricini* to be associated with a water-soluble, relatively heat-stable, nondialyzable material, which could be precipitated by ammonium sulfate, TCA, and acetone, and inactivated by subtilisin and pronase, but not by trypsin or pepsin.

In the last decade, the application of more advanced methods for separation of proteins has allowed considerable purification and characterization of the prothoracicotropic hormones, which turned out to be more than one. Ishizaki and Ichikawa (1967), using the biological test already mentioned, separated from aqueous, heated extracts of *Bombyx mori* brains three active fractions with approximate molecular weights of 31,000 (BH I), 12,000 (BH II), and 9000 (BH III). The separation steps were: ammonium sulfate precipitation, gel filtration on Sephadex G-100, chromatography on DEAE-cellulose, and again gel filtration. As little as 2 ng of the purified material was sufficient to cause development in a brainless pupa. Yamazaki and Kobayashi (1969) used as biological test material brainless pupae of *Bombyx mori*. The active substance they separated from aqueous extracts of acetone-dried brains of *Bombyx mori* pupae could not be identified with BHI, BH II, or BH III because of the completely different behavior of the substances on columns of DEAE- and of CM-cellulose. Its molecular weight, as estimated from gel-filtration on Sephadex G-100, is about 20,000; its isoelectric point, as determined by isoelectric focusing, is between 8.35 and 8.65. The hormone is heat-stable, and is inactivated by treatment with trypsin, pronase, and subtilisin, but not by chymotrypsin or sialidase. It contains sugar residues, and is therefore probably a glycoprotein. A quantity of 20 ng is sufficient to cause adult development in one brainless pupa.

No data are available for a comparison of these four factors isolated by the Japanese authors with the "activation factors I and II" that Gersch and Stürzebecher (1970) separated from aqueous extracts of brains of *Periplaneta americana* or of *Antherea pernyi* by means of gel filtration on Sephadex G-100 and biological testing on prothoracic glands of *Periplaneta americana* larvae (cytological test). Their molecular weights are reported to be 50,000 (I) and between 10,000 and 20,000 (II). Later, Gersch *et al.* (1973) and Gersch and Bräuer (1974) studied the effects of these two factors on isolated prothoracic glands of various insects *in vitro*. Factor II was found to increase the membrane potential of the prothoracic gland cells, whereas faction I significantly increased the incorporation into tissue RNA of [³H]uridine added to the medium.

Truman and Riddiford (1974*b*) studied the time of secretion of the prothoracicotropic hormone by the brain of tobacco hornworm larvae by means of ligaturing experiments at various stages. Secretion occurs on two occasions: the first lasts about 3.5 hr, and triggers transformation of the 5th instar larvae into the migrating form; the second occurs 2 days later, lasts 7 hr, and causes initiation of the pupal molt. In the first case, the prothoracic glands require the continuous presence of the hormone; in the second, the cells are "turned on" and continue to secrete ecdysone even after cessation of prothoracicotropic hormone secretion.

4.5. Silkworm Embryonic Diapause Hormone

As demonstrated by Hagesawa (1957, 1963), a diapause hormone (DH) secreted by the subesophageal ganglion of the silkworm *Bombyx mori* is responsible for the embryonic diapause in the eggs. It is released into the hemolymph and acts on oocytes, determining their ability to undergo hibernation after oviposition. In some strains of *Bombyx mori,* the eggs do not undergo hibernation (which is recognized from the development of a dark color due to accumulation of an ommochrome). Pupae of one of these strains are used as biological tests; i.e., injection of the hormone induces their eggs to diapause. Sonobe and Ohnishi (1971) purified DH from whole heads of adult female silkmoths, and found it to be heatstable, nondialyzable, precipitated by ammonium sulfate, and inactivated by incubation with proteolytic enzymes (pronase, subtilisin, trypsin, and chymotrypsin). Gel filtration on Sephadex G columns allowed separation of two factors with DH activity, the molecular weights of which were between 5000 and 10,000. The authors therefore consider them to be either proteins or complex molecules containing peptide linkages. Isobe *et al.* (1973) confirmed that the DH activity of head extracts is abolished by treatment with proteolytic enzymes and that two factors are present. For one, the molecular weight was estimated to be between 2000 and

4000 on the basis of gel filtration on Sephadex LH-20. This purified factor was found to contain several amino acids, as well as some other components.

4.6. Peptides That Act on Spontaneous Electrical Activity of Neurons

CC extracts are active on electrophysiological preparations of insect nerve cords. Experiments performed on cockroaches showed that such extracts depress the rate of spontaneous firing of neurons when applied at high concentrations (Ozbas and Hodgson, 1958), and increase this activity at lower concentrations (Milburn et al., 1960; Milburn and Roeder, 1962). The latter authors found that the active factor is heat-stable and resists incubation with chymotrypsin. Attempts at separation of this factor by means of paper-partition chromatography were made by Gersch and Richter (1963) and by Brown (1965), and both found that it cochromatographed with one of the heart-accelerating factors (neurohormone D/P_1; see Section 4.1). According to Brown, however, it differed from P_1, since it was not inactivated by incubation with chymotrypsin. By means of column ion-exchange chromatography on CM-cellulose and gel filtration on Sephadex G-15 and G-10, Natalizi et al. (1970) separated four heat-stable factors with this activity from CC extracts of *Periplaneta americana*. They were all inactivated by treatment with trypsin.

4.7. Bursicon and Other Tanning Factors

Cuticular tanning in insects involves primarily the oxidation of dihydrophenols to quinones, which then react with free amino groups of cuticular proteins. The subsequent cross-linking yields an extremely hard and insoluble exocuticle. The phenolic tanning agent appears to be *N*-acetyldopamine, which is formed through acetylation of dopamine (Karlson et al., 1962). Cottrell (1962a,b) and Fraenkel and Hsiao (1962) demonstrated independently that the process of hardening and darkening, which the cuticle of newly emerged blowflies undergoes following emergence, is under hormonal control. A ligature placed on the neck of a newly emerged fly prevents the cuticle from tanning, and injection of blood from a fly in the process of tanning into a ligatured fly causes the latter to tan. Cottrell (1962b) found that the active factor was nondialyzable, inactivated by alcohol and by incubation with subtilisin. Fraenkel and Hsiao (1965) proposed for this hormone the term "bursicon" and found it in extracts of both the brain neurosecretory cells and the combined ganglion of the thorax. As shown by ligaturing experiments, activation by the brain is essential in blowflies, and represents a control of the initiation of tanning modulated via the brain according to external stimuli. In fact, tanning is delayed by many hours if the fly is prevented from digging itself out of the medium (e.g., sand) from which it has emerged. The hormone was shown by these authors to be relatively heat-labile and to be precipitated by typical protein precipitants—with loss of activity in the case of TCA, alcohol, acetone, but with full retention of activity in the case of ammonium sulfate. It is non-dialyzable, and is inactivated by treatment with trypsin or pronase. They also showed that it was not species-specific. Blood from newly molted *Periplaneta americana* larvae, or adults in the process of tanning, was active when injected into blowflies. In cockroaches, ligaturing of the neck did not prevent tanning. Presumably, this insect does not require a mechanism to delay tanning until the establishment of favorable conditions, as is the case for flies.

Mills *et al.* (1965) extended these observations on *Periplaneta americana* and found that ligatures placed between thorax and abdomen and between various abdominal segments in freshly ecdysed cockroaches always confined the cuticular darkening to the posterior portions. The main source of the hormone was found to be the 6th abdominal ganglion (later, Vandenberg and Mills, 1974, reported it to be more exactly the neurohemal organs located posteriorly to the last three abdominal ganglia). Vincent (1971) also found that in locusts the sites of release of bursicon are the abdominal ganglia.

Mills and Lake (1966) and Mills and Nielsen (1967) purified bursicon to a certain extent, starting from ventral nerve cords or from whole newly molted American cockroaches, using ammonium sulfate fractionation, gel filtration, and column chromatography on DEAE cellulose, and estimated its molecular weight to be approximately 40,000. This agreed with the figure that Fraenkel *et al.* (1966) had obtained for blowfly bursicon. The latter authors also found that the electrophoretic characteristics of hormone preparations from the brain, the hemolymph, and the ganglia of blowflies and cockroaches were similar, but not identical.

Bursicon appears to be not merely a tanning agent, but to affect cuticular differentiation *in toto*. By means of ligaturing experiments, Fogal and Fraenkel (1969) have shown that purified bursicon controls both the formation of melanin and the endocuticle deposition that follows it. Injection of actinomycin D or of puromycin into newly molted flies inhibited the formation of endocuticle, but did not affect melanization. The action of bursicon on endocuticle formation is therefore mediated through stimulation of RNA and protein synthesis, but synthesis of the enzyme DOPA-decarboxylase (the presence of which is necessary for melanization) does not seem to be under its control. Bursicon caused enhanced uptake into the epidermis of [^{3}H]leucine injected in flies, but did not affect the [^{3}H]uridine uptake. The site of action of bursicon on RNA synthesis related to endocuticle formation is evidently a different tissue, perhaps the fat body.

When purified preparations of bursicon were added *in vitro* to homogenates of freshly ecdysed American cockroaches, no effect could be demonstrated by Mills *et al.* (1967) on any of the enzymatic reactions involved in the metabolism of tyrosine to *N*-acetyldopamine. Therefore, Mills and Whitehead (1970) suggested that the mode of action of the hormone was instead to increase membrane permeability (e.g., of the hemocytes), thus making tyrosine (the concentration of which in the hemolymph was found to rise before ecdysis) available to the enzymes present inside the hemocytes. On the basis of the observation that diuretic hormone and bursicon rise simultaneously in the hemolymph after ecdysis, and of the similarity in molecular weights and in sites of release, the authors suggest that bursicon and diuretic hormone (see section 4.3) are the same substance.

Cyclic AMP can mimic the action of bursicon in increasing cuticular darkening in the ligatured cockroach thorax (Vandenberg and Mills, 1974). Cyclic AMP could therefore act as a second messenger, the synthesis of which would be stimulated by bursicon, and in turn would increase membrane permeability to tyrosine and other precursors.

Some other protein factors have been shown to be active in modifying the chitinous integument of insects. Zdarek and Fraenkel (1969) reported the existence of a hormone, derived from neurosecretory brain cells of blowfly larvae, that accelerates puparium formation. The factor was also present in the hemolymph of pupariating larvae. It was later recognized (Fraenkel et al., 1972) that two factors were actually involved, one active on tanning of the puparium (PTF), and one on its retraction (ARF). They were both inactivated by alcohol and acetone, precipitated by TCA and ammonium sulfate, nondialyzable, and destroyed by trypsin or pronase. In an effort to characterize the factors by means of gel filtration on Biogel P-300 and SDS gel electrophoresis, Sivasubramanian et al. (1974) found that ARF, which is present mainly in the hemolymph and is heat-labile, has a molecular weight of approximately 180,000 (with subunits of 90,000 each), whereas PTF, which is heat-stable and has a molecular weight of 312,000, exists in two forms—one, found in the nervous system, with subunits of 26,000; the other, in the hemolymph, with subunits of 80,000.

4.8. Hyperglycemic Peptides

Insect hemolymph contains glucose, but much higher concentrations of the nonreducing disaccharide trehalose. Its concentration was found by Steele (1961) to be strikingly increased in *Periplaneta americana* as a consequence of the injection of an extract of CC. The increase began 30 min after injection, reached a maximum of 150% after 5 hr, then decreased, reaching normality after approximately 48 hr. The extract was prepared in insect Ringer with heating at 100°C. This increase (Steele, 1963) was accompanied by a reduction of the glycogen reserves of the fat body (but not of muscle). The CC extracts were also active *in vitro*, increasing glycogenolysis of fat body tissue, and, at higher hormone concentrations, of ventral nerve cords as well. The author suggested that the hormone acts by increasing the activity of the enzyme phosphorylase, as do catecholamines and glucagon on liver, the only difference being that whereas liver dephosphorylates glucose-6-phosphate to glucose, the insect fat body has an enzymic mechanism for the conversion of glucose-phosphate to trehalose (Candy and Kilby, 1961). Hart and Steele (1973) provide further evidence to suggest that glycogenolysis in the ventral nerve cord is accelerated by the CC extract through an increase in the level of cAMP, which activates phosphorylase.

Hyperglycemic activity has also been found in CC extracts of blowflies (Friedman, 1967) and of locusts (Goldsworthy, 1969). The hormonal activity is not species-specific (Natalizi and Frontali, 1966; Friedman, 1967).

The hormone has been shown by Natalizi and Frontali (1966) to be completely inactivated by trypsin. At least two peptidic factors are actually present in CC extracts, and were separated by Brown (1965) by means of paper-partition chromatography and by Natalizi and Frontali (1966) by means of column chromatography on SE-Sephadex. Mordue and Goldsworthy (1969) found that one of the paper-chromatographically separated factors is extractable mainly from the glandular lobe of *Schistocerca gregaria* CC, the other from the storage lobe. Efforts toward their purification by means of gel filtration (Natalizi *et al.*, 1970; Traina *et al.*, 1976) showed that the active factors are retained by columns of Sephadex G-15, as are some of the cardioaccelerating factors, suggesting that they are relatively small peptides, containing aromatic amino acid residues.

Electrical stimulation of the blowfly brain was shown by Normann and Duve (1969) to elicit release of hyperglycemic hormone into the hemolymph, but only if the nerve connections between brain and CC were intact. After these connections are sectioned in locusts, the CC partially regenerate, but the new tissues lack the glandular component. Extracts of these regenerated CC were found by Highnam and Goldsworthy (1972) still to contain the hyperglycemic hormone. Gersch *et al.* (1970) have shown that electrical stimulation of the nervus CC II (but not of nervus CC I) causes the release of the hyperglycemic hormone in an *in vitro* preparation of cockroach CC. This finding was confirmed by *in vivo* experiments (Gersch, 1974*b*) in which the concentration of trehalose in the hemolymph was significantly reduced by sectioning of nervus CC II, but sectioning of nervus CC I had no such effect. It appears that the hyperglycemic hormone or hormones are produced by the brain medial neurosecretory cells and are transported through the nervus CC II to the CC, from which they are presumably released into the hemolymph.

4.9. Adipokinetic Hormone

The fatty acid (especially diglyceride) level in the hemolymph of locusts increases considerably during flight. Mayer and Candy (1969) have shown that the same happens when extracts of *Schistocerca gregaria* CC are injected into adult locusts. The CC extract also acts *in vitro*, stimulating the release of diglyceride from the fat body. Hemolymph from flown locusts contains significant amounts of hormone, whereas hemolymph from unflown locusts does not. The adipokinetic hormone is stable to boiling, but is inactivated by treatment with proteolytic enzymes; from a column of Sephadex G-15, it emerges between V_0 and V_t. The hormone, secreted by the CC, apparently contributes to the regulation of diglyceride concentration in locust hemolymph during flight, stimulating its release from the fat body.

4.10. Peptide and Protein Pheromones from Male Accessory Glands

Female insects after mating exhibit important changes in their behavior, consisting mainly in decreased receptivity toward successive matings (monogamy) and increased rate of oviposition. Both these changes have been shown in some species to be induced by humoral factors brought over to the female through the sperm or male accessory gland material, and therefore belonging to the category of pheromones. Some excellent reviews of the rich literature on this subject have been published (Truman and Riddiford, 1974*a;* de Wilde and de Loof, 1973). These factors are not neurosecretory, but they are mentioned in this chapter because they apparently affect the nervous system.

In a search for chemical markers for genetic studies in *Drosophila,* Fox *et al.* (1959) found that paper chromatograms of whole males contained a ninhydrin-positive spot that was absent in females and that was due to peptidic material ("sex peptide"). This material was later shown by Chen and Diem (1961) to be localized in the male paragonial glands and by Baumann and Chen (1973) to be actually made up of two substances, paragonial substance 1 (PS1) and PS2, with pheromone activity. PS1 was found to be a monogamic factor, while PS2 stimulated oviposition. The two substances have been further purified from methanolic extracts of *Drosophila funebris* paragonial glands (Baumann, 1974). PS1 was shown to be a peptide consisting of 27 amino acid residues. Two forms of PS1 were present in the ratio of 7:3, differing only in the presence of Val or Leu; fractions containing partially purified Val-PS1 and Leu-PS1 had the same female receptivity–lowering activity. The amino acid composition was found to be: Lys (1), Arg (1), Asp (6), Thr (1), Ser (3), Glu (3), Pro (2), Ala (9), and Val or Leu (1). The molecular weight, as estimated by means of Sephadex G-50, was found to be 2700 in 50% acetic acid and 5400 in 1 M acetic acid, suggesting dimerization. PS2 was found to be a low-molecular-weight substance containing GLY, and in addition carbohydrate. However, its chemical nature is not established.

Another group of workers (Fuchs *et al.,* 1969; Fuchs and Hiss, 1970) purified the monogamic factor (which they called "matrone") from whole adult males of the yellow fever mosquito *Aedes aegypti.* The factor was found to be non-dialyzable, readily precipitable by 60% ammonium sulfate, heat-labile, and inactivated by chymotrypsin. Gel filtration on Sephadex G-100 led to loss of activity because of separation of the protein into two subunits, alpha and beta, with molecular weights of 60,000 and 30,000, respectively, which needed to be combined for full development of the monogamic effect. The alpha subunit alone, however, was found by Hiss and Fuchs (1972) to induce the increased rate of oviposition behavior. On the basis of surgical experiments, Gwadz (1972) suggested that the copulatory behavior in the female of *Aedes aegypti* is regulated by the terminal abdominal ganglion, and that this ganglion is the site of action of matrone.

4.11. Neurosecretory Factors with As Yet Undemonstrated Peptidic Nature

There is a whole series of physiologically active factors the neurosecretory origin of which has been postulated on the basis of nerve transection experiments, e.g., a factor inducing "calling behavior" in virgin female adults of silkmoths (Riddiford and Williams, 1971). Severance of the connections of their CC with the brain has been shown to inhibit this behavior. For other factors, a humoral nature has been fully demonstrated through the preparation of extracts from nervous or neurosecretory tissue and the induction, following their injection, of definite physiological effects. Their neurosecretory origin suggests a peptidic or proteinaceous nature, but this nature has not yet been investigated (e.g., no experiments with proteolytic enzymes have been reported). Such is the "eclosion hormone" demonstrated by Truman (1971, 1973a) in silkmoths, the secretion of which by the brain and release through the CC into the hemolymph triggers the onset of preeclosion behavior and, in the tobacco hornworm (Truman, 1973b; Truman and Endo, 1974), the secretion of bursicon.

Another factor that may be mentioned in this section is active on mitochondrial oxidative metabolism. Removal of the CC was shown manometrically by Keeley and Friedman (1967, 1969) to depress respiratory metabolism of whole adult male cockroaches (*Blaberus discoidalis*) and of their excised fat body tissue *in vitro*. Mitochondria isolated from the fat body of CC-deprived animals showed a respiratory rate 40% lower than the controls, and this was due to depression of definite enzyme systems (Keeley, 1971). Addition of an extract of CC *in vitro* did not restore respiration, but (Keeley and Waddill, 1971) daily *in vivo* injections of CC extracts returned to normal the depressed respiratory quotient of operated cockroaches. The authors suggest that the neurosecretory factor active on mitochondrial oxidative metabolism is a protein because it is inactivated by heating at 100°C. No mention is made of inactivation experiments with proteolytic enzymes. Its lack of direct activity on mitochrondria *in vitro* suggests that it does not act as a cofactor to the electron-transport system, but affects other biochemical functions, which in turn act on electron transport.

4.12. Peptides of Unknown Function Found in the Nervous System

During ion-exchange chromatographic analysis of the free amino acids contained in deproteinized extracts of honeybee brains, an unknown peak appeared between Taur and Asp (Frontali, 1964). On acid hydrolysis, it resolved into the amino acids Asp, Glu, Ser, Ala, Gly, and His.

5. Mollusca

5.1. Cardioactive Peptides

Cardioaccelerator peptides (referred to as "substance X") have been extracted from ganglia and heart of gastropod mollusks (Hampe et al., 1969; Welsh, 1971). This activity was shown to be associated with neurosecretory granules from the ganglia of Venus (Cottrell and Maser, 1967). Gel-chromatographic analysis of substance X showed that it separated into four peaks of activity on Sephadex G-15, three of which (peaks A, B, and C) were inactivated by proteases (Frontali et al., 1967; Welsh, 1971). The activity of peaks B and C was similar to the cardioaccelerator action of serotonin, except that the activity was more persistent in the case of the peptides. A neurohormone with cardioexcitatory activity has also been reported in cephalopods (Blanchi et al., 1973).

5.2. Peptides That Regulate Neuronal Activity

A peptide factor extracted from the ganglia of gastropod mollusks induces bursting pacemaker potential activity in a specific neuron in Otala lactea (Ifshin et al., 1975). This peptide appears to have a molecular weight of about 1000 daltons, as determined by chromatography on Sephadex G-25, and is not inactivated by trypsin or chymotrypsin, but is totally inactivated by treatment with pronase. A similar factor from extracts of bag cell clusters in the parietovisceral ganglion of Aplysia that causes enhanced bursting pacemaker activity in R15 (found in the same ganglion) has been reported (Mayeri and Simon, 1975). See Chapter 11 for a discussion of these phenomena.

5.3. Peptides That Regulate Salt and Water

A number of morphological studies at the light- and electron-microscopic levels have implicated various neurosecretory cell groups in molluscan ganglia in the control of diuretic phenomena (see Golding, 1974, for a review). Unfortunately, no biochemistry has been done on these systems, and it is not known whether the hormones involved are peptides. Recent studies, however, have shown that one or more low-molecular-weight polypeptides extracted from the identified neuron R_{15} in the abdominal ganglion of Aplysia californica produce a rapid gain in weight of the animal when injected into the hemocele of Aplysia (Kupfermann and Weiss, 1976). These authors suggest that the peptide or peptides regulate water as opposed to salt in the animal, since the sensory input to R_{15} from the osphradium is specifically responsive to the osmotic pressure of the surrounding sea water (Jahan-Parwar et al., 1969; Stinnakre and Tauc, 1969). See Chapter 8 for a discussion of peptides synthesized in R_{15}.

5.4. Peptides Involved in Reproduction

A considerable literature exists showing that gametogenesis and reproductive activity in gastropod mollusks are under the regulation of neurohormones (see Golding, 1974). The caudodorsal cells in the cerebral ganglia of *Lymnaea stagnalis* would appear to be excellent candidates for a systematic biochemical study, but this work is only in its preliminary stages (Garaerts, 1975).

The most extensive studies in this area have been on the "egg-laying" hormone found in the bilateral bag cell clusters in the abdominal ganglion of *Aplysia californica*. These cells are normally inactive electrically, but are electrotonically connected and respond to input of relatively short duration with a synchronous, prolonged, and repetitive spike activity (Kupfermann and Kandel, 1970). This activity is assumed to be related to the secretion of a peptide hormone that induces egg-laying in the animal (Strumwasser *et al.*, 1969; Kupfermann, 1970, 1972). A low-molecular-weight "bag cell–specific protein" associated with this activity has been isolated (Toevs and Brackenbury, 1969), and biosynthetic studies on this system have been done (Arch, 1972*a,b;* Gainer and Wollberg, 1974). The peptide hormone is first synthesized as a precursor protein of about 25–29,000 daltons, which is then cleaved to various lower-molecular-weight polypeptides (Arch, 1972*a,b*). The actual identity of the peptide that contains the egg-laying activity is, at the time of this writing, still unclear. In radioisotopic Leu incorporation studies, Arch (1972*a*) showed that primarily low-molecular-weight proteins (i.e., less than 10,000 daltons) were released from the bag cell processes in the neurohemal area (sheath) on stimulation. Loh *et al.* (1975) found that the polypeptides released from the bag cell sheath by calcium-dependent potassium depolarization were heterogeneous. The released peptides were separated on SDS gels and stained with coomassie blue. At least three distinct stained bands, at about 12,000, 6000 and less than 3000 daltons, could be discerned to have been released. However, no bioassays were done to evaluate which, if any, of these peptides contained the biological activity. Recent studies by Arch *et al.*, (1976*a,b*) suggest that the active component is a basic peptide (isoelectric point about 9.3) with an approximate molecular weight of 6000 daltons. The mechanism of the hormone action may be by causing contraction of muscle fibers around the ovarian follicles (Coggeshall, 1970).

Several other marine gastropods have been studied with regard to "egg-laying" factors. Davis *et al.* (1974) have shown that crude extracts of whole nervous systems of egg-laying specimens of *Pleurobranchaea californica* induced egg-laying and also suppression of feeding in this animal. Either less activity or no activity at all was found in nervous system extracts from nonlaying specimens. Ram (1975) has reported that extracts of the parietal ganglia of *Busycon canaliculatum* and *B. carica* contain a peptide factor or peptide factors that cause laying of egg capsules. Members of both sexes of this dioecious prosobranch contain the "egg capsule–laying" substance.

6. Echinodermata: Radial Nerve Factor

Injection of extracts of radial nerves from echinoids (sea urchins) as well as asteroids (starfish) induces gamete maturation and release in these organisms (Chaet, 1967; Kanatani, 1969; Cochran and Englemann, 1972). The active factor has been termed the "radial nerve factor" (RNF) (Schuetz and Biggers, 1967), but has also been referred to as the "gonad (gamete) shedding substance" (GSS) by Kanatani (1967). The factor isolated from *Patiria miniata* is a peptide of about 4800 daltons molecular weight (Chaet, 1967), whereas the peptide factor isolated from *Asterias forbesi* has a molecular weight of about 2100 daltons (Kanatani, 1967; Kanatani *et al.*, 1971). These two peptides appear to be biologically distinct, since they are not interspecifically active.

The mechanism of action of the RNF is attributed, in part, to the synthesis and release of 1-methyladenine in the ovary (Schuetz and Biggers, 1967; Kanatani *et al.*, 1969; Hirai and Kanatani, 1971), which is not species-specific. Thus, the RNF peptide produces a "second" messenger, 1-methyladenine, which in turn may act through a "third" messenger produced by the cell membrane of the oocyte (Kanatani and Shirai, 1970; Harai *et al.*, 1971). The literature regarding this subject has been reviewed in detail by Golding (1974).

7. Conclusion

Invertebrate nervous systems have traditionally provided valuable experimental material for the analysis of neurosecretory phenomena (see the historical review of neurosecretion by Knowles, 1974). Many of these invertebrate neurosecretory systems are peptidergic (see Table 1), and considerable progress has been made in the identification of the peptide factors and their mechanisms of action. The relatively greater prominence of neurosecretion in the homeostatic mechanisms and adaptive behavior of invertebrates, as compared with vertebrates, is probably related to the paucity of true endocrine systems in the former. This may also reflect the lower phylogenetic position of the invertebrates with regard to neuronal evolution. Indeed, it has been suggested that the most primitive neurons during the evolution of the nervous system were highly specialized in *both* receptor and effector properties, and that the effector specialization involved secretory mechanisms (Parker, 1919; Grundfest, 1959). Since peptides are common constituents of all cells, neuronal and nonneuronal, it is conceivable that some of the first intercellular messengers that were released by these primitive neurons were peptides.

The study of peptidergic neurons in invertebrates, in addition to being an interesting subject on its own merit, will also be of value in the development of general principles that are transferable to vertebrate systems. For example, the work of Brown and Starratt (1975) on proctolin in the cockroach hindgut complements that of Otsuka and co-workers (see Chapter 5) on Substance P in the mammalian spinal cord, in that both indicate that peptides may be released as "neurotransmitters." Furthermore, the exciting behavioral effects of peptides discussed in Chapter 14 have comparable examples among the invertebrates (see Truman and Riddiford, 1974a). Finally, as mentioned earlier, specific invertebrate neurons are often large and easily identifiable from animal to animal, and thus provide model systems for experiments that are not feasible in vertebrate preparations.

8. References

Abramowitz, A.A., Hisaw, F.L., and Papandrea, D.N., 1944, The occurrence of a diabetogenic factor in the eyestalks of crustaceans, *Biol. Bull.* **86:**1–5.

Alexandrowicz, J.S., and Carlisle, D.B., 1953, Some experiments on the function of the pericardial organs in Crustacea, *J. Mar. Biol. Assoc. U.K.* **32:**175–192.

Arch, S., 1972a, Polypeptide secretion from the isolated parietovisceral ganglion of *Aplysia californica, J. Gen. Physiol.* **59:**47–59.

Arch, S., 1972b, Biosynthesis of the egg laying hormone (ELH) in the bag cell neurons of *Aplysia californica, J. Gen. Physiol.* **60:**102–119.

Arch, S., Early, P., and Smock, T., 1976a, Biochemical isolation and physiological identification of the egg-laying hormone in *Aplysia californica, J. Gen. Physiol.* **68:**197–210.

Arch, S., Smock, T., and Early, P., 1976b, Precursor and product processing in the bag cell neurons of *Aplysia californica, J. Gen Physiol.* **68:**211–225.

Aréchiga, H., Huberman, A., and Naylor, E., 1974, Hormonal modulation of circadian neural activity in *Carcinus maenas (L.), Proc. R. Soc. Lond. Ser. B.* **187:**299–313.

Aston, R.J., 1975, The role of AMP-cyclic monophosphate in relation to the diuretic hormone of *Rhodnius prolixus, J. Insect Physiol.* **21:**1873–1877.

Aston, R.J., and White, A.F., 1974, Isolation and purification of the diuretic hormone from *Rhodnius prolixus, J. Insect Physiol.* **20:**1673–1682.

Bauchau, A.G., Mengeot, J.C., and Olivier, M.A., 1968, Action de la sérotonine et de l'hormone diabétogène des crustacés sur la phosphorylase musculaire, *Gen. Comp. Endocrinol.* **11:**132–138.

Baumann, H., 1974, The isolation, partial characterization and biosynthesis of the paragonial substances PS1 and PS2 of *Drosophila funebris, J. Insect Physiol.* **20:**2181–2194.

Baumann, H., and Chen, P.S., 1973, Geschlechtsspezifische Ninhydrin-spezifische Substanzen in Adultmännchen von *Drosophila funebris, Rev. Suisse Zool.* **80:**865–890.

Baumann, E., and Gersch, M., 1973, Untersuchungen zur Stabilität des Neurohormons D., *Zool. Jahrb. Abt. Allg. Zool. Physiol. Tiere* **77:**153–160.

Beament, J.W.L., 1958, A paralysing agent in the blood of cockroaches, *J. Insect Physiol.* **2:**199–214.

Belamarich, F.A., 1963, Biologically active peptides from the periocardial organs of the crab, *Cancer borealis, Biol. Bull.* **124:**9–16.

Belamarich, F.A., and Terwilliger, R.C., 1966, Neurosecretion in invertebrates other than insects. I. The nature and localization of neurosecretory substances. Isolation and identification of cardio-excitor hormone from the pericardial organs of *Cancer borealis, Amer. Zool.* **6:**101–106.

Berlind, A., and Cooke, J.M., 1970, Release of a neurosecretory hormone as peptide by electrical stimulation of crab pericardial organs, *J. Exp. Biol.* **53:**679–686.

Blanchi, D., Noviello, L., and Libonati, M., 1973, A neurohormone of cephalopods with cardioexcitatory activity, *Gen. Comp. Endocrinol.* **21:**267–277.

Brown, B.E., 1965, Pharmacologically active constituents of the cockroach corpora cardiaca; resolution and some characteristics, *Gen. Comp. Endocrinol.* **5:**387–401.

Brown, B.E., 1967, Neuromuscular transmitter substance in insect visceral muscle, *Science* **155:**595–597.

Brown, B.E., 1975, Proctolin: A peptide transmitter candidate in insects, *Life Sci.* **17:**1241–1252.

Brown, B.E., and Starratt, A.N., 1975, Isolation of proctolin, a myotropic peptide, from *Periplaneta americana, J. Insect Physiol.* **21:**1879–1881.

Brown, F.A., Jr., 1935, Control of pigment migration within the chromatophores of *Palaemonetes vulgaris, J. Exp. Zool.* **71:**1–14.

Burnett, A.L., and Diehl, N.A., 1964, The nervous system of *Hydra*. III. The maturation of sexuality with special reference to the nervous system, *J. Exp. Zool.* **157:**237–235.

Burnett, A.L., Diehl, N.A., and Diehl, F., 1964, The nervous system of *Hydra*. II. The control of growth and regeneration by neurosecretory cells, *J. Exp. Zool.* **157:**227–235.

Cameron, M.L., 1953, Secretion of an orthodiphenol in the corpus cardiacum of the insect, *Nature London* **172:**349–350.

Candy, D.J., and Kilby, B.A., 1961, The biosynthesis of trehalose in the locust fat body, *Biochem. J.* **78:**531–536.

Cazal, M., 1967, Activité cardiaque de *Locusta migratoria in vitro* et influence des corpora cardiaca. Mise au point expérimentale et premiers résultats, *C. R. Acad. Sci. Paris* **264:**842–845.

Cazal, M., and Girardie, A., 1968, Controle humoral de l'équilibre hydrique chez *Locusta migratoria migratorioides, J. Insect Physiol.* **14:**655–668.

Chaet, A.B., 1967, Gamete release and shedding substance of seastars, *Symp. Zool. Soc. London* **20:**13–24.

Chen, P.S., and Diem, C., 1961, A sex-specific ninhydrin-positive substance found in the paragonia of adult males of *Drosophila melanogaster, J. Insect Physiol.* **7:**289–298.

Cochran, R.G., and Englemann, F., 1972, Echinoid spawning induced by a radial nerve factor, *Science* **178:**423–424.

Coggeshall, R.E., 1970, A cytologic analysis of the bag cell control of egg laying in *Aplysia, J. Morphol.* **132:**461–486.

Cook, B.J., and Holman, G.M., 1975, Sites of action of a peptide neurohormone that controls hindgut muscle activity in the cockroach *Leucophaea maderae, J. Insect Physiol.* **21:**1187–1192.

Cook, B.J., Holman, G.M., and Marks, E.P., 1975, Calcium and cyclic AMP as possible mediators of neurohormone action in the hindgut of the cockroach, *Leucophaea maderae, J. Insect Physiol.* **21:**1807–1814.

Cooke, J.M., and Hartline, D.K., 1975, Neurohormonal alteration of integrative properties of the cardiac ganglion of the lobster, *Homarus americanus, J. Exp. Biol.* **63:**33–52.

Cottrell, C.B., 1962*a*, The imaginal ecdysis of blowflies. The control of cuticular hardening and darkening, *J. Exp. Biol.* **39:**395–411.

Cottrell, C.B., 1962*b*, The imaginal ecdysis of blowflies. Detection of the blood-borne darkening factor and determination of some of its properties, *J. Exp. Biol.* **39:**413–430.

Cottrell, G.A., and Maser, M., 1967, Subcellular localization of 5-hydroxytryptamine and substance X in molluscan ganglia, *Comp. Biochem. Physiol.* **20:**901–906.

Davey, K.G., 1961*a*, Substances controlling the rate of beating of the heart of *Periplaneta, Nature London* **192:**284.

Davey, K.G., 1961*b*, The mode of action of the heart accelerating factor from the corpus cardiacum of insects, *Gen. Comp. Endocrinol.* **1:**24–29.

Davey, K.G., 1962, The mode of action of the corpus cardiacum on the hindgut in *Periplaneta americana, J. Exp. Biol.* **39:**319–324.

Davis, W.J., Mpitsos, G.J., and Pinneo, J.M., 1974, The behavioral hierarchy of the mollusk *Pleurobranchea*. II. Hormonal suppression of feeling associated with egg-laying, *J. Comp. Physiol.* **90:**225–243.

de Bessé, N., and Cazal, M., 1968, Action des extraits d'organes périsympathiques et de *corpora cardiaca* sur la diurèse de quelques Insectes, *C.R. Acad. Sc. Paris* **266:**615–618.

de Wilde, J., and de Loof, A., 1973, Reproduction, endocrine control, in: *The Physiology of Insecta* (M. Rockstein, ed.), pp. 97–158, Academic Press, New York and London.

Edman, P., Fänge, R., and Östlund, E., 1958, Isolation of the red pigment concentration hormone of the crustacean eyestalk, in: *Zweites Internationales Symposium über Neurosekretion* (W. Bargmann, E. Hanström, and B. Scharrer, eds.), pp. 119–123, Springer-Verlag, Berlin and New York.

Evans, J.J.T., 1962, Insect neurosecretory material separated by differential centrifugation, *Science* **136:**314–315.

Fernlund, P., 1971, Chromactivating hormones from *Pandalus borealis*. Isolation and purification of a light adapting hormone, *Biochim. Biophys. Acta* **237**:519–529.

Fernlund, P., 1974, Synthesis of the red pigment concentrating hormone of the shrimp, *Pandalus borealis, Biochem. Biophys. Acta* **371**:312–322.

Fernlund, P., and Josefsson, L., 1968, Chromactivating hormones of *Pandalus borealis*. Isolation and purification, *Biochim. Biophys. Acta* **158**:262–273.

Fernlund, P., and Josefsson, L.H., 1972, Crustacean color-change hormone: Amino acid sequence and chemical synthesis, *Science* **177**:173–175.

Fingerman, M., 1970, Dual control of the leucophores in the prawn, *Palaemonetes vulgaris*, by pigment dispersing and pigment concentrating substances, *Biol. Bull.* **138**:26–34.

Fingerman, M., 1973, Comparison of the effects of partially purified eyestalk extracts of the shrimp *Crangon septemspinosa* on its black, red and white chromatophores, *Physiol. Zool.* **46**:173–179.

Fingerman, M., and Connell, P.M., 1968, The role of cations in the actions of the hormones controlling the red chromatophores of the prawn, *Palaemonetes vulgaris, Gen. Comp. Endocrinol.* **10**:392–398.

Fingerman, M., and Mobberly, W.C., Jr., 1960, Investigation of the hormones controlling the distal retinal pigment of the prawn, *Palaemonetes, Biol. Bull.* **118**:393–406.

Fingerman, M., and Rao, K.R., 1969, A comparative study of leucophore-activating substances from the eyestalk of two crustaceans, *Palaemonetes vulgaris* and *Uca pugilator, Biol. Bull.* **136**:200–215.

Fingerman, M., Lowe, M.E., and Sundararaj, B.I., 1959, Dark-adapting and light-adapting hormones controlling the distal retinal pigment of the prawn, *Palaemonetes vulgaris, Biol. Bull.* **116**:30–36.

Fingerman, M., Krasnow, R.A., and Fingerman, S.W., 1971, Separation, assay and properties of the distal retinal pigment light-adapting and dark-adapting hormones in the eyestalks of the prawn, *Palaemonetes vulgaris, Physiol. Zool.* **44**:119–128.

Flattum, R.F., Watkinson, I.A., and Crowder, L.A., 1973, The effect of insect "autoneurotoxin" on *Periplaneta americana* and *Schistocerca gregaria* Malpighian tubules, *Pestic. Biochem. Physiol.* **3**:237–242.

Florey, E., and Florey, E., 1954, Über die mögliche Bedeutung von Enteramin (5-Oxytryptamin) als nervöser Aktionssubstanz bei cephalopoden und dekapoden Crustaceen, *Z. Naturforsch.* **9B**: 58–68.

Fogal, W., and Fraenkel, G., 1969, Role of bursicon in melanization and endocuticle formation in the adult fleshfly *Sarcophaga bullata, J. Insect Physiol.* **15**:1235–1240.

Fox, A.S., Mead, C.G., and Munyon, I.L., 1959, Sex peptide of *Drosophila melanogaster, Science* **129**:1489, 1490.

Fraenkel, G., and Hsiao, C., 1962, Hormonal and nervous control of tanning in the fly, *Science* **138**:27–29.

Fraenkel, G., and Hsiao, C., 1965, Bursicon: A hormone which mediates tanning of the cuticle in the adult fly and other insects, *J. Insect Physiol.* **11**:513–556.

Fraenkel, G., Hsiao, C., and Seligman, M., 1966, Properties of bursicon: An insect protein hormone that controls cuticular tanning, *Science* **151**:91–93.

Fraenkel, G., Zdarek, J., and Sivasubramanian, P., 1972, Hormonal factors in the central nervous system and haemolymph of pupariating fly larvae which accelerate puparium formation and tanning, *Biol. Bull.* **143**:127–139.

Freeman, A.R., Connell, P.M., and Fingerman, M., 1968, An electrophysiological study of the red chromatophore of the prawn *Palaemonetes:* Observations on the action of red pigment concentrating hormone, *Comp. Biochem. Physiol.* **26**:1015–1029.

Friedman, S., 1967, The control of trehalose synthesis in the blowfly *Phormia regina, J. Insect Physiol.* **13**:397–405.

Frontali, N., 1964, Brain glutamic acid decarboxylase and synthesis of gamma-aminobutyric acid in vertebrate and invertebrate species, in: *Comparative Neurochemistry* (D. Richter, ed.), pp. 185–192, Pergamon Press, Oxford.

Frontali, N., Williams, L., and Welsh, J.H., 1967, Heart excitatory and inhibitory substances in molluscan ganglia, *Comp. Biochem. Physiol.* **22**:833–841.

Fuchs, M.S., and Hiss, E.A., 1970, Partial purification and separation of the protein components of matrone from *Aedes aegypti, J. Insect Physiol.* **16:**931–940.

Fuchs, M.S., Craig, G.B., and Despommier, D.D., 1969, The protein nature of the substance inducing female monogamy in *Aedes aegypti, J. Insect Physiol.* **15:**701–709.

Gainer, H., and Wollberg, Z., 1974, Specific protein metabolism in identifiable neurons of *Aplysia californica, J. Neurobiol.* **5:**243–261.

Garaerts, W.P.M., 1975, Studies on the endocrine control of growth and reproduction in the hermaphrodite pulmonate snail *Lymnea stagnalis,* Ph.D. thesis, Free University of Amsterdam.

Gersch, M., 1958, Neurohormonale Beeinflussung der Herztätigkeit bei der Larve von *Corethra, J. Insect Physiol.* **2:**281–297.

Gersch, M., 1972, Experimentelle Untersuchungen zum Freisetzung-mechanismus von Neurohormonen nach elektrischer Reizung der Corpora cardiaca von *Periplaneta americana in vitro, J. Insect Physiol.* **18:**2425–2439.

Gersch, M., 1974a, Experimentelle Untersuchungen zur Ausschüttung von Neurohormonen aus Ganglien des Bauchmarks von *Periplaneta americana* nach elektrischer Reizung *in vitro, Zool. Jahrb. Abt. Allg. Zool. Physiol. Tiere* **78:**138–149.

Gersch, M., 1974b, Selektive Freisetzung des hyperglycaemischen Faktors aus den Corpora cardiaca von *Periplaneta americana in vivo, Experientia* **30:**767.

Gersch, M., 1975, *Principien neurohormonaler und neurohumoraler Steuerung physiologischer Prozesse,* Friedrich-Schiller-Universität, Jena, 151 pp.

Gersch, M., and Bräuer, R., 1974, *In vitro* Stimulation der Prothorakaldrüsen von Insekten als Testsystem (Prothorakaldrüsentest), *J. Insect Physiol.* **20:**735–741.

Gersch, M., and Richter, K., 1963, Auslösung von Nervenimpulsen durch ein Neurohormon bei *Periplaneta americana, Zool. Jahrb. Abt. Allg. Zool. Physiol. Tiere* **70:**301–308.

Gersch, M., and Stürzebecher, J., 1967, Zur Frage der Identität und des Vorkommens von Neurohormon D in verschiedenen Bereichen des Zentralnervensystems von *Periplaneta americana, Z. Naturforsch* **22b:**563.

Gersch, M., and Stürzebecher, J., 1970, Experimentelle Stimulierung der zellulären Aktivität der Prothorakaldrüsen von *Periplaneta americana* durch den Aktivations-faktor, *J. Insect Physiol.* **16:**1813–1926.

Gersch, M., and Stürzebecher, J., 1971, Further studies on chemical characterization of neurohormone D, *Proceedings of the International Symposium on Insect Endocrinology,* Brno, 1966, pp. 71–80.

Gersch, M., Fischer, F., Unger, H., and Koch, H., 1960, Die Isolierung neurohormonaler Faktoren aus dem Nervensystem der Küchenschabe *Periplaneta americana, Z. Naturforsch.* **15b:**319–322.

Gersch, M., Unger, H., Fischer, F., and Kapitza, W., 1963, Identifizierung einiger Wirkstoffe aus dem Nervensystem der Crustaceen und Insekten, *Z. Naturforsch.* **18b:**587, 588.

Gersch, M., Richter, K., Stürzebecher, J., and Fabian, B., 1969, Experimentelle Untersuchungen zum Wirkungsmechanismus des Neurohormons D von *Periplaneta americana* am ''biologischen Modell'' der Harnblase von *Bufo bufo, Gen. Comp. Endocrinol.* **12:**40–50.

Gersch, M., Richter, K., Böhm, G.A., and Stürzebecher, J., 1970, Selektive Ausschüttung von Neurohormonen nach elektrischer Reizung der Corpora cardiaca von *Periplaneta americana in vitro, J. Insect Physiol.* **16:**1991–2013.

Gersch, M., Richter, K. and Stürzebecher, J., 1971, Weitere experimentelle Untersuchungen zum Wirkungsmechanismus des Neurohormons D von *Periplaneta americana* am ''biologischen Modell'' der Harnblase von *Bufo bufo:* Beeinflussung des Na^+ Transportes, *Gen. Comp. Endocrinol.* **17:**281–286.

Gersch, M., Bräuer, R., and Birkenbeil, H., 1973, Experimentelle Untersuchungen zum Wirkungsmechanismus der beiden entwicklungs-physiologisch aktiven Fraktionen des ''Gehirnhormons'' der Insekten (Aktivationsfaktor I und II) auf die Prothorakaldrüse, *Experientia* **29:**425–427.

Goldbard, G.A., Sauer, J.R., and Mills, R.R., 1970, Hormonal control of excretion in the American cockroach. II) Preliminary purification of a diuretic and antidiuretic hormone, *Comp. Gen. Pharmacol.* **1:**82–86.

Golding, D.W., 1974, A survey of neuroendocrine phenomena in non-arthropod invertebrates, *Biol. Rev.* **49:**161–224.

Goldsworthy, G.J., 1969, Hyperglycaemic factors from the corpus cardiacum of *Locusta migratoria,* *J. Insect Physiol.* **15:**2131–2140.

Goldsworthy, G.J., and Mordue, W., 1974, Neurosecretory hormones in insects, *J. Endocrinol.* **60:**529–558.

Grundfest, H., 1959, Evolution of conduction in the nervous system, *Evolution of Nervous Control,* American Association for the Advancement of Science, Washington, D.C., p. 43–86.

Gwadz, R.W., 1972, Neuro-hormonal regulation of sexual receptivity in female *Aedes aegypti, J. Insect Physiol.* **18:**259–266.

Hagesawa, K., 1957, The diapause hormone of the silkworm *Bombyx mori, Nature London* **179:**1300, 1301.

Hagesawa, K., 1963, Studies on the mode of action of the diapause hormone in the silkworm *Bombyx mori.* I) The action of diapause hormone injected into pupae of different ages, *J. Exp. Biol.* **40:**517–529.

Hallahan, C., and Powell, B.L., 1971, Purification and characterization of the melanin-dispersing hormone from the sinus gland of the shore crab *Carcinus maenas, Gen. Comp. Endocrinol.* **17:**451–457.

Hampe, M.M.V., Tesch, A., and Jaeger, C.P., 1969, On the nature of cardio-active substances from *Strophocheilus oblongus* ganglia, *Life Sci.* **8:**827–835.

Hart, D.E., and Steele, J.E., 1973, The glycogenolytic effect of the corpus cardiacum on the cockroach nerve cord, *J. Insect Physiol.* **19:**927–939.

Highnam, K.C., and Goldsworthy, G.J., 1972, Regenerated corpora cardiaca and hyperglycaemic factor in *Locusta migratoria, Gen. Comp. Endocrinol.* **18:**83–88.

Highnam, K.C., and Hill, L., 1969, *The Comparative Endocrinology of the Invertebrates,* American Elsevier Publishing Co., New York, 270 pp.

Hirai, S., and Kanatani, H., 1971, Site of production of meiosis-inducing substance in ovary of starfish, *Exp. Cell Res.* **67:**224–229.

Hirai, S. Kubota, J., and Kanatani, K., 1971, Induction of cytoplasmic motivation by 1-methyl adenine in starfish oocytes after removal of the germinal vesicle, *Exp. Cell Res.* **68:**137–143.

Hiss, E.A., and Fuchs, M.S., 1972, The effect of matrone on oviposition in the mosquito *Aedes aegypti, J. Insect Physiol.* **18:**2217–2227.

Holman, G.M., and Cook, B.J., 1972, Isolation, partial purification and characterization of a peptide which stimulates the hindgut of the cockroach *Leucophaea maderae, Biol. Bull.* **142:**446–460.

Holman, G.M., and Marks, E.P., 1974, Synthesis, transport and release of a neurohormone by cultured neuroendocrine glands from the cockroach, *Leucophaea maderae, J. Insect Physiol.* **20:**479–484.

Ichikawa, M., and Ishizaki, H., 1963, Protein nature of the brain hormone of insects, *Nature London* **198:**308, 309.

Ifshin, M.S., Gainer, H., and Barker, J.L., 1975, Peptide factor extracted from molluscan ganglia that modulates bursting pacemaker activity, *Nature London* **254:**72–74.

Ishizaki, H., and Ichikawa, M., 1967, Purification of the brain hormone of the silkworm *Bombyx mori, Biol. Bull.* **133:**355–368.

Isobe, M., Hagesawa, K., and Goto, T., 1973, Isolation of the diapause hormone from the silkworm *Bombyx mori, J. Insect Physiol.* **19:**1221–1239.

Jahan-Parwar, B., Smith, M., and von Baumgarten, R., 1969, Activation of neurosecretory cells in *Aplysia* by osphradial stimulation, *Amer. J. Physiol.* **216:**1246–1251.

Johnson, B., 1966, Fine structure of the lateral cardiac nerves of the cockroach *Periplaneta americana, J. Insect Physiol.* **12:**645–653.

Johnson, B., and Bowers, B., 1963, Transport of neurohormones from the corpora cardiaca in insects, *Science* **141:**264–266.

Josefsson, L., 1967, Separation and purification of distal retinal hormone and red pigment concentrating hormone of the crustacean eyestalks, *Biochim. Biophys. Acta* **148:**300–303.

Josefsson, L., and Kleinholz, L.G., 1964, Isolation and purification of hormones of the crustacean eyestalk, *Nature London* **201:**301, 302.

Kanatani, H., 1967, Mechanism of starfish spawning with special reference to gonad-stimulating substance (GSS) of nerve and meiosis inducing substance (MIS) of gonad, *Jpn. J. Exp. Morphol.* **21**:61–78.

Kanatani, H., 1969, Mechanism of starfish spawning: Action of neural substance on the isolated ovary, *Gen. Comp. Endocrinol. Suppl.* **2**:582–589.

Kanatani, H., and Shirai, H., 1970, Mechanism of starfish spawning. III. Properties and action of meiosis-inducing substance produced in gonad under influence of gonad-stimulating substance, *Dev. Growth Differ.* **12**:119–140.

Kanatani, H., Shirai, H., Nakanishi, K., and Kurokawa, T., 1969, Isolation and identification of meiosis-inducing substance in starfish, *Asterias amurensis, Nature London* **221**:273, 274.

Kanatani, H., Ikegami, S., Shirai, H., Oide, H., and Tamura, S., 1971, Purification of gonad-stimulating substance obtained from radial nerves of the starfish, *Asterias amurensis, Dev. Growth Differ.* **13**:151–164.

Karlson, P., Sekeris, C.E., and Sekeris, K.E., 1962, Zum Tyrosinstoffwechsel der Insekten. VI. Identifizierung von *N*-acetyl-3,4-dihydroxy-*β*-phenaethylamin (*N*-acetyl-dopamin) als Tyrosinmetabolit, *Z. Physiol. Chem.* **327**:86–94.

Kater, S.B., 1968, Cardioaccelerator release in *Periplaneta americana, Science* **160**:765–767.

Keeley, L.L., 1971, Endocrine effects on the biochemical properties of fat body mitochondria from the cockroach, *Blaberus discoidalis, J. Insect Physiol.* **17**:1501–1515.

Keeley, L.L., and Friedman, S., 1967, Corpus cardiacum as a metabolic regulator in *Blaberus discoidalis.* I. Long-term effects of cardiatectomy on whole body and tissue respiration and on trophic metabolism, *Gen. Comp. Endocrinol.* **8**:129–134.

Keeley, L.L., and Friedman, S. 1969, Effects of long-term cardiatectomy-allatectomy on mitochondrial respiration in the cockroach *Blaberus discoidalis, J. Insect Physiol.* **15**:509–518.

Keeley, L.L., and Waddill, V.H., 1971, Insect hormones: Evidence for a neuroendocrine factor affecting respiration metabolism, *Life Sci.* **10**:737–745.

Keller, R., 1969, Untersuchungen zur Artspezifität eines Crustaceenhormons, *Z. Vergl. Physiol.* **63**:137–145.

Keller, R., and Andrew, E.M., 1973, The site of action of the crustacean hyperglycaemic hormone, *Gen. Comp. Endocrinol.* **20**:572–578.

Kleinholz, L.H., 1934, Eyestalk hormone and the movement of the distal retinal pigment in *Palaemonetes, Proc. Nat. Acad. Sci. U.S.A.* **20**:659–661.

Kleinholz, L.H., 1936, Crustacean eyestalk hormone and retinal pigment migration, *Biol. Bull.* **70**:159–184.

Kleinholz, L.H., 1961, Pigmentary effectors, in: *The Physiology of Crustacea* (T.H. Waterman, ed.), Vol. II, pp. 133–170, Academic Press, New York and London.

Kleinholz, L.H., 1972, Comparative studies of crustacean melanophore-stimulating hormones, *Gen. Comp. Endocrinol.* **19**:473–483.

Kleinholz, L.H., 1975, Purified hormones from the crustacean eyestalk and their physiological specificity, *Nature London* **258**:256, 257.

Kleinholz, L.H., and Keller, R., 1973, Comparative studies in crustacean neurosecretory hyperglycaemic hormones. I. The initial survey, *Gen. Comp. Endocrinol.* **21**:554–564.

Kleinholz, L.H., and Kimball, F., 1965, Separation of neurosecretory pigmentary-effector hormones of the crustacean eyestalk, *Gen. Comp. Endocrinol.* **5**:336–341.

Kleinholz, L.H., Kimball, F., and McGravey, M., 1967, Initial characterization and separation of hyperglycaemic (diabetogenic) hormone from the crustacean eyestalk, *Gen. Comp. Endocrinol.* **8**:75–81.

Knowles, F., 1974, Twenty years of neurosecretion, in: *Neurosecretion—The Final Endocrine Pathway* (F. Knowles and L. Vollrath, eds.), pp. 3–14, Springer-Verlag, Berlin.

Kupfermann, I., 1970, Stimulation of egg laying by extracts of neuroendocrine cells (bag cells) of abdominal ganglion of *Aplysia, J. Neurophysiol.* **33**:877–881.

Kupfermann, I., 1972, Studies on the neurosecretory control of egg laying in *Aplysia, Amer. Zool.* **12**:513–519.

Kupfermann, I., and Kandel, E.R., 1970, Electrophysiological properties and functional connections of two symmetrical neurosecretory clusters (bag cells) in abdominal ganglion of *Aplysia, J. Neurophysiol.* **33:**865–876.

Kupfermann, I., and Weiss, R., 1976, Water regulation by a presumptive hormone contained in identified neurosecretory cell R$_{15}$ of *Aplysia, J. Gen. Physiol.* **67:**113–123.

Lentz, T. L., 1965, *Hydra:* Induction of supernumerary heads by isolated neurosecretory granules, *Science* **150:**633–635.

Loeb, M.J., 1972, Strobilation in the Chesapeake Bay sea nettle, *Chrysaora quinquecirrha.* I. The effects of environmental temperature changes on strobilation and growth, *J. Exp. Zool.* **180:**279–292.

Loeb, M.J., 1974*a*, Strobilation in the Chesapeake Bay sea nettle, *Chrysaora quinquecirrha.* II. Partial characterization of the neck-inducing factor from strobilating polyps, *Comp. Biochem. Physiol.* **47A:**291–301.

Loeb, M.J., 1974*b*, Strobilation in the Chesapeake Bay sea nettle, *Chrysaora quinquecirrha.* III. Dissociation of the neck-inducing factor from strobilating polyps, *Comp. Biochem. Physiol.* **49A:**423–432.

Loeb, M.J., 1975, Strobilation in the Chesapeake Bay sea nettle, *Chrysaora quinquecirrha.* IV. Tissue levels of iodinated high molecular weight component and NIF in relation to temperature change–induced behavior, *Comp. Biochem. Physiol.* **51A:**37–42.

Loh, Y.P., Sarne, Y., and Gainer, H., 1975, Heterogeneity of proteins synthesized, stored and released by bag cells of *Aplysia californica, J. Comp. Physiol.* **100:**283–295.

Maddrell, S.H.P., 1963, Excretion in the blood-sucking bug *Rhodnius prolixus.* I. The control of diuresis, *J. Exp. Biol.* **40:**247–256.

Maddrell, S.H.P., 1966, The site of release of the diuretic hormone in *Rhodnius*—a new neurohaemal system in insects, *J. Exp. Biol.* **45:**499–508.

Maddrell, S.H.P., and Gee, J.D., 1974, Potassium-induced release of the diuretic hormones of *Rhodnius prolixus* and *Glossina austeni.* Calcium dependence, time course and localization of neurohaemal areas, *J. Exp. Biol.* **61:**155–171.

Maddrell, S.H.P., Pilcher, D.E.M., and Gardiner, B.O.C., 1969, Stimulatory effect of 5-hydroxytryptamine on secretion by Malpighian tubules of insects, *Nature London* **222:**784, 785.

Marks, E.P., and Holman, G.M., 1974, Release from the brain and acquisition by the corpora cardiaca of a neurohormone *in vitro, J. Insect Physiol.* **20:**2087–2093.

Mayer, R.J., and Candy, D.J., 1969, Control of haemolymph lipid concentration during locust flight. An adipokinetic hormone from the corpora cardiaca, *J. Insect Physiol.* **15:**611–620.

Mayeri, E., and Simon, S., 1975, Modulation of synaptic transmission and burster neuron activity after release of neurohormone in *Aplysia, Neurosci. Abstr.* **1:**584.

Maynard, D.M., and Welsh, J.H., 1959, Neurohormones of the pericardial organs of brachiuran crustacea, *J. Physiol.* **149:**215–227.

Milburn, N.S., and Roeder, K.B., 1962, Control of efferent activity in the cockroach terminal abdominal ganglion by extracts of corpora cardiaca, *Gen. Comp. Endocrinol.* **2:**70–76.

Milburn, N.S., Weiant, E. A., and Roeder, K.B., 1960, The release of efferent nerve activity in the roach *Periplaneta americana* by extracts of the corpus cardiacum, *Biol. Bull.* **118:**111–119.

Mills, R.R., 1967, Hormonal control of excretion in the American cockroach. I. Release of a diuretic hormone from the terminal abdominal ganglion, *J. Exp. Biol.* **46:**35–41.

Mills, R.R., and Lake, C.R., 1966, Hormonal control of tanning in the American cockroach. IV. Preliminary purification of the hormone, *J. Insect Physiol.* **12:**1395–1401.

Mills, R.R., and Nielsen, D.J., 1967, Hormonal control of tanning in the American cockroach. V. Some properties of the purified hormone, *J. Insect Physiol.* **13:**273–280.

Mills, R.R., and Whitehead, D.L., 1970, Hormonal control of tanning in the American cockroach: Changes in blood cell permeability during ecdysis, *J. Insect Physiol.* **16:**331–340.

Mills, R.R., Mathur, R.B., and Guerra, A.A., 1965, Studies on the hormonal control of tanning in the American cockroach. I. Release of an activation factor from the terminal abdominal ganglion, *J. Insect Physiol.* **11:**1047–1053.

Mills, R.R., Lake, C.R., and Alworth, W.L., 1967, Biosynthesis of *N*-acetyldopamine by the American cockroach, *J. Insect Physiol.* **13:**1539–1546.

Mordue, W., 1969, Hormonal control of Malpighian tubule and rectal function in the desert locust *Schstocerca gregaria, J. Insect Physiol.* **15**:273–285.

Mordue, W., and Goldsworthy, G.J., 1969, The physiological effects of corpus cardiacum extracts in locusts, *Gen. Comp. Endocrinol.* **12**:360–369.

Natalizi, G.M., and Frontali, N., 1966, Purification of insect hyperglycaemic and heart accelerating hormones, *J. Insect Physiol.* **12**:1279–1287.

Natalizi, G.M., Pansa, M.C., d'Ajello, V., Casaglia, O., Bettini, S., and Frontali, N., 1970, Physiologically active factors from corpora cardiaca of *Periplaneta americana, J. Insect Physiol.* **16**:1827–1836.

Normann, T.C., 1965, The neurosecretory system of the adult *Calliphora erythrocephala*. I. The fine structure of the corpus cardiacum with some observations on adjacent organs, *Z. Zellforsch.* **67**:461–501.

Normann, T.C., 1972, Heart activity and its control in the adult blowfly *Calliphora erythrocephala, J. Insect Physiol.* **18**:1793–1810.

Normann, T.C., and Duve, H., 1969, Experimentally induced release of a neurohormone influencing haemolymph trehalose level in *Calliphora erythrocephala* (Diptera), *Gen. Comp. Endocrinol.* **12**:449–459.

Ozbas, S., and Hodgson, E.S., 1958, Action of insect neurosecretion upon central nervous system *in vitro* and upon behavior, *Proc. Nat. Acad. Sci. U.S.A.* **44**:825–830.

Parker, G.H., 1919, *The Elementary Nervous System,* Lippincott, Philadelphia.

Passano, L.M., 1960, Molting and its control, in: *The Physiology of Crustacea* (T.H. Waterman, ed.), Vol. I, pp. 473–536, Academic Press, New York and London.

Perez-Gonzalez, M.D., 1957, Evidence for hormone-containing granules in sinus glands of the fiddler crab, *Uca pugilator, Biol. Bull.* **113**:426–441.

Pilcher, D.E.F., 1970, Hormonal control of the Malpighian tubules of the stick insect, *Carausius morosus, J. Exp. Biol.* **52**:653–665.

Pilcher, D.E.M., 1971, Stimulation of movements of Malpighian tubules of *Carausius* by pharmacologically active substances and tissue extracts, *J. Insect Physiol.* **17**:463–470.

Raabe, M., Cazal, M., Chalaye, D., and de Bessé, N., 1966, Action cardioaccéleratrice des organes neurohémaux périsympathiques ventraux de quelques insectes, *C. R. Acad. Sci. Paris* **263**:2002–2005.

Ralph, C.L., 1962, Heart accelerators and decelerators in the nervous system of *Periplaneta americana, J. Insect Physiol.* **8**:431–439.

Ram, J.J., 1975, Laying of egg capsules in *Busycotypus* induced by nervous system extracts, *Biol. Bull.* **149**:443.

Rao, K.R., 1965, Isolation and partial characterization of the moult-inhibiting hormone of the crustacean eyestalk, *Experientia* **21**:593, 594.

Richter, K., 1967, Untersuchungen zum Wirkungsmechanismus von Neurohormon D am Herzen von *Periplaneta americana, Zool. Jahrb. Abt. Allg. Zool. Physiol. Tiere* **73**:261–275.

Riddiford, L.M., and Williams, C.M., 1971, Role of the corpora cardiaca in the behavior of saturniid moths. Release of a sex pheromone, *Biol. Bull.* **140**:1–7.

Schaller, H.C., 1973, Isolation and characterization of a low molecular weight substance activating head and bud formation in *Hydra, J. Embryol. Exp. Morphol.* **29**:27–38.

Schaller, H.C., 1975, A neurohormone from *Hydra* is also present in rat brain, *J. Neurochem.,* **25**:187, 188.

Schaller, H.C., and Gierer, A., 1973, Distribution of the head activating substance in *Hydra* and its localization in membranous particles in nerve cells, *J. Embryol. Exp. Morphol.* **29**:39–52.

Scharrer, E., and Scharrer, B., 1937, Über Drusën-Nervenzellen und neurosekretorische Organe bei Wirbellosen und Wirbeltieren, *Biol. Rev.* **12**:185–216.

Schuetz, A.W., and Biggers, J.D., 1967, Regulation of germinal vesicle breakdown in starfish oocytes, *Exp. Cell Res.* **46**:624–628.

Sivasubramanian, P., Friedman, S., and Fraenkel, G., 1974, Nature and role of proteinaceous hormonal factors acting during puparium formation in flies, *Biol. Bull.* **147**:163–185.

Skorkowski, E.F., 1971, Isolation of three chromatographic hormones from the eyestalks of the shrimp *Crangon crangon, Mar. Biol. Berlin* **8**:220–223.

Skorkowski, E.F., 1972, Separation of three chromatographic hormones from the eyestalk of the crab *Rhithropanopeus harrisi, Gen. Comp. Endocrinol.* **18:**329–334.

Skorkowski, E.F., and Kleinholz, L.H., 1973, Comparison of white pigment concentrating hormone from *Crangon* and *Pandalus, Gen. Comp. Endocrinol.* **20:**595–597.

Sonobe, H., and Ohnishi, E., 1971, Silkworm *Bombyx mori:* Nature of diapause factor, *Science* **174:**835–838.

Sowa, B.A., and Borg, T.K., 1975, Density gradient centrifugation isolation of hormone-containing neurosecretory granules from the cockroach *Leucophaea maderae, J. Insect Physiol.* **21:**511–516.

Starratt, A.N., and Brown, B.E., 1975, Structure of the pentapeptide proctolin. A proposed neurotransmitter in insects, *Life Sci.* **17:**1253–1256.

Steele, J.E., 1961, Occurrence of a hyperglycaemic factor in the corpus cardiacum of an insect, *Nature London* **192:**680, 681.

Steele, J.E., 1963, The site of action of insect hyperglycaemic hormone, *Gen. Comp. Endocrinol.* **3:**46–52.

Stephens, G.C., Friedl, F., and Guttman, B., 1956, Electrophoretic separation of chromatophorotropic principles of the fiddler crab, *Uca, Biol. Bull.* **111:**312, 313.

Sternburg, J., 1960, Effect of insecticides on neurophysiological activity in insects, *J. Agric. Food Chem.* **8:**257–261.

Stinnakre, J., and Tauc, L., 1969, Central neuronal response to the activation of osmoreceptors in the osphradium of *Aplysia, J. Exp. Biol.* **51:**347–361.

Strumwasser, F., Jacklet, J.W., and Alvarez, R.B., 1969, A seasonal rhythm in the neural extract induction of behavioral egg-laying in *Aplysia, Comp. Biochem. Physiol.* **29:**197–206.

Stürzebecher, J., 1969, Differenzierung und Charakterisierung verschiedener neurohormonaler Factoren aus dem Nervensystem der Schabe *Periplaneta americana* L. mit Hilfe der Gelfiltration, Ph.D. dissertation, Jena, unpublished.

Terwilliger, R.C., Terwilliger, U.B., Clay, G.A., and Belamarich, F.A., 1970, The subcellular localization of a cardioexcitatory peptide in the pericardial organs of the crab, *Cancer borealis, Gen. Comp. Endocrinol.* **15:**70–79.

Toevs, L.A., and Brackenbury, R.W., 1969, Bag cell–specific proteins and the humoral control of egg laying in *Aplysia californica, Comp. Biochem. Physiol.* **29:**207–216.

Traina, M.E., Bellino, M., and Frontali, N., 1974, A comparison between neurohormone D and "peak 1" extracted from cockroach corpora cardiaca, *Zool. Jahrb. Abt. Allg. Zool. Physiol. Tiere* **78:**424–428.

Traina, M.E., Bellino, M., Serpietri, L., Massa, A., and Frontali, N., 1976, Heart accelerating peptides from cockroach corpora cardiaca, *J. Insect Physiol.* **22:**323–329.

Truman, J.W., 1971, Physiology of insect ecdysis. I. The eclosion behaviour of saturniid moths and its hormonal release, *J. Exp. Biol.* **54:**805–814.

Truman, J.W., 1973*a*, Physiology of insect ecdysis. II. The assay and occurrence of the eclosion hormone in the Chinese oak silkmoth *Antheraea pernyi, Biol. Bull.* **144:**200–211.

Truman, J.W., 1973*b*, Physiology of insect ecdysis. III. Relationship between the hormonal control of eclosion and of tanning in the tobacco hornworm, *Manduca sexta, J. Exp. Biol.* **58:**821–829.

Truman, J.W., and Endo, P.T., 1974, Physiology of insect ecdysis: Neural and hormonal factors involved in wing spreading behavior of moths, *J. Exp. Biol.* **61:**47–55.

Truman, J.W., and Riddiford, L.M., 1974*a*, Hormonal mechanisms underlying insect behaviour, *Adv. Insect Physiol.* **10:**297–352.

Truman, J.W., and Riddiford, L.M., 1974*b*, Physiology of insect rhythms. III. The temporal organization of the endocrine events underlying pupation of the tobacco hornworm, *J. Exp. Biol.* **60:**371–382.

Unger, H., 1956, Neurohormonale Steuerung der Herztätigkeit bei Insekten, *Naturwissenschaften* **43:**66, 67.

Unger, H., 1957, Untersuchungen zur neurohormonalen Steuerung der Herztätigkeit bei Schaben, *Biol. Zentralbl.* **76:**204–225.

Unger, H., 1965, Der Einfluss der Neurohormone C und D auf die Farbstoffabsorptions-fahigkeit der Malpighischen Gefässe (und des Darmes) der Stabheuschrecke *Carausius morosus in vitro, Zool. Jahrb. Abt. Allg. Zool. Physiol. Tiere* **71:**710–717.

Vandenberg, R.D., and Mills, R.R., 1974, Hormonal control of tanning by the American cockroach: cyclic AMP as a probable intermediate, *J. Insect Physiol.* **20:**623–627.

Vincent, J.F.V., 1971, The effects of bursicon on cuticular proteins in *Locusta migratoria migratorioides*, *J. Insect Physiol.* **17:**625–636.

Wall, B.J., 1967, Evidence for antidiuretic control of rectal water absorption in the cockroach *Periplaneta americana*, *J. Insect Physiol.* **13:**565–578.

Wall, B.J., and Ralph, C.L., 1964, Evidence for hormonal regulation of Malpighian tubule excretion in an insect, *Periplaneta americana*, *Gen. Comp. Endocrinol.* **4:**452–556.

Welsh, J.H., 1971, Neurohumoral regulation and the pharmacology of a molluscan heart, *Comp. Gen. Pharmacol.* **2:**423–432.

Wigglesworth, V.B., 1940, The determination of characters at metamorphosis in *Rhodnius prolixus* (Hemiptera), *J. Exp. Biol.* **17:**201–222.

Williams, C.M., 1947, Physiology of insect diapause. II. Interaction between the pupal brain and prothoracic glands in the metamorphosis of the giant silkworm, *Phylosamia cecropia, Biol. Bull.* **93:**89–98.

Yamazaki, M., and Kobayashi, M., 1969, Purification of the proteinic brain hormone of the silkworm, *Bombyx mori, J. Insect Physiol.* **15:**1981–1990.

Zdarek, J., and Fraenkel, G., 1969, Correlated effects of ecdysone and neurosecretion in puparium formation (pupariation) of flies, *Proc. Nat. Acad. Sci. U.S.A.* **64:**565–572.

Zeleny, C., 1905, Compensatory regulation, *J. Exp. Zool.* **2:**1–102.

11

Physiological Roles of Peptides in the Nervous System

Jeffery L. Barker

1. Introduction

The possibility that peptides play physiological roles in neuronal function has long been suggested but only recently begun to be supported by various lines of evidence. For example, subcutaneous and intrahypothalamic injections of the hypothalamic peptide luteinizing hormone–releasing hormone (LH–RH) in ovariectomized, estrogen-pretreated female rats facilitates the induction of the lordosis reflex within an hour of injection (Moss and McCann, 1973, 1975; Foreman and Moss, 1975; Moss and Foreman, 1976), while intracranial injection of renin, angiotensin I, or angiotensin II leads rapidly to drinking behavior in rats (Epstein et al., 1970; Fitzsimons, 1971; Simpson and Routtenberg, 1973, 1975). That a single substance can organize and integrate the activity of precisely the population of neurons required for such complicated behaviors as mating and drinking strongly indicates an important role in the nervous system for such a substance. The cellular correlates and molecular mechanisms underlying these complicated, peptide-induced behaviors have not been established. The purpose of this chapter is to review recent investigations into the physiological roles of several species of peptide. Most of the material discussed will be from observations made at the cellular level, primarily as these events may relate to the elaboration of endocrine, autonomic, or motor behaviors.

Although relatively little rigorous research has been carried out thus far, several hypotheses regarding the physiological roles of peptides in the nervous system

Jeffery L. Barker · Behavioral Biology Branch, National Institute of Child Health and Human Development, National Institutes of Health, Bethesda, Maryland.

have emerged. One hypothesis is that peptides may function as neurotransmitters, mediating rapid cell–cell interactions in the CNS (Nicoll and Barker, 1971a; Konishi and Otsuka, 1974a,b; Phillis and Limacher, 1974a; Dyer and Dyball, 1974; Renaud et al., 1975; Saito et al., 1975). This notion is based mainly on demonstrations of pharmacological effects of peptides on neuronal excitability (for a review, see Barker, 1976). Another idea that has emerged concerns the possibility that peptides may function as neurohormones, modulating neuronal excitability in a long-term manner (Nicoll and Barker, 1971b; Barker and Gainer, 1974; Barker et al., 1975a; Henry et al., 1975a; Renaud et al., 1975; Mayeri and Simon, 1975; Barker and Smith, 1976).

The distinctions that can presently be drawn between neurotransmitter and neurohormone will be considered briefly in this introduction in order to provide some background for better evaluation of whether peptides are neurotransmitters or neurohormones or both.

1.1. Neurotransmitters

Much evidence has accumulated to indicate that nerve cells in vertebrate and invertebrate CNS can interact through chemical and electrical forms of transmission at specialized junctions, or synapses (for recent and thorough reviews, see Gerschenfeld, 1973; Hubbard, 1973; Krnjević, 1974). Electrical transmission is instantaneous, the electrical signal being transmitted to the follower (postsynaptic) cell through low-resistance junctions between the cells (for more detailed information, see M. V. L. Bennett, 1966). Electrical transmission is therefore limited to contiguous elements, and is mediated by ionic currents. Chemical transmission involves the release of a chemical (transmitter) from the presynaptic cell into a specialized 200-Å space between cells (synaptic cleft). Transmitter release is the conclusion to a complicated series of events, which are minimally understood. Transmitter release takes two forms, resting and evoked—the former referring to release that occurs in the absence of discernible electrical activity, the latter describing release that accompanies electrical events (action potentials that invade and excite the presynaptic terminal). Excitation of the terminal transiently depolarizes the terminal, increasing the permeability of the membrane to Ca^{2+}, allowing entry of Ca^{2+} into the terminal. This last step is in some way coupled to the exit of transmitters (and hormones) "excitation–secretion coupling" (for details, see Douglas and Poisner, 1964; Miledi, 1973). The frequency of electrical events, the duration of terminal excitability, and the amount of Ca^{2+} entering are all important determinants of the magnitude and kinetics of evoked transmitter release.

Following release, transmitter molecules move across the synaptic cleft to interact with localized receptor sites on the postsynaptic cells. Activation of these sites leads to receptor-coupled changes in the conductance of the postsynaptic membrane to one or more of the predominant ions (Na^+, K^+, Cl^-, Ca^{2+}). These conductance changes are largely independent of the membrane potential of the cell. Conductance changes transiently increase or decrease the excitability of the postsynaptic element, the extent depending on the location of the synaptic receptors and the passive electrical properties of the cell. The polarity of the postsynaptic response depends on which ionic conductances are activated and what their concentration gradients across the cell are. All the electrically observable synaptic events are completed in a period of milliseconds to seconds. There are also longer-lasting consequences of receptor activation of postsynaptic cells that are only recently becoming appreciated, including selective changes in cAMP and cGMP levels (Kebabian *et al.*, 1975) and in protein synthesis (Gainer and Barker, 1974, 1975). Inactivation of synaptic transmission is accomplished through active enzymatic degradation of the transmitter molecule, through reuptake and degradation, or through passive diffusion. Other factors that have been shown to be important to the quality of synaptic transmission include availability of releasable transmitter, desensitization of postsynaptic receptors following continued release, and hypersensitivity of receptors after a period of diminished release.

A major goal of research on synaptic transmission is to identify the naturally occurring transmitter molecules that mediate various synaptic transmissions. Some basic requirements for the establishment of this identity include (1) isolation of the molecule from presynaptic terminals; (2) pharmacological actions of the molecule on postsynaptic membrane identical to the physiological response observed during synaptic transmission; (3) sensitivity of both pharmacological action and physiological transmission to the same antagonists; and (4) physiological release from presynaptic terminals (see Werman, 1966, for a discussion of criteria). Additional requirements might include demonstration of synthetic and degradative enzymes, of reuptake mechanisms at the synapse, and of transmitter receptor and receptor protein. Some of the molecules proposed as putative synaptic transmitters include amino acids (Gly, Tau, Gln, GABA, Pro), catecholamines (norepinephrine, serotonin, dopamine), and acetylcholine. None of these molecules has clearly fulfilled all the basic requirements, although available evidence strongly supports GABA, norepinephrine, and acetylcholine as mediators of various peripheral synaptic events in vertebrates and invertebrates.

Identification of transmitters mediating central synaptic events is greatly hindered by the embedding of synapses within a maze of other elements (glial and neuronal). Therefore, evidence gathered in the CNS is contaminated and fragmentary, at best. Fractionation of the CNS to yield tissue rich in synaptic elements (synaptosomes) containing presynaptic terminals and postsynaptic membranes has been only partly successful in providing a preparation to identify central transmitters. The same basic requirements as stated above cannot yet be met by research on central transmitters with the same rigor as has been applied to peripheral transmitters. General evidence of a central transmitter includes (1) presence of the molecule in synaptosomal preparations; (2) histochemical or immunohistochemical demonstration of distinct pathways or endings containing the putative transmitter; (3) presence of synthetic, degradative, and reuptake processes specific for the molecule; (4) evoked release of the molecule in a Ca^{2+}-dependent manner; (5) demonstration of specific receptors for the transmitter (typically by specific antagonist binding studies); and (6) parallel sensitivity of synaptic potentials and postsynaptic pharmacological responses by specific antagonists. The specific conductance changes and biochemical events associated with synaptic transmission in the CNS have largely eluded observation thus far, since such observations are not possible in intact tissue. Dissociation of neural elements and growth in tissue culture should provide a preparation to study both these aspects of synaptic transmission (Nelson, 1975).

To summarize, neurotransmitters are a low-molecular-weight class of molecules synthesized by nerve cells that underlie a form of cell–cell interaction in the nervous system characterized as synaptic transmission. Transmitters function over relatively small and discrete areas for brief periods (milliseconds to seconds). The main function of the transmitter class of molecules appears to be mediation at specific locations of short-lived interactions between cellular elements. At the electrophysiological level, such interactions rapidly regulate postsynaptic cellular excitability.

1.2. Neurohormones

Distinctions among various forms of secretion and comparisons between neuronal and nonneuronal forms have been made repeatedly (Bargmann, 1966; E. Scharrer, 1966; B. Scharrer, 1970). It is clear that certain cells in the CNS can synthesize and secrete substances that have a target far removed from the cell of origin. For example, large cells clustered in the paraventricular and supraoptic (''magnocellular'') nuclei synthesize several classes of peptides that are transported to axon terminals to be released for export by the general (and portal) circulation to nonneural (and endocrine) target tissues (for a source book, see Berde, 1968). These substances are derived from elements that appear to have many of the physiological properties of nerve cells, including excitability (Kandel, 1964; Yagi et al., 1966; Kelly and Dreifuss, 1970; Barker et al., 1971b,c; Koizumi and Yamashita, 1972), synaptic input (Kandel, 1964; Dreifuss and Kelly, 1972a,b; Barker et al., 1971a; Nicoll and Barker, 1971a; Negoro and Holland, 1972; Koizumi and Yamashita, 1972), and chemosensitivity (Moss et al., 1971; Barker et al., 1971a; Nicoll and Barker, 1971a; Dreifuss and Kelly, 1972a,b; Moss et al., 1972a,b). Since the magnocellular terminals in the posterior pituitary do not synapse on other neurons, but rather end in vascular areas, the cells are ''neurosecretory'' (Bargmann, 1966; E. Scharrer, 1966), and their peptide products are ''neurohormones'' (B. Scharrer, 1969, 1972). Also included in these definitions are parvicellular neurosecretory cells, the axons of which terminate on the hypophyseal portal system of the median eminence to release peptide neurohormones that have as primary targets endocrine cells in the anterior pituitary. As presently defined, then, neurohormones are substances released by nerve cells at extrasynaptic sites (portal, general, or cerebrospinal circulations) to regulate the activity of both endocrine and nonendocrine target tissues.

Neurohormones differ from neurotransmitters in (1) the location of their release sites; (2) the distance between release and target sites; (3) their time-course of action; (4) the potentiality for multiple actions on a variety of nonneural and neural cellular elements; and (5) their peptide structure. Similarities between neurotransmitter and neurohormone physiology include (1) Ca^{++}-dependent release from membrane-bound stores (Douglas and Poisner, 1964; Daniel and Lederis, 1967; Douglas *et al.*, 1971; Hubbard, 1973; Peck *et al.*, 1975); (2) activation of adenylate or guanylate cyclase activity of target tissues (Orloff and Handler, 1967; Borgeat *et al.*, 1972; Kebabian *et al.*, 1975); and (3) direct or indirect regulation of product release from endocrine and neural elements.

The possibility that magnocellular, parvicellular, and other peptides may play functional roles in the CNS has been raised following evidence of (1) peptides in extrahypothalamic areas (Winokur and Utiger, 1974; Winters *et al.*, 1974; Leeman and Mroz, 1974; Hökfelt *et al.*, 1975*a,b;* Hughes, 1975; Hughes *et al.*, 1975; Brownstein *et al.*, 1975); (2) the induction of specific CNS-dependent behavioral patterns by peptides (Moss and McCann 1973, 1975; Epstein *et al.*, 1970; Fitzsimons, 1971; Simpson and Routtenberg, 1973); (3) the effects of peptides on neuronal excitability in a wide variety of preparations (Nicoll and Barker, 1971*a,b;* Moss *et al.*, 1972*a,b;* Barker and Gainer, 1974; Krnjević and Morris, 1974; Konishi and Otsuka, 1974*a,b;* Henry *et al.*, 1975; Phillis and Limacher, 1974*a,b;* Barker *et al.*, 1975*a;* Renaud and Martin, 1975*a,b;* Renaud *et al.*, 1975; Dyer and Dyball, 1974; Mayeri and Simon, 1975; Kawakami and Sakuma, 1974, 1976); and (4) recurrent collaterals of magnocellular neurosecretory cells (Kandel, 1964; Barker *et al.*, 1971*a;* Dreifuss and Kelly, 1972*a;* Koizumi and Yamashita, 1972; Novin and Durham, 1973; Koizumi *et al.*, 1973). Peptides may therefore not only be released into circulations to affect nonneural targets, but may also be released in the CNS to mediate short-term events via synaptic mechanisms or long-term events via hormonal mechanisms. Neurohormones may be redefined in terms of function, as substances synthesized by nerve cells with effects in the CNS that are operationally distinct from transmitter actions. By this definition, neurohormones may or may not be released at synapses (and "putative transmitters" with actions similar to peptides may be reconsidered as neurohormones). The remainder of the chapter is devoted to reviewing the evidence that peptides may play neurotransmitter and neurohormonal roles.

2. Substance P

Arg–Pro–Lys–Pro–Gln–Gln–Phe–Phe–Gly–Leu–Met (NH₂)

2.1. Distribution

Substance P is an undecapeptide that has received increasing investigative attention over the past 45 years, beginning with its isolation from equine brain and intestine and demonstration of actions on smooth muscle by von Euler and Gaddum (1931). Final characterization and purification, as well as synthesis, worked out by Leeman and co-workers (reviewed by Leeman and Mroz, 1974; see also Chapter 5), have accelerated research on the functions of the peptide. Biochemical, pharmacological, and immunohistochemical techniques have revealed a widespread distribution of Substance P or Substance P–like activity in both the mammalian central and peripheral nervous systems, including nerve endings in the skin, in fibers terminating on sweat glands and around vessels, in intestinal plexuses, in peripheral nerves, in sensory ganglia and axons, throughout the substantia gelatinosa, around the central spinal canal and in cells located in the ventral horn of the cord, and in mesencephalon, medulla oblongata, hypothalamus, and cortex (Pernow, 1953; Amin et al., 1954; Gaddum, 1960; Katoaka, 1962; Ryall, 1964; Shaw and Ramwell, 1968; Takahashi et al., 1974; Takahashi and Otsuka, 1975; Hökfelt et al., 1975a,b). The recent immunohistochemical methods of Hökfelt and colleagues (1975a,b) have further localized Substance P–like immunoreactive material to a small-sized population of sensory ganglion cells and their unmyelinated axons and terminals in cats and rats. The relatively greater resolution and selectivity of this method appear to have compromised its sensitivity, since Substance P–like immunohistofluorescence was observed only in sensory cells and their axons following chemical or mechanical block of axonal transport of the peptide material. The lack of fluorescence in the sensory cells of the untreated preparation contrasts with the radioimmunoassayable and bioassayable activity in sensory ganglia and axons (Takahashi et al., 1974; Takahashi and Otsuka, 1975). The different methods of assaying peptide activity could account for the different results.

Assayable activity is transported out to the periphery (Holton, 1960; Hökfelt et al., 1975b), as well as into the spinal cord (Takahashi and Otsuka, 1975; Hökfelt et al., 1975b). Although the rate of transport has not been determined, assayable activity accumulates within 24 hr (Hökfelt et al., 1975b), suggesting that the peptide is transported during the rapid phase of transport. Whether the material is transported membrane-bound and in final form remains to be determined. That it is transported intragranularly is suggested by the release of activity from a particulate fraction of peripheral nerves following treatment with hypotonic solutions (von Euler, 1963).

The distribution of activity in the spinal cord changes within 10 days of transection of sensory axons in dorsal roots: activity is greatly depressed in the substantia gelatinosa, but not in the remainder of the spinal cord (Takahashi and Otsuka, 1975; Hökfelt et al., 1975b). These findings, taken with those showing accumulation of activity after blocking transport, indicate that peptide activity is transported from cell body to afferent terminals. These results, then, provide evidence to fulfill one of the requirements to support either a neurotransmitter or neurohormone role for Substance P or a closely related peptide: presence in nerve cells and terminals.

2.2. Actions in the Central Nervous System

Functional roles for Substance P in the CNS are suggested by a number of pharmacological actions of the synthetic or extracted peptide. Substance P (and several structurally related peptides) cause excitation of cuneate neurons (Krnjević and Morris, 1974), dorsal horn cells (Henry et al., 1975; Henry, 1975), motoneurons (Konishi and Otsuka, 1974a,b; Takahashi et al., 1974; Takahashi and Otsuka, 1975), and Betz cells and unidentified cortical units (Phillis and Limacher, 1974*a,b*). Similar effects of the peptide on the different types of neurons have been observed, and are characterized by a gradual and sustained excitation (Fig. 1–4). The onset of action of the peptide is typically slower than that of putative transmitters, as can be seen in comparing Substance P and glutamate excitation of motoneurons and interneurons in the spinal cord. The response to bath application of glutamate is immediate (Figs. 1 and 2; see also Barker and Nicoll, 1973). The duration of the peptide excitation outlasts the period of application, in contrast to the amino acid response, which decays rapidly. The character and time-course of the responses on motoneurons in the isolated spinal cord presumably reflect postsynaptic actions, since the responses and their different kinetics remain the same after elimination of synaptic transmission (Barker and Nicoll, 1973; Konishi and Otsuka, 1974*a,b*). The relatively slower time-course of peptide action might be due to the kinetics of diffusion of the larger molecule to its site of action, or it might reflect the intrinsic activity of the peptide. The slow time-course *per se* should not immediately exclude the material from consideration as a neurotransmitter, since the effects of some putative transmitters have similarly slow time-courses: acetylcholine (ACh) excitation of cortical neurons (Fig. 4; Krnjević and Phillis, 1963; Phillis and Limacher, 1974*a*) and sympathetic ganglion cells (Weight and Votava, 1970), and certain depolarizing responses to serotonin in molluscan cells (Gerschenfeld and Paupardin-Tritsch, 1974; Barker, 1975). All these slow time-courses are thought to be due to a receptor-mediated decrease in resting K^+ conductance.

The peptide and amino acid responses on motoneurons in the isolated cord preparation differ not only in kinetics, but also in relative potency, with the peptide active over the 10^{-8}–10^{-6} M range (Konishi and Otsuka, 1974*a,b;* Otsuka and Konishi, 1975), while glutamate is active over the 10^{-4}–10^{-3} M range (Barker and Nicoll, 1973; Konishi and Otsuka, 1974*a,b*). A similar difference in potency may also be present in other areas of the nervous system in which

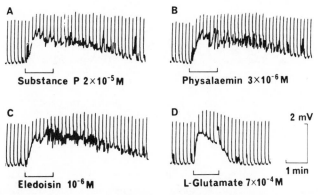

FIG. 1. Effects of Substance P, physalaemin, eledoisin, and L-glutamate on the isolated spinal cord of frog. Potential changes generated in the motoneurons were recorded from the 8th ventral root. Drugs were applied in the bath during the periods indicated in the figure. The corresponding dorsal root was stimulated maximally at the rate of 0.1 Hz. DC recording. Depolarization upward. From Konishi and Otsuka (1974*a*).

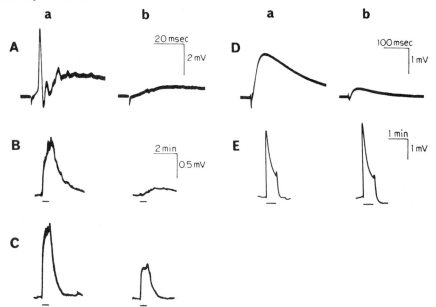

FIG. 2. Effects of Lioresal on the reflex activities and the responses to Substance P and amino acids. (A–C) Recordings from L3 ventral root: (A) mono- and polysynaptic reflexes; (B) responses to Substance P (2×10^{-7} M); (C) responses to L-glutamate (10^{-3}M). (D, E) Recordings from L5 dorsal root: (D) dorsal root potentials induced by a single volley in L4 dorsal root; (E) responses to GABA (3×10^{-5} M). Horizontal bars under tracings mark the periods of drug applications. In each pair of tracings: (a) obtained in normal Krebs solution; (b) obtained in the solution containing Lioresal (5×10^{-6} M). Scales for B also apply to C. Positivity at the recording electrode upward. From Saito et al. (1975).

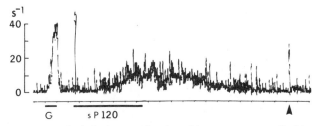

FIG. 3. Rate-meter record of discharge of a dorsal horn neuron excited by Substance P (SP) and glutamate (G). The unit was also activated following a jet of air to the peripheral receptive field (delivered at arrowhead). Ordinate in spikes/sec. Time marks: 10 sec. From Henry et al. (1975).

neuronal responses to Substance P and transmitters have been compared using microiontophoretic application. Responses to Substance P have occurred following iontophoresis from micropipettes containing 0.8–7 mM peptide solutions (Krnjević and Morris, 1974; Phillis and Limacher, 1974a,b; Henry et al., 1975). In contrast, 0.1–1.0 M solutions of putative transmitter have usually been employed in iontophoretic experiments to obtain comparable neuronal responses (Krnjević

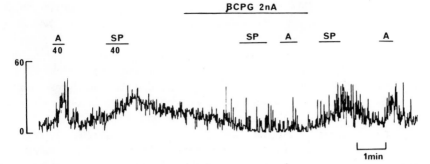

FIG. 4. Rate-meter recording of the firing of a single corticospinal neuron. Microiontophoresis of Substance P (SP) and ACh (A) increases the spontaneous activity, and both these excitatory effects are depressed by concomitant iontophoresis of β-chlorophenyl-GABA (βCPG). Recovery occurs rapidly following termination of the βCPG application. Periods of drug application are indicated by bars (with ejection currents below bars in nA). Ordinate: spikes/sec. From Phillis (unpublished observations).

and Morris, 1974; Phillis and Limacher, 1974a,b). Presumably, the 1000-fold greater concentrations of transmitter do not reflect *less* transport numbers for these molecules in an electric field. In fact, transport numbers recently established for several peptides are similar to those already developed for transmitters (Renaud, personal communication). It is thus possible that the requirement for different concentrations reflects different intrinsic biological activities. (Other possibilities include greater sensitivity to enzymatic or uptake processes of the transmitters relative to the peptide, thus requiring more transmitter molecules per unit time to produce comparable postsynaptic effects.)

The observed postsynaptic effects of Substance P and a group of closely related analogues (Konishi and Otsuka, 1974a) have been interpreted as evidence for (Konishi and Otsuka, 1974a,b; Takahashi et al., 1975; Saito et al., 1975) and against (Krnjević and Morris, 1974; Henry et al., 1975) the original hypothesis of Lembeck (1953) that Substance P or a closely related peptide is the transmitter mediating synaptic input from primary afferents into the CNS. Those who believe that the demonstrated slow time-course of peptide effect eliminates Substance P from consideration as a *transmitter* mediating rapid synaptic events (Krnjević and Morris, 1974; Henry et al., 1975) have suggested that the peptide might play a modulatory role acting over a longer period of time than individual synaptic events. Others, noting the presence of peptide only in small sensory cells, unmyelinated afferents, and their terminals (Hökfelt et al., 1975a, b), as well as clear excitatory effects on nociceptive neurons in the spinal cord (Henry, 1975), have proposed that the peptide might be either the transmitter (Hökfelt et al., 1975b) or a regulator (Henry, 1975) mediating nociceptive pathways. Another proposed synaptic function is that Substance P or a like peptide mediates excitatory transmission from primary afferents to motoneurons (Konishi and Otsuka, 1974a,b; Takahashi et al., 1974). However, other lines of evidence are required to support any of these notions.

One line of investigation that is important for better establishing the candidacy of Substance P has been developed by Konishi, Otsuka, and colleagues. They have recently reported that Lioresal, a chlorinated benzene derivative of GABA that is used therapeutically to reduce spasticity (Pinto *et al.*, 1972; Knutsson *et al.*, 1973), blocks both the depolarizing action of Substance P on motoneurons and mono- and polysynaptic activation of motoneurons (Otsuka and Konishi, 1975; Saito *et al.*, 1975). Although a thorough pharmacological analysis of Lioresal on synaptic physiology and pharmacology in the spinal cord has not been carried out, Davidoff and Sears (1974) have shown that the drug does not block either glutamate or aspartate depolarizations of motoneurons, but does reduce primary afferent–motoneuron transmission in the frog. Similar results have recently been reported by Saito *et al.* (1975) in the isolated rat spinal cord. (It should be noted that the latter authors did report some depression of glutamate excitation by Lioresal.) The foregoing findings suggest some specificity of the drug's antagonism of the peptide response in the spinal cord. Konishi and Otsuka (1974*b*) have also demonstrated that Substance P depresses, then facilitates, primary afferent–motoneuron transmission. The inhibitory component is blocked by coincident perfusion with picrotoxin, an antagonist of GABA responses in the amphibian (Tebēcis and Phillis, 1969; Davidoff, 1972; Barker and Nicoll, 1973; Barker *et al.*, 1975*b*) and mammalian spinal cord (Levy, 1974), leading Konishi and Otsuka (1974*b*) to suggest that the inhibitory effect of Substance P may be due to activation of a "GABA-operated inhibitory mechanism." GABA is considered a putative transmitter mediating inhibition of primary afferent–motoneuron transmission through depolarization of primary afferent terminals and inactivation of transmitter release–"presynaptic inhibition" (Eccles *et al.*, 1963; Barker and Nicoll, 1972, 1973; Davidoff, 1972; Barker *et al.*, 1975*c*). However, depolarizing synaptic events occurring on primary afferent terminals presynaptic to motoneurons are not completely antagonized by picrotoxin (Barker *et al.*, 1975*c*). In fact, the depolarizing potentials, which may last several seconds, are antagonized by Lioresal, which does not block the depolarizing effects of GABA on primary afferents (Davidoff and Sears, 1974; Saito *et al.*, 1975).

Although definitive conclusions cannot be drawn from these results, the relative specificity for Lioresal's antagonism of Substance P on motoneurons, as well as its depression of primary afferent transmissions to motoneurons and to other primary afferents (causing presynaptic inhibition), suggest that the peptide may be involved in *both* synaptic pathways. Thus, Substance P might be released from primary afferents synapsing directly on other afferents or on GABAergic interneurons that themselves release GABA to depolarize other afferents. This might provide a mechanism for enhancing the sensory input from one population of afferents relative to the inputs from other afferents. Perhaps a particular sensory modality is served by peptide-containing afferents (e.g., nociception) and this form of input is "tuned up" to the relative exclusion of others, which are "tuned down" through concomitant presynaptic inhibition. The possibility that Substance P may not act primarily as a transmitter either in presynaptic inhibitory transmissions or in primary afferent input into the CNS, but rather might act more as a modulator to set the bias of motoneuron and afferent membranes postsynaptic to afferents, is suggested by the observation of Davidoff and Sears (1974) that Lioresal causes a hyperpolarization of both motoneurons and primary afferents that is largely abolished by Mg^{2+}. These results may be interpreted as indicating that Substance P is tonically released from elements presynaptic to motoneurons and primary afferents. The rate of tonic (as well as phasic) release would be important in determining the level of presynaptic inhibition and the level of postsynaptic excitation. Unfortunately, Lioresal's relative specificity for Substance P effects in the spinal cord does not extend to cortical structures, where it blocks ACh responses equally as well as peptide responses (with glutamate responses remaining resistant) (Fig. 4).

At present, it is impossible to establish or eliminate either a transmitter or a modulator role for Substance P in primary afferent transmissions. The presence of this peptide, as well as its clear electrophysiological effects (and those of related analogues), warrant further study. For example, demonstration of its release would fulfill an important requirement for candidacy, thus far lacking. Factors that regulate release of the peptide my act indirectly to modulate the level of sensory input into the CNS. One intriguing possibility comes from consideration of the marked overlap in distribution of both immunohistochemical Substance P–like activity (Hökfelt *et al.*, 1975a,b) and opiate receptors (Snyder, 1975) in the substantia gelatinosa. The latter receptors, which are receptors for naturally occurring peptides (Hughes, 1975; Hughes *et al.*, 1975), might be placed on Substance P–containing afferents to allow peptide regulation of the release of another peptide, by analogy to releasing-factor peptide regulation of peptide hormone release from the pituitary. Or the opiate peptide receptors might be placed on cells postsynaptic to Substance P–containing afferents to interact postsynaptically with Substance P. In this regard, Substance P has been shown to antagonize the analgesic and respiratory depressant actions of morphine and to produce hyperalgesia and respiratory stimulation (Zetler, 1956; Keele and Armstrong, 1964).

Although sedation and stupor have been reported following intraventricular administration (von Euler and Pernow, 1954, 1956), and anti-convulsant activity has been observed against both picrotoxin and strychnine after subcutaneous injection (Zetler, 1956), clear, reproducible, and specific endocrine, autonomic, or motor effects using purified Substance P have not yet been demonstrated. Recent immunohistochemical techniques have confirmed the presence of peptidelike material unevenly distributed in CNS structures (Nilsson *et al.*, 1974; Höfelt, 1975a), and effects on excitability at the cellular level have been reported (Phillis and Limacher, 1974a,b; Fig. 4), but other aspects of the physiology of the peptide in the CNS (e.g., cell bodies of origin, synthesis, transport, release) are presently lacking. Therefore, serious conjectures about supraspinal roles of Substance P have not been offered.

2.3. Actions in the Peripheral Nervous System

Besides being transported into the spinal cord, Substance P is also transported to the periphery from sensory cell bodies, and is present in free nerve endings

(Hökfelt *et al.*, 1975*b*). The peptide has a multitude of peripheral actions, including contraction of the small intestine (Pernow, 1960), vasodilation and hypotension (von Euler and Gaddum, 1931; Pernow, 1960; Takahashi and Otsuka, 1975), and salivation (Chang and Leeman, 1970). Its presence in the periphery has led to the suggestion that it might mediate the phenomenon of antidromic vasodilation (Hökfelt *et al.*, 1975*b*), originally proposed by Lembeck (1953). Subcutaneous injection of the peptide in volunteers causes the sensation of itching (Hökfelt *et al.*, 1975*b*), while application in approximately micromolar concentration to a blister base produces severe pain (Armstrong *et al.*, 1954). Recently, Juan and Lembeck (1974) have used an isolated, perfused rabbit ear preparation to compare the effects of ACh and peptides on both the electrical activity of the sensory nerve and the reflex fall in blood pressure. Substances were injected into the arterial line, and electrical activity in sensory fibers was recorded extracellularly from the central end of the greater auricular nerve. ACh produced discharge of activity that was rapid in onset and brief in duration. Substance P, extracted from beef gut and brain, elicited marked action potential activity, the response being slower in onset and longer in duration. The peptide was about 700 times more potent in producing these effects than ACh. Apparent specificity of these pharmacological effects is suggested by the inactivity of the structurally related peptides physalaemin and eledoisin, as well as prostaglandin E_1, vasopressin, oxytocin, and adrenaline. When the greater auricular nerve was left intact, injection of Substance P and ACh into the injection line produced a reflex fall in blood pressure. The peptide and ACh responses again had long and short time-courses, respectively. Substances inactive in eliciting activity on sensory fibers were likewise inactive in causing hypotension. Juan and Lembeck concluded that Substance P and ACh may be playing excitatory roles in activating "paravascular pain receptors." That these receptors do not respond to the closely related peptide analogues physalaemin and eledoisin is remarkable, since the latter are active at the central terminals of primary afferents in the spinal cord (Konishi and Otsuka, 1974*a*) and elsewhere in the CNS (Phillis and Limacher, 1974*b*). (It should be noted that the *inactivity* of synthetic Substance P obtained commercially and the lower activity of Substance P obtained from peptide synthesis—by Dr. Susan Leeman's laboratory—suggest that the active factor may not have precisely the same structure as Substance P.)

The demonstrated presence and activity of peptide in the periphery raise the question of its function. One possibility is that the secretion of the peptide alters the membrane properties of, and sets the bias on the excitability of, free nerve endings, thus regulating, in either a general or a specific way, sensory input into the nervous system. Regulation of the secretion of the peptide would then be important in determining the level of sensory input. Factors involved in such regulation might well be correlated with various behavioral states in which sensory input is altered either in a general way (as in sleep or in hibernation) or in a specific way (as during estrus).

2.4. Conclusions

Future considerations of the roles of Substance P and related peptides in neuronal function will depend on a closer correlation between physiological pathways and pharmacological actions. Of equal importance in understanding the peptide's relationship to the nervous system will be a demonstration of its biochemistry (synthesis and metabolism). Enzymatic cleavage of select portions of the undecapeptide substrate might be not only a pathway of metabolism, but also a means of producing peptide metabolites that interact with different receptors and have different functions. Clearly, the investigation of the physiological roles of Substance P in the nervous system has just begun.

3. Angiotensin II

$$\text{Asp–Arg–Val–Tyr–Ile–His–Pro–Phe}$$
$$1 \quad 2 \quad 3 \quad 4 \quad 5 \quad 6 \quad 7 \quad 8$$

3.1. Distribution

Renin is a proteolytic enzyme synthesized and secreted primarily by renal tissue that cleaves its substrate, an α_2 globulin made in liver cells, to produce the decapeptide angiotensin I, the carboxyl-terminal dipeptide moiety of which is in turn removed by the action of converting enzyme to produce the active octapeptide angiotensin II (for a recent review, see Oparil and Haber, 1974). The primary peripheral actions of the active peptide are directed at vascular smooth muscle cells to cause vasoconstriction and at aldosterone-producing cells in the adrenal cortex, where the peptide stimulates the secretion of aldosterone. Aldosterone increases Na^+ retention by renal tubular cells, which leads to increased extracellular volume. Angiotensin II is deaspartylated to angiotensin III (an active metabolite; see Tsai et al., 1975). Both angiotensin II and III are inactivated in peripheral capillary beds by angiotensinases. Until recently, almost all research on the renin–angiotensin system focused on the peripheral components of the enzyme–peptide–steroid regulation of blood pressure and volume homeostasis. However, various lines of evidence have been put forward to indicate that the renin–angiotensin system also utilizes central and peripheral nervous systems in regulating circulatory pressure and volume (for recent reviews, see Severs and Daniels-Severs, 1973; Regoli et al., 1974). All the components of the renin–angiotensin system are present in the CNS (Ganten et al., 1971a,b; Fischer-Ferraro et al., 1971), the availability of the active product apparently being dependent on the availability of converting enzyme (Yang and Neff, 1972). The possibility that renin present in the CNS represents contamination from the periphery has been raised, but seems improbable, since renin cannot be detected in CSF in conditions that greatly elevate circulating levels of renin (Ganten et al., 1971a).

Radioimmunological and immunohistochemical methods have recently been utilized to reveal an uneven distribution of angiotensin I in the CNS (Changaris *et al.*, 1976). Angiotensin I is present in relatively high concentrations in the subfornical organ, pars intermedia, corpus callosum, hypothalamus, and choroid plexus. Specific receptors for angiotensin II have also been demonstrated throughout the CNS (McLean, *et al.*, 1975), the highest specific binding being in hypothalamic and midbrain–thalamic regions. Some but not all of these receptors are distributed in the circumventricular organs, CNS areas that interface with the circulatory system at leaky, fenestrated parts of the blood–brain barrier. These areas include the area postrema, subfornical organ, and supraoptic crest (Broadwell and Brightman, 1976). In this regard, [^{14}C]angiotensin II given intravenously appears to enter the CNS at the level of the third ventricle, and is found there and in the aqueduct (Volicer and Loew, 1971). More recent evidence, however, shows that very little radioactivity can be detected in CSF following intravenous administration of labeled peptide and, further, that what is detected is not angiotensin II by immunoreactivity (Ganten, 1976). It is thus possible that some of the effects of angiotensin II on the CNS may be attributed to peptide derived from a CNS renin–angiotensin system, while others may be due to circulating angiotensin II that has diffused across the more permeable parts of the blood–brain barrier.

3.2. Actions in the Central Nervous System

3.2.1. Drinking Behavior

Epstein *et al.* (1970) found that injection of picomoles of angiotensin II through cannulae inserted into anterior hypothalamus, preoptic region, and septum caused water-replete rats to drink. The behavior was elicited rapidly and specifically. Other forms of behavior (e.g., hyperactivity, eating) were not observed. Carbachol, a structural analogue of the putative transmitter acetylcholine, also caused drinking when injected at the angiotensin-sensitive sites, but the time-course of the effects was different, as was the specificity of the behavior pattern evoked. Drinking was not elicited by injections of bradykinin, vasopressin, catecholamines, or aldosterone. Fitzsimons (1971) carried out a structure–activity study to investigate the molecular requirements for activity. An absolute requirement for phenylalanine at the carboxyl terminal was found, while various other reductions in the number of amino acids depressed activity by varying degrees. Both renin substrate and angiotensin I were fully effective, as was renin. With the latter, a considerable latency to drinking was observed, presumably due to time required to cleave available substrate and produce angiotensin I, which would then be enzymatically converted to the active peptide. The ventricular site of peptide-induced dipsogenic activity appears to be the anterior portion of the third ventricle, and in particular the organum vasculosum area (Phillips, 1976).

These results have been confirmed and extended in the goat by Andersson and colleagues (Andersson *et al.*, 1970, 1975; Andersson and Eriksson, 1971). They have found that the dipsogenic activity of intraventricular injections of angiotensin II is dependent on CSF Na$^+$ concentration. Prehydration of the goat abolished dipsogenic responses to intraventricularly administered angiotensin II unless hypertonic NaCl was included in the infusion. Intraventricular infusions of either the peptide or hypertonic NaCl alone in goats in normal water balance resulted in drinking that was greatly enhanced by simultaneous NaCl infusion, demonstrating apparent synergism. These results have led to the suggestion that angiotensin II promotes the uptake of Na^+ into sensitive nerve cells involved in the drinking behavior (Andersson *et al.*, 1975). Severs *et al.* (1970) have also reported drinking activity after intraventricular administration of angiotensin II, but this activity was abolished by treatment with anticholinergic drugs or by prior hypophysectomy. It is possible that these treatments affected the behavior of the preparation in a general, not a specific, way.

Another angiotensin-sensitive area that appears to be involved in dipsogenic activity is the subfornical organ (Simpson and Routtenberg, 1973, 1975). Dipsogenic activity can also be initiated by injection of acetylcholine into the subfornical organ (Simpson and Routtenberg, 1975). Recent evidence indicates that angiotensin II–induced drinking mediated by the subfornical organ is initiated *only* by bloodborne peptide, and not by intraventricularly administered material (Phillips, 1976).

Prostaglandin (PG) participation in the (intraventricular) peptide-induced drinking is suggested by the fact that intraventricular PGE_1 supresses both drinking behavior and peptide-induced dipsogenic activity (Epstein and Kennedy, 1976). Low doses of PGE_1 did not block hypertonic saline- or carbachol-stimulated drinking, indicating a relatively specific interaction with angiotensin II–related dipsogenic activity. Performance of drinking behavior *per se* was unaffected, as were other ingestive behaviors. PGE_2 also suppressed peptide drinking, primarily by prolonging latency to drinking. Other prostaglandins (PGA_1 and $PGF_{2\alpha}$) were ineffective. These results suggest that E prostaglandins, which are endogenous to the CNS, may be natural satiators of thirst. In fact, peptide-induced drinking can be prolonged by coincident administration of protaglandin synthesis inhibitors (Phillips, 1976).

Investigations into the cellular basis for the dipsogenic effects of angiotensin II in the CNS have only recently been initiated. Microiontophoretic application of angiotensin II to neurons in the subfornical organ invariably caused potent excitatory responses (Felix and Akert, 1974; Felix, 1976) (Fig. 5). The peptide frequently caused bursts of action potential activity. Acetylcholine also evoked excitatory discharges from subfornical organs (discharges that were blocked by atropine). If we assume that the transport of peptides and of acetylcholine by current passage through the micropipette is similar, then angiotensin dispensed from 10^{-5}–10^{-3} M solutions would appear to be considerably more potent than acetylcholine dispensed from 5×10^{-1} M solutions. The difference in potency might relate to density of receptors or efficacy of receptor-coupled excitation. That mediation of the peptide excitation was not likely through a release of acetylcholine is suggested by the lack of atropine effect on peptide excitation. Half the cells tested for comparison of peptide and acetylcholine effects were excited by the peptide alone. Other peptides examined, including bradykinin, eledoisin, and physalae-

A

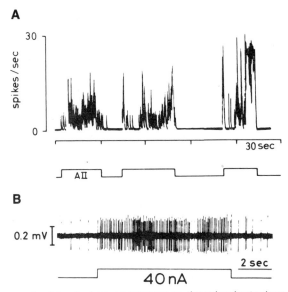

B

FIG. 5. Responses of subfornical organ neurons to the microiontophoresis of angiotensin II. (A) Integrated firing frequency of a neuron. Ejections of angiotensin II (+35 nA, +35 nA, and +50 nA), marked on bottom line, cause excitation after short delay. (B) Continuous recording of the neuronal discharge, showing enhancement of firing rate during the application of angiotensin II (+40 nA) over a period of 8 sec. From Felix (1976).

min, were ineffective. Angiotensin II microiontophoresed in the cortex and hippocampus did not produce significant excitatory effects. These results demonstrate the presence of neurons in the subfornical organ that are activated by angiotensin. Later observations by Phillips and Felix (1976), showing a block of peptide excitation by a structurally related analogue, suggest the presence of receptors on these neurons. Activation of these units would appear to be one of the cellular components of dipsogenic activity.

A study on the cellular effects of iontophoretic and intravenous angiotensin II has been carried out by Wayner et al. (1973) on other parts of the CNS considered to be involved in drinking. They found many neurons responsive to either iontophoretic or intravenous administration of angiotensin present in the lateral hypothalamus, zona incerta, ventromedial and dorsomedial hypothalamic nuclei, dentate gyrus, and thalamus. Some of the neurons excited by angiotensin II were also excited by either intravenous or iontophoretic application of Na^+. Coincident application of both Na^+ and peptide by both routes of administration often evoked a discharge of activity far greater than was evoked by either peptide or ion alone. These results, although obtained on unidentified neurons using extracellular recording techniques, may provide examples of cellular components of the observations made by Andersson and colleagues regarding synergism between Na^+ and angiotensin II in promoting dipsogenic activity. However, caution must be exercised in interpreting the excitatory effects of intravenous angiotensin II on hypothalamic neurons removed from the permeable areas of the blood–brain barrier, since they might well be due to effects secondary to the increases in blood pressure induced by intravenous angiotensin. Exactly how angiotensin II and Na^+ interact to initiate drinking remains for future investigation to determine.

3.2.2. Vasopressin Release

Intravenous, intracarotid, and intraventricular administration of angiotensin II have all been reported to increase circulating vasopressin levels (Bonjour and Malvin, 1970; Severs *et al.*, 1970; Mouw *et al.*, 1971; Keil *et al.*, 1975; Malvin, 1976). Stimulation of vasopressin release by intraventricular angiotensin II is potentiated by coincident infusion of hypertonic NaCl (Andersson *et al.*, 1970; Andersson and Westbye, 1970; Andersson and Eriksson, 1971). The precise mechanisms have not been fully worked out, although angiotensin II both excites the cell bodies of cat supraoptic neurosecretory cells (Nicoll and Barker, 1971*b;* Fig. 6) and organ-cultured dog neurosecretory cells (Sakai *et al.*, 1974) and releases vasopressin from the isolated terminals of neurosecretory cells in the posterior pituitary (Gagnon *et al.*, 1973). The excitatory effects on supraoptic neurosecretory cell bodies and releasing effects on isolated terminals were antagonized by angiotensin II analogues with substitutions for phenylalanine at the 8-position (Sakai *et al.*, 1974; Gagnon *et al.*, 1973). Angiotensin I also stimulated vasopressin release,

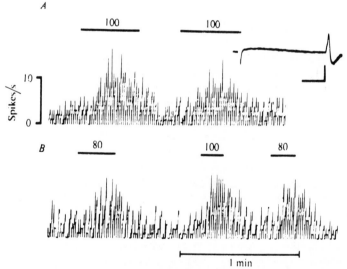

FIG. 6. Responses of neurons in the supraoptic nucleus to the microiontophoresis of angiotensin II. (A) Polygraph record of 1-sec integrated activity from a supraoptic neurosecretory cell electrophysiologically activated by antidromic stimulation (inset). Ejection of angiotensin II at various cationic currents (in nA, solid lines above tracing) causes an excitation of activity after a short delay. Withdrawal of the ejecting current results in a return to baseline activity. (B) A similar response from another neuron in the supraoptic nucleus (exhibiting recurrent inhibition). Inset calibration: 2 mV, 2 msec. From Nicoll and Barker (1971*b*).

mainly following conversion to angiotensin II, since the effect of the former was blocked by converting enzyme inhibitor (Sirois and Gagnon, 1975). Application of 10^{-9} M 1-Sar-8-Leu angiotensin II also depressed *resting* vasopressin release (Gagnon *et al.*, 1973).

More recent work from several laboratories has not supported the notion of peptide-sensitive sites in the posterior pituitary (Malvin, 1976; Hoffman, 1976). Angiotensin II failed to initiate vasopressin release from isolated neuronal lobes (Malvin, 1976), while systemic administration of angiotensin II caused the release of vasopressin only indirectly (Hoffman, 1976; Phillips, 1976). Nicoll and Barker (1971*b*), recording from antidromically activated neurosecretory cells in the supraoptic nucleus, failed to observe any change in activity of these cells following systemic administration of angiotensin II. This finding would also suggest little immediate activation of neurosecretory cells by this route of administration.

These conflicting results do not permit firm conclusions regarding the sites or sites of action of angiotensin II in evoking vasopressin release. If other components of the renin–angiotensin system also produced supraoptic neurosecretory cell excitation and vasopression release, then a central angiotensin II regulation of vasopressin would need to be strongly considered. Release of vasopressin would help to maintain normovolemia and normotension by causing antidiuresis and vasoconstriction. Presumably, angiotensin II specifically releases vasopressin and not oxytocin, since the latter plays important roles unrelated to the circulatory system and salt-water balance. This specificity has yet to be investigated.

3.2.3. Pressor Response

That intraventricular and intracarotid administration of angiotensin II reliably evokes pressor responses has been established over the past 15 years (Bickerton and Buckley, 1961; Yu and Dickenson, 1966; Smookler *et al.*, 1966; Severs *et al.*, 1967; Ueda *et al.*, 1969; Deuben and Buckley, 1970; Severs *et al.*, 1970). Angiotensin III given intraventricularly also produces pressor responses (Solomon *et al.*, 1976). Various lines of evidence indicate that the pressor response is composed of autonomic and endocrine components. The autonomic part consists of central activation of sympathetic activity (Gildenberg, *et al.*, 1973; Ferrario *et al.*, 1970) at several sites in the CNS—the subnucleus medialis (Deuben and Buckley, 1970), the area postrema (Joy and Lowe, 1970; Ueda *et al.*, 1972; Ferrario *et al.*, 1972), and the anteroventral aspect of the third ventricle (Phillips, 1976). The endocrine aspect, which is delayed in onset, is the release of vasopressin already mentioned. The possibility that serotoninergic pathways play a role in peptide-induced pressor responses is suggested by block of pressor effects to intraventricular angiotensin II following pretreatment with *p*-chlorophenylalanine (PCPA), which inhibits tryptophan hydroxylase (Finkielman *et al.*, 1976). In PCPA-treated animals, the peptide produced a profound, long-lasting *fall* in both systolic and diastolic blood pressure. Thus, the peptide may be exerting multiple central actions on cardiovascular tone.

Little work has been carried out on the cellular bases underlying central angiotensin-mediated changes in sympathetic activity. Ueda *et al.* (1972) have demonstrated that intravertebral artery administration of angiotensin II activates units in the area postrema, an area covered by relatively permeable blood–brain barrier (Broadwell and Brightman, 1976). Excitatory effects of angiotensin II in the isolated spinal cord have also been reported (Konishi and Otsuka, 1974*a*). Whether these latter effects are of physiological significance remains to be determined.

3.3. Actions in the Peripheral Nervous System

Circulating angiotensin II affects peripheral excitable tissue associated with the circulatory system in many ways. Besides directly activating smooth muscle cells in the vasculature (reviewed by Page and Bumpus, 1961; Park *et al.*, 1973), angiotensin II and analogues have direct, positive inotropic effects on myocardial cells, prolonging the plateau phase of the action potential and increasing tension (Dempsey *et al.*, 1971; Bonnardeaux and Regoli, 1974). Veno- and vasoconstrictor responses to sympathetic nerve stimulation in cutaneous, renal, and mesenteric vascular beds are potentiated by angiotensin II (Zimmerman, 1967; Panisset and Bourdois, 1968; Zimmerman and Whitmore, 1967; Hughes and Roth, 1971; Kadowitz *et al.*, 1971). These effects have been attributed to both blockade of norepinephrine reuptake (Panisset and Bourdois, 1968; Palaic and Khairallah, 1967; Peach *et al.*, 1969) and facilitation of norepinephine release (Zimmerman and Whitmore, 1967; Hughes and Roth, 1971; Kadowitz *et al.*, 1971; Zimmerman *et al.*, 1972). Angiotensin II also stimulates sympathetic nerve cell bodies directly (Reit, 1972), and indirectly by increasing ACh release from preganglionic terminals (Panisset, 1967). The peptide thus acts in an almost ubiquitous manner to modulate ganglionic and neuromuscular transmission in the sympathetic nervous system and alter vascular tone.

Angiotensin II also activates catecholamine synthesis in a variety of intact smooth muscle tissues innervated by the sympathetic nervous system (Boadle-Biber *et al.*, 1972). The activating effect of the peptide is directed at the rate-limiting enzyme, tyrosine hydroxylase, since no increase in catecholamine synthesis is observed if 3,4-dihydroxyphenylalanine (the catalytic product of tyrosine hydroxylase) is used as substrate rather than tyrosine. The peptide also stimulates protein synthesis in a dose-dependent manner, threshold being about 10^{-8} M. Synthesis of both catecholamine and protein induced by angiotensin II is inhibited by puromycin and cycloheximide (Roth and Hughes, 1975; Boadle-Biber and Roth, 1976). Electrical stimulation of sympathetically innervated tissues also results in increased catecholamine synthesis due to an increase in tyrosine hydroxylase activity (Morgenroth *et al.*, 1974). Angiotensin II and the hepapeptide des-Asp^1–angiotensin II both elevate tyrosine hydroxylase in isolated rabbit portal vein tissue (Boadle-Biber and Roth, 1976). Of additional interest is that cell-free preparations of the enzyme exhibit a dependency on Ca^{2+} in the micromole range (Boadle-Biber and Roth, 1976). One hypothesis that may account for some of the peptide-induced catecholamine and protein synthesis in these tissues is that the peptide depolarizes the sympathetic terminals sufficiently to increase Ca^{2+} influx. A test of this hypothesis would be to determine whether angiotensin II increases synthesis in the absence of Ca^{2+}. The relationship between peptide stimulation of protein synthesis and increased tyrosine hydroxylase activity remains to be clarified. These effects of angiotensin in peripheral tissues may provide some insight into actions of the peptide in CNS tissue.

3.4. Conclusions

Considerations of data published thus far leads first to the conclusion that CNS effects of the peptide could result partly from circulating peptide that activates nerve cells in CNS areas strategically placed outside the main blood–brain barrier. Angiotensin or angiotensins derived from peripheral precursors synthesized by hepatic and renal tissue fall outside the definitions and requirements for a *neuro*chemical. Central actions of the angiotensin or angiotensins that are due to materials endogenously synthesized by CNS tissue may be considered to reflect the actions of a neurochemical. Evidence to support such a neurochemical origin includes (1) presence in CSF and uneven distribution in the CNS; (2) specific CNS receptors distant from circumventricular organs; and (3) demonstrated effects on aggregates of CNS cells also lying well inside the blood–brain barrier. Presently lacking are (1) specific nerve cells of origin; (2) demonstrated intracellular synthesis of the various components; and (3) demonstration of release of neuronally synthesized material(s). Without such evidence, it is too premature yet to describe the peptide as a neurochemical.

Another important consideration that has not been fully investigated is the role or roles played by the peripheral or central renin–angiotensin system or both, the physiological regulation of salt and water balance and cardiovascular tone. Peripheral and central administration of structurally related peptide analogues with specific antagonistic activity does *not* alter drinking behavior, blood pressure, or salt and water balance in normotensive–normovolemic animals (Pals *et al.*, 1971; Ganten, 1976). Central, but not peripheral, administration of antagonist does, however, significantly reduce blood pressure in one strain of spontaneously hypertensive rats (SHRs) (Ganten, 1976), and peripheral administration of antagonist does reduce blood pressure in a small percentage of undiagnosed hypertensive patients (Streeten *et al.*, 1975). Furthermore, the concentration of angiotensin II in the CSF of SHRs is elevated, while peripheral levels remain normal (Ganten, 1976). Centrally and peripherally administered antagonist also reduces blood pressure and depresses drinking behavior in animals in which salt and water balance has been disturbed by Na^+-deficient diets or dehydration (Samuels, 1976). These results suggest that the central and peripheral renin–angiotensin systems may *not* contribute significantly to the maintenance of endocrine, autonomic, and drinking behaviors under conditions of adequate salt and water balance and blood pressure, but rather operate to influence the expression of these behaviors when the organism is stressed.

4. Parvicellular Peptides:
Thyrotropin-Releasing Hormone
Luteinizing Hormone–Releasing Hormone

$$(\text{Pyro})\text{Glu–His–Pro(NH}_2)$$
$$(\text{Pyro})\text{Glu–His–Trp–Ser–Tyr–Gly–Leu–Arg–Pro–Gly(NH}_2)$$

4.1. Distribution

A number of hypothalamic peptides that have well-established functions in the release of hormones from anterior pituitary cells and hence are called "releasing factor" peptides (reviewed by Blackwell and Guillemin, 1973; Schally *et al.*, 1973; Grant *et al.*, 1973) have recently been found distributed outside the hypothalamus (Winokur and Utiger, 1974; Brownstein *et al.*, 1974, 1975; Oliver *et al.*, 1974; Jackson and Reichlin, 1974) and in species devoid of pituitary tissue (Grimm-Jørgensen *et al.*, 1975). Furthermore, these peptides have been demonstrated in synaptosomal preparations of medial basal hypothalamus (Pelletier *et al.*, 1974a,b; G.W. Bennett *et al.*, 1975; Peck *et al.*, 1975) and of extra-eminence hypothalamus (G.W. Bennett *et al.*, 1975). They are thought to be derived from parvicellular hypothalamic neurons, which project from the periventricular region and ventromedial and arcuate nuclei to portal vessels originating in the median eminence (Szentagothai *et al.*, 1972), although precise localization is still lacking (see Chapter 6). The possibility that parvicellular neurosecretory cells participate in neuronal as well as endocrine events is suggested by electrophysiological and morphological evidence showing tuberoinfundibular projections to local and distant sites other than the median eminence (Millhouse, 1973; Barry *et al.*, 1974; Yagi and Sawaki, 1974; Renaud and Martin, 1975b; Conrad and Pfaff, 1975; Kawakami and Sakuma, 1976; Renaud, 1976a). Indirect support for neuronal roles also comes from the observation of resting and evoked parvicellular peptide release from hypothalamic "synaptosomes" (G.W. Bennett and Edwardson, 1975; Peck *et al.*, 1975), although definitive demonstration of parvicellular peptidergic neuron participation in neuronal events has not yet been reported.

4.2. Actions in the Central Nervous System

Both TRH and LH–RH have effects in the CNS. TRH, for example, antagonizes many of the pharmacological actions of barbiturates, including sleeping time, depressed respiration and heart rate, and hypothermia, without altering plasma levels of the drug (Brown and Vale, 1975; Kraemer *et al.*, 1975). These effects are still present following hypophysectomy, excluding an endocrine contribution. Other studies indicate that TRH potentiates the anticonvulsant effects of low doses of barbiturates, while antagonizing the sedative actions of high doses (Nemeroff *et al.*, 1974). TRH has also been reported to have anorectic actions independent of its endocrine effects and independent of catecholamine pathways in the CNS (Barlow *et al.*, 1975). More evidence of an extraendocrine function for a parvicellular peptide comes from the observation that LH–RH can induce mating behavior in ovariectomized female rats pretreated with estrone (Moss and McCann, 1973, 1975). The peptide-induced behavior is also present in adrenalectomized rats, thus excluding adrenal progesterone as a contributing factor. Similar findings have been reported in the estradiol-primed, hypophysectomized, ovariectomized rat by Pfaff (1973). That the behavioral effects of the peptide are mediated by direct actions on the CNS is indicated by recent experiments showing intrahypothalamic infusion of LH–RH, but not TRH, potentiates lordosis and mating behavior (Foreman and Moss, 1975; Herrenkohl and Verhulst, 1975; Moss and Foreman, 1976). LH–RH-sensitive sites in the hypothalamus are confined to the medial preoptic and arcuate nucleus areas; the peptide is not active when infused into either the lateral hypothalamus or the cerebral cortex. Further work has established that the lordosis reflex is optimally elicited approximately 1–2 hr following intrahypothalamic injection, causing Moss and Foreman (1976) to conclude that LH–RH is probably not acting like a transmitter. Presumably, part of the delay is due to diffusion of the peptide to target neurons, while part may be due either to slowness of onset or to the induction of biochemical events.

The cellular mechanisms underlying the effects of LH–RH and TRH on the CNS have not been studied in detail. Since the behavioral actions of the peptides are quite complicated, and the peptide-sensitive areas are not well localized, it is impossible to know which sets of neurons are important and how they function in the overall peptide effects. An initial approach has been to identify peptide-sensitive neurons in the CNS. Microiontophoresis of peptide (from micropipettes containing 0.001–0.01 M solutions) has revealed both excitatory and depressant effects of LH–RH and TRH on cellular excitability of hypothalamic and extrahypothalamic neurons (Figs. 7–9; Dyer and Dyball, 1974; Renaud and Martin, 1975*a;* Kawakami and Sakuma, 1974, 1976; Renaud *et al.,* 1975; Moss and Foreman, 1976; Renaud, 1976*b*). Moss and colleagues have carried out an extensive study on the chemosensitivity of antidromically activated and unactivated medial preoptic and arcuate–ventromedial neurons (located in the two areas of LH–RH sensitivity as measured with the lordosis reflex assay). The results show that LH–RH excites a majority of activated and unactivated units in these areas. TRH inhibits a majority of activated medial preoptic neurons, but excites most of the unactivated cells in this area. Somewhat in contrast, Renaud and colleagues have observed an almost uniform depression of excitability by TRH and LH–RH in both hypothalamic and extrahypothalamic areas (Renaud and Martin, 1975*a;* Renaud *et al.,* 1975; Renaud, 1976*b*).

These results provide only a catalogue of effects. At best, they indicate chemosensitivity at the cellular level. The relatively short time-course of action on

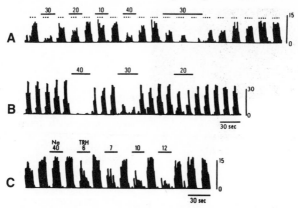

FIG. 7. Action of TRH on glutamate-evoked activity. The polygraph records taken from individual neurons in the ventromedial hypothalamus (A), cerebral cortex (B), and cuneate nucleus (C) indicate a dose-related depression of firing frequency of glutamate-evoked activity (· · ·) during administration of TRH (———). The upper two records indicate some prolongation in recovery to control excitability levels after larger and longer TRH applications. The vertical bar on the right indicates the number of spike counts per single bin. From Renaud and Martin (1975a).

excitability would appear to suggest transmitterlike roles for these peptides. That parvicellular peptides may play roles in the hypothalamus as neurotransmitters is supported by accumulation of evidence showing (1) parvicellular projections to areas in the hypothalamus other than the median eminence (see above) and (2) recurrent excitatory and inhibitory pathways from median eminence to tuberoinfundibular cell bodies (Renaud, 1975a). If parvicellular neurons showing multiple projections and recurrent forms of synaptic physiology can be convincingly demonstrated to be "peptidergic," then their neurosecretory products should be strongly considered for synaptic as well as endocrine functions. It is possible that parvicellular elements have a short-circuit feedback system similar to that possessed by magnocellular elements (see below).

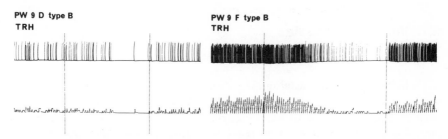

FIG. 8. Left: Polygraph record obtained from a preoptic neuron that was inhibited by TRF (5 nA), but not by LRF or oxytocin. Right: Trace from another unit inhibited by TRF (7 nA) and unaffected by oxytocin. The vertical dotted lines mark 40-sec periods of TRF application. Upper traces are reflections of single spikes; bottom traces are integrated activity records. From Dyer and Dyball (1974).

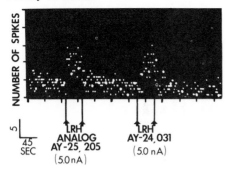

FIG. 9. Extracellular recording from a neuron in the septal–preoptic area. Excitatory respones to microiontophoresis of LR–RH and analogue (ejected with 5 nA currents). Activity as "number of spikes." From Moss (unpublished observations).

4.3. Conclusions

Although "releasing factor" peptides have yet to be definitely localized to hypothalamic parvicellular elements, all evidence suggests that these peptides are parvicellular in origin. From what little has been done, it appears that these peptides may play physiological roles in the CNS both as neurotransmitters in hypothalamic circuits and as neurohormones in modulating specific behavior patterns. Clearly, there is too little data yet to give anything more than an indication of neuronal functions apart from endocrine effects.

5. Magnocellular Peptides: Antidiuretic Hormone (Lysine Vasopressin) and Oxytocin

$$
\begin{array}{c}
\overline{\text{CyS}-\text{Tyr}-\text{Phe}-\text{Glu(NH}_2)-\text{Asp(NH}_2)-\text{CyS}}-\text{Pro}-\text{Lys}-\text{Gly(NH}_2) \\
1 \quad\; 2 \quad\;\; 3 \quad\;\; 4 \quad\quad\;\; 5 \quad\quad\;\; 6 \quad\; 7 \quad\; 8 \quad\; 9 \\
\overline{\text{CyS}-\text{Tyr}-\text{Ile}-\text{Glu(NH}_2)-\text{Asp(NH}_2)-\text{CyS}}-\text{Pro}-\text{Leu}-\text{Gly(NH}_2)
\end{array}
$$

5.1. Vertebrate Studies

5.1.1. Distribution and Peripheral Actions

The magnocellular nuclei in the vertebrate hypothalamus contain large-sized neurons that synthesize both large ($\sim$10,000 mol. wt.) and small ($\sim$1000 mol. wt.) peptides (for details, see Berde, 1968; Hayward, 1975). The former have been labeled *neurophysins,* of which there are two major and one minor species (Watkins, 1972; Zimmerman *et al.,* 1974), while the latter have long been known as *oxytocin* and *vasopressin* or *antidiuretic hormone.* One of the two major neurophysin species is apparently synthesized, and becomes associated with either oxytocin (neurophysin I) or vasopressin (neurophysin II) (Robinson *et al.,* 1971; Pickup *et al.,* 1973; Zimmerman *et al.,* 1973; Evans and Watkins, 1973). Both large and small peptides are transported rapidly to neurosecretory terminals, probably in intragranular form (Jones and Pickering, 1972; Nörstrom and Sjöstrand, 1971), and both have been found in CSF, hypophyseal blood, and plasma (Vorherr et al., 1968; Cheng and Friesen, 1970; Zimmerman et al., 1973; Robinson and Zimmerman, 1973; Robertson et al., 1970).

Functional roles for oxytocin and vasopressin have been established for peripheral peptide, and include renal tubule cell resorption of water (antidiurectic function); vascular (vasopressin function), epididymal, and uterine smooth muscle and myoepithelial cell contraction; and an increase in Na^+ and water permeability of amphibian (bladder) mucosal cells (Hays and Leaf, 1962; Civan and Frazier, 1968; Berde, 1968; Hib, 1974). Portal peptide has been implicated in the release of ACTH (reviewed by Yates et al., 1971). While vasopressin (and oxytocin) are usually undetectable at baseline plasma levels (Vorherr et al., 1968; Forsling et al., 1973) or in the nanomolar range (Robertson et al., 1970), neurophysin is detectable in plasma (Forsling et al., 1973). Neurophysin II release parallels that of vasopressin following hemorrhage (Fawcett et al., 1968), apparently in a 1:1 ratio (Forsling et al., 1973). Evidence to establish functional roles for portal and peripheral neurophysins has not yet been reported.

The presence of both large and small magnocellular peptides in the CSF following the appropriate stimuli suggests that the peptides are secreted not only into the portal and peripheral circulations, but also into the CSF, either directly from terminals abutting on ependymal cells or indirectly from intrahypothalamic terminals releasing peptide into the interstitium. CSF vasopressin is unlikely to be due to the leakage of plasma peptide across the blood–brain barrier, since artificial elevation of plasma levels does not result in detectable CSF peptide (Vorherr et al., 1968). In this regard, vasopressin has recently been released from synaptosomal fractions of the hypothalamus (including the median eminence) (Mulder et al., 1970; Bennett and Edwardson, 1975). It is thus likely that some magnocellular neurons project to the plexus of portal vessels in the median eminence (Parry and Livett, 1973); it is also possible that magnocellular projections exist that end in the hypothalamus (and not on ependymal cells or portal vessels or in the neurophypophysis). Functional roles for peptide released into the CSF have not been established.

5.1.2. Actions in the Central Nervous System

Recordings from magnocellular neurons in a variety of vertebrates have revealed the presence of synaptic modulation of magnocellular activity following electrical stimulation of the neurosecretory terminals in the neurohypophysis (Kandel, 1964; Barker et al., 1971b,c; Koizumi and Yamashita, 1972; Koizumi et al., 1973; Dreifuss and Kelly, 1972a; Negoro and Holland, 1972; Negoro et al., 1973; Hayward and Jennings, 1973a,b). The modulation of magnocellular activity could be mediated by activation of neurosecretory axon collaterals synapsing either directly on magnocellular somata or indirectly on interneurons that would then synapse on somata. Presumably, an axon collateral of the neurosecretory cell must

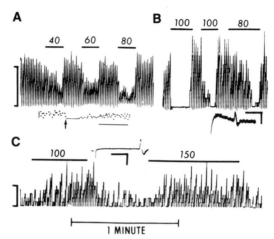

FIG. 10. Responses of supraoptic neurosecretory cells to microiontophoresis of vaso-pressin. (A) Dose–response depression of activity in a cell exhibiting recurrent inhibition (histogram beneath record; calibration: 5 counts/address; 100 msec) by vasopressin (solid lines above trace, currents in nA). Calibration bracket at left: 0–20 spikes/sec. (B) Vasopressin depression of neurosecretory cell activity. Polygraph calibration same as in (A). Antidromic spike below trace. Calibration: 5 msec/100 μV. (C) Vasopressin excitation of a neurosecretory cell. Calibration bracket: 0–5 spikes/sec. Antidromic spike centered above trace. Calibration: 2 msec/4 mV. From Nicoll and Barker (1971a).

be involved in the synaptic modulation of magnocellular activity. Therefore, if Dale's Principle (1934)—that the same substance is released from all branches of a particular neuron—is valid for neurosecretory cells, then either the larger or the smaller peptides synthesized by magnocellular neurons and secreted into portal and peripheral circulations must also be released from terminals at central synapses involved in recurrent synaptic modulation. A direct form of recurrent inhibition (without interneurons) has been suggested by a number of laboratories (Kandel, 1964; Barker et al., 1971c; Dreifuss and Kelly, 1972a; Koizumi and Yamashita, 1972; Vincent et al., 1972; Negoro and Holland, 1972; Hayward and Jennings, 1973a). This type of synapse, involving the placement of axon collaterals near the cell body of origin, would require that the peptide also inhibit magnocellular activity. Nicoll and Barker (1971a) tested this hypothesis by microiontophoresing vasopressin in the vicinity of antidromically activated supraoptic neurosecretory cells and recording their activity with extracellular electrodes. A majority (80%) of activated cells were inhibited by the peptide (Fig 10), while nearby cells that could not be antidromically activated showed both inhibitory (35%) and excitatory (65%) responses. Cortical neurons were predominantly excited (90%). These results indicate that vasopressin can alter neuronal activity in both the hypothalamus and the cortex. The predominance of inhibitory responses in activated supraoptic neurons is suggestive evidence in support of the aforementioned hypothesis. A variety of amino acids and catecholamines, as well as ACh, also inhibited antidromically activated supraoptic neurons (Barker et al., 1971c; Nicoll and Barker, 1971a), but while the pharmacological responses to these substances could be reliably inhibited by concurrent iontophoresis of the appropriate antagonists, recurrent synaptic inhibition could not be reliably blocked by iontophoretic or systemic administration of the same antagonists. An independent series of experiments carried out on antidromically activated neurons in the paraventricular nucleus has also shown chemosensitivity to both putative transmitters and peptides (Moss et al., 1972a,b). In this nucleus, however, oxytocin excited the neurosecretory cells, thus excluding it from consideration as a mediator of a monosynaptic form of recurrent inhibition.

Serious doubt that vasopressin participates in inhibitory synaptic events has been raised by Cross (1974), following recent demonstrations of recurrent inhibition in the Brattleboro strain of rats (Dyball, 1974; Dreifuss et al., 1974), which do not synthesize vasopressin with peripheral biological activity (Valtin and Schroeder, 1964). Brattleboro rats homozygous for the trait have diabetes insipidus, and also do not synthesize vasopressin-associated neurophysin (Burford et al., 1974). Recently, radioimmunoassay and immunohistochemical techniques for the localization of vasopressin have been applied to neurosecretory systems in the Brattleboro strain. The results using immunoperoxidase methods have been contradictory: one group found evidence of reaction product in both magnocellular nuclei of Brattleboro rats following incubation with anti-[8-lysine]–vasopressin serum (Watkins, 1975), while another did not (LeClerc and Pelletier, 1974). Vasopressin has also been demonstrated in Brattleboro rat neurophypophyses by radioimmunoassay using antibodies directed against arginine-8-vasopressin, although the amount present is approximately 5% of the normal content (Greidanus et al., 1974). Although the authors of these papers consider their techniques specific for vasopressin, cross-reactivity with oxytocin has not been entirely eliminated. In fact, using immunofluorescent methods, Swaab and Pool (1975) have abolished any "false" positive reaction for vasopressin in Brattleboro rats by preincubating their antibodies with oxytocin beads. Whether Brattleboro rats synthesize an analogue of vasopressin that has biological activity in the CNS but cross-reacts poorly with present-day assay techniques remains to be investigated.

The presence of recurrent inhibition, necessitating the involvement of neurosecretory axon collaterals in *both* normal and Brattleboro rats, still eludes easy explanation. Since these magnocellular systems may serve as models of similar events in parvicellular systems, several possibilities will be briefly considered. They include (1) transport of a putative transmitter along with neurophysin II and vasopressin, with release of all three at all terminals (a logical extension of Dale's Principle); (2) synthesis of all three substances and differential transport, so that putative transmitter alone migrates to synapses (in violation of Dale's Principle); (3) synthesis and transport of peptides to all terminals, with differential structural requirements for activity. Firm evidence to support the first explanation is lacking, though neurohypophyseal nerve-ending fractions containing elementary granules and rich in vasopressin also appear to contain small (ACh) vesicles (Lederis and Livingston, 1970). These authors concluded, however, that ACh was present mainly in endings separate from those with peptides. Differential transport of material in axons has not been reported; rather, in sensory cells with several processes, peptide (Hökfelt, 1975*b*) and protein (Barker *et al.*, 1976) appear to be transported in both axonal processes. Structural requirements for synaptic activity different from those for peripheral effects of vasopressin might allow recurrent inhibition to persist in the absence of assayable activity (as Brattleboro rats). (A different structural requirement for peptide activity at synapses might entail the need for "converting" enzymes localized in terminals or clefts.) Another explanation for recurrent inhibition is that electrical stimulation of the neurohypophysis activates (multipolar) cholinergic axons in the neurosecretory pathway that synapse onto magnocellular neurons.

Physiological activation of, and the functional significance of, recurrent pathways in the magnocellular neurosecretory systems remain to be elucidated. The apparent presence of peptide receptors (for angiotensin, vasopressin, and oxytocin) on antidromically activated magnocellular neurons suggests that these peptides may play some functional role as neurotransmitter or neurohormone in the regulation of activity in these systems. It is clearly too early to discount the possibility that vasopressin and oxytocin (or structurally related analogues) do participate in neuronal, as well as endocrine, functions. Although effects of magnocellular peptides on CNS function unrelated to recurrent inhibition have been reported (Walter *et al.*, 1975), specific patterns of behavior (as are released by angiotensin) have not thus far been reported.

5.2. Invertebrate Studies

5.2.1. Distribution

Considerable evidence has accumulated recently to establish that molluscan nervous systems contain magnocellular-sized neurons that synthesize a variety of low-molecular-weight proteins and peptides that are transported from cell body and released (Wilson, 1971; Arch, 1972; Gainer, 1972*b;* Gainer and Wollberg, 1974; Loh and Peterson, 1974; Loh *et al.*, 1975; Loh and Gainer, 1975*a,b*). Some of these cells contain large granules reminiscent of those found in vertebrate magnocellular systems, and it has been suggested that operationally, the cells are neurosecretory (Frazier *et al.*, 1967; Gainer, 1972*b;* Gainer and Wollberg, 1974; Loh and Gainer, 1975*a,b*). Specific functions for the various cells thus far described are largely lacking except for (1) a cluster known as "bag cells," the neurosecretory products of which appear to be involved in egg-laying (Strumwasser *et al.*, 1969; Kupfermann, 1970; Toevs, 1970; Arch, 1972), and (2) a cell that generates endogenous bursting pacemaker activity (see Section 5.2.2.a), the extracts of which appear to regulate water content in a sea mollusk (Kupfermann and Weiss, 1975). Despite the relative paucity of knowledge regarding the functions of these neurosecretory cells, their large size and easy identifiability have made them increasingly useful model systems to study neuronal function. Although these molluscan species precede primates by some 300–400 million years in evolution, specific and dramatic effects of vasopressin and several related peptides on these cells have recently been established (Barker and Gainer, 1974; Barker *et al.*,

1975a; Barker and Smith, 1976), suggesting that some peptide structures may have been largely conserved through evolution.

5.2.2. Actions in the Central Nervous System

a. Vertebrate Magnocellular Peptides. The two cells responsive to vasopressin that have been described thus far have similar physiological and biochemical properties (Gainer, 1972a,b; Strumwasser, 1973; Loh, et al., 1976). The cells are capable of synthesizing low-molecular-weight proteins and of generating endogenous bursting pacemaker potential (BPP) activity characterized by slow membrane potential oscillations coupled to bursts of spikes (Carpenter and Gunn, 1970; Carpenter, 1973; Gainer 1972b; Strumwasser, 1973; Barker and Gainer, 1975a). Generation of the burst of spikes is predicated on sufficient and sustained depolarization of the membrane by the slow oscillation. The latter can occur in the absence of the former, indicating that the slow events are relatively independent of the fast ones (Strumwasser, 1973; Mathieu and Roberge, 1971; Barker and Gainer, 1975a). Vertebrate magnocellular neurosecretory cells with apparently similar electrical activity ("bursters" and "phasic" cells) have recently been reported (Hayward and Jennings, 1973a,b; Arnauld et al., 1974, 1975; Harris et al., 1975; Wakerley et al., 1975), suggesting that such activity may not be peculiar to invertebrates. The mechanisms underlying BPP activity have been studied in the invertebrate, and appear to involve the interplay of a time- and voltage-dependent K^+ conductance with slow inactivation kinetics, coupled to a voltage-dependent Na^+ conductance with little or no activiation (Smith et al., 1975). The membrane continuously generates BPPs as it cycles through the depolarizing effect of the Na^+ current and the hyperpolarizing action of the K^+ current.

The effect of vasopressin and related peptides is to initiate or enhance BPP activity in two specific cells (Barker and Gainer, 1974; Barker et al., 1975a). One of these cells, present in the land snail *Otala lactea*, is relatively inactive physiologically and biochemically when the snail is dormant; the cell cannot generate BPPs or produce low-molecular-weight proteins under these conditions (Gainer, 1972a,b). Addition of 10^{-9}–10^{-7} M peptide in the bathing solution causes rapid changes in steady-state membrane properties that long outlast the period of application (Fig. 11). Associated with the initiation of BPP activity is a change in the current-voltage characteristic of the cell from linear to nonlinear. Similar results can be obtained using low-current (1–10 nA) microiontophoresis from pipettes containing 1 mM solutions of peptide (Fig. 12). This latter technique—which allows unknown, but presumably minute, quantities of peptide to be dispensed—has permitted demonstration of the voltage-dependent character of the peptide effect as well as localized the responsive area to the axon hillock. Iontophoresis of the peptide while the cell's membrane is hyperpolarized is not associated with a measurable change in membrane properties at that potential, but after the membrane is returned to control potential, the presence of BPP activity is revealed (Fig. 12D). In contrast, the amplitude of a response to iontophoresced ACh on the

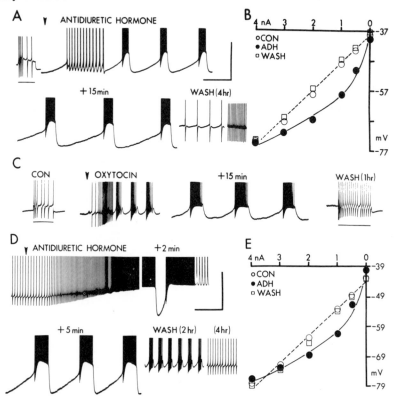

FIG. 11. Effects of ADH (vasopressin) and oxytocin on membrane properties of aestivated cell 11. (A) Membrane potential traces of cell 11 from an aestivated snail. Under control conditions, the cell is silent and cannot generate BPPs on injection of depolarizing current (applied during bar under left hand trace, top row). Addition of 2×10^{-8} M vasopressin at arrow causes a rapid depolarization of the membrane potential and initiation of BPP activity, which increases in amplitude with time (+15 min). Washing for 4 hr with vasopressin-free saline restores the cell to its inability to generate BPPs, although it is now spontaneously active. (B) I–V relationships of the same cell in (A), before, during, and after addition of vasopressin. Plots were constructed by measuring voltage response to 2-sec current pulses, using threshold for firing as the origin. (C) Another cell from an aestivated snail exposed to 10^{-8} M oxytocin (at arrow). Under control conditions (CON), cell is silent and cannot produce BPPs on injection of depolarizing current (applied during the bar beneath trace). Oxytocin initiates BPP activity, which increases in amplitude with time. Washing with oxytocin-free saline for 1 hr returns the cell to its control condition. Calibration: 40 mV, 12 sec (2 min during initial course of hormone applications and in far right-hand trace, 2nd row in A). (D) Membrane potential traces of cell 11 taken from a semiaestivated snail. Application of 10^{-8}M vasopressin (at arrow) both causes a depolarization of the membrane potential and increases the rate of repetitive activity. Spontaneous BPP cycle occurred (as in right hand trace, top row), but after 5 min, sustained injection of hyperpolarizing current was required to demonstrate the BPP activity. Washing for 4 hr in vasopressin-free saline was necessary to restore the cell to control activity. (E) I–V relationships of the same cell illustrated in (A), before, during, and after exposure to vasopressin. The plot was constructed as in (B). Calibration: 40 mV, 12 sec. From Barker et al. (1975a).

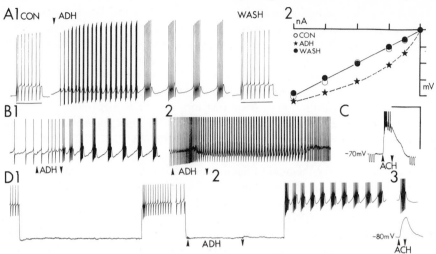

FIG. 12. Effects of ADH (vasopressin) on the membrane properties of cell 11. Membrane potential traces of cell 11s taken from dormant and semidormant snails. (A1) Resting membrane potential, -48 mV; (CON) control. Injection of depolarizing current (during the period marked by a bar beneath the trace) does not produce BPP activity. Bath applica-tion of 10^{-9} M vasopressin (at arrow) induces BPP rhythm. The first half of the trace is recorded at ¼ the speed of the last half. Washing in vasopressin-free saline for 1 hr (WASH) restores the cell's membrane properties. (A2) Current–voltage relationships of the cell before, during, and after treatment with vasopressin. The marks on the axes rep-resent 1-nA and 10-mV intervals. The origin is the threshold for firing (-42 mV). Vaso-pressin reversibly induces marked nonlinearity in the curve. (B1) Iontophoresis of vaso-pressin leads to BPP activity. Vasopressin was ejected between the arrows by using a 4-nA cationic current. The resting membrane potential was -46 mV. (B2) Effects of vasopressin (iontophoresed between arrows) long outlast the application period. (C) Bath application of 10^{-5}M ACh depolarizes the membrane potential and increases mem-brane conductance. Downward deflections are voltage responses to constant-current pulses (1 nA). (D1) Sustained hyperpolarization of a cell does not induce BPP activity once the hyperpolarizing current is removed. The resting membrane potential was -48 mV. (D2) Iontophoresis of vasopressin (between the arrows) when the cell is hyperpo-larized does not cause an observable voltage response, but on removal of the hyperpo-larizing current, well-developed BPP activity is evident. (D3) Similar hyperpolarization of the cell increases the size of the depolarizing ACh response. Calibration: 40 mV; 12 sec in A1, B1, and D1–3; 48 sec in A1 and B2; 6 min in C. From Barker and Gainer (1974).

same cell increases as the cell's membrane is hyperpolarized, and the response can be extrapolated to an inversion potential of about 0 mV (Barker, 1975). The latter result, taken together with ion-change experiments, indicates that ACh transiently increases the voltage-independent conductances of the membrane to Na$^+$ and K$^+$ ions, as has been described for this and other transmitters at other synapses (see the reviews by Gerschenfeld, 1973; Krnjević, 1974).

More recently, voltage-clamp analysis has been applied to the responses of the cell-to-bath application of 10^{-7}–10^{-6} M vasopressin. Voltage clamp of the cell exposed to the depolarizing effects of vasopressin reveals an increase in inward holding current and a decrease in membrane conductance, suggesting that the depolarization might be due to a decrease in voltage-independent K$^+$ conductance relative to voltage-independent Na^+ conductance (Barker and Smith, unpublished observations). After several minutes bathing in peptide, the membrane exhibits changes in steady-state properties, including the appearance of a voltage-dependent Na^+ conductance, which inactivates slowly or not at all, and a voltage-dependent K^+ conductance, which inactivates slowly (Barker and Smith, 1976; Fig. 13). Thus, the steady-state current–voltage curve of the untreated cell changes from one with an all-positive slope to one that has an area of negative slope and is N-shaped. The negative slope, present over the same potential range as the nonlinear part of the current–voltage curve observed under current clamp, is due primarily to the voltage-dependent Na^+ conductance. The appearance of the latter is thus largely responsible for the marked nonlinear electrical qualities of membrane. The peptide also appears to alter the kinetics of the voltage-dependent pacemaker K^+ conductance, which regulates the period of BPP (Barker and Smith, unpublished observations). Similar effects of vasopressin on unclamped membrane potential behavior and steady-state membrane properties have been observed on the other vasopressin-sensitive cell ("R_{15}"), which is found in the sea slug $Aplysia$ $californica$. These results clearly indicate that peptides can cause changes in neuronal

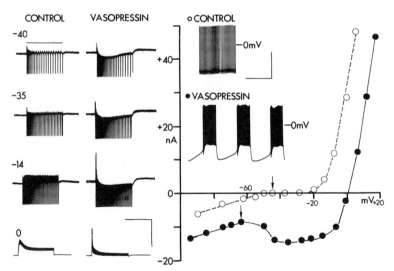

FIG. 13. Alteration by vasopressin (ADH) of the steady-state I–V curve. Recordings from cell 11 taken from an aestivating snail before (CONTROL) and after bath application of 1 μM Lys-vasopressin. Membrane potential activity illustrated in insets of I–V plot on right. CONTROL (before ADH): beating pacemaker activity. VASOPRESSIN (after ADH): bursting pacemaker activity. Zero membrane potential: 0 mV. Left: Membrane of cell voltage clamped and 5-sec voltage steps imposed (during time shown by bar above uppermost left-hand trace). Current resulting from voltage steps to potentials indicated by numbers above traces under control conditions and in presence of vasopressin. Right: I–V curve derived from these currents using most negative or least positive current evoked after 1 sec during command. Cell's membrane potential held at −45 mV in control and −62 mV in vasopressin (downward arrows). Calibrations: (left) 10 nA (upper three pairs of traces) and 40 nA (lowermost pair of traces), 5 sec; (right) 50 mV, 20 sec. From Barker and Smith (1976).

membrane properties that are quite different from those produced by putative neurotransmitters. Specifically, putative transmitters act on voltage-independent membrane conductances, whereas vasopressin acts primarily on voltage-dependent conductances. Whether these differences in mode of action will prove to be consistent distinguishing properties of neurotransmitters and neurohormones remains to be determined. Whether the long-lasting nature of the respone is due to peptide remaining bound to the cell or to long-lasting changes induced by the peptide also remains to be determined. It is conceivable that these several differences may become additional distinguishing characteristics between neurotransmitters and neurohormones.

Although all the appropriate studies to demonstrate the presence of a peptide receptor on the peptide-sensitive cells have not been performed, a structure–activity study of various peptide analogues and their responses on an identified cell in the land snail has been carried out (Barker *et al.*, 1975*a*). The results show that activity appears to require nearly the complete structure, since either removal of the glycinamide terminal or acetylation of cysteine at the 1-position abolishes activity. Neither the hexapeptide ring nor the tripeptide tail alone is active. Minor alterations in the amino acid structure at the 8-position can change activity. While specific antagonism of peptide activation has not yet been observed, the results indicate specific structural requirements for activity, which implies the existence of a receptor. The presumed receptor is unlike vertebrate magnocellular peptide receptors in that it cannot distinguish between oxytocin and vasopressin.

Attempts to isolate the naturally occurring peptide or peptides from molluscan nervous systems have been partly successful (Ifshin *et al.*, 1975). Extracts of *Aplysia* and *Otala* nervous systems can initiate or enhance BPP activity in the peptide-sensitive cells. The extracted factor or factors appear to have a molecular weight between 700 and 5000. Enzymatic treatment of the factor or factors indicates that the structure is not precisely the same structure as that of either vasopressin or oxytocin. Further localization of the active factor to a single cell type should allow study of the physiology and biochemistry of the active factor or factors. Interestingly, extracts of the bag cell clusters (in the *Aplysia* visceral ganglion) have qualitatively similar BPP-enhancing effects on the vasopressin-sensitive cell (R_{15}) in *Aplysia* (Mayeri and Simon, 1975). The naturally occurring peptide that activates peptide receptors on R_{15} may thus be synthesized by this cluster of cells.

The molecular mechanisms by which peptide initiates or enhances BPP activity remain to be determined. A possible clue to unraveling the mechanism may come from consideration of the effects of vasopressin on toad bladder epithelium, where the peptide displaces Ca^{2+} with a time-course similar to its induction of transepithelial Na^+ transport (Cuthbert and Wong, 1974). In fact, lowering of extracellular Ca^{2+} concentration initiates or enhances BPP activity in the peptide-sensitive cells (Barker and Gainer, 1975*b*). More experiments are needed to clarify the possibility that the peptide response is mediated through a change in the affinity of select parts of the cell's membrane for Ca^{2+}.

b. Bag Cell Extract. The "bag cells" of *Aplysia* are a cluster of cells containing neurosecretory granules with processes terminating in the connective tissue sheath surrounding the remainder of the parietovisceral ganglion (Frazier et al., 1967). Injection of bag cell extracts produces egg-laying behavior (Kupfermann, 1970, 1972), in part by stimulating muscle cells in the ovotestis to extrude eggs (Coggeshall, 1972). The neurohormone or neurohormones secreted by bag cells appear to be of low molecular weight (Toevs and Brackenbury, 1969; Arch, 1972). The precise molecular weight or weights are not certain, but release of more than one species has been reported (Loh et al., 1975).

Recently, the cellular bases of the egg-laying and associated behavioral effects of bag cell neurohormones have been investigated by examining the effects of extracts and bag cell stimulation on the electrophysiological behavior of identified neurons in the parietovisceral ganglion of *Aplysia* (Mayeri and Simon, 1975). These authors have observed several distinct effects: (1) an increase in the amplitude of excitatory postsynaptic potentials recorded in some identified cells that are (monosynaptically) postsynaptic to a cell called L_{10} (Fig. 14A); (2) a change in the membrane potential activity of L_{10} from beating to bursting pacemaker (Fig. 14B); and (3) an increase in the amplitude of BPP activity in R_{15}. These effects typically take minutes to become discernible, become most apparent within an hour, and last for another hour. Since there is no evidence for a synaptic input from bag cells to any of the cell types sensitive to the extracted material, it is probable that the effects observed are due to the release of neurohormone into neurohemal areas for distribution via extrasynaptic venues (hemolymph).

The location of the bag cell extract receptors is unknown, as is the mechanism (or mechanisms) of inactivation of the active material. The facilitation of synaptic potentials from L_{10} to follower cells may be partly due to the induction of BPP behavior in L_{10}, with subsequent facilitated release of transmitter (following propagation of clustered action potentials to L_{10}'s terminals); however, inhibitory synaptic potentials simultaneously generated by L_{10} do not always show a similar, parallel facilitation of amplitude (Fig. 14A). Furthermore, facilitated synaptic potentials can occur when L_{10} is not generating BPPs (Mayeri, personal communication). The mechanism or mechanisms of neurohormone-induced synaptic facilitation, whether pre- or postsynaptic or both, have not yet been elucidated. Several possibilities include (1) a change in excitation–secretion coupling, so that more transmitter is released; (2) a change in the efficacy of postsynaptic receptor-coupled conductance mechanisms, so that either the affinity of the receptor for transmitter increases or the conductance change itself is increased; (3) a change in the passive membrane properties of the follower cell to produce more voltage response of the membrane for a given receptor-coupled conductance change; and (4) a change in the inactivation kinetics of the transmitter, so as to allow more transmitter to be effective over a longer period.

The BPP-enhancing effects of bag cell neurohormone and its general timecourse are similar to the effects of the vertebrate magnocellular peptides discussed in Section 5.2.2a. These results suggest that the structure of the active bag cell neurohormone may overlap with that of magnocellular peptides. The possible physiological consequences of enhancement of BPP activity in R_{15} by the bag cell

A

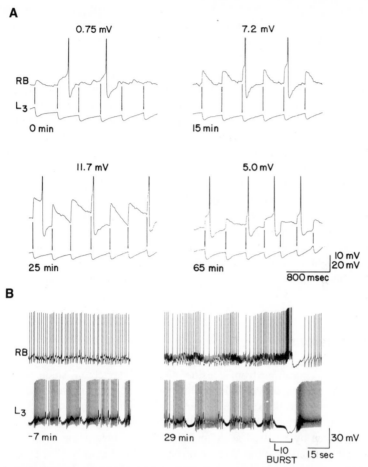

FIG. 14. Effects of bag cell extract on membrane potential and synaptic activity. Facilitation of excitatory postsynaptic potentials (EPSPs) in the isolated abdominal ganglion of *Aplysia* following application of crude extract of the neurosecretory bag cells to the bathing medium. (A) Vertical lines denote simultaneous, spontaneously occurring EPSPs in cell RB and IPSPs in cell L_3 produced directly in these cells by interneuron L_{10}. Following application of the extract at 0 min, the amplitude of the EPSPs increased to 11.7 mv at 25 min and steadily declined thereafter. (B) Lower gain, compressed time scale recordings from same cells showing the effect of synaptic facilitation of RB spike activity. At 29 min, a burst of L_{10} activity, also induced by the extract, strongly excites the RB cell. The L_{10} bursts, which are likely due to intrinsic properties of the interneuron, continued to occur periodically over the next 30 min. From Mayeri and Simon (1975).

neurohormone are presumably similar to those resulting from activation by magnocellular peptides. The induction or enhancement of BPP activity raises the general level of excitability of the cell, generating more action potentials per unit time. More important, the clustering of action potentials is associated with a progressive prolongation of action potential duration during the bursts (Strumwasser, 1973; Stinnakre and Tauc, 1973), which in turn is coupled to a progressively greater influx of Ca^{2+} into the cell (Stinnakre and Tauc, 1973; Stinnakre, 1975). If a similar progression of action potential duration and Ca^{2+} influx occurs at the terminals of cells generating BPPs, then a facilitated, pulsatile release of transmitter or neurosecretory product should occur.

Another possible physiological consequence derives from the alteration in steady-state membrane properties from a relatively linear current–voltage relationship of the membrane (over a physiological range of potential) to a markedly nonlinear relationship in the presence of peptide. This change should effectively alter the input–output relationships of the cell, so that inhibitory or excitatory synaptic input no longer causes changes in the cell's output that are simply a linear function of the input. Rather, output will now be related to input in a nonlinear manner, with small changes in synaptic input producing large changes in output.

A third possible consequence of BPP activity and facilitated Ca^{2+} entry at the soma may relate to the role Ca^{2+} might play in the synthesis and transport of proteins by nerve cells. Extracellular Ca^{2+} appears to be required for axonal transport in vertebrate sensory nerves (Hammerschlag et al., 1975) and for transport of low-molecular-weight proteins in R_{15} (unpublished observations). More research is needed to investigate these and other possible physiological consequences of BPP activity.

6. Conclusions

Conclusions resulting from consideration of invertebrate studies will be included in this section on general conclusions to the chapter, since these studies have provided observations that may yield insight into some of the ways peptides may alter neuronal excitability in the vertebrate nervous system. The salient observations produced by research at the cellular level in invertebrate systems include (1) cell–cell interaction in the absence of physiological evidence for synaptic contacts between the cells; (2) relatively slower action of onset; (3) long-term duration of action; (4) relatively greater potency; and (5) alterations in voltage-independent and voltage-dependent membrane properties. Contrasting observations in these systems have been made on putative neurotransmitter effects and include (1) cell–cell interaction mediated through specialized areas (synapses); (2) more rapid onset of action; (3) shorter duration of action; (4) relatively weaker potency; and (5) alterations in voltage-independent membrane properties only. The effects of peptides in these systems are to regulate neuronal excitability at both nerve cell body and nerve cell terminal levels in a long-term manner. Transmitters also regulate neuronal excitability at both levels, but with a much shorter time-course and essentially through different mechanisms of action on the cell's membrane. Peptides would thus appear to be important primarily in determining long-term levels of neuronal excitability (minutes to hours), while transmitters would be important in determining short-term changes in excitability (milliseconds to seconds). Whether peptide neurohormones also serve as another form of transmitter, mediating short- or long-term changes in excitability solely through changes in voltage-dependent membrane properties, has yet to be established or eliminated in either vertebrate or invertebrate nervous systems. If the distinction between neurohormones and neurotransmitters rests primarily on mechanisms of action, rather than on loci or duration of action, then peptides released at synapses that activate voltage-dependent membrane properties should be considered as locally active neurohormones. Neurohormones with neolocal or distant actions may be those that are released at extrasynaptic sites to activate voltage-dependent membrane properties on nearby or distant neurons. By these definitions, then, putative neurotransmitters that activate voltage-dependent membrane properties through either synaptic or extrasynaptic receptors should also be considered as neurohormones. Such actions have yet to be demonstrated. The various similarities and distinctions that can presently be drawn between molecules that function as neurotransmitters and those that may act as neurohormones are summarized in Table 1. Other possible mechanisms of neurohormone action in the nervous system besides direct effects on neuronal membranes to alter excitability have not been discussed because there is little phenomenological evidence to warrant such speculations. Future research will presumably determine how peptides can facilitate certain synaptic transmissions, whether it be presynaptically by acting on the processes of excitation–secretion coupling or postsynaptically by altering either transmitter affinity or the mechanisms underlying receptor-coupled conductance changes, or both.

In conclusion, although most investigations into the behavioral and cellular effects of peptides on the nervous system reported thus far have been either preliminary or primitive, and certainly not of the depth or breadth required to strongly support candidacy for either neurotransmitter or neurohormone, the evidence suggests that some peptides may belong to a class of neurochemical mediating cell–cell interactions that is operationally different from neurotransmitters. Peptide

TABLE 1. Comparison of Neurotransmitters and Neurohormones

	Neurotransmitter	Neurohormone
Substance	Amino acid, catecholamine, acetylcholine, peptide?	Peptide, putative "transmitters"?
Synthesis	Cell body, terminal	Cell body, terminal?
Transport	Rapid, intravesicular	Rapid, intragranular
Release	Synaptic terminals	Neurocirculatory (extrasynaptic) terminals, synaptic terminals?
	Ca^{2+}-dependent (evoked)	Ca^{2+}-dependent (evoked)
	Ca^{2+}-independent (resting)	Ca^{2+}-independent (resting)?
Actions	Rapid onset (msec–sec)	Slow onset (sec–min)
	Change in voltage-independent membrane properties	Change in voltage-dependent membrane properties
	Short duration (msec–sec)	Long duration (min–hr)
	Change in protein synthesis	Change in protein synthesis
Inactivation	Enzyme and reuptake	Enzyme, reuptake?
Function	Involved in momentary mediation of single cell–cell interactions	Involved in sustained modulation of cell aggregates and specific behaviors

neurohormones appear to be operating on a broader plane of integration of neuronal function, regulating the excitability of a complex circuit of cells, over an extended period, through engagement of receptor-coupled membrane mechanisms unlike those utilized by putative transmitters. Physiological roles of neurohormones in the nervous system may be somewhat analogous to those played by neurohormones and nonneurohormones in the periphery: simultaneous activation of discrete, functionally distinct elements with complementary endocrine, autonomic, and motor outputs, thus allowing a concerted, appropriate response. Just as hormones regulate the physiology of their target tissues, so also might neurohormones modulate the activity of specific sets of target neurons that underlie the expression of the relatively immutable basic drives (drinking, eating, mating, grooming, sleeping, aggression, hibernation, territoriality).

The teleology of neurohormones mediating an integrative class of operations in the nervous system apart from individual cell–cell interactions is not entirely clear. Neurohormones may have been selected for and preserved through evolution because they constitute a relatively simple method of activating a complex program. While public secretion of a substance allows for widespread and distant effects, the privacy and appropriateness of the effects are conveyed by the distribution of receptors. Thus, one substance can rapidly and discretely command many diverse elements in the nervous system that have as their final output a response of established physiological value. Neurohormones may play similar roles in the production of behaviors more subtle than instincts that are not of immediate, but rather of continuing, survival value. That is, specific neurohormones may be elaborated by various species to mediate some prescribed higher-order functions of adaptive value in the survival of the species (e.g., banding together, forms of communication and conditioning).

7. References

Amin, A.H., Crawford, T.B.B., and Gaddum, J.H., 1954, The distribution of substance P and 5-hydroxytryptamine in the central nervous system of the dog, *J. Physiol. London* **126:**596–618.

Andersson, B., and Eriksson, L., 1971, Conjoint action of sodium and angiotensin on brain mechanisms controlling water and salt balance, *Acta Physiol. Scand.* **81:**18–29.

Andersson, B., and Westbye, O., 1970, Synergistic action of sodium and angiotensin on brain mechanisms controlling fluid balance, *Life Sci.* **9:**601–608.

Andersson, B., Eriksson, L., and Oltner, R., 1970, Further evidence for angiotensin sodium interaction in central control of fluid balance, *Life Sci.* **9:**1091–1096.

Andersson, B., Leksell, L.G., and Rundgren, M., 1975, Duration of central action of angiotensin II estimated by its interaction of CSF Na$^+$, *Acta Physiol. Scand.* **93:**472–476.

Arch, S.W., 1972, Biosynthesis of the egg-laying hormone (ELH) in the bag cell neurons of *Aplysia californica, J. Gen. Physiol.* **60:**102–119.

Armstrong, D., Jepson, J.B., Keele, C.A., and Stewart, W.A., 1954, Development of pain producing substance in human plasma, *Nature London* **174:**791, 792.

Arnauld, E., Vincent, J.D., and Dreifuss, J.J., 1974, Firing patterns of hypothalamic supraoptic neurons during water deprivation in monkeys, *Science* **185:**535–537.

Arnauld, E., Dufy, B., and Vincent, J.D., 1975, Hypothalamic supraoptic neurons: Rates and patterns of action potential firing during water deprivation in the unanesthetized monkeys, *Brain Res.* **100:**315–325.

Bargmann, W., 1966, Neurosecretion, *Int. Rev. Cytol.* **19:**183–201.

Barker, J.L., 1975, CNS depressants: Effects on post-synaptic pharmacology, *Brain Res.* **92:**35–55.

Barker, J.L., 1976, Peptides: Roles in neuronal excitability, *Physiol. Rev.* **56:**435–452.

Barker, J.L., and Gainer, H., 1974, Peptide regulation of bursting pacemaker activity in a molluscan neurosecretory cell, *Science* **184:**1371–1373.

Barker, J.L., and Gainer, H., 1975a, Studies on bursting pacemaker potential activity in molluscan neurons, I. Membrane properties and ionic contributions, *Brain Res.* **84:**461–477.

Barker, J.L., and Gainer, H., 1975b, Studies on bursting pacemaker potential activity in molluscan neurons, II. Regulation by divalent cations, *Brain Res.* **84:**479–500.

Barker, J.L., and Nicoll, R.A., 1972, GABA: Role in primary afferent depolarization, *Science* **176:**1043–1045.

Barker, J.L., and Nicoll, R.A., 1973, The pharmacology and ionic dependency of amino acid responses in the frog spinal cord, *J. Physiol. London* **228:**259–277.

Barker, J.L., and Smith, T.G., 1976, Peptide regulation of neuronal membrane properties, *Brain Res.* **103:**167–170.

Barker, J.L., Crayton, J.W., and Nicoll, R.A., 1971a, Noradrenaline and acetylcholine responses of supraoptic neurosecretory cells, *J. Physiol. London* **218:**19–32.

Barker, J.L., Crayton, J.W., and Nicoll, R.A., 1971b, Supraoptic neurosecretory cells: Autonomic modulation, *Science* **171:**206–207.

Barker, J.L., Crayton, J.W., and Nicoll, R.A., 1971c, Antidromic and orthodromic responses of paraventricular and supraoptic neurosecretory cells, *Brain Res.* **33:**353–366.

Barker, J.L., Ifshin, M., and Gainer, H., 1975a, Studies on bursting pacemaker potential activity in molluscan neurons. III. Effects of hormones, *Brain Res.* **84:**501–513.

Barker, J.L., Nicoll, R.A., and Padjen, A., 1975b, Studies on convulsants in the isolated frog spinal cord. I. Antagonism of amino acid responses *J. Physiol. London* **245:**521–536.

Barker, J.L., Nicoll, R.A., and Padjen, A., 1975c, Studies on convulsants in the isolated frog spinal cord. II. Effects on root potentials, *J. Physiol. London* **245:**536–548.

Barker, J.L., Neale, J.H., and Gainer, H., 1976, Rapidly transported proteins in sensory, motor and sympathetic nerves of the isolated frog nervous system, *Brain Res.* **105:**497–515.

Barlow, J.L., Cooper, B.R., and Breese, A.J., 1975, Effects of thyrotropin releasing hormone on behavior: Evidence for an anorexic-like action, *Neurosci. Abstr.* **1:**334.

Barry, J., Dubois, M.P., and Poulain, P., 1974, Immunofluorescence study of the preoptico-infundibular LRF neurosecretory pathway in the normal, castrated or testosterone-treated male guinea pig, *Endocrinology* **95:**1416–1423.

Bennett, G.W., and Edwardson, J.A., 1975, Release of corticotrophin releasing factor and other hypophysiotropic substances from isolated nerve endings (synaptosomes), *J. Endocrinol.* **65:**35–44.

Bennett, G.W., Edwardson, J.A., Holland, D., Jeffcoate, S.L., and White, N., 1975, Release of immunoreactive luteinizing hormone–releasing hormone and thyrotropin-releasing hormone from hypothalamic synaptosomes, *Nature London* **257:**323–325.

Bennett, M.V.L., 1966, Physiology of electrotonic junctions, *Ann. N. Y. Acad. Sci* **188:**242–269.

Berde, B. (ed.), 1968, Neurohypophyseal hormones and similar polypeptides, in: *Handbook of Experimental Pharmacology,* Vol. 23, Springer-Verlag, New York, 967 pp.

Bickerton, R.K., and Buckley, J.P., 1961, Evidence for a central mechanism in angiotensin-induced hypertension, *Proc. Soc. Exp. Biol. Med.* **106:**834–836.

Blackwell, R.E., and Guillemin, R., 1973, Hypothalamic control of adenohypophyseal secretion, *Annu. Rev. Physiol.* **35:**357–390.

Boadle-Biber, M.C., and Roth, R.H., 1976, Catecholamine synthesis: Effect of angiotensin II and III, in: *International Symposium on the Central Effects of Angiotensin and Related Hormones* (J.P. Buckley, ed.), Pergamon Press, Elmsford, New York.

Boadle-Biber, M.C., Hughes, J.C., and Roth, R.H., 1972, Acceleration of catecholamine biosynthesis in sympathetically innervated tissue by angiotensin-II amide, *Br. J. Pharmacol.* **46:**289–299.

Bonjour, J.P., and Malvin, R.L., 1970, Stimulation of ADH release by the renin–angiotensin system, *Amer. J. Physiol.* **218:**1555–1559.

Bonnardeaux, J.L., and Regoli, D., 1974, Action of angiotensin and analogues on the heart, *Can. J. Physiol. Pharmacol.* **52:**50–60.

Borgeat, P, Chavancy, G., Dupont, A., Labrie, R., Arimura, A., and Schally, A.V., 1972, Stimulation of adenosine 3′:5′-cyclic monophosphate accumulation in anterior pituitary gland *in vitro* by synthetic luteinizing hormone–releasing hormone, *Proc. Nat. Acad. Sci. U.S.A.* **69:**2677–2681.

Broadwell, R.D., and Brightman, M.W., 1976, Entry of peroxidase into neurons of the central and peripheral nervous systems from extracerebral and cerebral blood, *J. Comp. Neurol.* in press.

Brown, M., and Vale, W., 1975, Central nervous system effects of hypothalamic peptides, *Endocrinology* **96:**1333–1336.

Brownstein, M., Palkovits, M., Saaverda, J.M., Bassiri, R.M., and Utiger, R.D., 1974, Thyrotropin-releasing hormone in specific nuclei of the brain, *Science* **185:**267–269.

Brownstein, M., Arimura, A., Sato, H., Schally, A.V., and Kizer, J.S., 1975, The regional distribution of somatostatin in the rat brain, *Endocrinology* **96:**1456–1461.

Burford, G.D., Dyball, R.E.J., Moss, R.L., and Pickering, B.T., 1974, Synthesis of both neurohypophysial hormones in both paraventricular and supraoptic nuclei of the rat, *J. Anat. London* **117:**261–269.

Carpenter, C.O., 1973, Ionic mechanisms and models of endogenous discharge of *Aplysia* neurons, in: *Neurobiology of Invertebrates,* pp. 35–58, Tihany, Hungary.

Carpenter, D.O., and Gunn, R., 1970, The dependence of pacemaker discharge of *Aplysia* neurons upon Na$^+$ and Ca^{++}, *J. Cell. Physiol.* **75:**121–128.

Chang, M.M., and Leeman, S.E., 1970. Isolation of a sialogogic peptide from bovine hypothalamic tissue and its characterization as substance P, *J. Biol. Chem.* **245:**4784–4790.

Changaris, D.G., Demers, L., Keil, L.C., and Severs, W., 1976, Immunopharmacology of angiotensin I in rat brain, in: *International Symposium on the Central Actions of Angiotensin and Related Hormones* (J.P. Buckley, ed.), Pergamon Press, Elmsford, New York.

Cheng, K.W., and Friesen, H.G., 1970, Physiological factors regulating secretion of neurophysin, *Metabolism* **19:**876–890.

Civan, M.M., and Frazier, H.S., 1968, The site of the stimulatory action of vasopressin on sodium transport in toad bladder, *J. Gen. Physiol.* **51:**589–605.

Coggeshall, R.E., 1972, The muscle cells of the follicle of the ovotestis in *Aplysia* as the probable target organ for bag cell extract, *Amer. Zool.* **12:**521–524.

Conrad, L.C.A., and Pfaff, D.W., 1975, Axonal projections of medical preoptic and anterior hypothalamic neurons, *Science* **190:**1112–1114.

Cross, B.A., 1974, The neurosecretory impulse, in: *Neurosecretion—The Final Neuroendocrine Pathway,* 6th International Symposium on Neurosecretion, London (1973), pp. 115–128, Springer-Verlag, New York.

Cuthbert, A.W., and Wond, P.Y.D., 1974, Calcium release: Relation to permeability changes in toad bladder epithelium following antidiuretic hormone, *J. Physiol. London* **241:**407–422.

Dale, H.H., 1934, Pharmacology and nerve endings, *Proc. Roy. Soc. Med.* **28:**319–322.

Daniel, A.R., and Lederis, K., 1967, Release of neurohypophysial hormones *in vitro, J. Physiol. London* **190:**171–187.

Davidoff, R.A., 1972, Gamma-aminobutyric acid antagonism and presynaptic inhibition in the frog spinal cord, *Science* **175:**331–333.

Davidoff, R.A., and Sears, E.S., 1974, The effects of Lioresal on synaptic activity in the isolated spinal cord, *Neurology* **24:**957–963.

Dempsey, P.J., McCallum, Z.T., Kent, K.M., and Cooper, T., 1971, Direct myocardial effects of angiotensin II, *Amer. J. Physiol.* **220:**477–481.

Deuben, R.R., and Buckley, J.P., 1970, Identification of a central site of action of angiotension II, *J. Pharmacol. Exp. Ther.* **175:**139–146.

Douglas, W.W., and Poisner, A.M., 1964, Stimulus–secretion coupling in a neurosecretory organ and the role of calcium in the release of vasopressin from the neurohypophysis, *J. Physiol. London* **172:**1–18.

Douglas, W.W., Nagasawa, J., and Schulz, R., 1971, Electron-microscopic studies on the mechanism of secretion of posterior pituitary hormones and significance of microvesicles ("synaptic vesicle"): Evidence of secretion by exocytosis and formation of microvesicles as a by-product of this process, in: *Subcellular Organization and Function in Endocrine Tissues, Mem. Soc. Endocrinol. No. 19* (H. Heller and K. Lederis, eds.) pp. 353–378, Cambridge University Press, London.

Dreifuss, J.J., and Kelly, J.S., 1972a, Recurrent inhibition of antidromically identified rat supraoptic neurons, *J. Physiol. London* **220:**87–103.

Dreifuss, J.J., and Kelly, J.S., 1972b, The activity of identified supraoptic neurones and their response to acetylcholine applied by iontophoresis, *J. Physiol. London* **222:**105–118.

Dreifuss, J.J., Nordmann, J.J., and Vincent, J.D., 1974, Recurrent inhibition of supraoptic neurosecretory cells in homozygous Brattleboro rats, *J. Physiol. London* **237:**25, 26.

Dyball, R.E.J., 1974, Single unit activity in the hypothalamo–neurohypophysial system of Brattleboro rats, *J. Endocrinol.* **60:**135–143.

Dyer, R.G., and Dyball, R.E.J., 1974, Evidence for a direct effect of LRF and TRF on single unit activity in the rostral hypothalamus, *Nature London* **252:**486–488.

Eccles, J.C., Schmidt, R.F., and Willis, W.D., 1963, Pharmacological studies on presynaptic inhibition, *J. Physiol.* **168:**500–530.

Epstein, A., and Kennedy, N.J., 1976, Suppression of thirst induced by the E prostaglandins, in: *International Symposium on the Central Actions of Angiotensin and Related Hormones* (J.P. Buckley, ed.), Pergamon Press, Elmsford, New York.

Epstein, A.N., Fitzsimons, J.T., and Rolls, B.J., 1970, Drinking induced by injection of angiotensin into the brain of the rat, *J. Physiol. London* **210:**457–474.

Evans, J.J., and Watkins, W.B., 1973, Localization of neurophysin in the neurosecretory elements of the hypothalamus and neurohypophysis of the normal and osmotically stimulated guinea-pig as demonstrated by immunofluorescence histochemical techniques, *Z. Zellforsch.* **145:**39–55.

Fawcett, C.P., Powell, A.E., and Sachs, H., 1968, Biosynthesis and release of neurophysins, *Endocrinology* **83:**1299–1310.

Felix, D., 1976, Peptide and acetylcholine actions on neurons of the cat subfornical organ, *Naunyn-Schmeideberg's Arch. Pharmacol.* **292:**15–20.

Felix, D., and Akert, K., 1974, The effect of angiotensin II on neurons of the cat subfornical organ, *Brain Res.* **76:**350–353.

Ferrario, C.M., Dickinson, C.J., and McCubbin, J.W., 1970, Central vasomotor stimulation by angiotensin, *Clin. Sci. London* **39:**239–245.

Ferrario, C.M., Gildenberg, P.L., and McCubbin, J.W., 1972, Cardiovascular effects of angiotensin mediated by the central nervous system, *Circ. Res.* **30:**259–262.

Finkielman, S., Goldstein, D.J., and Nahmod, V.E., 1976, Introduction of angiotensin II and serotonin in the central nervous system, in: *International Symposium on the Central Effects of Angiotensin and Related Hormones* (J.P. Buckley, ed.), Pergamon Press, Elmsford, New York.

Fischer-Ferraro, C., Nahmod, V.E., Goldstein, D.J., and Finkielman, S., 1971, Angiotensin and renin in rat and dog brain, *J. Exp. Med.* **133:**353–361.

Fitzsimons, J.T., 1971, The effect on drinking of peptide precursors and of shorter chain peptide fragments of angiotensin II injected into the rat's diencephalon, *J. Physiol. London* **214:**295–303.

Foreman, M., and Moss, R.L., 1975, Enhancement of lordotic behavior by intrahypothalamic infusion of luteinizing hormone–releasing hormone, *Neurosci Abstr.* **1:**435.

Forsling, M.L., Martin, M.J., Sturdy, J.C., and Burton, A.M., 1973, Observations on the release and clearance of neurophysin and the neurohypysial hormones in the rat, *J. Endocrinol.* **57:**307–315.

Frazier, W.T., Dandel, E.R., Kupfermann, I., Waziri, R., and Coggeshall, R.E., 1967, Morphological and functional properties of identified neurons in the abdominal ganglion of *Aplysia californica, J. Neurophysiol.* **30:**1288–1351.

Gaddum, J.H., 1960, Substance P distribution, in: *Polypeptides Which Affect Smooth Muscles and Blood Vessels* (M. Schachter, ed.), pp. 163–170, Pergamon Press, Elmsford, New York.

Gagnon, D.J., Cousineau, D., and Boacher, P.J., 1973, Release of vasopressin by angiotensin II and prostaglandin E_2 from the rat neurohypophysis *in vitro, Life Sci.* **12:**487–497.

Gainer, H., 1972a, Effects of experimentally induced diapause on the electrophysiology and protein synthesis of identified molluscan neurons, *Brain Res.* **39**:387–402.

Gainer, H., 1972b, Electrophysiological behavior of an endogenously active neurosecretory cell, *Brain Res.* **39**:403–418.

Gainer, H., and Barker, J.L., 1974, Synaptic regulation of specific protein synthesis in an identified neuron, *Brain Res.* **78**:314–319.

Gainer, H., and Barker, J.L., 1975, Selective modulation and turnover of proteins in identified neurons of *Aplysia*, *Comp. Biochem. Physiol.* **51B**:221–227.

Gainer, H., and Wollberg, Z., 1974, Specific protein metabolism in identifiable neurons of *Aplysia californica*, *J. Neurobiol.* **5**:243–261.

Ganten, D., 1976, Endogenous brain angiotensin: Its possible role in central mechanisms of blood pressure regulation and its relationship to circulating plasma angiotensin, in: *International Symposium on the Central Effects of Angiotensin and Related Hormones* (J.P. Buckley, ed.), Pergamon Press, Elmsford, New York.

Ganten, D., Marquez-Julio, A., Granger, P., Hayduk, K., Karsunky, K.P., Boucher, R., and Genest, J., 1971a, Renin in dog brain. *Amer. J. Physiol.* **221**:1733–1737.

Ganten, D., Minnich, J., Granger, P., Hayduk, K., Brecht, H.M., Barbeau, A., Boucher, R., and Genest, J., 1971b, Angiotensin-forming enzyme in brain tissue, *Science* **173**:64, 65.

Gerschenfeld, H., 1973, Chemical transmission in invertebrate central nervous systems and neuromuscular junctions, *Physiol. Rev.* **53**:1–119.

Gerschenfeld, H., and Paupardin-Tritsch, D., 1974, Ionic mechanisms and receptor properties underlying the responses of molluscan neurons to 5-hydroxytryptamine, *J. Physiol. London* **243**:427–456.

Gildenberg, P.L., Ferrario, C.M., and McCubbin, J.W., 1973, Two sites of cardiovascular action of angiotensin II in the brain of the dog, *Clin. Sci. London* **44**:417–420.

Grant, G., Vale, W., Brazeau, P., Rivier, J., Monahan, M., Gilon, C., Amoss, M., Rivier, C., Ling, N., Burgus, R., and Guillemin, R., 1973, *Recent Stud. Hypothalamic Function,* pp. 180–195.

Greidanus, Tj., Van Wiersma, B., Buys, R.M., Hollemans, H.J.G., and de Jong, W., 1974, A radioimmunoassay of vasopressin. A note on pituitary vasopressin content in Brattleboro rats, *Experientia* **30**:1217, 1218.

Grimm-Jørgensen, Y., McKelvy, J.K., and Jackson, I.M.D., 1975, Immunoreactive throtropin releasing factor in gastropod circumoesophageal ganglia, *Nature London* **254**:620.

Hammerschlag, R., Dravid, A.R., and Chin, A.Y., 1975, Mechanism of axonal transport: A proposed role for calcium ions, *Science* **188**:273–275.

Harris, M.C., Dreifuss, J.J., and Legros, J.J., 1975, Excitation of phasically firing supraoptic neurones during vasopressin release, *Nature London* **258**:80–82.

Hays, R.M., and Leaf, A., 1962, Studies on the movement of water through isolation toad bladder and its modification by vasopressin, *J. Gen. Physiol.* **45**:905–919.

Hayward, J.N., 1975, Neural control of the posterior pituitary, *Annu. Rev. Physiol.* **37**:191–210.

Hayward, J.N., and Jennings, D.P., 1973a, Activity of magnocellular neuroendocrine cells in the hypothalamus of unanesthetized monkeys, I. Functional cell types and their anatomical distribution in the supraoptic nucleus and the internuclear zone, *J. Physiol. London* **232**:515–543.

Hayward, J.N., and Jennings, O.P., 1973b, Activity of magnocellular neuroendocrine cells in the hypothalamus of unanesthetized monkeys. II. Osmosensitivity of functional cell types in the supraoptic nucleus and the internuclear zone, *J. Physiol. London* **232**:545–572.

Henry, J.L., 1975, Substance P excitation of spinal nociceptive neurons, *Neurosci. Abstr.* **1**:390.

Henry, J.L., Krnjević, K., and Morris, M.E., 1975, Substance P and spinal neurons, *Can. J. Physiol. Pharmacol.* **53**:423–432.

Herrenkohl, L.R., and Verhulst, I.M., 1975, Intracerebral infusions of luteinizing hormone releasing factor induce lordosis in rats, *Neurosci. Abstr.* **1**:436.

Hib, J., 1974, The *in vitro* effects of oxytocin and vasopressin on spontaneous contractility of the mouse cauda epididymis, *Biol. Reprod.* **11**:436–439.

Hoffman, W.E., 1976, Antidiuretic hormone release by angiotensin II, in: *International Symposium on the Central Effects of Angiotensin and Related Hormones* (J.P. Buckley, ed.), Pergamon Press, Elmsford, New York.

Hökfelt, T., Kellerth, J.O., Nilsson, G., and Pernow, B., 1975a, Substance P: Localization in the central nervous system and in some primary sensory neurons, *Science* **190**:889, 890.

Hökfelt, T., Kellerth, J.O., Nilsson, G., and Pernow, B., 1975b, Experimental immunohistochemical studies on the localization and distribution of substance P in cat primary sensory neurons, *Brain Res.* **100**:235–252.

Holton, P., 1960, Substance P concentration in degenerating nerve, in: *Polypeptides Which Affect Smooth Muscles and Blood Vessels* (M. Schachter, ed.), pp. 192–194, Pergamon Press, Oxford.

Hubbard, J., 1973, Microphysiology of vertebrate neuromuscular transmission, *Physiol. Rev.* **53**:674–723.

Hughes, J., 1975, Isolation of an endogenous compound from the brain with pharmacological properties similar to morphine, *Brain Res.* **88**:295–308.

Hughes, J., and Roth, R.H., 1971, Evidence that angiotensin enhances transmitter release during sympathetic nerve stimulation, *Br. J. Pharmacol.* **41**:239–255.

Hughes, J., Smith, T., Kosterlitz, H., Fothergill, L., Morgan, B., and Morris, H., 1975, Identification of two related pentapeptides from the brain with potent opiate agonist activity, *Nature London,* **258**:577–579.

Ifshin, M., Gainer, H., and Barker, J.L., 1975, Peptide factor extracted from molluscan ganglia that modulates bursting pacemaker activity. *Nature London* **254**:72–74.

Jackson, I.M.D., and Reichlin, J., 1974, TRH: Distribution in hypothalamic and extra-hypothalamic brain tissues of mammalian and submammalian chordates, *Endocrinology* **95**:854–862.

Jones, C.W., and Pickering, B.T., 1972, Comparison of the effects of water and sodium chloride inhibition on the hormone content of the neurohypophysis of the rat, *J. Physiol. London* **203**:449–458.

Joy, M.D., and Lowe, R.D., 1970, Evidence that the area postrema mediates the central cardiovascular response to angiotensin II, *Nature London* **228**:1303, 1304.

Juan, H., and Lembeck, F., 1974, Action of peptides and other algesic agents on paravascular pain receptors of the isolated perfused rabbit ear, *Naunyn-Schmeideberg's Arh. Pharmacol.* **283**:151–164.

Kadowitz, P.J., Sweet, C.S., and Brody, M.J., 1971, Potentiation of adrenergic venomotor responses by angiotensin, prostaglandin $F_{2\alpha}$ and cocaine, *J. Pharmacol. Exp. Ther.* **176**:167–173.

Kandel, E.R., 1964, Electrical properties of hypothalamic neurosecretory cells, *J. Gen. Physiol.* **12**:81–96.

Kataoka, K., 1962, The subcellular distribution of substance P in the nervous tissues, *Jpn. J. Physiol.* **12**:81–96.

Kawakami, M., and Sakuma, Y., 1974, Responses of hypothalamic neurons to the microiontophoresis of LH–RH, LH and FSH under levels of circulating ovarian hormones, *Neuroendocrinology* **15**:290–307.

Kawakami, M., and Sakuma, Y., 1976, Electrophysiological evidences for possible participation of periventricular neurons in anterior regulation, *Brain Res.* **101**:79–94.

Kebabian, J.W., Bloom, F.E., Steiner, A.L., and Greengard, P., 1975, Neurotransmitters increase cyclic nucleotides in postganglionic neurons: Immunocytochemical demonstration, *Science* **190**:157–159.

Keele, C.A., and Armstrong, D., 1964, *Substance Producing Pain and Itch,* Arnold, London.

Keil, L.C., Summy-Long, J., and Severs, W.B., 1975, Release of vasopressin by angiotensin II, *Endocrinology* **96**:1063–1065.

Kelly, J.J., and Dreifuss, J.J., 1970, Antidromic inhibition of identified rat suraoptic neurons, *Brain Res.* **22**:406–409.

Knutsson, E., Lindblom, V., and Martensson, A., 1973, Differences in effects in gamma and alpha spasticity induced by GABA derivative Baclofen (Lioresal), *Brain* **96**:29–46.

Koizumi, K., and Yamashita, H., 1972, Studies of antidromically identified neurosecretory cells of the hypothalamus by intracellular and extracellular recordings, *J. Physiol. London* **221**:683–705.

Koizumi, K., Ishikawa, T., and Brooks, McC.C., 1973, The existence of facilitatory axon collaterals in neurosecretory cells of the hypothalamus, *Brain Res.* **63**:408–413.

Konishi, S., and Otsuka, M., 1974a, The effects of Substance P and other peptides on spinal neurons of the frog, *Brain Res.* **65**:397–410.

Konishi, S., and Otsuka, M., 1974b, Excitatory action of hypothalamic substance P on spinal motoneurons of newborn rats, *Nature London* **252**:734, 735.

Kraemer, G.W., Mueller, R.A., Breese, G.R., Cooper, B.R., McKinney, W.B., and Prange, A.J., 1975, Reversal of pentobarbital sleep by thyrotropin releasing hormone in the rhesus monkey, *Neurosci. Abstr.* **1**:334.

Krnjević, K., 1974, Chemical nature of synaptic transmission in vertebrates, *Physiol. Rev.* **54**:418–540.

Krnjević, and Morris, M.E., 1974, An excitatory action of substance P on cuneate neurons, *Can. J. Physiol. Pharmacol.* **52**:736–744.

Krnjević, K., and Phillis, J.W., 1963, Acetylcholine sensitive cells in the cerebral cortex, *J. Physiol. London* **166**:296–327.

Kupfermann, I., 1970, Stimulation of egg laying by extracts of neuroendocrine cells (Bag cells) of abdominal ganglion of *Aplysia, J. Neurophysiol.* **33**:377–381.

Kupfermann, I., 1972, Studies on the neurosecretory control of egg-laying in *Aplysia, Amer. Zool.* **12**:513–516.

Kupfermann, I., and Weiss, K.R., 1975, Water regulation by a presumptive hormone contained in identified neurosecretory cell R_{15} of *Aplysia, J. Gen. Physiol.* **67**:113–123.

LeClerc, R., and Pelletier, G., 1974, Electron microscopic immunohistochemical localization of vasopressin in the hypothalamus and neurohypophysis of the normal and Brattleboro rat, *Amer. J. Anat.* **140**:583–588.

Lederis, K., and Livingston, A., 1970, Neuronal and subcellular localisation of acetylcholine in the posterior pituitary of the rabbit, *J. Physiol.London* **210**:187–204.

Leeman, S., and Mroz, E.A., Substance P, 1974, *Life Sci.* **15**:2033–2044.

Lembeck, F., 1953, Zur Frage der zentralen Übertragung afferenter Impulse III. Das Vorkommen und die Bedeutung der Substanz P in den dorsalen Wurseln des Ruckenmarks, *Arch. Exp. Pathol. Pharmakol.* **219**:197–213.

Levy, R.A., 1974, GABA: A direct depolarizing action at the mammalian primary afferent terminal, *Brain Res.* **76**:155–160.

Loh, Y.P., and Gainer, H., 1975a, Low molecular weight specific proteins in identified molluscan neurons. I. Synthesis and storage, *Brain Res.* **92**:181–192.

Loh, Y.P., and Gainer, H., 1975b, Low molecular weight specific proteins in identified molluscan neurons. II. Processing, turnover and transport, *Brain Res.* **92**:193–205.

Loh, Y.P., and Peterson, R.P., 1974, Protein synthesis in phenotypically different, single neurons of *Aplysia, Brain Res.* **78**:83–98.

Loh, Y.P., Sarne, Y., and Gainer, H., 1975, Heterogeneity of proteins synthesized, stored and released by the Bag Cells of *Aplysia californica, J. Comp. Physiol.* **100**:283–295.

Loh, Y.P., Barker, J.L., and Gainer, H., 1976, Neurosecretory cell protein metabolism correlated with diapause in the land snail, *Otala lactea, J. Neurochem.* **26**:25–30.

Malvin, R.L., 1976, Hypothalamic stimulation of ADH release by angiotensin II, in: *International Symposium on the Central Effects of Angiotensin and Related Hormones* (J.P. Buckley, ed.), Pergamon Press, Elmsford, New York.

Mathieu, P.A., and Roberge, F.A., 1971, Characteristics of pacemaker oscillations in *Aplysia* neurons, *Can. J. Physiol. Pharmacol.* **49**:787–795.

Mayeri, E., and Simon, S., 1975, Modulation of synaptic transmission and burster neuron activity after release of a neurohormone in *Aplysia, Neurosci. Abstr.* **1**:584.

McLean, A.S., Sirett, N.E., Bray, J.J., and Hubbard, J.I., 1975, Regional distribution of angiotensin II receptors in the rat brain, *Proc. Univ. Otago Med. Sch.* **53**:19, 20.

Miledi, R., 1973, Transmitter release induced by injection of calcium ions into nerve terminals, *Proc. Roy. Soc. London Ser. B* **183**:421–425.

Millhouse, O.E., 1973, The organization of the ventromedial hypothalamic nucleus, *Brain Res.* **55**:71–87.

Morgenroth, V.H., Boadle-Biber, M., and Roth, R.H., 1974, Tyrosine hydroxylase: Activation by nerve stimulation, *Proc. Nat. Acad. Sci. U.S.A.* **71**:4283–4287.

Moss, R.L., and Foreman, M.M., 1976, Potentiation of lordosis behavior by intrahypothalamic infusion of synthetic luteinizing hormone-releasing hormone, *Neuroendocrinology* in press.

Moss, R.L., and McCann, S. M., 1973, Induction of mating behavior in rats by luteinizing hormone releasing factor, *Science* **181**:177–179.

Moss, R.L., and McCann, S.M., 1975, Action of luteinizing hormone releasing factor (LRF) in the initiation of lordosis behavior in the estrone-primed ovariectomized female rat, *Neuroendocrinology* **17**:309–318.

Moss, R.L., Dyball, R.E.J., and Cross, B.A., 1971, Responses of antidromically identified supraoptic and paraventricular units to acetylcholine, noradrenaline and glutamate applied iontophoretically, *Brain Res.* **35**:573–575.

Moss, R.L., Dyball, R.E.J., and Cross, B.A., 1972a, Excitation of antidromically identified neurosecretory cells of the paraventricular nucleus by oxytocin applied iontophoretically, *Exp. Neurol.* **34**:95–102.

Moss, R.L., Urban, I., and Cross, B.A., 1972b, Microiontophoresis of cholinergic and aminergic drugs on paraventricular neurons, *Amer. J. Physiol.* **223**:310–318.

Mouw, D., Bonjour, J.P., Malvin, R.L., and Vander, A., 1971, Central action of angiotensin in stimulating ADH release, *Amer. J. Physiol.* **220**:239–242.

Mulder, A.H., Genze, J.J., and de Wied, D., 1970, Studies on the subcellular localization of corticotrophin releasing factor (CRF) and vasopressin in the median eminence of the rats, *Endocrinology* **87**:61–79.

Negoro, H., and Holland, R.C., 1972, Inhibition of unit activity in the hypothalamic paraventricular nucleus following antidromic activation, *Brain Res.* **42**:385–402.

Negoro, H., Visessuwan, S., and Holland, R.C., 1973, Inhibition and excitation of units in paraventricular nucleus after stimulation of the septum, amygdala and neurohypophysis, *Brain Res.* **57**:479–483.

Nelson, P.G., 1975, Nerve and muscle cells in culture, *Physiol. Rev.* **55**:1–61.

Nemeroff, C.B., Prange, A.J., Bissette, G., and Lipton, M.A., 1974, Thyrotropin releasing hormone (TRH): Potentiation of the anticonvulsant properties of phenobarbital sodium, *Fed. Proc. Fed. Amer. Soc. Exp. Biol. Abstr.*, **33**:322.

Nicoll, R.A., and Barker, J.L., 1971a, The pharmacology of recurrent inhibition in the supraoptic neurosecretory system, *Brain Res.* **35**:501–511.

Nicoll, R.A., and Barker, J.L., 1971b, Excitation of supraoptic neurosecretory cells by angiotensin II, *Nature London New Biol.* **233**:172–174.

Nilsson, G., Hökfelt, T., and Pernow, B., 1974, Distribution of substance P–like immunoreactivity in the rat central nervous system as revealed by immunohistochemistry, *Med. Biol.* **52**:424–427.

Nörstrom, A., and Sjöstrand, J., 1971, Effect of haemorrhage on the rapid axonal transport of neurohypophysial proteins of the rat, *J. Neurochem.* **18**:2017–2026.

Novin, D., and Durham, R., 1973, Orthodromic and antidromic activation of the paraventricular nucleus of the hypothalamus in the rabbit, *Exp. Neurol.* **41**:418–430.

Oliver, C., Charvet, J.P., Codaecioni, J.L., Vague, J., and Porter, J.C., 1974, TRH in human CSF, *Lancet* **i**:873.

Oparil, S., and Haber, E., 1974, The renin–angiotensin system, *N. Engl. J. Med.* **291**:389–401.

Orloff, J., and Handler, J., 1967, The role of adenosine 3′,5′-phosphate in the action of antidiuretic hormone, *Amer. J. Med.* **42**:757–770.

Otsuka, M., and Konishi, S., 1975, Substance P and excitatory transmitter of primary sensory neurons, *Cold Spring Harbor Symp. Quant. Biol.* **40**:44 (abstract).

Page, I.H., and Bumpus, F.M., 1961, Angiotensin, *Physiol. Rev.* **41**:331–390.

Palaic, D., and Khairallah, P.A., 1967, Inhibition of norepinephrine uptake by angiotensin, *J. Pharm. Pharmacol.* **19**:396, 397.

Pals, D.T., Masucci, F.D., Denning, G.S., Sipos, F., and Fessler, D.C., 1971, Role of angiotensin II in experimental hypertension, *Circ. Res.* **29**:673–681.

Panisset, J.C., 1967, Effect of angiotensin on the release of acetylcholine from preganglionic and postganglionic nerve endings, *Can. J. Physiol. Pharmacol.* **45**:313–317.

Panisset, J., and Bourdois, P., 1968, Effect of angiotensin on the response to noradrenaline and sympathetic nerve stimulation and on [³H]noradrenaline uptake in cat mesenteric blood vessels, *Can. J. Physiol. Pharmacol.* **46:**125–131.

Park, W.K., Regoli, D., and Rioux, F., 1973, Characterization of angiotensin receptors in vascular and intestinal smooth muscles, *Br. J. Pharmacol.* **48:**288–301.

Parry, H.B., and Livett, B.G., 1973, A new hypothalamic pathway to the median eminence containing neurophysin and its hypertrophy in sheep with natural scrapie, *Nature London* **242:**63–65.

Peach, M.J., Bumpus, F.M., and Khairallah, P.A., 1969, Inhibition of norepinephrine uptake in hearts by angiotensin II and analogues, *J. Pharmacol. Exp. Ther.* **167:**291–299.

Peck, E.J., Tytell, M., and Clark, J.H., 1975, Hypothalamic synaptosomes and *in vitro* LRF secretion, *Endocrine Soc. Abstr.*, p. 95.

Pelletier, G., Labrie, F., Puviani, R., Arimura, A., and Schally, A.V., 1974*a*, Immunohistochemical localization of luteinizing hormone releasing hormone in the rat median eminence, *Endocrinology* **95:**314–317.

Pelletier, G., Labrie, F., Arimura, A., Schally, A.V., 1974*b*, Electron microscopic immunohistochemical localization of growth hormone release inhibiting hormone in the rat median eminence, *Amer. J. Anat.* **140:**445–450.

Pernow, B., 1953, Studies on substance P. Purification, occurrence and biological actions, *Acta Physiol. Scand.* **79:**Suppl. 104.

Pernow, B., 1960, Effect of substance P on smooth muscle, in: *Polypeptides Which Affect Smooth Muscles and Blood Vessels* (M. Schachter, ed.), pp. 163–170, Pergamon Press, Elmsford, New York.

Pfaff, D.W., 1973, Luteinizing hormone–releasing factor potentiates lordosis behavior in hypophysectomized ovariectomized female rats, *Science* **182:**1148, 1149.

Phillips, M.I., 1976, Sensitive sites in the ventricular system for drinking and blood pressure responses to angiotensin, in: *International Symposium on the Central Effects of Angiotensin and Related Hormones* (J.P. Buckley, ed.), Pergamon Press, Elmsford, New York.

Phillips, M.I., and Felix, D., 1976, Specific angiotensin II receptive neurons in the cat subfornical organ, *Brain Res.* **109:**531–540.

Phillis, J.W., and Limacher, J.J., 1974*a*, Substance P excitation of cerebral cortical Betz cells, *Brain Res.* **69:**158–163.

Phillis, J.W., and Limacher, J.J., 1974*b*, Excitation of cerebral cortical neurons by various polypeptides, *Exp. Neurol.* **53:**414–423.

Pickup, J.C., Johnston, C.I., Nakamura, S., Uttenthal, L.O., and Hope, D.B., 1973, Subcellular organisation of neurophysins, oxytocin, (8-lysine)-vasopressin and adenosine triphosphatase in porcine posterior pituitary lobes, *Biochem. J.* **132:**361–371.

Pinto, O.DeS., Polikar, M., and Debono, G., 1972, Results of international clinical trials with Lioresal, *Postgrad. Med. J.* **48:**18–23.

Regoli, D., Park, W.K., and Rioux, F., 1974, Pharmacology of angiotensin, *Pharm. Rev.* **26:**69–129.

Reit, E., 1972, Actions of angiotensin on the adrenal medulla and autonomic ganglia, *Fed. Proc. Fed. Amer. Soc. Exp. Biol.* **31:**1338–1343.

Renaud, L.P., 1976*a*, Tuberoinfundibular neurons in the basomedial hypothalamus of the rat: Electrophysiological evidence for axon collaterals to hypothalamic and extrahypothalamic areas, *Brain Res.* in press.

Renaud, L.P., 1976*b*, Response of identified ventromedial hypothalamic nucleus neurons to putative neurotransmitters applied by microiontophoresis, *Br. J. Pharmacol.* in press.

Renaud, L.P., and Martin, J.B., 1975*a*, Thyrotropin releasing hormone: Depressant action on central neuronal activity, *Brain Res.* **86:**150–154.

Renaud, L.P., and Martin, J.B., 1975*b*, Electrophysiological studies of connections of hypothalamic ventromedial nucleus neurons in the rat: Evidence for a role in neuroendocrine regulation, *Brain Res.* **93:**145–151.

Renaud, L.P., Martin, J.B., and Brazeau, P., 1975, Depressant action of TRH, LHRH and somatostatin on activity of central neurons, *Nature London* **255:**233–235.

Robertson, G., Klein, L.A., Roth, J., and Gorden, P., 1970, Immunoassay of plasma vasopressin in man, *Proc. Nat. Acad. Sci. U.S.A.* **66:**1298–1305.

Robinson, A.G., and Zimmerman, E.A., 1973, Cerebrospinal fluid and ependymal neurophysin, *J. Clin. Invest.* **52:**1260–1267.

Robinson, A.G., Zimmerman, E.A., Engleman, E.G., and Frantz, A.G., 1971, Radioimmunoassay of bovine neurophysin: Specificity of neurophysin I and neurophysin II, *Metabolism* **20:**1138–1147.

Roth, R.H., and Hughes, J., 1975, Acceleration of protein synthesis by angiotensin—correlation with angiotensin's effect on catecholamine biosynthesis, *Biochem. Pharmacol.* **21:**3182–3187.

Ryall, R.W., 1964, The subcellular distributions of acetylcholine, Substance P. 5-hydroxytriptamine, γ-aminobutyric acid and glutamic acid in brain homogenates, *J. Neurochem.* **11:**131–145.

Saito, K., Konishi, S., and Otsuka, M., 1975, Antagonism between Lioresal and substance P in rat spinal cord, *Brain Res.* **97:**177–180.

Sakai, K.K., Marks, B.H., George, J., and Koestner, A., 1974, Specific angiotensin II receptors in organ-cultured canine supra-optic nucleus cells, *Life Sci.* **14:**1337–1344.

Samuels, A.I., 1976, Studies on the role of central angiotensin in Na⁺ deficient dogs, in: *International Symposium on the Central Effects of Angiotensin and Related Hormones* (J.P. Buckley, ed.), Pergamon Press, Elmsford, New York.

Schally, A.V., Arimura, A., and Kostin, A.J., 1973, Hypothalamic regulatory hormones, *Science* **179:**341–350.

Scharrer, B., 1969, Neurohumors and neurohormones: Definitions and terminology, *J. Neuro-Visc. Relat. Suppl. 9,* pp. 1–20.

Scharrer, B., 1970, General principles of neuroendocrine communication in: *The Neurosciences: Second Study Program* (F.O. Schmitt, ed.), pp. 519–529, The Rockefeller University Press, New York.

Scharrer, B., 1972, Neuroendocrine communication (neurohormonal, neurohumoral and intermediate), in: *Topics in Neuroendocrinology, Prog. Brain Res.* **38:**7–18 (J. Arienskappers and J.P. Schade, eds.), Elsevier, New York.

Scharrer, E., 1966, Principles of neuroendocrine integration, in *Endocrines and the Central Nervous System, Res. Publ. Assoc. Res. Nerv. Ment. Dis.* **43:**1–40.

Severs, W.B., and Daniels-Severs, A.E., 1973, Effects of angiotensin on the central nervous system, *Pharmacol. Rev.* **25:**415–449.

Severs, W.B., Daniels, A.E., and Buckley, J.P., 1967, On the central hypertensive effect of angiotensin II, *Neuropharmacology* **6:**199–205.

Severs, W.B., Summy-Long, J., Taylor, J.S., and Connor, J.D., 1970, A central effect of angiotensin: Release of pituitary pressor material, *J. Pharmacol. Exp. Ther.* **174:**27–34.

Shaw, J.E., and Ramwell, P.W., 1968, Release of a Substance P polypeptide from the cerebral cortex, *Amer. J. Physiol.* **215:**262–267.

Simpson, J.B., and Routtenberg, A., 1973, Subfornical organ: Site of drinking elicitation by angiotensin II, *Science* **181:**1172–1175.

Simpson, J.B., and Routtenberg, A., 1975, Subfornical lesions reduce intravenous angiotensin-induced drinking, *Brain Res.* **88:**154–161.

Sirois, P., and Gagnon, D.J., 1975, Increase in cyclic AMP levels and vasopressin release in response to angiotensin I in neurohypophyses: Blockade following inhibition of the converting enzyme, *J. Neurochem.* **25:**727–729.

Smith, T.G., Barker, J.L., and Gainer, H., 1975, Requirements for bursting pacemaker potential activity in molluscan neurons, *Nature London* **253:**450–452.

Smookler, H.H., Severs, W.B., Kinnard, W.J., and Buckley, J.P., 1966, Centrally mediated cardiovascular effects of angiotensin II, *J. Pharmacol. Exp. Ther.* **153:**485–494.

Snyder, S., 1975, The opiate receptor, *Neurosci. Res. Prog. Bull.* **13:**1–27.

Solomon, T.A., Herzig, M.F., and Cameron, E.A., 1976, Direct and centrally mediated cardiovascular activity of angiotensin III, in: *International Symposium on the Central Effects of Angiotensin and Related Hormones* (J.P. Buckley, ed.), Pergamon Press, Elmsford, New York.

Stinnakre, J., 1975, Etude des variations d'activite intracellulaire du calcium dans des neurones de Mollusques injectes d'aequorine, Ph.D. thesis, University of Paris, 97 pp.

Stinnakre, J., and Tauc, L., 1973, Calcium influx in active *Aplysia* neurones detected by injected aequouin, *Nature London New Biol.* **242:**113–115.

Streeten, D.H.P., Anderson, G.H., Fueiberg, J.M., and Dalakos, T.G., 1975, Use of angiotensin II antagonist (Saralasin) in the recognition of "angiotensinogenic" hypertension, *N. Engl. J. Med.* **292:**657–662.

Strumwasser, F., 1973, Neural and humoral factors in the temporal organization of behavior, *Physiologist* **16:**9–42.

Stumwasser, F., Jacklet, J.W., and Alvarez, R.B., 1969, A seasonal rhythm in the neural extract induction of behavioral egg-laying in *Aplysia, Comp. Biochem. Physiol.* **29:**197–206.

Swaab, D.F., and Pool, C.W., 1975, Specificity of oxytocin and vasopressin immunofluorescence, *J. Endocrinol.* **66:**263–272.

Szentagothai, J., Flerko, B., Mess, B., and Halasz, B., 1972, *Hypothalamic Control of the Anterior Pituitary,* pp. 22–109, Akademiai Kaido, Budapest.

Takahashi, T., and Otsuka, M., 1975, Regional distribution of substance P in the spinal cord and nerve roots of the cat and the effect of dorsal root section, *Brain Res.* **87:**1–11.

Takahashi, T., Konishi, S., Powell, D., Leeman, S.E., and Otsuka, M., 1974, Identification of the motoneuron depolarizing peptide in bovine dorsal root as hypothalamic substance P, *Brain Res.* **73:**59–69.

Tebecis, A.K., and Phillis, J.W., 1969, The use of convulsants in studying possible functions of amino acids in the toad spinal cord, *Comp. Biochem. Physiol.* **28:**1303–1315.

Toevs, L., 1970, Identification and characterization of the egg-laying hormone from the neurosecretory bag cells in *Aplysia,* Ph.D. dissertation, California Institute of Technology.

Toevs, L., and Brackenbury, R., 1969, Bag cell–specific proteins and the humoral control of egg-laying in *Aplysia californica, Comp. Biochem. Physiol.* **29:**207–216.

Tsai, B.S., Peach, M.J., Khosla, M.C., and Bumpus, F.M., 1975, Synthesis and evaluation of [des-Asp']angiotensin I as a precursor for [Des-Asp']angiotensin II ("Angiotensin III"), *J. Med. Chem.* **18:**1180–1183.

Ueda, H., Vchida, Y., Veda, K., Gondaira, T., and Katayama, S., 1969, Centrally mediated vasopressor effect of angiotensin II in man, *Jpn. Heart J.* **10:**243–247.

Ueda, H., Katayama, S., and Kato, P., 1972, Area postrema angiotensin-sensitive site in brain, *Adv. Exp. Biol. Med.* **17:**109–116.

Valtin, H., and Schroeder, H.A., 1964, Familial hypothalamic diabetes insipidus in rats (Brattleboro strain), *Amer. J. Physiol.* **206:**425–436.

Vincent, J.D., Arnauld, E., and Nicolesen-Catargi, A., 1972, Osmoreceptors and neurosecretory cells in the supraoptic complex of the unanesthetized monkey, *Brain Res.* **45:**278–281.

Volicer, L., and Loew, C.G., 1971, Penetration of angiotensin II into the brain, *Neuropharmacology* **10:**631–636.

von Euler, U.S., 1963, Substance P in subcellular particles in peripheral nerves, *Ann. N. Y. Acad. Sci.* **104:**449–463.

von Euler, U.S., and Gaddum, J.H., 1931, An unidentified depressor substance in certain tissue extracts, *J. Physiol. London* **72:**74–87.

von Euler, U.S., and Pernow, B., 1954, Effects of intraventricular administration of substance P, *Nature London* **174:**184.

von Euler, U.S., and Pernow, B., 1956, Neurotropic effects of Substance P, *Acta Physiol. Scand.* **36:**265–275.

Vorherr, H., Bradburg, M.W.B., Hoghoughi, M., and Kleeman, C.R., 1968, Antidiuretic hormone in cerebrospinal fluid during endogenous and exogenous changes in its blood level, *Endrocrinology* **83:**246–250.

Wakerley, J.B., Poulain, D.A., Dyball, R.E.J., and Cross, B.A., 1975, Activity of phasic neurosecretory cells during haemorrhage, *Nature London* **25:**82–84.

Walter, R., Hoffman, P.L., Flexner, J.B., and Flexner, L.B., 1975, Neurohypophysial hormones analogues and fragments: Their effect on puromycin-induced amnesia, *Proc. Nat. Acad. Sci. U.S.A.* **72:**4180–4184.

Watkins, W.B., 1972, The tentative identification of three neurophysins from the rat posterior pituitary gland, *J. Endocrinol.* **55:**577–589.

Watkins, W.B., 1975, Presence of neurophysin and vasopressin in the hypothalamic magnocellular nuclei of rats homozygous and heterozygous for diabetes insipidus (Brattleboro strain) as revealed by immuniperoxidase histology, *Cell Tissue Res.* **157**:101–113.

Wayner, M.J., Ono, T., and Nolley, D., 1973, Effects of angiotensin II on central neurons, *Pharmacol. Biochem. Behav.* **1**:679–691.

Weight, F.W., and Votava, J., 1970, Slow synaptic excitation in sympathetic ganglion cells: Evidence for synaptic inactivation of potassium conductance, *Science* **170**:755–758.

Werman, R., 1966, Criteria for identification of a central nervous system transmitter, *Comp. Biochem. Physiol.* **18**:745–766.

Wilson, D.L., 1971, Molecular weight distribution of proteins synthesized in single, identified neurons of *Aplysia, J. Gen. Physiol.* **57**:26–40.

Winokur, A., and Utiger, R.D., 1974, Thyrotropin-releasing hormone regional distribution in rat brain, *Science* **185**:265–267.

Winters, A.J., Eskay, R.L., and Porter, J.C., 1974, Concentration and distribution of TRF and LRF in human fetal brain, *J. Clin. Endocrinol. Metab.* **39**:960–963.

Yagi, K., and Sawaki, Y., 1974, Recurrent inhibition and facilitation: Demonstration in the tuberoinfundibular system and effects of strychnine and picrotoxin, *Brain Res.* **84**:155–159.

Yagi, K., Azuma, T., and Matsuka, K., 1966, Neurosecretory cell: Capable of conducting impulse in rats. *Science* **154**:778, 779.

Yang, H.Y., and Neff, N.H., 1972, Distribution and properties of angiotensin converting enzyme of rat brain, *J. Neurochem.* **19**:2443–2450.

Yates, F.E., Russell, D.M., and Maran, J.W., 1971, Brain–adenohypophysial communication in mammals, *Annu. Rev. Physiol.* **33**:393–444.

Yu, R., and Dickinson, C.J., 1966, Neurogenic effects of angiotensin, *Lancet* **2**:1276, 1277.

Zetler, G., 1956, Substanz P, ein Polypeptid ans Darm und Gehirn mit depressiven, hyperalgetischen und Morphin-antagonistischen Wirkungen auf das Zentralnervensystem, *Arch. Exp. Pathol. Pharmakol.* **228**:513–538.

Zimmerman, B.G., 1967, Evaluation of central and peripheral sympathetic components of action of angiotensin on the sympathetic nervous system, *J. Pharmacol. Exp. Ther.* **158**:1–10.

Zimmerman, B.G., and Whitmore, L., 1967, Effect of angiotensin and phenoxy-benzamine on release of norepinephrine in vessels during sympathetic nerve stimulation, *Int. J. Neuropharmacol.* **6**:27–38.

Zimmerman, B.G., Gomer, S.K., and Liao, J.C., 1972, Action of angiotensin on vascular adrenergic nerve endings, *Fed. Proc. Fed. Amer. Soc. Exp. Biol.* **31**:1344–1350.

Zimmerman, E.A., Hsu, K.C., Robinson, A.G., Carmel, P.W., Frantz, A.G., and Tannenbrum, M., 1973, Studies of neurophysin secreting neurons with immunoperoxidase techniques employing antibody to bovine neurophysin. I. Light microscopic findings in monkey and bovine tissue, *Endocrinology* **92**:931–940.

Zimmerman, E.A., Robinson, A.G., Husain, M.K., Acosta, M., Frantz, A.G., and Sawyer, W.H., 1974, Neurohypophysial peptides in the bovine hypothalamus: Relationship of neurophysin I to oxytocin, and neurophysin II to vasopressin in supraoptic and paraventricular regions, *Endocrinology* **95**:931–936.

12

Electrical Activity of Neurosecretory Terminals and Control of Peptide Hormone Release

Ian M. Cooke

1. Introduction

Nerve cells specialized for the release of peptide hormones to the circulation, i.e., classic neurosecretory cells, retain their full complement of neuronal properties. Their activity is under control of the CNS through excitatory and inhibitory synaptic mediation. They integrate these influences with their own capabilities for endogenous activity to ultimately generate action potentials propagated to the secretory terminals. They represent the "final neuroendocrine pathway" (Knowles, 1974; E. Scharrer, 1965). In this chapter, I wish to examine the relationship between those action potentials that are propagated to the secretory terminals and release of peptide hormones from them. How much of the extensive, detailed knowledge of the mechanisms governing the release of transmitters at synapses (for reviews, see, for example, Katz, 1969; Gerschenfeld, 1973) is applicable to release of peptides from neurosecretory terminals? Are there modifications of the electrical activity of neurosecretory cells, particularly their terminals, that are related to peptide secretion? For example, is the longer duration of action potentials, well documented for the neuron somata, also a feature of neurosecretory axons and terminals, and what is its significance? What is the significance for hormone release of the "spontaneous" activity often recorded from neurosecretory cells, and of firing in bursts or patterned activity?

Ian M. Cooke · Laboratory of Sensory Sciences and Department of Zoology, University of Hawaii, Honolulu, Hawaii.

This chapter will not attempt a comprehensive review of the literature. A number of extensive reviews have appeared recently (e.g., Cross *et al.*, 1975; Hayward, 1975; Finlayson and Osborne, 1975) or are soon to appear (e.g., Berlind, 1976; Bern and Mason, 1976; Normann, 1976). Rather, I will briefly mention a few of the studies that have provided information bearing on mechanisms governing hormone release from terminals, particularly the role of electrical activity in controlling release.* I will then describe experiments on the crab sinus gland neurosecretory system that are ongoing in my laboratory. In reintroducing this classic invertebrate preparation, I hope to show that it offers special advantages as a model system for addressing these questions.

1.1. Correlation of Electrical Activity and Hormone Release

A correlation between electrical activity and directly assayed hormone release has been demonstrated in the vertebrate neurohypophysis (e.g., Mikiten and Douglas, 1965; Ishida, 1968; Dyball and Dyer, 1971; and others cited later), in the insect brain–corpora cardiaca system (e.g., Kater, 1968; Normann and Duve, 1969; Gersch *et al.*, 1970), in *Aplysia* bag cells (Kupfermann, 1970), and in two crustacean systems, the pericardial organs (e.g., Cooke, 1964; Berlind and Cooke, 1970) and the sinus gland (Haylett *et al.*, 1975; see Section 2.3).

Less direct evidence, based on changes observed histologically or with electron microscopy, for hormone release in response to electrical stimulation is available for a number of neurosecretory systems. These systems include the neurohypophysis (for a review, see, for example, Douglas, 1973), fish urophysis (e.g., Fridberg *et al.*, 1966), insect corpora cardiaca (e.g., Hodgson and Geldiay, 1959; B. Scharrer and Kater, 1969; Normann, 1974), and crayfish sinus gland (Bunt and Ashby, 1968).

There are observations in several systems suggesting that modulation of hormone secretion may involve changes in spontaneous rates of firing of neurosecretory cells. Is there evidence for facilitation of peptide secretion by repetitive stimulation analogous to facilitation of transmitter release at synapses (e.g., del Castillo and Katz, 1954; Mallart and Martin, 1967; Gillary and Kennedy, 1969; Gerschenfeld, 1973)? A study of hypothalamic neurosecretory units in suckling rats showed periodic bursts of high frequency firing that were followed by the appearance of oxytocic activity in the blood (Lincoln, 1974; Lincoln and Wakerley, 1975). In isolated rat neurohypophyses, there was previous evidence that significant release of oxytocin required stimulation at repetitive rates greater than a threshold frequency of the order of 20/sec (Ishida, 1970; Dreifuss *et al.*, 1971; see Dreifuss and Ruf, 1972, Fig. 6). There is similar information for rabbits and cats (Harris *et al.*, 1969; Cross, 1974). Hypothalamic cells responsive to osmotic stimuli also frequently show spontaneous activity, sometimes in bursts (Hayward and

---

* Literature references are indicative only; where many have contributed to a point, I have tried to indicate an early and a recent publication. I regret the omission of many important papers.

Jennings, 1973a,b; Dyball and Pountney, 1973). The relationship between the activity of such units and appearance of antidiuretic hormone (ADH, vasopressin) in the circulation is not simple; hormone release continues for too long after cessation of electrical activity (e.g., Dyball, 1971; see also Vincent et al., 1972a,b).

In Aplysia and several species of snails, there are a number of identified neurons, presumed neurosecretory on morphological grounds, that show truly endogenous activity (e.g., Alving, 1968; Gainer, 1972a–c). In some cells, this activity is highly regular, and in cell R_{15}—the "hyperbolic burster"—of the Aplysia abdominal ganglion and its homologous cell in snails, organized into recurring bursts (e.g., Frazier et al., 1967; Kater and Kaneko, 1972; Barker and Gainer, 1975a,b; Barker et al., 1975). Cell R_{15} has its bursting modulated by environmental osmotic changes via the osphradial nerve (Jahan-Parwar et al., 1969; Stinnakre and Tauc, 1969). This cell has recently been shown to contain a water balance hormone (Kupfermann and Weiss, 1976). The bag cell clusters of neurosecretory cells in Aplysia present another case for which there is evidence that secretion of the egg-laying hormone is triggered by periods of high-frequency, repetitive firing (Kupfermann, 1970; Kupfermann and Kandel, 1970).

There are observations of patterned or bursting activity of neurons recorded from a number of other neurosecretory systems, though secretion of hormone has not been correlated with the activity. These systems include leech neurosecretory cells (Yagi et al., 1963), fish urophysis (e.g., Yagi and Bern, 1965), insect corpus cardiacum cells (Normann, 1973) and protocerebral cells (e.g., Wilkens and Mote, 1970; Cook and Milligan, 1972), and X-organ cells of the crayfish sinus gland (Iwasaki and Satow, 1969, 1971). It is a fair summary to say that presumptive neurosecretory cells have been observed to show "spontaneous" activity in every major animal group in which electrical recording has been tried.

1.2. The Calcium Hypothesis

The essential role of calcium for "excitation–secretion coupling" was recognized by Douglas (Douglas and Poisner, 1964a,b; Douglas, 1968; see the review by Rubin, 1974), and a detailed role in synaptic transmitter release was proposed by Katz and Miledi (e.g., 1967a,b). The hypothesis, in brief, is that depolarization of the terminal (presynaptic or neurosecretory) leads to an increased permeability to calcium, and extracellular calcium enters and, on reaching the interior, promotes the release of material. For neurosecretion, considerable evidence has accumulated suggesting release by a process of fusion of granules with the membrane and exocytosis of the contents (see, for example, the reviews by Douglas, 1973; Poisner, 1973; Normann, 1976).

Evidence supporting the applicability of the calcium hypothesis to the mechanisms governing the control of hormone release from neurosecretory terminals has been obtained in all the systems in which it has been examined. Tests of the calcium hypothesis in neurosecretory systems have for the most part involved observing increased rates of hormone release when the terminals are exposed to elevated levels of potassium, this being an effective means of depolarizing nerve membranes, and finding that release fails to be stimulated by high potassium if calcium is absent or reduced. These tests have been successfully applied to isolated neurohypophyses (Douglas and Poisner, 1964a, and a number of subsequent investigators, e.g., Dicker, 1966; Daniel and Lederis, 1967; Ishida, 1967; Dreifuss et al., 1971), to fish urophyses (e.g., Berlind, 1972), to insect diuretic hormone release (Maddrell and Gee, 1974), to the egg-laying hormone of Aplysia bag cells (Arch, 1972), and (to be described below) to crab sinus gland hormones. Effectiveness of calcium withdrawal in blocking electrically elicited release of hormone has been demonstrated for the neurohypophysis (e.g., Mikiten and Douglas, 1965; Ishida, 1968), and for cardioexcitor release from cockroach corpus cardiacum (Gersch et al., 1970) and crab pericardial organs (Berlind and Cooke, 1968, 1971), and is described in Section 2.3.2 for the crab sinus gland system. Only for the neurohypophysis has uptake of calcium during stimulation been demonstrated (e.g., Douglas and Poisner, 1964b; Ishida, 1967).

In summary, there is now abundant evidence that neurosecretory cells retain their neuronal properties, including integrative abilities and the ability to propagate action potentials, and that the release of hormone is in turn controlled by the propagated electrical activity. It is clear that hormone secretion is controlled with great exactitude, but there is a paucity of information about how this electrical activity (which can be patterned) governs the release of hormone from the terminals. Unfortunately, the relationship between electrical and secretory events at neurosecretory terminals is not amenable to the intimate, fine time-resolution analysis that has been accomplished at certain synapses, because there is no closely applied, sensitive bioassay system equivalent to the postsynaptic membrane of a synapse that provides an immediate measure for secretion of exquisite sensitivity and rapid (millisecond) time-course. By contrast, experiments for peptide release generally require repetitive stimulation for a period, collection of perfusate or blood, and subsequent assay. Assay methods are for the most part cumbersome, slow, and frequently of limited quantitative accuracy.

2. The Crustacean X-Organ Sinus Gland Neurosecretory System

The remainder of this chapter will be devoted to reintroducing the crustacean X-organ–sinus gland neurosecretory system as a model preparation for studies of the mechanisms by which electrical activity controls hormone release (as well as for other problems of neurosecretory physiology). It provides a preparation that can be easily isolated and maintained in a viable condition, a neurohemal structure that is discrete and in which there are no complicating nonpeptidergic neurons or other cells present. There is a rapid, convenient, and semiquantitative assay system for released hormonal products. Finally, and perhaps unique to this system, the neurosecretory terminals reach sizes large enough to permit investigation of their physiology with intracellular microelectrodes.

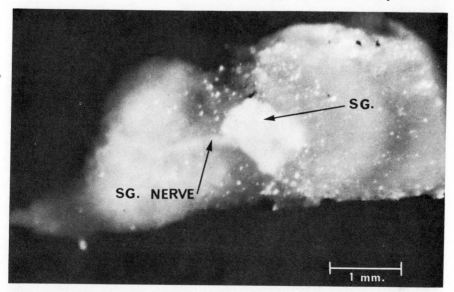

FIG. 1. Crab eyestalk freshly dissected to show the sinus gland and sinus gland nerve. The carapace, muscle, and connective tissue overlying the optic ganglia have been removed. The sinus gland (SG) is at the surface; it is a dense aggregation of neurosecretory terminals. Their axons form the sinus gland nerve. The iridescent white appearance is attributed to refraction from neurosecretory granules. The optic peduncle (proximal) is to the left. For experiments, the sinus gland and sinus gland nerve are dissected and placed in a perfused chamber. *Portunus sanguinolentus.*

2.1. Background

The *sinus gland* of crustaceans (see Fig. 1) was described and so named by Hanström in the 1930's (see the review by Gabe, 1966) because the appearance in histological sections of cellular profiles palisaded against blood sinuses suggested an endocrine function. Hanström also named a cluster of glandlike cells in the eyestalk the X-organ. The recognition of the X-organ and sinus gland as parts of a neurosecretory system (e.g., Bliss and Welsh, 1952; Passano, 1953; Durand, 1956) occurred at about the same time that the neurosecretory nature of the hypothalamic–neurohypophyseal system was gaining acceptance. The same selective staining procedures are effective for these systems. The histology revealed a tract of axons connecting the X-organ with the sinus gland. Thus, the sinus gland represents the bulbous final nerve terminals of neurosecretory cells packed around a blood sinus, i.e., a neurohemal organ. These terminals are derived primarily from the X-organ cells, but also include those of other cell groups in the eyestalk, brain, and even as far distant as the thoracic ganglion (Potter, 1956). Histological (Potter, 1956, 1958; Rehm, 1959) and electron-microscopic studies (Bunt and Ashby, 1967; Andrews et al., 1971) of the sinus gland are consistent in distinguishing five types of terminals by their staining characteristics, and by the inclusion of morphologically distinct neurosecretory granules. There is a literature presenting endocrinological evidence extending back to the 1920's (e.g., Perkins, 1928) that provides adequate numbers of hormonal activities mediated by principles from the sinus gland to make attractive the assignment of a different hormone to each morphological type of terminal. In crabs, such activities include control of two colors of pigmentary effectors, retinal pigment migration, molt inhibition, control of growth, regeneration, gonadal development, water and ionic balance, and blood sugar levels. The hormones producing these effects have been sufficiently characterized to say that the two chromatophorotropins and retinal pigment hormones are low-molecular-weight peptides and are distinct entities; the other effects have been shown to be attributable to at least two different small proteins (for review, see Kleinholz, 1966; Berlind, 1976; Fernlund and Josefsson, 1972). There is good evidence for the storage of the chromatophorotropins in granules (e.g., Pérez-González, 1957). It may be noted that the functional role of the crustacean sinus gland is highly analogous to that of the vertebrate neurohypophysis.

2.2. A Bioassay System

In order to take advantage of this preparation for studies on the mechanisms of hormone release, it was essential to have a convenient technique for assay of hormones appearing in the perfusate. We find that local beach crabs (*Ocypode laevis*, "ghost" crab) provide a satisfactory solution.*

These crabs, when they are just emerging from their burrow, are often a bright pink color, but blanch in less than 2 min. This rate of change is an order of magnitude faster than responses of most hormonally mediated chromatophores. Eyestalkless crabs, or perfused leg segments, become pink due to expansion of erythrophores. Injection of eyestalk extracts causes erythrophore contraction. If the state of contraction of at least five chromatophores is staged (5 = stellate, 1 = punctate), the average change of stage at a given time (e.g., 2 min) is a linear function of the log of the concentration of extract over a range of about 2 log units. Response variation between individual leg segments leads to quantitative uncertainty, if only a single sample can be assayed, of 0.5 log unit. Our standard assay volume is 0.2 ml; the homogenate of a single sinus gland in 0.6 ml can be diluted 10,000-fold and still provide a reliable chromatophore change of a half-stage. The equivalent change is obtained with 1×10^{-10} gm/ml synthetic shrimp red chromatophore-concentrating hormone (Fernlund and Josefsson, 1972).†
The synthetic hormone would thus appear to be about $^1/_{10}$ as effective on chromatophores of this species as on the original species. We have tried extracts of sinus glands of several crab species on the assay system, and have found little difference in the potency per single sinus gland. An important question in considering the data to be presented is the degree to which the assay system is specific for

* Development and quantification of the assay system is mainly the work of Tina Weatherby. The participation of James Kanz in initial work on the assay is acknowledged.
† I am grateful to Dr. Per Fernlund for providing a sample of synthetic hormone.

the red-concentrating hormone. Data in the literature do not offer a clear answer. However, in view of the degree of overlap of effects of various vertebrate neurohypophyseal peptides, it would not be surprising if a number of the hormones present in the sinus gland are capable of producing the response, if in sufficient concentration.

Another convenient characteristic of the ghost crab leg assay system, in addition to its sensitivity and speed of reaction, is that the chromatophores and their ability to respond to sinus gland material are little affected by reducing the calcium in the saline to half-normal, or increasing potassium to 5 times normal, or making both changes at once.

2.3. Correlation of Electrical Responses to Stimulation with Hormone Release

2.3.1. Electrical Stimulation in Normal Saline

The availability of a convenient assay system has facilitated a series of experiments on the release of hormones from the isolated sinus gland–sinus gland nerve preparation of *Cardisoma carnifex*.* This species has been used because it consistently has the largest sinus gland terminals of any species examined. Thus, in parallel experiments, the effect of various experimental procedures on electrical responses of individual terminals can be observed by intracellular recording. The stump of the sinus gland nerve is drawn with saline into a capillary ("suction") electrode, and another capillary electrode (tip diameter, 50–70 μm) is placed on the sinus gland to monitor electrical responses propagated into the terminal area as a result of stimulation. The preparation is continuously perfused† except during experimental periods, usually 5 or 7 min, after which the fluid that remained in the depression with the preparation (total volume about 0.1 ml) is partially removed (so as not to bring the preparation through the fluid–air interface) and two rinses applied and collected. The collected material is made up to a standard volume, usually 0.2 ml, and assayed on an *Ocypode* leg segment. Figure 2 illustrates an experiment to observe the correlation of the amount of hormone or hormones released with the size of compound electrical potentials recorded from the surface of the sinus gland at different stimulus intensities. Stimuli, in all the experiments to be described, were applied in a standard pattern of 5 pulses (0.3 msec duration) at 5/sec, every 10 sec, for a total of 100 or 175 stimuli. This stimulus routine was chosen because it gave optimal repetitive responses as recorded intracellularly from terminals (see Section 2.7). Unstimulated controls were considered satisfactory if on assay the collected material gave less than a 1-stage chromatophore change. "Leaky" preparations often failed to show good electrical responses, and probably were damaged during dissection. As illustrated in Fig. 2, a significant amount of assayable material appears in the fluid when the stimulus intensity is

* The collaboration of Beverley Haylett and Tina Weatherby is gratefully acknowledged.
† The normal saline used had the following composition (mM/1iter): Na, 468; K, 17.6; Ca, 25; Mg, 17; Cl, 552; SO_4, 8.8; HBO_3, 9. It was adjusted to pH 7.4 with NaOH.

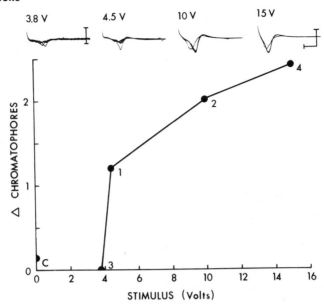

FIG. 2. Sinus gland electrical activity and hormone release in response to stimulation. The electrical responses shown were recorded with a 70μm–pore electrode applied to the surface of the sinus gland in response to electrical stimuli (0.3-msec duration) applied to the sinus gland nerve with a suction electrode. Each frame shows the 5 superimposed responses to the 10th train of stimulus pulses given at 5/train, 5/sec every 10 sec at the voltages shown. Perfusate was collected after 175 stimuli and assayed for red chromatophore–concentrating activity. The assay results (average change of stage of 5 chromatophores, here at 1 min rather than the usual 2 min) are plotted vs. stimulus voltage, and show hormone appearing at the voltage corresponding to threshold for propagated electrical responses. Numbers at graph points indicate the order of the trials; C is the average of the unstimulated controls. This figure and subsequent figures present data from isolated sinus gland–sinus gland nerve preparations of *Cardisoma carnifex* unless otherwise noted. Scales: 2 msec; 3.8 V, 500 μV; others, 2.5 mV.

sufficient to produce an electrical response; as the stimulus intensity is increased, the size of the electrical response increases markedly, and the amount of material released increases in parallel. We conclude that release of hormonal material is brought about by electrical activity propagated by the axons into the neurosecretory terminals.

During the periods between stimulus trains and for a brief time (about 1 min) after a stimulated test period, the extracellular recording electrode detects heightened frequency of spontaneous unit potentials and an increased number of different units observable, as identified by their recurrent, unique waveform (see Fig. 3). The equivalent observation of heightened frequency of spontaneous activity after repetitive stimulation has been made during intracellular recording from sinus gland terminals. The heightened spontaneous activity, however, is not sufficient to clearly account for the frequently observed higher hormone release levels in the unstimulated control period following a period of effective stimulation than before it. It is possible that the hormone enters the blood sinuses present in the structure during stimulation, and is slowly released to the perfusate later. The preparation is routinely left undisturbed for 1 min after completion of the stimulus routine, and another minute is required to collect the perfusate and the several rinses of the chamber. An alternative possibility is that this delayed release represents a process analogous to the increased frequency of spontaneous postsynaptic potentials (min-epps) following repetitive stimulation observed at various synapses (e.g., del Castillo and Katz, 1954; Miledi and Thies, 1971).

We have not yet explored in detail how many periods of such repetitive stimulation can be given before hormone release begins to decline. It is certainly more than the 4 or 5 we have used in our experiments thus far. The isolated preparations survive in a condition in which electrical responses to stimulation and accompanying hormone release continue for up to 2 days without application of any special organ culture techniques beyond slow perfusion with a simple saline (no organic constituents).

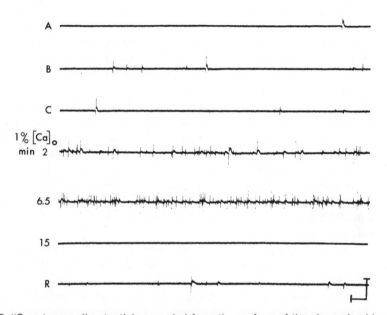

FIG. 3. "Spontaneous" potentials recorded from the surface of the sinus gland by a 70 μm–pore electrode. Rapid biphasic followed by monophasic potentials as in (A) are interpreted as an individual axonal action potential giving rise to a terminal action potential. Records during pauses between stimulus trains show increasing activity following repetitive stimulation: (A) between 1st and 10th trains; (B) between 20th and 30th trains; (C) 30 sec after cessation of stimulation, 35 trains (175 stimuli, 5/train at 5/sec, every 10 sec), shows decline of spontaneous potentials. Spontaneous activity as a result of introducing 0.25 mM calcium saline (1% normal) is shown as sampled at 2 and 6.5 min. Note the lack of slow, monophasic potentials at 6.5 min. Cessation of spontaneous activity (15) generally occurs between 12 and 18 min after exposure begins. Electrically evoked responses are still obtainable (not shown). Spontaneous activity reappears within minutes after returning to normal saline, first as fast, later as fast with slow, monophasic potentials. (R) is taken about 1 hr after return to normal. Scales: 100 msec, 500 μV; each record represents about 2 sec. Same sinus gland, same electrode position throughout. Compare Figs. 6 and 8.

2.3.2. Electrical Stimulation in Reduced-Calcium Salines

We have performed a series of experiments to determine the relationship between the extracellular concentration of calcium and neurally evoked hormone release from isolated sinus gland–sinus gland nerve preparations of *C. carnifex*. In every successfully completed experiment, reducing the calcium resulted in a reduction of hormone release in response to stimulation. An experiment was considered valid only if, on return to normal saline, stimulated levels of hormone release were close to those originally obtained in normal saline and significantly greater than in reduced calcium. All the preparations tested in low calcium salines afterward became more "leaky," i.e., released detectable amounts of hormone in the absence of stimulation. About half the preparations attempted were declared "unsuccessful" because they became too leaky during or after the exposure to low calcium. In Fig. 4, the results of 15 successful experiments in reduced calcium salines are shown plotted as the net chromatophore change (i.e., change in a stimulated less change in the preceding unstimulated period) vs. the log of the saline calcium concentration. It will be seen that the release of hormone is reduced as calcium is lowered. Converting the chromatophore changes to changes in concentration of hormone by comparison with dose–response curves, a 10-fold reduction in extracellular calcium is found to reduce hormone release about 10-fold. At a saline calcium concentration of 1% normal (0.25 mM), hormone release in response to stimulation is, on the average, not detectably greater than unstimulated release in the same period of time. The effects of low calcium on release are not paralleled, in this case, by effects on the propagated electrical activity. Propagated potentials recorded at the sinus gland appear with reduced latency. The amplitude of the compound response in 30% calcium saline is sometimes greater; in 10%, it is somewhat smaller (1 exception); in 1%, it is reduced. The electrical responses in reduced calcium are further discussed in Section 2.6 in connection with the intracellularly recorded observations.

A series of experiments in low-calcium salines has also been completed with the isolated sinus gland preparation from *Portunus sanguinolentus*, with comparable results (Haylett *et al.*, 1975).

These results are consistent with the finding of increased excitability in reduced calcium described in many other preparations, including another crustacean neurosecretory system, the pericardial organs (Berlind and Cooke, 1968, 1971). The failure of hormone release in salines of low extracellular calcium concentration would appear to bring this neurosecretory system into conformity with the calcium hypothesis. This comfortable conclusion, however, may be oversimplified in the light of observations described below that reduced calcium salines alter the electrical responses of terminals.

2.4. Hormone Release in High-Potassium Salines

Also in conformity with the calcium hypothesis and with observations on other neurosecretory systems is our finding of a massive release of chromatophore-active material from the sinus gland when the isolated preparation is placed in

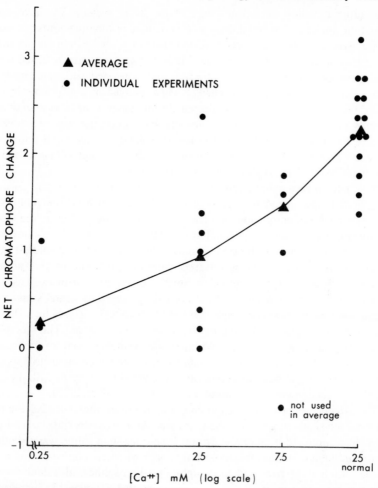

FIG. 4. Electrically elicited hormone release vs. log of saline calcium concentration. All observations in which net release after return to normal was greater than during exposure to low calcium are plotted (15 trials on 14 sinus glands). Ordinate is average chromatophore change produced by perfusate during stimulation (175 pulses in trains, as in Fig. 2) minus that from the preceding unstimulated control. The averaged points show that at lower extracellular calcium concentration, hormone release is reduced. At 1% of normal calcium, stimulated release is indistinguishable from unstimulated, though propagated electrical activity persists.

saline in which the normal (17.6 mM) concentration of potassium has been increased 10 times. This concentration may be calculated to depolarize the membrane potential of the axons and terminals to close to 0; as the high potassium is introduced to the experimental chamber, there is a brief period (about 10 sec) of heightened "spontaneous" unit activity recorded, and then the preparations become electrically silent, with respect to both unit activity and responsiveness to stimulation (as would be expected to result from depolarization). Within 5 min, sufficient material is released under these conditions to require 10- or 30-fold dilution to bring it within the linear range of the assay system. We have not yet explored the amount of material that can be obtained or the time-course of exhaustion; we have, on one occasion, collected material every 5 min for an hour in high-potassium saline, and, while a decline in the amount being released was apparent during the hour, the final collection still contained sufficient material to give a significant chromatophore change when assayed at a $1/30$ dilution of the standard 0.2 ml adjusted volume collected. In this case, as in all (three) other experiments in which response after high-potassium treatment was examined, the release of hormone in response to electrical stimulation an hour after return of the preparation to normal saline was at levels as high as or higher than it had been before application of the high-potassium saline.

It is hard to reconcile the apparent amount of material appearing during exposure of the sinus gland to high-potassium saline with high specificity of the assay system for red chromatophore–concentrating hormone. Rather, the observations add plausibility to the speculation that this assay system is responsive to a number of the hormones of the sinus gland. It would clearly be of interest to learn in detail what materials are being released under the conditions of our various experiments. The question whether other substances are present in extracts, or are released by high potassium, that are not normally released by neural stimulation is an important one to answer, because most of the chemical analyses have depended on extracts or material released by high-potassium salines. The question of whether there are proteins released with peptide hormones from invertebrate peptidergic neurons equivalent to the neurophysines of the vertebrate neurohypophysis has apparently not yet been examined in any invertebrate system with the degree of resolution needed (see, for example, Berlind and Cooke, 1970).

The effects of removing calcium on high potassium–stimulated release of material from the sinus gland preparation was studied in five experiments on 4 preparations. In all, release in normal-calcium, high-potassium saline was significantly higher (minimum of 1 chromatophore stage, approximately corresponding to a 10-fold concentration difference) than in high-potassium saline with calcium omitted. In all but one experiment, however, there was also a noticeable increase over release in nominally calcium-free, normal-potassium controls on exposing the preparation to calcium-free, high-potassium saline. Since these assays were made at 1/10 dilution of the collection perfusate, this increase represents large amounts of released material. In all but one of the preparations, there was detectable (assayed at 1/10 dilution) release of material in the ''0'' calcium saline with normal potassium. Increased ''leakiness'' of preparations in reduced-calcium salines, even 30% of normal ($= 7.5$ mM), has been consistently observed. Attempts to control calcium at low levels by addition of 1 mM EGTA to salines bathing the preparation led to high levels of release of material in the absence of stimulation, and so were abandoned. It may be that a high level of divalent cations in the saline is essential to maintenance of the hormones in stored form.

In terms of giving us confidence in claiming the universal applicability of the calcium hypothesis, it would have been more convenient to find that in the sinus gland preparation, removal of extracellular calcium would block all hormone release, as it did for the neurohypophysis (e.g., Douglas and Poisner, 1964a). As a possible explanation, we might consider the possibility that the blood sinuses present in the sinus gland do not equilibrate rapidly with the perfusate, and in the low-calcium, high-potassium salines, provide a source of calcium that sustains release. But why, then, is this inadequate to sustain release in response to electrical stimulation? These observations again raise the question: how physiological is the release induced by high-potassium salines?

2.5. Intracellularly Recorded Electrical Activity of Neurosecretory Terminals

The number of preparations in which it has been possible to use electrophysiological technology to follow electrical events and characterize the electrical properties of neuronal presynaptic terminals include two: the squid stellate ganglion (e.g., Hagiwara and Tasaki, 1958; Katz and Miledi, 1967a) and the chick ciliary ganglion (e.g., Martin and Pilar, 1964). Hence, it is of great interest to explore the responses and membrane properties of sinus gland neurosecretory terminals (two abstracts have appeared: Cooke, 1967, 1971).

The tropical terrestrial crabs *Cardisoma guanhumi* and *C. carnifex* have the largest sinus gland terminals of any crustacean I have had an opportunity to examine; individual bulbs reach dimensions of 30 μm. The observations to be described refer specifically to these species, although enough stable recordings from terminals of other species of crabs (*Grapsus, Ocypode, Gecarcinus, Podopthalmus, Scylla, Portunus*) have been obtained to have confidence that the observations can be somewhat generalized. The sinus gland and sinus gland nerve were either exposed but left in place on the isolated eyestalk ganglia, or completely isolated. Since individual terminals can be visualized under the dissecting microscope and there is nothing else in the sinus gland large enough to penetrate, there can be little ambiguity about the location of the recording electrode. In early experiments, ferricyanide deposits were made with the electrode and later located in terminals histologically; more recently, terminals have been injected with Procion yellow.

2.5.1. "Spontaneous" and Stimulated Responses Reflect Complex Morphology

Some 138 terminals have been recorded during stable penetrations. Resting potentials have varied from -30 to -80 mV. About half the units exhibited "spontaneous" activity, which usually took the form of overshooting (up to as much as $+30$ mV) action potentials arising from a slowly depolarizing baseline. Action potentials show a rapid rising phase, a slower falling phase, and in some cases, a prominent, hyperpolarizing afterpotential lasting more than 100 msec. The duration (measured at half-amplitude) of these potentials varies from 5 to 20 msec, and is related to the prevailing resting potential (see Section 2.7). It is thus at least 5 times longer than action potentials recorded from axons of the optic peduncle of the same animal. Terminal potentials often appear "spontaneously", usually at slow repetition rates (less than 1/sec, but never faster than 4/sec). Their frequency is not closely correlated with the prevailing membrane potential. However, terminals with resting potentials more negative than -45 mV rarely show spontaneous activity. It is difficult to decide to what extent "spontaneous" activity arises as a result of damage from the penetration. Some terminals hyperpolarize some minutes after penetration, and then cease to show unstimulated potentials. Others maintain a stable resting potential and show continued activity for hours. Simultaneous extracellular recordings from relatively undissected eyestalks with two electrodes detect "spontaneous" activity both in the sinus gland nerve and in the sinus gland. Some of the responses, identifiable by their form, can be seen to be time-locked, the nerve potential preceding the terminal potential. This indicates that not all "spontaneous" activity is due to injury of the terminal.

Terminals sometimes show spontaneous activity that takes two or three distinct waveforms (see Fig. 5). Only one of these forms appears as a full-sized action potential. Evidence that none of the potentials represents interactions between terminals or a postsynaptic potential comes from observing responses to passing current through the recording microelectrode using a bridge circuit. Very small hyperpolarizing currents (often less than 1 nA) suffice to suppress spontaneous activity, while depolarizing current will accelerate it and will evoke all the distinct waveforms (Fig. 5). It is noteworthy that the rate of recurrence of the potentials shows no adaptation during long (minutes) application of current. Preliminary results from examination of Procion-injected terminals indicate that there are a number of dilations interconnected by thin (0.5 μm) processes. Therefore, I suggest that each discrete waveform represents a regenerative response (action potential) occurring in a different dilation or terminal bulb and recorded by electrotonic conduction in the bulb in which the electrode is placed; only when the recorded bulb itself supports an action potential is a full overshooting response observed.

2.5.2. Responses to Sinus Gland Nerve Stimulation

The multiterminal response suggestion is supported by observations obtained during electrical stimulation of the neurosecretory axons by means of an extracellular electrode on the sinus gland nerve. The intraterminal responses show a distinct threshold at which an overshooting action potential appears; there is no change in the waveform with further increase in stimulus intensity, though latency may decrease. If the axon is stimulated repetitively, a fractionation of the terminal response to discrete, fluctuating, smaller responses often occurs. Many terminals show this fractionation at repetition rates higher than 5/sec. The form of the individual potentials is closely comparable with that of those seen to occur "spontaneously" or as a result of depolarizing current passed through the electrode (Fig. 5). Sometimes, this fractionation of the response bears close resemblance to the isolation of axon and soma responses of a vertebrate central neuron on repetitive antidromic stimulation (see, for example, Kandel, 1964, and Hayward, 1974, for goldfish; Yamashita *et al.*, 1970, for cat neuroendocrine cells), and indeed the geometric situation is analogous. Often, one element of the fractionated response will follow high rates of stimulation (20 or more/sec) and show a much briefer

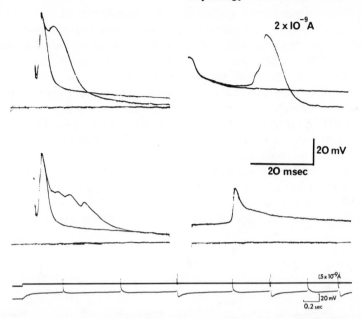

FIG. 5. Responses to depolarizing current and axon stimulation recorded intracellularly from a sinus gland terminal. The moving film record at the bottom shows the onset of a 2-nA depolarizing current passed through the recording electrode: two or more distinct forms of depolarizing responses occur at different repetitive rates. The rates show no accommodation. Initial resting potential about 50 mV, current trace not at zero. The waveforms may be seen in more detail in the fast time base frames above, right, obtained from tape replay of the same recording period. Frames at left show responses of the same terminal to repetitive stimulation of the sinus gland nerve: similar waveforms are seen in the fluctuating, slower responses. The initial, fast response followed high stimulus rates (to 20/sec), and is interpreted as the axonal action potential; it is rarely seen this prominently. Slower waveforms are attributed to responses of different terminal dilations of a single axon as recorded electrotonically from the penetrated terminal. *C. guanhumi.*

time-course (less than 2 msec duration); it is interpreted to represent the electrotonically recorded axonal action potential. These observations suggest that the duration of axonal action potentials is similar to that of "ordinary" axons. Extracellular recordings (see Fig. 3) with a pore (about 70 μm diameter) applied to the surface of the sinus gland show spontaneously recurring unitary potentials with a rapid initial biphasic deflection followed by a much slower monophasic deflection. The time-courses are consistent with interpretation of such potentials as an axonal action potential of ordinary time-course followed by the much slower terminal action potential.

2.6. The Ionic Bases of Terminal and Axonal Potentials

Experiments to examine the ionic basis of the sinus gland responses indicate further differences in the characteristics of axonal and terminal electrical activity. In brief, perfusion of the isolated sinus gland–sinus gland nerve preparation with saline having half the normal sodium (Tris substituted; Fig. 6), or with tetrodo-toxin (TTX, 10^{-7} M; Fig. 7), leads to abrupt failure of terminal responses elicited by stimulation of the axon. This proves to be a failure of axonal conduction, because regenerative responses occur to depolarizing current passed through the recording electrode. The threshold and rise time are increased, and overshoot is reduced (use of a bridge makes absolute membrane potentials somewhat uncertain during current-passing). Regenerative responses have persisted in 5% normal so-dium, and in TTX up to 10^{-6} M. These observations are very similar to those ob-tained in studies on a number of neurons in which the experimental evidence led to the conclusion that both sodium and calcium ions contributed to carrying regen-erative inward current. Of particular interest are the very similar changes in the presence of half-normal sodium or TTX observed by Iwasaki and Satow (1971) in the neurosecretory somata of the crayfish X-organ.

Intracellular terminal responses during perfusion with salines having reduced

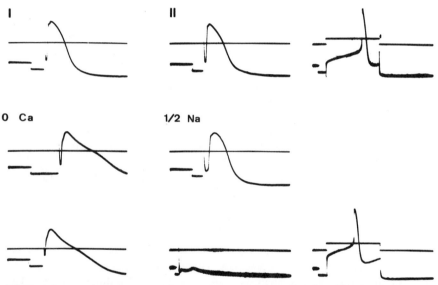

FIG. 6. Changes in responses of a sinus gland terminal during perfusion with calcium-free saline and with half-sodium saline. (I) Top: normal; middle and bottom: after 8 min perfusion with nominally calcium-free saline. In the middle record, the resting potential is held at approximately normal value (about −57 mV) by hyperpolarizing current through the electrode. In the bottom trace, the resting potential, unaltered, is about −35 mV. The falling phase is markedly slowed and overshoot is reduced (from 10 to 6 mV); hyperpolarization did not reverse the effects of reduced calcium. Responses to stimula-tion were not blocked. Compare Figs. 8 and 9. (II) Top: responses of same terminal 13 min after return to normal saline; middle: 6 min; bottom: 8 min perfusion with half-sodium, Tris-substituted saline. Response initiated by axonal stimulation fails abruptly before major reduction in response amplitude in the terminal. Response persists to depolarizing current (lower-right frame); depolarization to reach threshold has increased (from about 20 to 30 mV), and spike size has decreased (threshold to peak from about 60 to 50 mV). The effects were rapidly reversed following commencement of normal perfusion (not shown). Current monitor at −20 mV; current passed, about 1 nA; commencement of trace to downward deflection marks 10 msec; deflection is 10 mV. *C. guanhumi.*

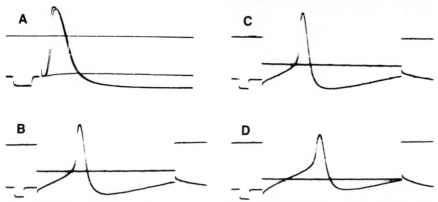

FIG. 7. Effects of tetrodotoxin on sinus gland terminal responses. (A) Responses to superthreshold stimulation of the sinus gland nerve, 3 superimposed sweeps: The 1st, fastest, response was recorded before application of TTX; the 2nd, 4 min 10 sec after switching perfusion to 10^{-7} M TTX; the 3rd, in which the response has failed, 10 sec later. There has been little change in terminal response at the time of conduction failure in the axon. (B) Response to intracellular depolarizing current, 1 nA, recorded immediately after last sweep of (A) (note slower time base). (C) Response to current before application of TTX for comparison with: (D) after 12 min in TTX. Current has been increased to 1.3 nA. There has been an increase in threshold, decrease in overshoot, and increase in duration, but responses did not fail. Current monitor trace marks 0 potential. Calibrations: 10 msec, 10 mV.

or nominally zero calcium have been studied. As the low-calcium saline begins to reach the experimental chamber, there is a noticeable decrease in threshold, and usually commencement or increase in spontaneous firing (Fig. 8). There is usually some, and in some cases a substantial (about 20 mV), decrease in resting potential. As exposure to the low calcium continues, there may be fractionation of the "spontaneous" potentials into several distinct waveforms (as seen during repetitive stimulation); the potentials become progressively longer in duration and smaller in amplitude. Responses to sinus gland nerve stimulation or to intracellular depolarizing currents show parallel changes. In 30% and 10% calcium saline, spontaneous firing and the lengthened, reduced responses to stimulation may continue; in some cases, they fail. In 1% or nominally zero-calcium salines, spontaneous activity usually ceases within 15 min of the first noticeable effects, often leaving a slightly oscillating, depolarized membrane potential in evidence. Previous to cessation, if the membrane potential is artificially repolarized to the normal resting potential, spontaneous activity slows, and the potentials return to a faster and larger form (see below). Even in terminals that do not depolarize significantly, there is increased spontaneity and decreased threshold, and the form of responses becomes more rounded, slower, and smaller. Hyperpolarization will partially reverse these changes. In many cases, when a terminal has become quiescent, sinus gland nerve stimulation elicits a small potential that can be interpreted as the electrotonically recorded axonal action potential. This interpretation is supported by the persistence of conducted potentials recorded extracellularly from the sinus gland nerve.

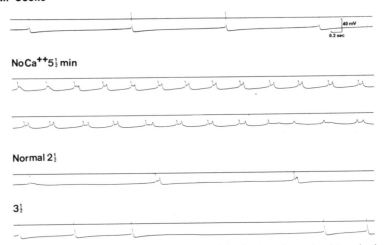

FIG. 8. Intracellular recording of spontaneous activity from a sinus gland terminal during perfusion with nominally calcium-free saline and recovery. Each trace represents about 7 sec. In the top trace, "spontaneous" potentials occur at about 0.5/sec; note prominent overshoot and afterhyperpolarization. The next two traces (No Ca⁺⁺ 5½ min) are continuous; the responses have increased in frequency to about 2/sec, and are fractionated into two or three distinct waveforms. These waveforms begin to fail, and shortly after the records shown, activity ceased. Response to stimulation was also blocked (not shown). During recovery, here sampled at 2½ and 3½ min, the sequence of changes recurs in reverse order. Note that there has not been a change in resting potential. Compare Figs. 6 and 9. *C. guanhumi.*

Extracellular recordings with a pore electrode on the sinus gland during exposure of an isolated preparation to reduced calcium salines (see Fig. 3) show increasing rates of "spontaneous" unit potentials. The level of activity reaches a maximum after 6–8 min, and then often suddenly ceases. A normal or slightly diminished compound propagated response to stimulation of the sinus gland nerve can still be evoked. Close examination of certain unit responses characterized by a rapid initial biphasic deflection followed by a slower monophasic potential indicates that the slow potential is progressively reduced in amplitude as low-calcium exposure continues, and often fails some minutes before general quiescence occurs. In view of observations with intracellular recording, the fast potential is interpreted to represent the axonal action potential, and the slow, the terminal response.

In 3 of 8 preparations tested in "0" calcium saline, terminals have continued to show large, overshooting action potentials (see Fig. 6). In these terminals, it is the falling phase of the action potential that is most markedly altered: the overshooting peak of the potential is much sharper; the initial fall, just after the peak, is more rapid; when the potential has returned to a little below the zero potential line, there is a slowing of rate; hyperpolarizing afterpotentials are not evident.

In some cases, the effect of changing the resting potential on the form of action potentials was examined during perfusion with reduced-calcium salines (see Fig. 9). The increase in duration and reduction of afterhyperpolarization produced by low calcium could be reversed by hyperpolarizing. A lengthening of the action

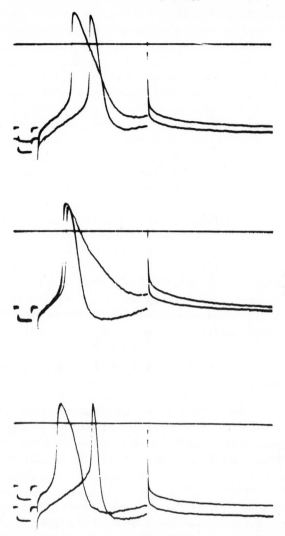

FIG. 9. Changes in sinus gland terminal responses in reduced calcium saline: interaction with membrane potential. Each frame superimposes two responses to depolarizing current (1 nA) through the electrode. Top: Responses after 8 min perfusion with 10% calcium saline (2.5 mM). First spike with long falling phase, at unaltered resting potential (about −55 mV); there is "spontaneous" firing at about 2/sec (not shown). The 2nd response occurs during hyperpolarization by about 10 mV to normal resting potential; spontaneous activity is suppressed. Note decrease in spike duration. Middle: First, long-duration spike, recorded after 10 min in 10% calcium saline; resting potential about −50 mV, spontaneous firing about 3/sec. Second trace, taken as normal saline reenters; resting potential is beginning to recover, spontaneous activity at about 1/sec. Bottom: After 20 min in normal saline; 1st, long-duration spike is response while holding resting membrane potential at about −40 mV; this produced spontaneous firing at about 1/sec. Note continued afterhyperpolarization. Second trace, response at normal resting potential (−60 mV), no spontaneous firing. Compare Figs. 6 and 11. Calibration: 10 msec, 10 mV.

potential in normal saline results from depolarization, although this treatment does not eliminate the hyperpolarizing afterpotential. The changes in the action potential (in low calcium) are interpretable as shifts in the voltage-dependent conductance changes equivalent to maintained depolarization, as described from voltage clamp experiments on squid axons (Fraenkenhauser and Hodgkin, 1957). They are also consistent with a contribution of calcium-inward current to the action potentials, as described in molluscan neurons (e.g., Geduldig and Junge, 1968; Standen, 1975a,b). The reduction of afterhyperpolarization observed in low calcium may also imply an effect of that calcium that reaches the interior of the terminal in activation of potassium permeability increase (Meech and Standen, 1975).

It should be emphasized that the changes described above occur rapidly (within 3–8 min). An experiment was considered valid only if, on return to normal saline, the unit's response returned to the pre-low-calcium form.

At this point, the studies of intracellular terminal responses in salines of altered ionic composition or containing TTX lead to the conclusion that terminals support regenerative depolarizing potentials in which the depolarizing current represents inward movement of both sodium and calcium ions. The neurosecretory axons, by contrast, have an action potential dominated by a classic sodium-mediated depolarizing current. Iwasaki and Satow (1971) also concluded that crayfish sinus gland axons showed sodium action potentials, though the action potentials of the X-organ somata were dual sodium–calcium spikes.

2.7. Changes of Terminal Responses During Repetitive Stimulation

A number of neurosecretory systems, including X-organ cells of crayfish (Iwasaki and Satow, 1969), provide examples in which the activity of neurosecretory elements is patterned in bursts or periods of high-frequency activity. Data on the effect of stimulus rate and pattern on hormone release from the isolated sinus gland system are not yet in hand, but observations on the intracellularly recorded responses of the terminals lead to the prediction that grouped stimuli will release more hormone than an equal number of stimuli uniformly spaced. This prediction is based on the observation that the duration of the terminal action potentials increases as a result of repetitive stimulation. In the experiment from which Fig. 10 is taken, for example, some change was visible at rates of 0.5/sec (not shown), and a maximal lengthening of about 36% occurs after 5 stimuli at a rate of 2/sec. At 4/sec, the same maximal duration is attained, and requires 10 stimuli to reach maximum. Longer trains, if the frequency is much higher than 4/sec, show progressive slowing of the rising phase of the responses and reduction of the overshoot.

In other experiments, I found that the duration of the action potential was highly sensitive to the prevailing membrane potential (see Fig. 11). Holding a terminal hyperpolarized decreases the duration mainly by increasing the rate of repolarization, but in some cases, the rate of rise and the amount of overshoot are also increased (hyperpolarization can suppress the response of the terminal to

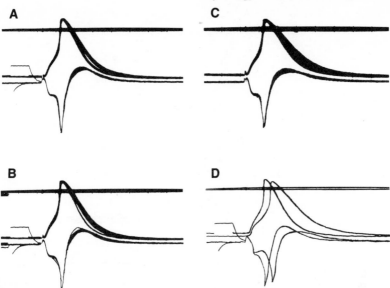

FIG. 10. Changes of terminal action potential duration with repetitive stimulation. Each frame shows superimposed responses of a terminal to stimulation of the sinus gland nerve (middle baselines) and the simultaneous passive differentiation of the response (retouched, time constant, 2.5 msec). (A) 5 stimuli at 1/sec. (B) 5 stimuli at 2/sec. (C) train of 10 stimuli at 2/sec, 1st response not recorded. (D) 1st and 10th response to train of 10 stimuli at 4/sec. Note increase in duration from minimum of about 11 to 18 msec (at half amplitude); the increase for any single train is about 36%. Differentiation shows little change in rate of rise except at 4/sec. Calibration: 10 msec, 10 mV. Compare Fig. 11.

stimulation of the sinus gland nerve, leaving a small response attributable to the axonal action potential). Depolarization of the terminal will slow the falling phase of the action potential, but also reduce rise time and overshoot. However, the degree of experimental manipulation possible is limited by the tendency of the depolarizing current to give rise to "spontaneous" action potentials. Changes in the duration of the action potential following an abruptly imposed change in the resting potential require some tens of seconds to half a minute to reach their full, maintained extent. The changes are reversed along a similar time course on releasing the imposed hyperpolarization or depolarization. Thus, it seems that the series of depolarizations represented by a train of action potentials has a cumulative effect similar to an artificially imposed small, maintained depolarization.

In speculating on the significance of the increases in duration of terminal action potentials with repetitive stimulation for the release of peptide hormones, we have recourse to the only preparation for which precise information exists relating electrical events at the terminal to secretion: the squid stellate ganglion synapse. There, the size of the postsynaptic response, which is taken as a measure (though nonlinear) of the amount of transmitter released, is sharply (10-fold for 7.7 mV) dependent on the size of the presynaptic depolarization, once the latter has exceeded a threshold level (Katz and Miledi, 1967a). Postsynaptic response is also dependent on the concentration of extracellular calcium (log release vs. log extra-

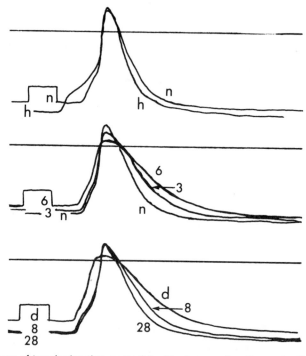

FIG. 11. Change of terminal action potential with change of resting potential. Responses to sinus gland nerve stimulation at normal and at altered resting potential have been traced to superimpose the peaks of the responses. Top: At normal (n) and (h) after being held about 6 mV hyperpolarized for 28 sec by passing 0.5 nA through the recording electrode. Note reduced fall time. Middle: At normal (n) and depolarized resting potentials (3) held about 3 mV depolarized for 30 sec by passing 0.2 nA and (6) held depolarized for 50 sec at 3 mV, 10 sec at 6 mV (0.5 nA). Spontaneous action potentials occurred at an average frequency of about 0.5/sec during depolarization. Note reduced overshoot and increased fall time. Bottom: Recovery from depolarization (d) is trace (6) of middle; (8) 8 sec and (28) 28 sec after resting potential was returned to normal (−48 mV). Note rapid recovery of overshoot, slower recovery in rate of fall. Calibrations: 10 msec, 10 mV. Compare Fig. 9, same terminal.

cellular calcium, about 2.5; Katz and Miledi, 1970), and finally on the duration of the depolarization. Uncertainty about the distribution of potential along the terminal and its time course caused Katz and Miledi to refrain from quantifying the relation. Their figure (1967a, Fig. 13) would suggest a 3-fold increase as a result of a 30% increase in the half-amplitude duration of a presynaptic depolarization of rounded form. While considering the squid preparation, it is interesting to note that these presynaptic terminals are found to be capable of regenerative calcium-mediated responses after treatment with TTX (and tetraethylammonium; Katz and Miledi, 1969a). These responses are limited to the presynaptic membrane area, and are not present in the preterminal axon, or in the postsynaptic structure. There is also circumstantial evidence for such responses at frog motor axon terminals (Katz and Miledi, 1969b).

In the light of the observations on the stellate ganglion synapse, the selective increase in duration of action potentials observed during repetitive stimulation provides a plausible mechanism, at membrane level, for facilitation of hormone release during higher frequency stimulation. For that matter, it may offer a new hypothesis for the explanation of facilitation at synapses during repetitive stimulation; no unified theory has yet emerged in explanation of this phenomenon. The sinus gland preparation provides a suitable system for correlating the effects of different stimulus patterns on terminal responses with those on hormone release.

It would be of great interest to know to what extent the detailed mechanisms by which the interaction of membrane potential changes and calcium movements that control transmitter release at a synapse are also applicable to peptide hormone release from neurosecretory terminals. Unfortunately, there is little prospect for impulse-by-impulse assay of release from individual terminals; mechanisms will have to be deduced by the combinations of experiments of the kind illustrated in this chapter: correlating the changes in responses of individual terminals to altered experimental conditions or routines, and following the effects on release from the neurohemal organ as a whole caused by the same experimental manipulations.

3. General Conclusions

The work on the X-organ–sinus gland system allows it to be added to the list of neurosecretory systems for which the direct correlation of hormone release with electrical activity propagated to the terminals by the axons of the neurosecretory cells themselves has been established. An essential role of calcium in linking the stimulus, i.e., action potentials, to hormone release has been demonstrated in this system. In this system, however, that role is seen to include major participation in the electrical response of the terminal. It seems reasonable to suggest that the long duration and major participation of calcium in the terminal depolarizing responses (action potentials) is adaptive for the admission of extracellular calcium to the terminal for its subsequent promotion of release of neurosecretory granule material.

It could be suggested that a dual sodium–calcium spiking mechanism is the more general and primitive form of regenerative electrical response, and the sodium-dominated action potential of axons a specialization for rapid conduction. There is increasing evidence for the existence of a component of membrane potential–dependent calcium permeability change and calcium current in action potentials of axons (for a review, see Reuter, 1973; see also, for example, Baker et al., 1971; Meves and Vogel, 1973), as well as the somata of neurons of several diverse major animal groups (e.g., annelids: Kleinhaus and Prichard, 1975; gastropod mollusks: Geduldig and Junge, 1968; Wald, 1972; Barker and Gainer, 1975a; Standen, 1975a,b; crustaceans: Iwasaki and Satow, 1971; amphibians: Koketsu and Nishi, 1969; Barrett and Barrett, 1976). The X-organ cells of crayfish are not the only ones shown to have different soma and axon spike mechanisms with the soma showing dual sodium–calcium-mediated spikes and the axon dominantly sodium-mediated spikes; the equivalent situation has been documented for molluscan neurons (e.g., Wald, 1972; Kado, 1973; Standen, 1975a), and would be anticipated in the amphibian neurons. There is evidence for axonal action potentials having a dominantly sodium-mediated mechanism in several neurosecretory systems based on failure of electrical conduction following TTX treatment. These systems include the vertebrate neurohypophysis (e.g., Dreifuss et al., 1971), the crab pericardial organs (Berlind and Cooke, 1971), and axons of the sinus gland nerve (Iwasaki and Satow, 1971; Cooke, 1971; this chapter).

Most intriguing is the question of how general the involvement of calcium-inward currents in active electrical responses of terminals will turn out to be. There is evidence for it in all the three widely different presynaptic terminals in which it has been sought: the squid stellate ganglion giant synapse (Katz and Miledi, 1969a; Llinás et al., 1972), frog nerve-muscle junctions (Katz and Miledi, 1969b), and the sinus gland neurosecretory terminals. The ability to elicit release from rat neurohypophyses in sodium-free, calcium sulfate saline by 1-msec electrical pulses applied directly to the terminal region suggests that they may be capable of calcium-mediated spikes (Douglas and Sorimachi, 1971). A report (Zeballos et al., 1975) on extracellular recording with small (2 μm) electrodes from cat neurohypophysis reveals slower potentials at the periphery (where terminals are dense) and faster potentials more centrally where axons pass. It is tempting to speculate on the basis of these observations that mammalian terminals will prove to have long, sodium–calcium-mediated spikes, while the axons have normal, sodium-mediated action potentials, as found in the crab sinus gland–sinus gland nerve.

Will it turn out to be a general principle that the membrane characteristics of the soma reflect those of the neuronal terminals, as they appear to do in the cells of the X-organ–sinus gland system? It is noteworthy that action potentials involving a major component of calcium-inward current are generally of longer duration than those dominated by a sodium-mediated spike, even in the case of a soma and axon of the same neuron. Whether the long-duration action potentials so often associated with neurosecretory cells reflect an action potential involving prominent calcium current remains to be examined. It seems clearly adaptive for control of hormone release that propagated action potentials be a means of setting off major inward calcium movement in view of the demonstrated role of calcium in stimulus–secretion coupling.

Peptidergic neurons, in all cases in which their electrical activity has been examined, have proved to have the full complement of neuronal activities. They receive inhibitory and excitatory synaptic input, initiate action potentials, and release hormone as a result of these electrical activities. Their sites of synaptic interaction lie within, and their activities appear to be controlled by, the CNS. There is as yet no compelling evidence that peptidergic neurons function as their own receptor. This role has been suggested in the case of osmosensitive neurosecretory cells in the vertebrate hypothalamus (for a review, see Cross et al., 1975) and in the brain neurosecretory cells of insects the activities of which are related to photoperiod. Peptide secretion takes place at discrete axonal terminals that are often organized into well-ordered neurohemal structures. There has so far been no evidence that secretion can take place from the soma or axon; the possibility has been suggested but not tested. Except in the case of the vertebrate median eminence, there is no evidence to support the suggestion that peptide secretion is controlled by other neurons accompanying them. In the case of release of oxytocin and vasopressin by the posterior pituitary (e.g., Douglas and Poisner, 1964a; Daniel and Lederis, 1967), and of the cardioaccelerator peptide from the crab pericardial organ (Berlind et al., 1970), the possibility has been specifically examined and ruled out. The question whether peptide secretion occurs spontaneously from terminals in the absence of commanding action potentials is hard to resolve unequivocally. The amounts to be detected are sufficiently low, and the available bioassays sufficiently cumbersome, to make such measurements difficult. In so many respects the physiology of peptide secretion appears to closely parallel the release of neurotransmitter at synapses that some equivalent to spontaneous quantal release might be anticipated. It is hard to imagine that such spontaneous release would reach physiologically significant levels. If this were the case, control of secretion would require modulation of the synthesis or transport of peptides from the neurons. In view of the dramatic effect of hypertonic saline in increasing the frequency of miniature endplate potentials at frog neuromuscular junctions (Fatt and Katz, 1952), it would seem at least conceivable that the release of antidiuretic hormones could result from a direct effect of increased osmotic pressure of the circulating fluid on peptidergic terminals. This effect might account for some of the lack of direct correspondence between electrical activity and the amount of oxytocin and vasopressin release in response to hypertonic saline stimuli (e.g., Dyball, 1971; Dyball and Dyer, 1971). However, there do occur centrally mediated responses of hypothalamic neurons to osmotic changes too small to be expected to initiate such hormone release directly from terminals. By contrast with lack of evidence for any of these other mechanisms, there is a great deal of evidence that, in a number of peptidergic neuronal systems, propagated electrical activity reaching the terminals signals the secretion of hormone from them. As at ordinary synapses, calcium appears to form the necessary link coupling excitation to secretion. Both from the evidence at the synapse and from the evidence for a vital role of extracellular calcium on peptide secretion, adaptation of terminal membrane for a long duration of electrical depolarization and for major participation of calcium in the terminal responses would seem highly relevant modifications for peptide secretion of mechanisms generally present in neurons.

ACKNOWLEDGMENTS

Beverley Haylett and Tina Weatherby were collaborating investigators in much of the original work here described. I am grateful to Martha W. Goldstone and Beverley Haylett for unfailing technical assistance; to Dr. Martin Vitousek for collecting and transporting C. carnifex; to Jon Hayashi for help in reproduction of the figures; and to Carol Kosaki, Sally S. Oshiro, and Brenda Lum for preparation of the manuscript. Research in the author's laboratory was supported in part by National Science Foundation grant BMS72-02421A1, by a grant from the University of Hawaii Research Council, and by University of Hawaii Foundation grants to the Laboratory of Sensory Sciences.

4. References

Alving, B. O., 1968, Spontaneous activity in isolated somata of *Aplysia* pacemaker neurons, *J. Gen. Physiol.* **45:**29–45.

Andrews, P.M., Copeland, D.E., and Fingerman, M., 1971, Ultrastructural study of the neurosecretory granules in the sinus gland of the blue crab *Callinectes sapidus, Z. Zellforsch.* **113:**461–471.

Arch, S., 1972, Polypeptide secretion from the isolated parietovisceral ganglion of *Aplysia californica, J. Gen. Physiol.* **59:**47–59.

Baker, P.F., Hodgkin, A.L., and Ridgeway, E.B., 1971, Depolarization and calcium entry in squid giant axons, *J. Physiol. London* **218:**709–755.

Barker, J.L., and Gainer, H., 1975*a,* Studies on bursting pacemaker potential activity in molluscan neurons. I. Membrane properties and ionic contributions, *Brain Res.* **84:**461–477.

Barker, J.L., and Gainer, H., 1975*b,* Studies on bursting pacemaker potential activity in molluscan neurons. II. Regulation by divalent cations, *Brain Res.* **84:**479–500.

Barker, J.L., Ifshin, M.S., and Gainer, H., 1975, Studies on bursting pacemaker potential activity in molluscan neurons. III. Effects of hormones, *Brain Res.* **84:**501–513.

Barrett, E., and Barrett, J.N., 1976, Separation of two voltage-sensitive potassium currents, and demonstration of a tetrodotoxin-resistant calcium current in frog motoneurones, *J. Physiol.* **255:**737–774.

Berlind, A., 1972, Teleost caudal neurosecretory system; release of urotensin II from isolated urophyses, *Gen. Comp. Endocrinol.* **18:**557–560.

Berlind, A., 1976, Cellular dynamics in invertebrate neurosecretory systems, *Int. Rev. Cytol.* in press.

Berlind, A., and Cooke, I.M., 1968, Effect of calcium omission on neurosecretion and electrical activity of crab pericardial organs, *Gen. Comp. Endocrinol.* **11:**458–463.

Berlind, A., and Cooke, I.M., 1970, Release of a neurosecretory hormone as peptide by electrical stimulation of crab pericardial organs, *J. Exp. Biol.* **53:**679–686.

Berlind, A., and Cooke, I.M., 1971, The role of divalent cations in electrically elicited release of neurohormone from crab pericardial organs, *Gen Comp. Endocrinol.* **17:**60–72.

Berlind, A., Cooke, I.M., and Goldstone, M.W., 1970, Do the monoamines in crab pericardial organs play a role in peptide secretion? *J. Exp. Biol.* **53:**669–677.

Bern, H., and Mason, C., 1976, Cellular biology of the neurosecretory neuron, in: *Handbook of Physiology,* American Physiology Society, Washington, D.C., in press.

Bliss, D.E., and Welsh, J.H., 1952, The neurosecretory system of brachyuran Crustacea, *Biol. Bull.* **103:**157–169.

Bunt, A.H., and Ashby, E.A., 1967, Ultrastructure of the sinus gland of the crayfish *Procambarus clarkii, Gen. Comp. Physiol.* **9:**334–342.

Bunt, A., and Ashby, E., 1968, Ultrastructural changes in the crayfish sinus gland following electrical stimulation, *Gen. Comp. Endocrinol.* **10:**376–382.

Cook, D.J., and Milligan, J.V., 1972, Electrophysiology and histology of the medial neurosecretory cells in adult male cockroaches, *Periplaneta americana, J. Insect. Physiol.* **18:**1197–1214.

Cooke, I.M., 1964, Electrical activity and release of neurosecretory material in crab pericardial organs, *Comp. Biochem. Physiol.* **13:**353–366.

Cooke, I.M., 1967, Potentials recorded intracellularly from neurosecretory terminals, *Amer. Zool.* **7:**732–733.

Cooke, I.M., 1971, Calcium dependent depolarizing responses recorded from crab neurosecretory terminals, *Proc. Int. Union Physiol. Sci.* **9:**119.

Cross, B.A., 1974, The neurosecretory impulse, in: *Neurosecretion—The Final Neuroendocrine Pathway* (F. Knowles and L. Vollrath, eds.), pp. 115–128, Springer-Verlag, New York.

Cross, B.A., Dyball, R.E.J., Dyer, R.G., Jones, C.W., Lincoln, D.W., Morris, J.F., and Pickering, B.T., 1975, Endocrine neurons, *Recent Prog. Horm. Res.* **31:**243–286.

Daniel, A.R., and Lederis, K., 1967, Release of neurohypophysial hormones *in vitro, J. Physiol.* **190:**171–187.

del Castillo, J., and Katz, B., 1954, Statistical factors involved in neuromuscular facilitation and depression, *J. Physiol.* **124:**574–585.

Dicker, S.E., 1966, Release of vasopressin and oxytocin from isolated pituitary glands of adult and new-born rats, *J. Physiol.* **185:**429–444.

Douglas, W.W., 1968, Stimulus–secretion coupling: The concept and clues from chromaffin and other cells, *Br. J. Pharmacol.* **34:**451–474.

Douglas, W.W., 1973, How do neurons secrete peptides? Exocytosis and its consequences, including "synaptic vesicle" formation, in the hypothalamo–neurohypophyseal system, *Recent Prog. Brain Res.* **39:**21–38.

Douglas, W.W., and Poisner, A.M., 1964*a,* Stimulus–secretion coupling in a neurosecretory organ and the role of calcium in the release of vasopressin from the neurohypophysis, *J. Physiol.* **172:**1–18.

Douglas, W.W., and Poisner, A.M., 1964*b,* Calcium movement in the neurohypophysis of the rat and its relation to the release of vasopressin, *J. Physiol.* **172:**19–30.

Douglas, W.W., and Sorimachi, M., 1971, Electrically evoked release of vasopressin from isolated neurohypophyses in sodium-free media, *Br. J. Pharmacol.* **42:**647P.

Dreifuss, J.J., and Ruf, K.B., 1972, A transpharyngeal approach to the rat hypothalamus, in: *Experiments in Physiology and Biochemistry* (G. Kerkut, ed.), pp. 213–228, Academic, London.

Dreifuss, J., Kalnins, I., Kelly, J.S., and Ruf, K.B., 1971, Action potentials and release of neurohypophysial hormones *in vitro, J. Physiol.* **215:**805–817.

Durand, J.B., 1956, Neurosecretory cell types and their secretory activity in the crayfish, *Biol. Bull.* **111:**62–76.

Dyball, R.E., 1971, Oxytocin and ADH secretion in relation to electrical activity in antidromically identified supraoptic and paraventricular units, *J. Physiol.* **214:**245–256.

Dyball, R.E., and Dyer, R.G., 1971, Plasma oxytocin concentration and paraventricular neurone activity in rats with diencephalic islands and intact brains, *J. Physiol.* **216:**227–235.

Dyball, R.E., and Pountney, P.S., 1973, Discharge patterns of supraoptic and paraventricular neurones in rats given a 2 per cent NaCl solution instead of drinking water, *J. Endocrinol.* **56:**91–98.

Fatt, P., and Katz, B., 1952, Spontaneous subthreshold activity at motor nerve endings, *J. Physiol.* **117:**109–128.

Fernlund, P., and Josefsson, L., 1972, Crustacean color-change hormone; amino acid sequence and chemical synthesis, *Science* **177:**173–175.

Finlayson, L.H., and Osborne, M.P., 1975, Secretory activity of neurons and related electrical activity, in: *Adv. Comp. Physiol. Biochem.* **6:**165–258 (O. Lowenstein, ed.), Academic Press, New York.

Fraenkenhauser, B., and Hodgkin, A.L., 1957, The action of calcium on the electrical properties of squid axons, *J. Physiol.* **137:**218–244.

Frazier, W.T., Kandel, E.R., Kupfermann, I., Waziri, R., and Coggeshall, R.E., 1967, Morphological and functional properties of identified neurons in the abdominal ganglion of *Aplysia californica, J. Neurophysiol.* **30:**1288–1351.

Fridberg, G., Iwasaki, S., Yagi, K., Bern, H., Wilson, D.M., and Nishioka, R., 1966, Relation of impulse conduction to electrically induced release of neurosecretory material from the urophysis of the teleost fish *Tilapia mossambica, J. Exp. Zool.* **161:**137–150.

Gabe, M., 1966, *Neurosecretion,* Pergamon Press, Oxford, 872 pp.

Gainer, H., 1972*a,* Patterns of protein synthesis in individual, identified, molluscan neurons, *Brain Res.* **39:**369–386.

Gainer, H., 1972*b,* Effects of experimentally induced diapause on the electrophysiology and protein synthesis of identified molluscan neurons, *Brain Res.* **39:**387–402.

Gainer, H., 1972*c,* Electrophysiological behavior of an endogenously active neurosecretory cell, *Brain Res.* **39:**403–418.

Geduldig, D., and Junge, D., 1968, Sodium and calcium components of action potentials in *Aplysia* giant neurone, *J. Physiol.* **199**:347–365.

Gersch, M., Richter, K., Böhm, G.-A., and Stürzebecher, J., 1970, Selektive Ausschüttung von Neurohormonen nach elektrischer Reizung der Corpora Cardiaca von *Periplaneta americana in vitro*, *J. Insect Physiol.* **16**:1991–2013.

Gerschenfeld, H.M., 1973, Chemical transmission in invertebrate nervous systems and neuromuscular junctions, *Physiol. Rev.* **53**:1–119.

Gillary, H.L., and Kennedy, D., 1969, Neuromuscular effects of impulse pattern in a crustacean motor neuron, *J. Neurophysiol.* **32**:607–612.

Hagiwara, S., and Tasaki, I., 1958, A study of the mechanism of impulse transmission across the giant synapse of the squid, *J. Physiol.* **143**:114–137.

Harris, G.W., Manabe, Y., and Ruf, K.B., 1969, A study of the parameters of electrical stimulation of unmyelinated fibres in the pituitary stalk, *J. Physiol.* **203**:67–81.

Haylett, B.A., Weatherby, T.M., and Cooke, I.M., 1975, Electrically elicited release of neurohormone from isolated crab sinus gland and its dependence on Ca^{++}, *Physiologist* **18**:242.

Hayward, J.N., 1974, Physiological and morphological identification of hypothalamic magnocellular neuroendocrine cells in goldfish preoptic nucleus, *J. Physiol.* **239**:103–124.

Hayward, J.N., 1975, Neural control of the posterior pituitary, *Annu. Rev. Physiol.* **37**:191–210.

Hayward, J.N., and Jennings, D.P., 1973*a*, Activity of magnocellular neuroendocrine cells in the hypothalamus of unanaesthetized monkeys. I. Functional cell types and their anatomical distribution in the supraoptic nucleus and the internuclear zone, *J. Physiol.* **232**:515–543.

Hayward, J.N., and Jennings, D.P., 1973*b*, Activity of magnocellular neuroendocrine cells in the hypothalamus of unanaesthetized monkeys II. Osmosensitivity of functional cell types in the supraoptic nucleus and the internuclear zone, *J. Physiol.* **232**:545–572.

Hodgson, E., and Geldiay, S., 1959, Experimentally induced release of neurosecretory materials from roach corpora cardiaca, *Biol. Bull.* **117**:275–283.

Ishida, A., 1967, The effect of tetrodotoxin on calcium-dependent link in stimulus–secretion coupling in neurohypophysis, *Jpn. J. Physiol.* **17**:308–320.

Ishida, A., 1968, Stimulus–secretion coupling on the oxytocin release from the isolated posterior pituitary lobe, *Jpn. J. Physiol.* **18**:471–480.

Ishida, A., 1970, The oxytocin release and the compound action potential evoked by electrical stimulation of the isolated neurohypophysis of the rat, *Jpn. J. Physiol.* **20**:84–96.

Iwasaki, S., and Satow, Y., 1969, Spontaneous grouped discharge of secretory neuron soma in X-organ of crayfish, *Procambarus clarkii*, *J. Physiol. Soc. Jpn.* **31**:629–630.

Iwasaki, S., and Satow, Y., 1971, Sodium- and calcium-dependent spike potentials in the secretory neuron soma of the X-organ of the crayfish, *J. Gen. Physiol.* **57**:216–238.

Jahan-Parwar, B., Smith, M., and von Baumgarten, R., 1969, Activation of neurosecretory cells in *Aplysia* by osphradial stimulation, *Amer. J. Physiol.* **216**:1246–1257.

Kado, R.T., 1973, *Aplysia* giant cell: Soma–axon voltage clamp differences, *Science* **182**:843.

Kandel, E.R., 1964, Electrical properties of hypothalamic neuroendocrine cells, *J. Gen. Physiol.* **47**:691–717.

Kater, S., 1968, Cardioaccelerator release in *Periplaneta americana* (L.) *Science* **160**:765–767.

Kater, S., and Kaneko, C., 1972, An endogenously bursting neuron in the gastropod mollusc, *Helisoma trivolvis*, *J. Comp. Physiol.* **79**:1–14.

Katz, B., 1969, *The Release of Neural Transmitter Substances*, Liverpool University Press, Liverpool, 60 pp.

Katz, B., and Miledi, R., 1967*a*, A study of synaptic transmission in the absence of nerve impulses, *J. Physiol.* **192**:407–436.

Katz, B., and Miledi, R., 1967*b*, The release of acetylcholine from nerve endings by graded electric pulses, *Proc. R. Soc. London Ser. B.* **167**:23–38.

Katz, B., and Miledi, R., 1969*a*, Tetrodotoxin-resistant electric activity in presynaptic terminals, *J. Physiol.* **203**:459–487.

Katz, B., and Miledi, R., 1969*b*, Spontaneous and evoked activity of motor nerve endings in calcium Ringer, *J. Physiol.* **203**:689–706.

Katz, B., and Miledi, R., 1970, Further study of the role of calcium in synaptic transmission, *J. Physiol.* **207**:789–801.

Kleinhaus, A., and Prichard, J., 1975, Calcium dependent action potentials produced in leech Retzius cells by tetraethylammonium chloride, *J. Physiol.* **246**:351–361.

Kleinholz, L.H., 1966, Separation and purification of crustacean eyestalk hormones, *Amer. Zool.* **6**:161–167.

Knowles, F., 1974, Twenty years of neurosecretion, in: *Neurosecretion—The Final Neuroendocrine Pathway* (F. Knowles and L. Vollrath, eds.), pp. 3–11, Springer-Verlag, New York.

Koketsu, K., and Nishi, S., 1969, Calcium and action potentials of bullfrog sympathetic ganglion cells, *J. Gen. Physiol.* **53**:608–628.

Kupfermann, I., 1970, Stimulation of egg laying by extract of neuroendocrine cells (bag cells) of abdominal ganglion of *Aplysia*, *J. Neurophysiol.* **33**:877–881.

Kupfermann, I., and Kandel, E.R., 1970, Electrophysiological properties and functional interconnections of two symmetrical neurosecretory clusters (bag cells) in abdominal ganglion of *Aplysia*, *J. Neurophysiol.* **33**:865–876.

Kupfermann, I., and Weiss, K.R., 1976, Water regulation by a presumptive hormone contained in identified cell R15 of *Aplysia*, *J. Gen. Physiol.* **67**:113–123.

Lincoln, D.W., 1974, Dynamics of oxytocin secretion, in: *Neurosecretion—The Final Neuroendocrine Pathway* (F. Knowles and L. Vollrath, eds.), pp. 129–133, Springer-Verlag, New York.

Lincoln, D.W., and Wakerley, J.B., 1975, Factors governing the periodic activation of supraoptic and paraventricular neurosecretory cells during suckling in the rat, *J. Physiol.* **250**:443–461.

Llinás, R., Blinks, J.R., and Nicholson, C., 1972, Calcium transient in presynaptic terminals in squid giant synapse: Detection with aequorin, *Science* **176**:1127–1129.

Maddrell, S., and Gee, J., 1974, Potassium-induced release of the diuretic hormones of *Rhodnius prolixus* and *Glossina austeni:* Ca dependence, time course and localization of neurohaemal areas, *J. Exp. Biol.* **61**:155–171.

Mallart, A., and Martin, A.R., 1967, An analysis of facilitation of transmitter release at the neuromuscular junction of the frog, *J. Physiol.* **193**:679–694.

Martin, A.R., and Pilar, G., 1964, An analysis of electrical coupling at synapses in the avian ciliary ganglion, *J. Physiol.* **171**:454–475.

Meech, R.W., and Standen, N.B., 1975, Potassium activation in *Helix aspersa* neurones under voltage clamp: A component mediated by calcium influx, *J. Physiol.* **249**:211–239.

Meves, H., and Vogel, W., 1973, Calcium inward currents in internally perfused giant axons, *J. Physiol.* **235**:225–266.

Mikiten, T.M., and Douglas, W., 1965, Effect of calcium and other ions on vasopressin release from rat neurohypophysis stimulated electrically *in vitro, Nature London* **207**:302.

Miledi, R., and Thies, R., 1971, Tetanic and post-tetanic rise in frequency of miniature end-plate potentials in low-calcium solutions, *J. Physiol.* **212**:245–257.

Normann, T., 1973, Membrane potential of the corpus cardiacum neurosecretory cells of the blowfly, *Calliphora erythrocephala, J. Insect Physiol.* **19**:303–318.

Normann, T., 1974, Calcium-dependence of neurosecretion by exocytosis, *J. Exp. Biol.* **61**:401–409.

Normann, T., 1976, Neurosecretion by exocytosis, *Int. Rev. Cytol.* **45**: in press.

Normann, T., and Duve, H., 1969, Experimentally induced release of a neurohormone influencing hemolymph trehalose level in *Calliphora erythrocephala* (Diptera), *Gen. Comp. Endocrinol.* **12**:449–459.

Passano, L.M., 1953, Neurosecretory control of molting in crabs by the X-organ sinus gland complex, *Physiol. Comp. Oecol.* **3**:155–189.

Pérez-González, M.D., 1957, Evidence of hormone containing granules in sinus glands of the fiddler crab *Uca pugilator, Biol. Bull.* **113**:426–441.

Perkins, E.B., 1928, Color changes in Crustaceans, especially in *Palaemonetes, J. Exp. Zool.* **50**:71–195.

Poisner, A.M., 1973, Stimulus–secretion coupling in the adrenal medulla and posterior pituitary gland, in: *Frontiers in Neuroendocrinology* (W. Ganong and L. Martini, eds.), pp. 33–59, Oxford, New York.

Potter, D.D., 1956, Observations on the neurosecretory system of portunid crabs, Ph.D. thesis, Harvard University.

Potter, D.D., 1958, Observations on the neurosecretory system of portunid crabs, in: 2, *Internationales Symposium über Neurosekretion* (W. Bargmann, B. Hanström, E. Scharrer, and B. Scharrer, eds.), pp. 113–118, Springer-Verlag, Berlin.

Rehm, M., 1959, Observations on the localization and chemical constitution of neurosecretory material in nerve terminals in *Carcinus maenas*, *Acta Histochem.* **7:**88–106.

Reuter, H., 1973, Divalent cations as charge carriers in excitable membranes, *Prog. Biophys. Mol. Biol.* **26:**3–43.

Rubin, R.P., 1974, *Calcium and the Secretory Process*, Plenum Press, New York, 189 pp.

Scharrer, B., and Kater, S., 1969, Neurosecretion XV. An electron microscopic study of the corpora cardiaca of *Periplaneta americana* after experimentally induced hormone release, *Z. Zellforsch.* **95:**177–186.

Scharrer, E., 1965, The final common pathway in neuroendocrine integration, *Arch. Anat. Microsc. Morphol. Exp.* **54:**359–370.

Standen, N.B., 1975*a*, Calcium and sodium ions as charge carriers in the action potential of an identified snail neurone, *J. Physiol.* **249:**241–252.

Standen, N.B., 1975*b*, Voltage-clamp studies of the calcium inward current in an identified snail neurone: Comparison with the sodium inward current, *J. Physiol.* **249:**253–268.

Stinnakre, J., and Tauc, L., 1969, Central neuronal response to activation of osmoreceptors in the osphradium of *Aplysia*, *J. Exp. Biol.* **51:**347–361.

Vincent, J.D., Arnauld, E., and Bioulac, B., 1972*a*, Activity of osmosensitive single cells in the hypothalamus of the behaving monkey during drinking, *Brain Res.* **44:**371–384.

Vincent, J.D., Arnauld, E., and Nicolescu-Catargi, A., 1972*b*, Osmoreceptors and neurosecretory cells in the supraoptic complex of unanesthetized monkey, *Brain Res.* **45:**278–281.

Wald, F., 1972, Ionic differences between somatic and axonal action potentials in snail giant neurones, *J. Physiol.* **220:**267–281.

Wilkens, J.L., and Mote, M.I., 1970, Neuronal properties of the neurosecretory cells in the fly *Sarcophaga bullata*, *Experientia* **26:**275–276.

Yagi, K., and Bern, H.A., 1965, Electrophysiological analysis of the response of the caudal neurosecretory system of *Tilapia mossambica* to osmotic manipulations, *Gen. Comp. Endocrinol.* **5:**509–526.

Yagi, K., Bern, H.A., and Hagadorn, I.R., 1963, Action potentials of neurosecretory neurons in the leech *Theromyzon rude*, *Gen. Comp. Endocrinol.* **3:**490–495.

Yamashita, H., Koisumi, K., and Brooks, C., 1970, Electrophysiological studies of neurosecretory cells in the cat hypothalamus, *Brain Res.* **20:**462–466.

Zeballos, G.A., Thornborough, J.R., and Rothballer, A.B., 1975, Neurohypophysial electrical activity in the anesthetized cat, *Neuroendocrinology* **18:**104–114.

13

Endogenous Opiate Peptides

Werner A. Klee

> The receptor for a foreign drug is really the receptor for a humoral substance, with which the foreign molecule also interacts. . . . We do not yet know the natural function of the macromolecule(s) with which morphine interacts.
>
> —H.O.J. Collier (1973)

1. Introduction

The discovery, within the past year or so, of the endogenous opiate peptides (Hughes *et al.*, 1975*b*) has captured the imagination of pharmacologists and chemists alike and led to a remarkable burst of scientific productivity. I try in this chapter to review this work in a coherent, if highly personal, way, with the sad realization that much of what I say may be outdated by the time this volume is in print. The statement by Collier quoted above is to my knowledge the first explicit recognition of the likelihood that endogenous ligands for the opiate receptor must exist, and was made at the International Congress of Pharmacology in San Francisco in 1972. Within three years, the endogenous opiate peptides were isolated and characterized, and further progress will undoubtedly be even more rapid. Nevertheless, a review of the field at this time may be of some value to those who are not specialists in opiates. For this reason, I have also chosen, by way of introduction, to review briefly some aspects of the mechanism of action of morphine as it has developed over the past several years.

The mechanism of action of the opiates—morphine and its thousands of synthetic relatives—has been under study for almost 200 years, and perhaps even

Werner A. Klee · Laboratory of General and Comparative Biochemistry, National Institute of Mental Health, Bethesda, Maryland.

longer (Eddy and May, 1973). Derosne in 1803 and Serturner in 1805 both isolated crystalline materials from crude opium, which Serturner called *morphine* and showed to have opiumlike effects in dogs. The early, and continued, interest in this class of drugs is the result of its unrivaled clinical utility as a pain-killer, anti-diarrheal, sleep-inducer, reliever of anxiety, and promoter of a general feeling of well-being. In addition, and particularly in recent years, it has been the hope of many that an understanding of the mechanism of opiate action may lead to real insight into some uncharted areas of brain function. It is one of the goals of this chapter to show that this latter hope is on its way to being fulfilled as the result of the discovery of the endogenous opiate peptides.

2. The Opiate Receptor

2.1. Pharmacological Evidence

A large body of evidence has been accumulated over the past 50 years showing that opiates act by virtue of interactions with specific receptors. This evidence, which has been amply reviewed (Beckett and Casy, 1954; Portoghese, 1965; Jacobson, 1972), may be summarized as follows: (1) There is a general structural similarity among those substances that have opiatelike pharmacology. Over the years, as more synthetic analgesics have been discovered, the minimal structural requirements for opiate activity have, apparently, been reduced to two in number. These are the presence of an aromatic ring structure and a nitrogen atom, usually as a tertiary amine, that is located at a distance of 2 saturated carbon atoms from the aromatic moiety (Eddy and May, 1973). (2) Morphine is an asymmetric molecule. The naturally occurring, levorotatory, isomer is the active form; the opposite enantiomer is, in the case of almost all the opiates, inactive or nearly so. (3) Morphine congeners (such as etorphine) have been synthesized that are 500–1000 times as potent as morphine, and yet differ in structure in only minor ways. (4) Perhaps the single most decisive fact requiring that opiate receptors exist is the observation, first made by Pohl (1915), that morphine derivatives may act as antagonists of the opiates. Small changes in structure, e.g., replacement of the methyl group of morphine by allyl to produce nalorphine, converts a potent analgesic to a drug that can completely antagonize morphine analgesics.

2.2. Biochemical Studies

The explosive progress made in the past few years in our understanding of the mechanism of opiate action is due, in no small part, to the direct biochemical demonstration of opiate receptors. Goldstein *et al.* (1971) reported experiments that established a set of criteria for stereospecificity of binding of opiates to their receptors. Subsequent work by Pert and Snyder (1973*a,b*), Terenius (1972, 1973), and Simon *et al.* (1973), using opiates of very high specific radioactivity, successfully demonstrated the presence of high-affinity, stereospecific opiate receptors in membrane preparations from brain and guinea pig ileum. These receptors were shown to interact with narcotic agonists and antagonists and, in general, to show a spectrum of binding affinites that correspond well with the pharmacological potencies of a large series of opiates derived from many chemical classes. The regional distribution of receptors in the brain of man and rhesus monkey has been studied in detail by Hiller *et al.* (1973) and Kuhar *et al.* (1973). These workers have shown that opiate receptors, although found throughout the brain, are concentrated in areas associated with the limbic system and the periaqueductal gray areas. The latter have been shown to be highly responsive sites of analgesia after direct application of morphine by microinjection techniques (Jacquet and Lajha 1974). Recently, radioautographic procedures, which allow the demonstration of opiate receptors at the light-microscope level of resolution, have been developed by Pert *et al.* (1975) and Schubert *et al.* (1975). These studies have served to emphasize the clustered nature of the distribution of opiate receptors even within regions of the brain that are well endowed with them. The detailed brain maps that should result from this work will doubtless lead to important new insights into the neurophysiology of pain, and perhaps of affective phenomena as well.

An apparently important observation, although not yet understood, is the specific effect of Na^+ on the properties of the opiate receptor. Inhibition of agonist binding by Na^+ was observed by Simon *et al.* (1973), whereas antagonist binding had been found by Pert and Snyder (1973*b*) not to be affected by Na^+. Subsequent studies (Pert and Snyder, 1974; Simon *et al.*, 1975) showed that Na^+ (and to some extent Li^+) specifically increases the binding affinity of antagonists and decreases that of opiate agonist. An allosteric model has been proposed in which a postulated equilibrium of the receptor between a state favoring agonist binding and one favoring antagonist binding is dependent on the concentration of Na^+. The intriguing and unanswered question here is the relationship, if any, between this phenomenon and Na^+ transport through the membrane.

Opiate receptors have been found to be present in easily measurable amounts in the brain of all vertebrate species examined (Pert *et al.*, 1974), but not in nervous tissue from any invertebrate species. Receptors are also found in some peripheral neurons, and have been studied by binding experiments in the myenteric plexus of the guinea pig ileum (Terenius, 1972; Pert and Snyder, 1973*a*). Pharmacological evidence for their existence in the mouse vas deferens, rabbit heart, and isolated cat nictitating membrane has been collected, but chemical demonstrations have not yet been reported for these tissues. A cultured cell line that is abundantly endowed with opiate receptors has been found by Klee and Nirenberg (1974). The cells in question, NG108-15, are somatic hybrids of a neuroblastoma clonal cell line and rat glioma line. The receptors present are very similar in specificity to those found in rat brain, and these cells have proved to be useful in studies of the biochemical mechanism of action of opiates.

2.3. Coupling with Adenylate Cyclase

Binding of a ligand to its receptor is only the first step in its action, and from a functional point of view, it is important to identify the next step. In the case of the opiates, there is now good evidence that the receptors are coupled to adenylate cyclase in the cell membrane as inhibitory regulators. The earliest biochemical experiments successfully linking opiates with adenylate cyclase were performed with rat brain homogenates by Collier and Roy (1974a). These workers found that the conversion of [^{3}H]ATP to [^{3}H]cAMP is stimulated approximately 2-fold by prostaglandin E_1, and that this stimulated activity can be completely prevented by micromolar concentrations of morphine. The inhibitory effect of morphine was prevented by naloxone, a specific and essentially pure opiate antagonist. Furthermore, the potency of a series of opiates in inhibiting cAMP formation correlated well with their pharmacological activity as analgesics (Collier and Roy, 1974b). Perhaps because of the extreme heterogeneity of brain and also the lability of many brain enzyme systems, a number of laboratories have been unsuccessful in their attempts to demonstrate receptor-mediated inhibition of adenylate cyclase. There have been some reports of successful replication of the work of Collier and Roy (1974a), and work with cell culture systems described below leaves little doubt of their validity and importance.

A neuroblastoma X glioma hybrid cell line, NG108-15, has already been referred to, since it was shown to contain approximately 300,000 morphine receptors/cell (Klee and Nirenberg, 1974). Homogenates prepared from this cell line contain an adenylate cyclase activity that is inhibited by opiates in a stereospecific, naloxone-reversible, receptor-mediated fashion (Sharma et al., 1975a). The parental cell lines, neuroblastoma N18TG-2 and glioma C6BU-1, which contain few and no morphine receptors, respectively, contain adenylate cyclase activities that are only slightly or not at all inhibited by opiates. The potency of a series of opiates as inhibitors of adenylate cyclase correlates well with their pharmacological activity and the inactive dextrorotating enantiomer of levorphanol, dextrophan, is without effect. Interestingly, basal as well as prostaglandin E_1- or adenosine-stimulated adenylate cyclase activity is inhibited by morphine in these cells (Sharma et al., 1975a,b). Work performed simultaneously and independently by Hamprecht and his colleagues using many of the same cell lines has also shown that cAMP levels in NG108-15 are lowered by opiates in a receptor-mediated process (Traber et al., 1974, 1975a–c). The interesting observation of Gullis et al. (1975) that GMP levels are raised by opiates in NG108-15 cells has not yet been confirmed by others. The vast bulk of the evidence so far compiled indicates that opiate receptors are functionally linked to adenylate cyclase, and that this inhibitory coupling may account for the acute effects of morphine and its relatives (Klee et al., 1975; Roy and Collier, 1975; Francis et al., 1975).

3. Adenylate Cyclase and the Mechanism of Addiction

More than 30 years ago, Himmelsbach (1943) postulated that opiate dependence is the result of a homeostatic increase in the intrinsic activity of brain centers that are depressed by the action of the drugs. On the biochemical level, such an increase might be expected to occur in the amount of adenylate cyclase activity found in morphine sensitive neurons. Such a compensatory change has now been observed in NG108-15 cells cultured for many hours in the presence of morphine (Sharma *et al.*, 1975*b;* Traber *et al.*, 1975*b;* Klee *et al.*, 1975), and leads to a simple biochemical model of opiate dependence and tolerance (Sharma *et al.*, 1975*b*). Morphine produces an immediate fall in cellular cAMP levels, due to inhibition of adenylate cyclase activity. The cAMP levels return toward normal as adenylate cyclase activity expands due to the late regulatory control step until the tolerant state is reached and cellular cAMP is once again that of untreated cells. Withdrawal of morphine from the tolerant cells lead to an immediate and large increase in cAMP levels, well above those normally found. These abnormally high cAMP levels may be the biochemical correlate of the abstinence syndrome in animals, and result from the release of an abnormally large amount of adenylate cyclase activity from morphine inhibition. Finally, after a number of hours in the continued absence of morphine, the cells recover their normal complement of adenylate cyclase activity, and cAMP levels also return to normal. The evidence that this type of dual regulation of adenylate cyclase is a major factor in addiction of animals to opiates is less direct, but nevertheless indicates that this can account for dependence and tolerance in animals as well (Francis *et al.*, 1975; Collier *et al.*, 1975).

4. Endogenous Opiates

4.1. Pharmacological Evidence

A natural question to ask, once the existence of specific opiate receptors had been clearly demonstrated, was, "Why are there morphine receptors?" Clearly, opiates are not normally consumed by most animals, and clearly also, therefore, the opiate receptor must ordinarily interact with substances other than opiates, as Collier (1973) had pointed out. This line of reasoning led many, over the past few years, to initiate a search for the endogenous opiates.

In addition, a number of experiments had been performed with the potent and essentially pure opiate antagonist naloxone, which are best understood if it is assumed that naloxone is counteracting the action of an endogenous opiate. The earliest such experiment is also perhaps the most dramatic and convincing. Akil *et al.* (1972, 1976) showed that the analgesia produced by electrical stimulation of the periaqueductal gray region of the brain stem of rats can be reversed, at least in part, by the administration of naloxone. Jacob *et al.* (1974) found that mice could be made more sensitive to pain by injection of small amounts of naloxone. These observations were recognized by Jacob and co-workers to indicate that there exists an endogenous opiate the action of which is being blocked by naloxone. Blockade of acupuncture analgesia by naloxone has been reported by Mayer *et al.* (1976). The implications of this interesting experiment are clearly in accord with the present discussion.

4.2. Steps in Their Isolation and Characterization

Progress toward isolation of the endogenous opiates was reported by Hughes (1975*a*) at a meeting held in May 1974. He was able to assay crude brain extracts with the mouse vas deferens and show that such extracts contain peptides of molecular weight between 300 and 700 that inhibit electrically stimulated contractions of this smooth muscle preparation, but are inactive if naloxone is also present. Terenius and Wahlström (1974) reported in an abstract that human cerebrospinal fluid contains a material, probably peptide in nature, that binds to the opiate receptor. By May 1975, a number of laboratories had obtained partially purified preparations of endogenous opiate peptides from mammalian brain (Pasternak *et al.*, 1975), pituitary (Teschemacher *et al.*, 1975; Cox *et al.*, 1975), and human CSF (Terenius and Wahlström, 1975). Hughes *et al.* (1975*a,b*) were clearly well ahead of the field in that the peptide they had isolated from porcine brain, which they chose to name *enkephalin,* was in an essentially pure state.

4.3. The Enkephalins

4.3.1. Endogenous Opiate Pentapeptides of Defined Structure

The excitement generated by the announcement by Hughes *et al.* (1975*b*) of the structure of the enkephalins, pentapeptides from brain with opiate activity, is probably unique in the history of pharmacology. These peptides, which differ by a single amino acid replacement in position 5, have the following structures: Tyr–Gly–Gly–Phe–Met (methionine–enkephalin) and Tyr–Gly–Gly–Phe–Leu (leucine–enkephalins). They were isolated from porcine brain extracts in a 5:1 ratio by Hughes and his colleagues. Due to their simple structure, they were quickly synthesized and studied in an unbelievably large number of laboratories, as were hundreds of other peptides with related structures. Confirmation of the structure deduced by Hughes and his colleagues for the enkephalins has also resulted from the work of Simantov and Snyder (1976), who isolated the same pair of peptides, although in a very different ratio, from extracts of beef brain. As was pointed out by Hughes *et al.* (1975*b*), a particularly fascinating aspect of enkephalin structure is that the enkephalin sequence is contained within the structure of the pituitary polypeptide β-lipotropin. It had been recognized earlier that β-lipotropin also contains within its structure the amino acid sequence of β-MSH, a peptide hormone the function of which in mammals is still obscure, but that in lower vertebrates such as frogs is responsible for adaptive changes in skin color. Thus, a single polypeptide, β-lipotropin, has the property of being, at least potentially, the precursor of two quite different polypeptide hormones. This interesting relationship will be discussed further below.

4.3.2. Biological Activity

The enkephalins were shown by Hughes *et al.* (1975*a,b*) to be potent inhibitors of electrically stimulated contraction of smooth muscle in the walls of the mouse vas deferens and the guinea pig ileum. Methionine–enkephalin has approximately 20 times the potency of morphine in the vas deferens assays, and is approximately equipotent with morphine in the guinea pig ileum assay. Leucine–enkephalin exhibits only approximately $^1/_2-^1/_5$ the potency of methionine–enkephalin in these (Hughes *et al.*, 1975*a*) and other assay systems (Klee and Nirenberg, 1976; Simantov and Snyder, 1976). These effects are naloxone reversible, and therefore presumably, receptor-mediated. The activity of the enkephalins is short-lived, but particularly so in the guinea pig ileum system, perhaps due to rapid degradation (Hughes *et al.*, 1975*a*). All attempts to show analgesic properties of enkephalins by systemic administration of even large amounts to animals have been unsuccessful due, in all likelihood, either to rapid degradation or the inability to cross the blood–brain barrier. On the other hand, many laboratories have been able to show that direct administration of enkephalins to the brain by intracerebroventricular injection lessens the sensitivity of rats or mice to pain (Belluzzi *et al.*, 1976; Büscher *et al.*, 1976). Very large doses of the pentapeptides are necessary to demonstrate analgesia, and the effects are extinguished within 2–10 minutes, again because of rapid degradation. Significantly, however, analgesic effects of enkephalin administration can be prevented by prior systemic application of low doses of naloxone. Therefore, the enkephalin may be assumed to interact with morphine receptors in the brain and thereby reduce analgesia.

The enkephalins have been shown to bind to opiate receptors present in particulate fractions of brain. The affinities of the peptides for the receptors is relatively low when compared with morphine (Simantov and Snyder, 1976; Chang *et al.*, 1976; Bradbury *et al.*, 1976). Binding measurements made in the presence and absence of Na^+ have led Simantov and Snyder (1976) to postulate that the enkephalins may have mixed agonist/antagonist action. No direct evidence supporting this proposal has yet been presented, and experiments with adenylate cyclase inhibition discussed below suggest that the enkephalins are primarily agonists as do the guinea pig ileum assays of Cox *et al.* (1976*a*).

The enkephalins have been found to be highly potent, opiate receptor–mediated inhibitors of adenylate cyclase activity in homogenates of neuroblastoma × glioma NG108-15 cells (Klee and Nirenberg, 1976). Inhibition of enzyme activity is receptor-mediated, since it is reversed by naloxone, and the amount of inhibition, in several cell lines, is correlated with the number of receptors present. Moreover, naloxone inhibition of methionine–enkephalin action takes place with an apparent binding constant of 30nM, which is, within experimental error, the same as the binding constant calculated for naloxone reversal of morphine action (20 nM), and that measured by direct binding experiments in this system (Sharma *et al.*, 1975*b*). Methionine–enkephalin has approximately 100 times the potency of morphine as an inhibitor of adenylate cyclase, and leucine enkephalin is approximately $^1/_5$ as potent as is the methionine peptide (Klee and Nirenberg, 1976). Thus, in neuroblastoma × glioma hybrid cells, as in the mouse vas deferens (Hughes *et al.*, 1975*a*), the enkephalins are much more potent than morphine. In brain and the guinea pig ileum, their potency is much reduced. It is perhaps no coincidence that the enkephalins are also relatively stable in the hybrid cells and vas deferens and produce relatively long-lived effects, while they are apparently very unstable in brain and the ileum and produce only short-lived effects. True potency may therefore be difficult to measure in these latter systems due to rapid degradation of the enkephalins.

One of the hopes of many workers interested in the enkephalins is that these substances may lead to the development of the ideal analgesic—one that is potent, but does not lead to tolerance or dependence. Because of the rapid inactivation of the enkephalins in most test systems, only a few experiments have been reported to date that examine this question. Thus, Waterfield *et al.* (1976) have found that the vas deferens of mice made tolerant to morphine is also tolerant to methionine–enkephalin. A more direct test of the tolerance and dependence liability of methionine–enkephalin was performed by Lampert *et al.* (1976), who found that neuroblastoma × glioma hybrid cells exposed to the methionine–enkephalin for periods of 12 hr or longer develop a 2–3-fold increase in adenylate cyclase activity, which, according to the dual regulation model of Sharma *et al.* (1975*b*), accounts for opiate addiction. Thus, the evidence to date indicates that the enkephalins have addictive properties, and will therefore not fulfill the pharmacologist's dream of the ideal analgesic. This judgment is based on only a small amount of experimental work; obviously, the important question of the addictive properties of endogenous opiate peptides is being examined experimentally in a large number of laboratories as this volume is being printed. Lampert *et al.* (1976) have also found that a partially purified peptide prepared from pituitary by the procedure of Teschemacher *et al.* (1975) also has addictive properties in neuroblastoma × glioma hybrid cell cultures, which suggests that all the endogenous opiate peptides may show this property. An alternate view of the possible role of enkephalin in addiction is presented in the interesting paper of Kosterlitz and Hughes (1975).

4.4. Are Morphine and the Enkephalins Structurally Related?

We have seen that enkephalins function by virtue of interaction with the opiate receptor. Indeed, the existence of morphine receptors led to the search for the endogenous ligands. Clearly, therefore, the opiate must be postulated to mimic the structure of the enkephalins in some ways so that they may bind to the same receptor. Morphine and its congeners may be regarded as derivatives of tyramine that have been locked into a particular conformation by virtue of substituents on the phenol ring, the α carbon atom, and the nitrogen moiety. Evidence that this may be true has been provided by binding experiments to the receptor of a series of tyramine derivatives with increasing degrees of substitution and conformational rigidity (Klee *et al.*, unpublished observations), and Horn and Rodgers (1976) have advanced similar arguments for the critical importance of a tyramine skeleton. The structure–activity relationships for a series of enkephalin derivatives shown in Table 1 provide some insight into this problem. Note first that there is a good, albeit not perfect, correlation between the order of potency in binding to receptors in rat brain and in inhibition of adenylate cyclase in neuroblastoma × glioma hybrid cells, even though the absolute values differ by an order of magnitude or more. Possibly, some of the discrepancies noted are due to degradation by peptidases present in brain during the binding assay, but real differences in the

TABLE 1. Receptor Binding and Adenylate Cyclase Inhibition of Enkephalins and Some Analogues

Peptide	ID_{50} receptor binding, ID_{50} (nM) [a]	Inhibition of adenylate cyclase, K_i (nM) [b]
Tyr–Gly–Gly–Phe–Met (methionine–enkephalin)	400	12
Tyr–Gly–Gly–Phe–Leu	1,400	40
Tyr–Gly–Gly–Phe–Met–Thr	2,000	30
Tyr–Gly–Gly–Phe–Met–NH$_2$	200	200
Tyr–Gly–Gly–Phe–Leu–NH$_2$	4,000	200
Arg–Tyr–Gly–Gly–Phe–Met	500	1,000
Ser–Tyr–Gly–Gly–Phe–Met	40,000	2,000
Tyr–Gly–Gly–Phe–Thr	20,000	400
Phe–Gly–Gly–Phe–Met	200,000	10,000
Tyr–Ser–Gly–Phe–Leu	—	10,000

[a] Data of Chang et al. (1976) measured with rat brain receptors in the presence of 0.1 M Na$^+$.
[b] Data of Klee et al. (1976) measured with homogenates of neuroblastoma × glioma hybrid cells.

receptors present in the two assay systems cannot be ruled out at present. The data show that an unsubstituted amino-terminus of the tyrosine residue is important for activity, as is the presence of a hydrophobic residue at position 5 and a glycine at position 2. These results support the hypothesis that enkephalin, when bound to the receptor, assumes a morphinelike conformation, with the tyrosine moiety corresponding to the tyramine portion of the morphine molecule and the side chains of residue 5 and perhaps 4 as well interacting, by hydrophobic forces, with portions of the receptor that also interact with the C and D rings of opiates. It is to be hoped that studies on the conformation of endogenous opiate peptides in solution and in the crystalline state, currently in progress in many laboratories, will lead to more precise answers to the questions of the relationship of structure to the function of these substances.

4.5. β-Lipotropin and the Endorphins

β-Lipotropin was first isolated from sheep pituitary glands by Birk and Li (1964) and found to have lipolytic activity, and to be at least a potential precursor of β-MSH (Gráf et al., 1971; Bradbury et al., 1975). The discovery by Hughes et al. (1975a) that residues 61–65 of this 91 amino acid residue peptide have the amino acid sequence of methionine–enkephalin led to an immediate flurry of renewed interest in the properties of β-lipotropin and its fragments. The amino acid sequence of human β-lipotropin (Li and Chung, 1976) is shown in Fig. 1. Amino acid residues 36–91 are, with the exceptions of conservative amino acid replacements at positions 42, 46, 83, and 87, identical with those of ovine and porcine β-lipotropin. Residues 1–35, for which there is no known function, are much more highly variable and may be, as is the C peptide of proinsulin, primarily important in the biosynthesis of this molecule.

H–Glu–Leu–Thr–Gly–Gln–Arg–Leu–Arg–Gln–Gly–Asp–Gly–Pro–Asn–Ala–Gly–Ala–Asn–Asp–
Gly–Glu–Gly–Pro–Asn–Ala–Leu–Glu–His–Ser–Leu–Leu–Ala–Asp–Leu–Val–Ala–Ala–Glu–Lys–
Lys–Asp–Glu–Gly–Pro–Tyr–Arg–Met–Glu–His–Phe–Arg–Trp–Gly–Ser–Pro–Pro–Lys–Asp–Lys–
Arg–Tyr–Gly–Gly–Phe–Met–Thr–Ser–Glu–Lys–Ser–Gln–Thr–Pro–Leu–Val–Thr–Leu–Phe–Lys–
Asn–Ala–Ile–Ile–Lys–Asn–Ala–Tyr–Lys–Lys–Gly–Glu–OH

FIG. 1. Amino acid sequence of human β-lipotropin.

Intact β-lipotropin has not usually been found to display opiatelike activity in any assay system (but see Graf *et al.*, 1976). On the other hand, all fragments of β-lipotropin that have tryosine 61 at their amino-terminus and are 5 or more amino acid residues in length do display opiatelike activity (Guillemin *et al.*, 1976; Bradbury *et al.*, 1976; Cox *et al.*, 1976b). It seems highly likely that β-lipotropin serves as a prohormone that, as the result of suitable proteolytic cleavage, is the precursor of both β-MSH and endogenous opiate peptides. Opiate peptide fragments of β-lipotropin (61–65 . . . 91) that are larger than enkephalin (61–65), such as 61–69, 61–79, 61–89, and 61–91, have, by general agreement, been called *endorphins*, which in this review will be used as a generic term for all endogenous opiate peptides as originally suggested by Eric Simon.

The larger endorphins derived from β-lipotropin have biological activity comparable to that of the enkephalins in the guinea pig ileum assay system (Guillemin *et al.*, 1976; Cox *et al.*, 1976b), and appreciably greater than that of the enkephalins when assayed by direct administration to the brain (Feldberg and Smyth, 1976) or by binding to brain receptors (Bradbury *et al.*, 1976). The experiments of Feldberg and Smyth (1976) are particularly instructive, in that they demonstrate that the endorphin corresponding to β-LPH 61–91 is not only highly potent in inducing analgesia after administration directly into the third ventricle of cat brain, but that the effect lasts for more than 2 hr. These experiments suggest that β-lipotropin 61–91 is not subject to rapid degradation in cat brain. Further evidence for long-term stability of β-lipotropin 61–91 has been provided by our unpublished experiments, which show that this peptide inhibits the adenylate cyclase of neuroblastoma × glioma hybrid cell homogenates for at least 2 hr, whereas, in parallel experiments, the inhibitory effects of enkephalin are lost after 15–20 min.

Lipotropin fragment binding to brain opiate receptors has been studied in detail by Bradbury *et al.* (1976). Some of their data are reproduced in Table 2. The data show that whereas fragment 61–91 is indeed 4 times more potent than enkephalin in this assay, fragments of intermediate length have reduced potency. This result suggests that the marked stability and high potency of fragment 61–91 may be primarily due to conformational factors. Thus, fragment 61–91 may acquire a stable conformation similar to the conformations of the globular or fibrous proteins that renders it resistant to proteolysis and that is not accessible to peptides of

TABLE 2. Binding of β-Lipotropin Fragments
to Brain Opiate Receptors

β-lipotropin fragment (residue number)	$ID_{50}{}^a$ (nM)
61–91	2
61–89	2
61–87	42
61–69	21
61–68	45
61–65	8
61–64	300
1–91	>1,000
Morphine	4.1

^{a} Experiments performed with [^{3}H]dihydromorphine in the
absence of added Na$^+$. The data are taken from Bradbury
et al. (1976).

shorter chain length. Guillemin et al. (1976) and Cox et al. (1976b) have both shown that a number of endorphin fragments of β-lipotropin are approximately equipotent with enkephalin in the guinea pig ileum assay. The clear implication of all this work is that a large number of different degradation products of β-lipotropin have opiate activity that is morphinelike or greater in potency. Furthermore, the stability and therefore the duration of action of these peptides varies widely, from the very short half-life of the enkephalins to the very long half-life of the complete 61–91 fragment. It seems both unnecessary and unwise to argue from these facts that one or the other type of peptide represents the physiologically significant principle. Both short-acting peptides, serving as classic neurotransmitters, and longer-acting ones, serving as modulators of neural pathways, can be physiologically important. These considerations suggest that specific proteolytic activation steps may play critical roles in the regulation of brain function (see Chapter 9 for a full discussion of the role of proteolysis in peptide function).

4.6. Other Types of Endorphin

Although it has been clearly established, as discussed above, that the enkephalins and a large number of other endogenous opiate peptides are structurally related to β-lipotropin, some endorphins may be of quite different structure. Goldstein and his colleagues (Cox et al., 1976a) have purified a group of peptides from pituitary that seem not to be identical with the β-lipotropin fragments just described, even though they are not yet completely pure or fully characterized chemically. These pituitary endorphins, in contrast to β-lipotropin fragments, are largely inactivated by trypsin and stable to cyanogen bromide cleavage. Conceivably, they are fragments of a β-lipotropin molecule containing Leu in place of Met at position 65; however, the very high potency of even the not completely pure material may argue that these peptides are structurally dissimilar to the β-lipotropin fragments. The pituitary endorphins have been shown to have opiatelike activity in a number of assay systems, including the guinea pig ileum and mouse vas deferens (Cox *et al.*, 1975), and as inhibitors of adenylate cyclase in rat brain and neuroblastoma × glioma homogenates (Goldstein *et al.*, 1976).

A possible third class of endorphin peptide has been found in blood by Pert *et al.* (1976). This material, although no larger than enkephalin, has very long-lasting analgesic effects. Elucidation of its structure is awaited with great interest.

5. Physiological Role of Endogenous Opiate Peptides

The physiological role of the endorphins is at present largely a matter for speculation. Certainly, these peptides function by virtue of their interaction with opiate receptors. Simantov *et al.* (1976) have studied the regional distribution of endorphin activity in monkey brain, and have found a good correlation between the concentration of endorphin and that of opiate receptors in all but three brain regions. The anterior hypothalamus, caudate, and globus pallidus are each much richer in endorphin activity than expected from the number of opiate receptors present in these regions. It may be that much of the endorphin produced in these regions is ultimately transported to other areas of the brain, or from the hypothalamus to the pituitary. Localization of at least a large fraction of pituitary endorphin to the posterior lobe (or pars intermedia) (Ross *et al.*, 1976) suggests that its site of synthesis may well be the hypothalamus.

The endorphins are clearly capable of inducing analgesia and probably other opiatelike effects as well. The biochemical mechanism of action of these peptides may well be to lower cAMP levels in neurons with the appropriate receptors. How this, or any other simple chemical change, can lead to the lessening of pain or a state of euphoria is one of the great unsolved problems in biology. To what extent the endorphins control feelings of pleasure or pain or states of arousal or sleep and how they do so will undoubtedly be the subject of research for many years to come.

6. References

Akil, H., Mayer, D.J., and Liebeskind, J.C., 1972, Comparason chez le rat entre l'analgesie induite par stimulation de la substance grise peri-aqueducale et l'analgesie morphinique, *C. R. Acad. Sci. Ser. D* **274**:3603.

Akil, H., Mayer, D.J., and Liebeskind, J.C., 1976, Antagonism of stimulation-produced analgesia by naloxone, a narcotic-antagonist, *Science* **191**:961, 962.

Beckett, A.H., and Casy, A.F., 1954, Synthetic analgesics: Stereochemical considerations, *J. Pharm. Pharmacol.* **6**:986–981.

Belluzzi, J.D., Grant, N., Garsky, V., Sarantakis, D., Wise, C.D., and Stein, L., 1976, Analgesia induced *in vivo* by central administration of enkephalin in rat, *Nature London* **260**:625, 626.

Birk, Y., and Li, C.H., 1964, β-Lipotropin, *J. Biol. Chem.* **239**:1048–1052.

Bradbury, A.F., Smyth, D.G., and Snell, C.R., 1975, Biosynthesis of β-MSH and ACTH, in: *Pep-*

tides: Chemistry, Structure and Biology (R. Walter and J. Meienhofer, eds.), pp. 609–615, Ann Arbor Science, Ann Arbor, Michigan.

Bradbury, A.F., Smyth, D.G., Snell, C.R., Birdsall, N.J.M., and Hulme, E.C., 1976, *C* fragment of lipotropin has a high affinity for brain opiate receptors, *Nature London* **260:**793–795.

Büscher, H.H., Hill, R.C., Römer, D., Cardinaux, F., Closse, A., Hauser, D., and Pless, J., 1976, Evidence for analgesic activity of enkephalin in the mouse, *Nature London* **261:**423–425.

Chang, J.-K., Fong, B.T.W., Pert, A., and Pert, C.B., 1976, Opiate receptor affinities and behavioral effects of enkephalin: Structure–activity relationship of ten synthetic peptide analogues, *Life. Sci.* **18:**1473–1481.

Collier, H.O.J., 1973, Pharmacological mechanisms of drug dependence, in: *Pharmacology and the Future of Man,* Proceedings of the 5th International Congress on Pharmacology (J. Cochin and E.L. Way, eds.), pp. 65–76, Karger, Basel.

Collier, H.O.J., and Francis, D.L., 1975, Morphine abstinence is associated with increased brain cyclic-AMP, *Nature London* **255:**159–162.

Collier, H.O.J., and Roy, A.C., 1974*a*, Morphine-like drugs inhibit stimulation by E prostaglandins of cyclic-AMP formation by rat-brain homogenate, *Nature London* **248:**24–27.

Collier, H.O.J., and Roy, A.C., 1974*b*, Hypothesis: Inhibition of E-prostaglandin sensitive adenyl cyclase as the mechanism of morphine analgesia, *Prostaglandins* **7:**361–376.

Collier, H.O.J., Francis, D.L., McDonald-Gibson, W.J., Roy, A.C., and Saud, S.A., 1975, Prostaglandins, cyclic-AMP and the mechanism of opiate dependence, *Life Sci.* **17:**85–90.

Cox, B.M., Opheim, K.E., Teschemacher, H., and Goldstein, A., 1975, A peptide-like substance from pituitary that acts like morphine. 2. Purification and properties, *Life Sci.* **16:**1777–1782.

Cox, B.M., Gentleman, S., Su. T.-P., and Goldstein, A., 1976*a*, Further characterization of morphine-like peptides (endorphins) from pituitary, *Brain Research* **115:**285–296.

Cox, B.M., Goldstein, A., and Li, C.H., 1976*b*, Opioid activity of a peptide, β-lipotropin-(61–91), derived from β-lipotropin, *Proc. Nat. Acad. Sci. U.S.A.* **73:**1821–1823.

Eddy, N.B., and May, E.L., 1973, The search for a better analgesic, *Science* **181:**407–414.

Feldberg, W., and Smyth, D.G., 1976, The *C*-fragment of lipotropin—a potent analgesic, *J. Physiol.* in press.

Francis, D.L., Roy, A.C., and Collier, H.O.J., 1975, Morphine abstinence and quasi-abstinence effects after phosphodiesterase inhibitors and naloxone, *Life Sci.* **16:**1901–1906.

Goldstein, A., Lowney, L.I., and Pol, B.K., 1971, Stereospecific and nonspecific interactions of the morphine congener levorphenol in subcellular fractions of mouse brain, *Proc. Nat. Acad. Sci. U.S.A.* **68:**1742–1747.

Goldstein, A., Cox, B.M., Klee, W.A., and Nirenberg, M., 1977, Endorphin from pituitary inhibits cyclic AMP formation in homogenates of neuroblastoma × glioma hybrid cells, *Nature London* **265:**362–363.

Gráf, L., Barát, E., Cseh, G., and Sajgó, M., 1971, Amino acid sequence of porcine β-lipotrophic hormone, *Biochim. Biophys. Acta* **229:**276–278.

Gráf, L., Ronai, A.Z., Bajusz, S., Cseh, G., and Szekely, J.F., 1976, Opiate agonist activity of β-lipotropin fragments: A possible biological source of morphine-like substances in the pituitary, *FEBS Lett.* **64:**181–184.

Guillemin, R., Ling, N., and Burgus, R., 1976, Endorphines, peptides d'origine hypothalamique et neurohypophysaire a activité morphinomimetique. Isolement et structure moleculaire d'α-endorphine, *C. R. Acad. Sci. Ser. D* **282:**783–785.

Gullis, R., Traber, J., and Hamprecht, B., 1975, Morphine elevates levels of cyclic-GMP in a neuroblastoma × glioma hybrid cell line, *Nature London* **256:**57–59.

Hiller, J.M., Pearson, J., and Simon, E.J., 1973, Distribution of stereospecific binding of potent narcotic analgesic etorphine in human brain—predominance in limbic system, *Res. Comm. Chem. Pathol. Pharmacol.* **6:**1052–1062.

Himmelsbach, C.K., 1943, The morphine abstinence syndrome, *Fed. Proc. Fed. Amer. Soc. Exp. Biol.* **2:**201–203.

Horn, A.S., and Rodgers, J.R., 1976, Structural and conformational relationships between the enkephalins and the opiates, *Nature London* **260:**795–797.

Hughes, J., 1975*a*, Search for the endogenous ligand of the opiate receptor, *Neurosci. Res. Prog. Bull.* **13:**55–58.

Hughes, J., 1975*b*, Isolation of an endogenous compound from the brain with pharmacological properties similar to morphine, *Brain Res.* **88:**295–308.

Hughes, J., Smith, T.W., Kosterlitz, H.W., Fothergill, L.A., Morgan, B.A., and Morris, H.R., 1975*a*, Identification of two related pentapeptides from the brain with potent opiate agonist activity, *Nature London* **258:**577–579.

Hughes, J., Smith, T., Morgan, B., and Fothergill, L., 1975*b*, Purification and properties of enkephalin—The possible endogenous ligand for the morphine receptor, *Life Sci.* **16:**1753–1758.

Jacob, J.J., Tremblay, E.C., and Colombel, M.-C., 1974, Facilitation de reactions nociceptives par la naloxone chez le sourit et chez le rat, *Psychopharmacologia* **37:**217–223.

Jacobson, A.E., 1972, Narcotic analgesics and antagonists, in: *Chemical and Biological Aspects of Drug Dependence* (S.J. Mule and H. Brill, eds.), pp. 101–118, Chemical Rubber Co. Press, Cleveland.

Jacquet, Y.F., and Lajtha, A., 1974, Paradoxical effects after microinjection of morphine in the periaqueductal gray matter of the rat, *Science* **185:**1055–1057.

Klee, W.A., and Nirenberg, M., 1974, A neuroblastoma × glioma cell line with morphine receptors, *Proc. Nat. Acad. Sci. U.S.A.* **71:**3474–3477.

Klee, W.A., and Nirenberg, M., 1976, The mode of action of endogenous opiate peptides, *Nature London* **263:**609–612.

Klee, W.A., Sharma, S.K., and Nirenberg, M., 1975, Opiate receptors as regulators of adenylate cyclase, *Life Sci.* **16:**1869–1874.

Klee, W.A., Pert, C.B., Nirenberg, M., Chang, J.K., and Fong, B., 1976, Inhibition of adenylate cyclase of neuroblastoma × glioma hybrid cells by enkephalin analogues (in preparation).

Kosterlitz, H.W., and Hughes, J., 1975, Some thoughts on the significance of enkephalin, the endogenous ligand, *Life Sci.* **17:**91–96.

Kubar, M.J., Pert, C.B., and Snyder, S.H., 1973, Regional distribution of opiate receptor binding in monkey and human brain, *Nature London* **245:**447–450.

Lampert, A., Nirenberg, M., and Klee, W.A., 1976, Tolerance and dependence evoked by an endogenous opiate peptide, *Proc. Nat. Acad. Sci. U.S.A.,* **73:**3165–3167.

Li, C.H., and Chung, D., 1976, Primary structure of human β-lipotropin, *Nature London* **260:**622–624.

Mayer, D.J., Price, D.D., Raffi, A., and Barber, J., 1976, Naloxone blockade of acupuncture analgesia, *Proc. 1st World Congress on Pain,* in press.

Pasternak, G.W., Goodman, R., and Snyder, S.H., 1975, An endogenous morphine-like factor in mammalian brain, *Life Sci.* **16:**1765–1769.

Pert, C.B., and Snyder, S.H., 1973*a*, Opiate receptor: Demonstration in nervous tissue, *Science* **179:**1011–1014.

Pert, C.B., and Snyder, S.H., 1973*b*, Properties of opiate-receptor binding in rat brain, *Proc. Nat. Acad. Sci. U.S.A.* **70:**2243–2247.

Pert, C.B., and Snyder, S.H., 1974, Opiate receptor-binding of agonists and antagonists affected differentially by sodium, *Mol. Pharmacol.* **10:**868–879.

Pert, C.B., Aposhian, D., and Snyder, S.H., 1974, Phylogenetic distribution of opiate receptor binding, *Brain Res.* **75:**356–361.

Pert, C.B., Kubar, M.J., and Snyder, S.H., 1975, Autoradiographic localization of the opiate receptor in rat brain, *Life Sci.* **16:**1850–1854.

Pert, C.B., Pert, A., and Tallman, J.A., 1976, Isolation of a novel endogenous opiate analgesic from human blood, *Proc. Nat. Acad. Sci. U.S.A.* **73:**2226–2230.

Pohl, J., 1915, Über das *N*-allylnorcodein, einen Antogonisten des Morphins, *Z. Exp. Pathol. Ther.* **17:**370–382.

Portoghese, P.S., 1965, A new concept on the mode of interaction of narcotic analgesia with receptors, *J. Med. Chem.* **8:**609–616.

Ross, M., Ehrenkranz, J., Cox, B.M., and Goldstein, A., 1976, Localization of pituitary morphine-like peptide (endorphin), *Fed. Proc. Fed. Amer. Soc. Exp. Biol.* **35:**(Abstract).

Roy, A.C., and Collier, H.O.J., 1975, Prostaglandins, cyclic-AMP and biochemical mechanism of opiate agonist action, *Life Sci.* **16:**1857–1862.

Schubert, P., Höllt, V., and Herz, A., 1975, Autoradiographic evaluation of the intracerebral distribution of [³H]etorphine in the mouse brain, *Life Sci.* **16:**1855, 1856.

Sharma, S.K., Nirenberg, M., and Klee, W.A., 1975a, Morphine receptors as regulators of adenylate cyclase activity, *Proc. Nat. Acad. Sci. U.S.A.* **72:**590–594.

Sharma, S.K., Klee, W.A., and Nirenberg, M., 1975b, Dual regulation of adenylate cyclase accounts for narcotic dependence and tolerance, *Proc. Nat. Acad. Sci. U.S.A.* **72:**3092–3096.

Simantov, R., and Snyder, S.H., 1976, Isolation and structure identification of a morphine-like peptide "encephalin" in bovine brain, *Life Sci.* **18:**781–788.

Simantov, R., Kuhar, M.J., Pasternak, G.W., and Snyder, S.H., 1976, The regional distribution of a morphine-like factor enkephalin in monkey brain, *Brain Res.* **106:**189–197.

Simon, E.J., Hiller, J.M., and Edelman, I., 1973, Stereospecific binding of the potent narcotic analgesic [³H]etorphine to rat-brain homogenate, *Proc. Nat. Acad. Sci. U.S.A.* **70:**1947–1949.

Simon, E.J., Hiller, J.M., Groth, J., and Edelman, I., 1975, Further properties of stereospecific opiate binding sites in rat-brain—Nature of sodium effect, *J. Pharmacol. Exp. Ther.* **192:**531–537.

Terenius, L., 1972, Specific uptake of narcotic analgesics by subcellular fractions of the guinea pig ileum, *Acta Pharmacol. Toxical.* **31:**50.

Terenius, L., 1973, Characteristics of the receptor for narcotic analgesics in synaptic plasma membrane fraction from rat brain, *Acta Pharmacol. Toxicol.* **33:**377–384.

Terenius, L., and Wahlström, A., 1974, Inhibitor(s) of narcotic receptor binding in brain extracts and cerebrospinal fluid, *Acta Pharmacol. Toxicol.* **35:**55 (abstract).

Terenius, L., and Wahlström, A., 1975, Morphine-like ligand for opiate receptors in human CSF, *Life Sci.* **16:**1759–1764.

Teschemacher, H., Opheim, K.E., Cox, B.M., and Goldstein, A., 1975, A peptide-like substance from pituitary that acts like morphine. 1. Isolation, *Life Sci.* **16:**1771–1776.

Traber, J., Fischer, K., Latzin, S., and Hamprecht, B., 1974, Morphine antagonizes action of prostaglandin in neuroblastoma cells but not of prostaglandin and noradrenaline in glioma and glioma × fibroblast hybrid cells, *Nature London* **248:**24.

Traber, J., Reiser, G., Fischer, K., and Hamprecht, B., 1975a, Measurements of adenosine 3>-5>-cyclic monophosphate and membrane-potential in neuroblastoma × glioma hybrid cells—opiates and adrenergic agonists cause effects opposite to those of prostaglandin-El, *FEBS Lett.* **52:**327–333.

Traber, J., Gullis, R, and Hamprecht, B., 1975b, Influence of opiates on levels of adenosine 3>-5>-cyclic monophosphate in neuroblastoma × glioma hybrid cells, *Life Sci.* **16:**1863–1868.

Traber, J., Fischer, K., Latzin, S., and Hamprecht, B., 1975c, Morphine antagonizes action of prostaglandin in neuroblastoma and neuroblastoma × glioma hybrid cells, *Nature London* **253:**120–122.

Waterfield, A.A., Hughes, J., and Kosterlitz, H.W., 1976, Cross tolerance between morphine and methionine–enkephalin, *Nature London* **260:**624, 625.

7. Addendum

Because of the rapid progress being made in the study of endogenous opiate peptides, this addendum was deemed essential. An indication of the amount of work being done in this area is that of 66 papers in the proceedings of the 1976 meeting of the International Narcotics Research Club (Kosterlitz, 1976), 28 are concerned with opiate peptides. I try, in this addendum, to organize the very recent literature (since June, 1976) around a few basic questions being asked of the endogenous opiate peptides.

1. How many are there and which are physiologically important?

These questions are still a source of confusion and controversy. All fragments of β-lipotropin (β-LPH) which have Tyr_{61} as the amino terminus and extend to residue 65 or longer have activity in one or more of the standard morphine assays. As seen in Table 3, however, the amount of activity found relative to that of morphine varies greatly depending upon the assay, the peptide, and, to some extent, the laboratory. Analgesia, after central administration, is seen primarily with the longest of the peptides, β-LPH 61–91 (also called β-endorphin or C fragment). However, the activity of the various β-LPH fragments 61–65 . . . 91 is fairly constant with chain length in the guinea pig ileum assay and in receptor binding assays. An interesting finding is the great difference in relative binding potency depending upon whether the radioactive ligand is naloxone, or enkephalin. Surprisingly, in both the mouse vas deferens and neuroblastoma adenylate cyclase assays, 61–65 is appreciably more potent than 61–91. These data do not single out one of the β-LPH fragments as a best candidate for physiological signif-

TABLE 3. Activity Relative to Morphine (or Normorphine) of β-Lipotropin Fragments 61–N [a]

Assay	65	68	69	76	77	78	79	87	89	91	Reference [b]
Analgesia, central administration											
Cat, 3rd ventricle	<0.01									100	a
Rat, periaqueductal gray	0			0						50	b
Rat, intraventricular	0				0	0				4–10	c
Mouse and rat, intraventricular	<0.02	<0.2	<0.2							18–33	d
Rat, intraventricular	<0.05						0.5			50	e
Rat, intraventricular	0.01–0.05										f
Guinea pig ileum											
	0.7									0.9	g
	1			0.6	0.36	0.23				4.5	h
	1						0.42			1.4	e
	0.28									0.42	i
Mouse vas deferens	38									7	g
Neuroblastoma adenylate cyclase	100				10	10				10	j
Receptor Binding											
[3H]naloxone	0.5				0.3					0.5	k
	0.3	0.2	0.3				0.3	0.2	5	9	l
[3H]Leu-enkephalin	340									470	g

[a] Relative activity: morphine = normorphine = 1.
[b] Tail-pinch test, Smyth *et al.*, 1976; (b) toe and tail pinch, pin prick, hot plate, ice water, Jacquet and Marks, 1976; (c) tail pinch and pin prick, Bloom *et al.*, 1976; (d) mouse tail flick and hot plate, Loh *et al.*, 1976; (e) hot plate, Graf *et al.*, 1976; (f) Belluzzi *et al.*, 1976; Büscher *et al.*, 1976; (g) Lord *et al.*, 1976; (h) Ling and Guillemin, 1976; since no morphine data is given I have assumed 61–65 = 1; (i) Cox *et al.*, 1976; (j) Klee *et al.*, 1976; Klee, Nirenberg, Lis, Seidah, and Chretien, in preparation; (k) Pert *et al.*, 1976a; (l) Bradbury *et al.*, 1976.

icance. Perhaps several of these peptides have important roles to play *in vivo*. The pituitary endorphins studied by Goldstein and his colleagues (Goldstein, 1976; Gentleman *et al.*, 1976) as well as others (Kromer *et al.*, 1976; Teschemacher *et al.*, 1976) may form a second class of opiate peptides. The possibility exists that these may be the Leu-65 analogues of the β-LPH fragments. In view of the intriguing differences in regional distribution between Met-enkephalin and Leu-enkephalin (Smith *et al.*, 1976; Simantov and Snyder, 1976*a*) which may be indicative of functional differences between the two, such a possibility is an exciting one. Peptides with morphinelike activity have been found in many tissues, and even in casein hydrolysates (Wajda *et al.*, 1976).

2. What is the function of the endogenous opiates and how do they work?

In most studies to date, the β-LPH fragments have shown activities which differ only qualitatively from those exhibited by morphine and the other classical opiates. The work summarized in Table 3 is clear evidence for many similarities in mechanism. The enkephalins, in particular, have also been shown to have effects upon cyclic nucleotide levels in brain slices (Minneman and Iverson 1976) and neuroblastoma × glioma hybrid cells (Gullis *et al.*, 1976; Brandt *et al.*, 1976*a*), as well as upon adenylate cyclase activity in such cells (Klee *et al.*, 1976) which are qualitatively indistinguishable from those of morphine even though there appear to be large differences in potency. Neurophysiological studies of the actions of morphine, enkephalin, and other opiate peptides show similar effects on single neurons of many brain regions (Gent and Wolstencroft, 1976; Hill *et al.*, 1976; Bradbury *et al.*, 1976*a*,*b*; Zieglgänsberger and Fry, 1976; Frederickson *et al.*, 1976). These effects are sometimes inhibitory and sometimes excitatory, but the direction is usually the same with either morphine or the peptides.

Studies on the conformation of the enkephalins are proceeding in a number of laboratories (Roques *et al.*, 1976; Jones *et al.*, 1976; Bleich *et al.*, 1976; Schiller *et al.*, 1977). Although these are still in preliminary stages, it seems as if the enkephalins can assume conformations which resemble the structure of opiates, particularly that of the oripavines.

Morphine does not appear to mimic all actions of the endogenous opiates. Microinjection of β-LPH 61–91 into the periaqueductal gray region of the brain produces catalepsy characterized by a "waxy flexibility" along with the analgesia listed in Table 3 (Jacquet *et al.*, 1976; Jacquet and Marks, 1976). In contrast, analgesic doses of morphine injected into the same brain region produce paradoxical hyperactivity (Jacquet and Lajtha, 1974). In related studies, intracerebroventricular injection of β-LPH 61–91 in somewhat larger doses was found to produce catatonia (or catalepsy) characterized by a boardlike stiffness (Bloom *et al.*, 1976; Motomatsu *et al.*, 1976).

These actions of the peptide are blocked by naloxone and thus are probably mediated by opiate receptors. Jacquet and Marks (1976) as well as Bloom *et al.* (1976) have called attention to the possible relation between these phenomena and some aspects of schizophrenia. The observation that naloxone can halt the auditory hallucinations of schizophrenic patients (L.M. Gunne, L. Lindstrom, and L. Terenius, *J. Neurotransmission*, in press) has not yet been confirmed. It is nevertheless an intriguing possibility that these peptides may in one way or another play a role in some forms of mental illness (Byck, 1976). Studies of endorphines in CSF may help clarify this issue (Wahlstrom *et al.*, 1976).

A disappointing number of studies have shown minimal or no effects of naloxone on the behavior of naive subjects. Thus, naloxone has essentially no effect upon hypnotic analgesia in man (Goldstein and Hilgard, 1975), body temperature in rats (Goldstein and Lowery, 1975), shock escape threshold in the rat (Goldstein *et al.*, 1976), or pain perception in humans (El-Sobky *et al.*, 1976). More promising perhaps are studies showing naloxone antagonism of conditioned hyperthermia (Lal *et al.*, 1976) or of fatigue of the guinea pig ileum pressure reflex (Van Nueten *et al.*, 1976) which hint at functional roles for the endogenous opiates.

3. Are the endogenous opiates addictive?

The answer to this question seems to be clearly yes, at least in animal and cell culture systems. Loh *et al.* (1976) and Wei and Loh (1976) show that continuous infusion of β-LPH 61–65 or 61–91 into the periaqueductal gray–fourth ventricle region of rat brain over a period of 70 hr produces physical dependence as evidenced by the production of withdrawal signs after naloxone administration. Tolerance and dependence were also elicited by repeated injections of D-Ala^2-Met-enkephalin amide (Pert *et al.*, 1976*b*) intraventricularly into rats over a period of 9 days (Pert, 1976) and by pituitary endorphins as well (Blasig and Herz, 1976). Studies of the effects of culturing neuroblastoma $\times$ glioma hybrid cells for a number of hours in the presence of Met-enkephalin (Lampert *et al.*, 1976) or Leu-enkephalin (Brandt *et al.*, 1976*b*) show that the increased adenylate cyclase activity associated with opiate tolerance and dependence is elicited. Thus, the endogenous opiate peptides clearly have dependence liability in animals and will probably do so in humans as well if this can ever be tested. Reports of analgesia after systemic administration of β-LPH 61–91 Tseng *et al.*, 1976) if they can be confirmed (see Bloom *et al.*, 1976) indicate that these peptides may reach the brain from the blood and therefore the question of addiction liability becomes an important obstacle to the development of clinically useful drugs based on the enkephalins or other endogenous opiate peptides. The generally accepted picture of naloxone as an inert drug in the absence of prior opiate treatment may require modification in the light of the dependence liability of these peptides and of recent studies which show that regional glucose utilization in brain of naive animals is profoundly affected by naloxone (Sakurada *et al.*, 1976).

The possible role of opiate peptide turnover or levels in the development of tolerance and dependence is also being studied. Simantov and Snyder (1976*b*) have found elevated enkephalin activity in the brains of morphine dependent rats whereas Clouet and Ratner (1976) report a decreased incorporation of [^{3}H]glycine into enkephalin in morphine-tolerant rats. These observations are not necessarily contradictory in that they could both be consistent with decreased enkephalin turnover in the addicted state. Further studies of these interesting phenomena are clearly needed.

4. How are synthesis, release and degradation carried out and controlled?

These fascinating questions are just beginning to be approached experimentally and will doubtless account for an ever increasing share of the literature. Although much more opiate peptide activity is found in pituitary than in brain, hypophysectomy has been shown not to affect brain levels of endorphin activity (Cheung and Goldstein, 1976). Thus the brain, as well as the pituitary, must be capable of synthesizing opiate peptides. Indeed, the [^{3}H]glycine incorporation experiments of Clouet and Ratner (1976) were performed with intracisternal injections of labeled glycine. Although it still seems likely that β-lipotropin is the precursor of the opiate peptides (Bradbury *et al.*, 1976*a,b*; Chretien *et al.*, 1976; Ronai *et al.*, 1976; Lazarus *et al.*, 1976), direct evidence has not yet been provided.

Experiments showing release of opiate peptides from pituitary (Kromer *et al.*, 1976) or brain (Smith *et al.*, 1976) have been carried out. In brain preparations, release requiries Ca^{++} and is evoked by depolarization with high K$^+$. Whether specific release mechanisms exist in either brain or pituitary is not yet known. Stress may play a role in releasing these peptides (Akil *et al.*, 1976).

Although the endogenous opiate peptides are clearly susceptible to proteolytic degradation (Hambrock *et al.*, 1976; Jacquet *et al.*, 1976), there is scant information yet available on the mechanisms of control of this process or on the number of the enzymes involved.

8. Addendum References

Akil, H., Madden IV, J., Patrick, R. L., and Barchas, J. D., 1976, Stress-induced increase in endogenous opiate peptides: Concurrent analgesia and its partial reversal by naloxone, in *Opiates and Endogenous Opioid Peptides* (H. W. Kosterlitz, ed.), pp. 63–70, North Holland, Amsterdam.

Belluzzi, J. D., Grant, N., Garsky, V., Sarantakis, D., Wise, C. D., and Stein, L., 1976, Analgesia induced *in vivo* by central administration of enkephalin in rat, *Nature* **260**:625–626.

Blasig, J., and Herz, 1976, Tolerance and dependence induced by morphine-like pituitary peptides in rats, *N-S Arch. Physiol.* **294**:297–300.

Bleich, H. E., Cutnell, J. D., Day, A. R., Freer, R. J., Glasel, J. A., and McKelvy, J. F., 1976, Preliminary analysis of ^{3}H and ^{13}C spectral and relaxation behavior in methionine-enkephalin, *Proc. Nat. Acad. Sci. U.S.A.* **73**:2589–2593.

Bloom, F., Segal, D., Ling, N., and Guillemin, R., 1976, Endorphins: Profound behavioral effects in rats suggest new etiological factors in mental illness, *Science* **194**:630–632.

Bradbury, A. F., Feldberg, W. F., Smyth, D. G., and Snell, C. R., 1976*a*, Lipotropin C-fragment: An endogenous peptide with potent analgesic activity, in *Opiates and Endogenous Opioid Peptides* (H. W. Kosterlitz, ed.), pp. 9–18, North Holland, Amsterdam.

Bradbury, A. F., Smyth, D. G., and Snell, C. R., 1976*b*, Lipotropin: Precursor to two biologically active peptides, *Biochem. Biophys. Res. Commun.* **69**:950–956.

Brandt, M., Gullis, R. J., Fischer, K., Buchen, C., Hamprecht, B., Moroder, L., and Wunsch, E.,

1976a, Enkephalin regulates levels of cyclic nucleotides in neuroblastoma × glioma hybrid cells, *Nature* **262**:311–313.

Brandt, M., Fischer, K., Moroder, L., Wunsch, E., and Hamprecht, B., 1976b, Enkephalin evokes biochemical correlates of opiate tolerance and dependence in neuroblastoma × glioma hybrid cells, *FEBS Letters* **68**:38–40.

Büscher, H. H., Hill, R. C., Römer, D., Cardinaux, F., Closse, A., Hauser, D., and Pless, J., 1976, Evidence for analgesic activity of enkephalin in the mouse, *Nature* **261**:423–425.

Byck, R., 1976, Peptide transmitters: A unifying hypothesis for euphoria, respiration, sleep and the action of lithium, *Lancet* **2**:72—73.

Cheung, A. L., and Goldstein, A., 1976, Failure of hypophysectomy to alter brain content of opioid peptides (endorphins), *Life Sci.* **19**:1005–1006.

Chretien, M., Benjannet, S., Dragon, N., Seidah, N. G., and Lis, M., 1976, Isolation of peptides with opiate activity from sheep and human pituitaries: Relationship to beta-lipotropin. *Biochem. Biophys. Res. Commun.* **72**:472–478.

Clouet, D. H., and Ratner, M., 1976, The incorporation of ^{3}H-glycine into enkephalins in the brains of morphine treated rats, in *Opiates and Endogenous Opioid Peptides* (H. W. Kosterlitz, ed.), pp. 71–78, North Holland, Amsterdam.

Cox, B. M., Goldstein, A., and Li, C. H., 1976, Opioid activity of a peptide, β-lipotropin-(61–91), derived from β-lipotropin, *Proc. Nat. Acad. Sci. U.S.A.* **73**: 1821–1823.

El-Sobky, A., Dostrovsky, J. O., and Wall, P. D., 1976, Lack of effect of naloxone on pain perception in humans, *Nature* **263**:783–784.

Frederickson, R. C. A., Nickander, R., Smithwick, E. L., Shuman, R., and Norris, F. H., 1976, Pharmacological activity of Met-enkephalin and analogues *in vitro* and *in vivo*–depression of single neuronal activity on specified brain regions, in *Opiates and Endogenous Opioid Peptides* (H. W. Kosterlitz, ed.), pp. 239–246, North Holland, Amsterdam.

Gent, J. P., and Wolstencroft, J. H., 1976, Actions of morphine, enkephalin and endorphin on single neurones in the brain stem, including the raphe and the periaqueductal gray, of the cat, in *Opiates and Endogenous Opioid Peptides* (H. W. Kosterlitz, ed.), pp. 217–224, North Holland, Amsterdam.

Gentleman, S., Ross, M., Lowney, L. I., Cox, B. M., and Goldstein, A., 1976, Pituitary endorphins, in *Opiates and Endogenous Opioid Peptides* (H. W. Kosterlitz, ed.), pp. 27–34, North Holland, Amsterdam.

Goldstein, A., 1976, Opioid peptides (endorphins) in pituitary and brain, *Science* **193**:1081–1086.

Goldstein, A., and Hilgard, E. R., 1975, Failure of opiate antagonist naloxone to modify hypnotic analgesia. *Proc. Nat. Acad. Sci. U.S.A.* **72**:2041–2043.

Goldstein, A., and Lowery, P. J., 1975, Effect of opiate antagonist naloxone on body temperature in rats. *Life Sci.* **17**:927–931.

Goldstein, A., Pryor, G. T., Otis, L. S., and Larsen, F., 1976, Role of endogenous opioid peptides: Failure of naloxone to influence shock escape threshold in rat, *Life Sci.* **18**:599–604.

Graf, L., Ronai, A. Z., Bajusz, S., Cseh, G., and Szekely, J. F., 1976, Opiate agonist activity of β-lipotropin fragments: A possible biological source of morphine-like substances in the pituitary, *FEBS Lett.* **64**:181–184.

Gullis, R. J., Buchen, C., Moroder, L., Wunsch, E., and Hamprecht, B., 1976, Opiate-like effects of enkephalins on neuroblastoma × glioma hybrids, in *Opiates and Endogenous Opioid Peptides* (H. W. Kosterlitz, ed.), pp. 143–152, North Holland, Amsterdam.

Hambrock, J. M., Morgan, B. A., Rance, M. J., and Smith, C. F. C., 1976, Mode of deactivation of enkephalins by rat and human plasma and rat brain homogenates, *Nature* **262**:782–783.

Hill, R. G., Pepper, C. M., and Mitchell, J. F., 1976, The depressant action of iontophoretically applied Met-enkephalin on single neurones in rat brain, in *Opiates and Endogenous Opioid Peptides* (H. W. Kosterlitz, ed.), pp. 225–230, North Holland, Amsterdam.

Jacquet, Y. F., and Lajtha, A., 1974, Paradoxical effects after microinjection of morphine in the periaqueductal gray matter of the rat, *Science* **185**:1055–1057.

Jacquet, Y. F., and Marks, N., 1976, C-Fragment of beta-lipotropin: Endogenous neuroleptic or antipsychotogen, *Science* **194**:632–635.

Jacquet, Y. F., Marks, N., and Li, C. H., 1976, Behavioral and biochemical properties of "opioid" peptides, in *Opiates and Endogenous Opioid Peptides* (H. W. Kosterlitz, ed.), pp. 411–414, North Holland, Amsterdam.

Jones, C. R., Gibbons, W. A., and Garsky, V., 1976, Proton magnetic resonance studies of conformation and flexibility of enkephalin peptides, *Nature* 262:779–782.

Klee, W. A., Lampert, A., and Nirenberg, M., 1976, Dual regulation of adenylate cyclase by endogenous opiate peptides, in *Opiates and Endogenous Opioid Peptides* (H. W. Kosterlitz, ed.), pp. 153–160, North Holland, Amsterdam.

Kosterlitz, H. W. (ed.), 1976, *Opiates and Endogenous Opioid Peptides,* North Holland, Amsterdam, 456 p.

Kromer, W., Bläsig, J., Westenthanner, A., Haarmann, I., and Teschemacher, H., 1976, Characteristics of porcine pituitary peptides which act like opiates: Release from anterior lobes *in vitro*— relation between central effects and degradation by brain enzymes, in *Opiates and Endogenous Opioid Peptides* (H. W. Kosterlitz, ed.), pp. 1–8, North Holland, Amsterdam.

Lal, H., Miksic, S., and Smith, N., 1976, Naloxone antagonism of conditioned hyperthermia: An evidence for release of endogenous opioid, *Life Sci.* 18:971–976.

Lampert, A., Nirenberg, M., and Klee, W. A., 1976, Tolerance and dependence evoked by an endogenous opiate peptide, *Proc. Nat. Acad. Sci. U.S.A.* 73:3165–3167.

Lazarus, L. H., Ling, N., and Guillemin, R., 1976, Beta-lipotropin as a pro-hormone for morphinomimetic peptides endorphins and enkephalins, *Proc. Nat. Acad. Sci. U.S.A.* 73:2156–2159.

Ling, N., and Guillemin, R., 1976, Morphinomimetic activity of synthetic fragments of beta-lipotropin and analogs, *Proc. Nat. Acad. Sci. U.S.A.* 73:3308–3310.

Loh, H. H., Tseng, L. F., Wei, E., and Li, C. H., 1976, Beta-endorphin is a potent analgesic agent, *Proc. Nat. Acad. Sci. U.S.A.* 73:2895–2898.

Lord, J. A. H., Waterfield, A. A., Hughes, J., and Kosterlitz, H. W., 1976, Multiple opiate receptors, in *Opiates and Endogenous Opioid Peptides* (H. W. Kosterlitz, ed.), pp. 275–280, North Holland, Amsterdam.

Minneman, K. P., and Iversen, L. L., 1976, Enkephalin and opiate narcotics increase cyclic GMP accumulation in slices of rat neostriatum, *Nature* 262:313–314.

Motomatsu, T., Lis, M., Seidah, N., and Chretien, M., 1976, Cataleptic effect of 61–91 beta-lipotropic hormone in rats, *Can. J. Neurological Sci.,* in press.

Pert, A., 1976, Behavioral pharmacology of D-alanine[2]-methionine-enkephalin amide and other long-acting opiate peptides, in *Opiates and Endogenous Opioid Peptides* (H. W. Kosterlitz, ed.), pp. 87–94, North Holland, Amsterdam.

Pert, C. B., Bowie, D. L., Fong, B. T. W., and Chang, J. -K., 1976*a,* Synthetic analogues of Met-enkephalin which resist enzymatic destruction, in *Opiates and Endogenous Opioid Peptides* (H. W. Kosterlitz, ed.), pp. 79–86, North Holland, Amsterdam.

Pert, C. B., Pert, A., Chang, J. -K., and Fong, B. T. W., 1976*b,* D-Ala[2]-Met-Enkephalinamide: Potent, long-lasting synthetic pentapeptide analgesic, *Science* 194:330–332.

Ronai, A. Z., Szekely, J. I., Graf, L., Dunai-Kovács, Z., and Bajusz, S., 1976, Morphine-like analgesic effect of a pituitary hormone, beta-lipotropin, *Life Sci.* 19:733–738.

Roques, B. P., Garbay-Jaureguiberry, C., Oberlin, R., Anteunis, M., and Lala, A. K., 1976, Conformation of Met[5]-enkephalin determined by high field PMR spectroscopy, *Nature* 262:778–779.

Sakurada, D., Shinohara, M., Klee, W. A., Kennedy, C., and Sokoloff, L., 1976, Local cerebral glucose utilization following acute or chronic morphine administration and withdrawal, *Neuroscience Abstracts,* 6th Annual Meeting of the Society for Neuroscience, p. 613.

Schiller, P. W., Yam, C. F., and Lis, M., 1977, Evidence for topographical analogy between methionine-enkephalin and morphine derivatives, *Biochemistry,* in press.

Simantov, R., and Snyder, S. H., 1976*a,* Brain-pituitary opiate mechanisms: Pituitary opiate receptor binding, radioimmunoassays for methionine enkephalin and leucine enkephalin, and ^{3}H-enkephalin interactions with the opiate receptor, in *Opiates and Endogenous Opioid Peptides* (H. W. Kosterlitz, ed.), pp. 41–48, North Holland, Amsterdam.

Simantov, R., and Snyder, S. H., 1976*b,* Elevated levels of enkephalin in morphine dependent rats, *Nature* 262:505–507.

Smith, T. W., Hughes, J., Kosterlitz, H. W., and Sosa, R. P., 1976, Enkephalins: Isolation, distribution and function, in *Opiates and Endogenous Opioid Peptides* (H. W. Kosterlitz, ed.), pp. 57–62, North Holland, Amsterdam.

Teschemacher, H., Blasig, J., and Kromer, W., 1976, Porcine pituitary peptides with opiate-like activity: Partial purification and effects in rat after intraventricular injection, *N-S Arch. Physiol.* **294:**293–295.

Tseng, L. -F., Loh, H. H., and Li, C. H., 1976, β-Endorphin as a potent analgesic by intravenous injection, *Nature* **263:**239–240.

Van Nueten, J. M., Janssen, P. A. J., and Fontaine, J., 1976, Unexpected reversal effects of naloxone on the guinea pig ileum, *Life Sci.* **18:**803–810.

Wahlstrom, A., Johansson, L., and Terenius, L., 1976, Characterization of endorphines (endogenous morphine-like factors) in human CSF and brain extracts, in *Opiates and Endogenous Opioid Peptides* (H. W. Kosterlitz, ed.), pp. 49–56, North Holland, Amsterdam.

Wajda, I. J., Neidle, A., Ehrenpreis, S., and Manigault, I., 1976, Properties and distribution of morphine-like substances, in *Opiates and Endogenous Opioid Peptides* (H. W. Kosterlitz, ed.), pp. 129–136, North Holland, Amsterdam.

Wei, E., and Loh, H., 1976, Physical dependence on opiate-like peptides, *Science* **193:**1262–1263.

Zieglgänsberger, W., and Fry, J. P., 1976, Actions of enkephalin on cortical and striatal neurones of naive and morphine tolerant/dependent rats, in *Opiates and Endogenous Opioid Peptides* (H. W. Kosterlitz, ed.), pp. 231–238, North Holland, Amsterdam.

14

Behavioral Effects of Peptides

D. de Wied and W.H. Gispen

1. Introduction

The pituitary–adrenal system plays an essential role in homeostatic functions. Numerous aspects of stress-induced pituitary–adrenal activation in relation to peripheral mechanisms of adaptation have been studied since Selye's first observations on the general adaptation syndrome some 40 years ago (Selye, 1950). Little attention, however, has been paid to the brain as a target for these hormones. Clinical observations frequently commented on psychological changes in addition to electrophysiological alterations in hyper- as well as hypocorticism (Cleghorn, 1957; Von Zerssen, 1976). Many a laboratory experiment during the last decade, however, disclosed the implication of a number of pituitary and hypothalamic hormonal peptides on various brain functions. The importance of these entities was revealed by observations on behavioral disturbances following extirpation of the pituitary gland in rats (de Wied, 1969); in animals with hereditary diabetes insipidus, which lack the ability to synthesize vasopressin (de Wied *et al.*, 1975a); or in rats in which the action of vasopressin in the brain is neutralized by intraventricular administration of specific vasopressin antiserum (van Wimersma Greidanus *et al.*, 1975a). Impaired behavior associated with the measures mentioned above can be readily amended by treatment with ACTH, melanocyte-stimulating hormone (MSH), or vasopressin, but also with fragments of these polypeptides that in themselves are practically devoid of classic endocrine effects such as the stimulation of corticoidogenesis in the adrenal cortex in case of ACTH fragments, or antidiuretic, vasopressor, and other endocrine effects in case of vasopressin fragments. On the basis of these findings, it was postulated that the pituitary

D. de Wied and W.H. Gispen · Rudolf Magnus Institute for Pharmacology, Medical Faculty, University of Utrecht, Vondellaan 6, Utrecht, The Netherlands.

manufactures and releases peptides designated as *neuropeptides* that are involved in the formation and maintenance of new behavior patterns (de Wied, 1969). Of more recent date are observations on the CNS effects of various hypothalamic and other brain oligopeptides (Prange *et al.*, 1975) that suggest that the hypothalamic–pituitary complex employs a great variety of oligopeptides that carry specific information to the brain to facilitate the generation of behavior patterns in response to environmental needs.

2. Implication of the Pituitary Gland in Acquisition and Maintenance of Conditioned Avoidance Behavior

2.1. Adenohypophysectomy and Hypophysectomy on Acquisition of Active and Passive Avoidance Behavior

2.1.1. Effect of Pituitary Peptides

Removal of the pituitary gland in rats may induce a severe deficit in the rate of acquisition of a two-way active shuttle-box avoidance response (Applezweig and Baudry, 1955; Applezweig and Moeller, 1959; de Wied, 1969). Bélanger (1958), using a somewhat different situation, found that the performance of hypophysectomized rats was nearly as good as that of control rats, although the response of the hypophysectomized rats to the conditioned stimulus (CS) was significantly slower. We found that adenohypophysectomized rats are far inferior to sham-operated controls in acquiring the avoidance response in the shuttle box (de Wied, 1964). These findings confirmed those of Applezweig and Baudry (1955), and indicated that the deficiency in avoidance-learning is caused by the absence of anterior pituitary hormones. Indeed, the removal of the posterior lobe of the pituitary, which includes the intermediate lobe, does not interfere with avoidance acquisition (de Wied, 1965). Total hypophysectomy induced the same behavioral disturbance as adenohypophysectomy (de Wied, 1969). Debility appeared to be a factor in poor acquisition by hypophysectomized rats, although the rats were capable of making all bodily movements normally involved in performing the avoidance response in shuttle-box conditioning (Gispen *et al.*, 1973). Replacement therapy with adrenocortical steroids, thyroxin, and testosterone improves the physical condition and the rate of acquisition of a shuttle-box avoidance response, but not toward that of sham-operated controls (de Wied, 1964, 1971a). This improvement is due to effects of thyroxin and testosterone. Control of the physical ability of hypophysectomized rats by dietary means also allows hypophysectomized animals to acquire a shuttle-box avoidance response (Harris, 1973). Administration of natural ACTH as a long-acting zinc phosphate preparation in amounts that maintain adrenal activity in hypophysectomized rats nearly completely restores avoidance learning. This effect is not the result of ACTH-mediated adrenal activity, since dexamethasone failed to stimulate shuttle-box avoidance behavior of hypophysectomized rats (de Wied, 1971a). Thus, the behavioral effect of ACTH must be due to an extra target effect of this polypeptide hormone. That it is was demonstrated with the use of long-acting zinc phosphate complexes of structurally related peptides such as α-MSH, $ACTH_{1-10}$, and $ACTH_{4-10}$, which are virtually devoid of corticotropic activities. These peptides are as effective as $ACTH_{1-24}$ in restoring the rate of acquisition of the avoidance response in hypophysectomized rats (de Wied, 1969). Studies on the influence of $ACTH_{4-10}$ on endocrine and metabolic function, on motor and sensory capacities, and on the general health and condition of the hypophysectomized rat suggested that the effect of this polypeptide was not due to any of these functions. The sensitivity of the hypophysectomized rat to electric shock is even greater than that of sham-operated controls (Gispen *et al.*, 1970). Escape speed in a runway, however, which is markedly decreased in hypophysectomized rats, was significantly improved by treatment with $ACTH_{4-10}$, although not to the level of that of sham-operated rats. There might therefore be an effect of the heptapeptide on motor or sensory capacities or both (de Wied, 1969). This conclusion may be too simple, since motivational phenomena (see Section 3.1) might also be involved in runway performance (Barry and Miller, 1965). Furthermore, it was found that hypophysectomy also affects one-way active-avoidance behavior as studied in either a jump box (Gispen, 1970) or in a pole-jumping test. The rate of acquisition by hypophysectomized rats in these behavioral situations is not so poor as in two-way active avoidance behavior in the shuttle box. The performance in the jump box and in the pole-jumping test can be restored by treatment with $ACTH_{4-10}$. In the latter situation, the vasopressin fragment [des-Gly–NH_2^9,Arg^8]vasopressin (DG-AVP) also restores the performance, as depicted in Fig. 1. In view of these findings, it was postulated that peptides like $ACTH_{4-10}$ or closely resembling this heptapeptide normally operate in acquisition of new behavioral patterns. Such peptides, designated as

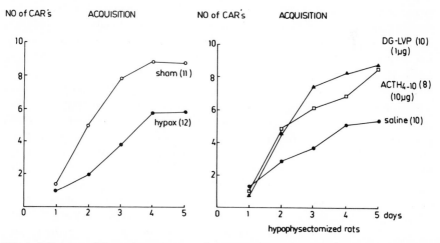

FIG. 1. Influence of hypophysectomy and treatment with neuropeptides on the rate of acquisition of a pole-jumping avoidance response. Hypophysectomy was performed via the transauricular route 3 days prior to the first training session. The training consisted of a daily session of 10 trails in 10 min on 5 consecutive days. The peptides were dissolved in saline and administered subcutaneously 1 hr prior to each acquisition session.

neuropeptides, may be synthesized by the pituitary gland or generated from precursor molecules and released on adequate stimulation to affect brain structures involved in learning processes.

In attempting to test the hypothesis that ACTH has a general excitatory effect that in a fear situation would augment fear-motivated responses, Weiss *et al.* (1970) employed a passive-avoidance test. Hypophysectomized rats showed attenuated avoidance behavior under active and passive conditions. ACTH enhanced the poor avoidance of hypophysectomized rats in both tests. Since the effect was obtained in both active and passive avoidance behavior, it could not be due to motor-activity differences. Deficient passive avoidance behavior of hypophysectomized rats was also found by Lissák and Bohus (1972) when low shock intensities were used, but not if animals were exposed to higher shock levels. These authors were able to restore passive avoidance behavior of the hypophysectomized rat with $ACTH_{1-24}$ and [Lys^8] vasopressin (LVP).

2.1.2. Presence of a Behaviorally Active Peptide in Pituitary Material

The availability of a porcine pituitary extract in large quantities afforded the opportunity to undertake a search for behaviorally active pituitary peptides (Lande *et al.*, 1971). Initial screening studies for activity were performed in hypophysectomized rats, using shuttle-box avoidance behavior. Fractions were obtained by ion-exchange chromatography on carboxylmethyl cellulose. The most potent fraction was further resolved by gel filtration on Sephadex G-25. Using a Sephadex column calibrated for molecular weight vs. elution volume, four peaks of biological activity corresponding to materials with a molecular size of about 40, 20, 13, and 8 amino acid residues were obtained. The smallest component was further purified by ion-exchange chromatography on DEAE-Sephadex. The product was identified as [des-Gly–NH_2^9, Lys^8] vasopressin (DG-LVP) (Lande *et al.*, 1971). In subsequent experiments, a pole-jumping avoidance test was used in which inhibition of extinction of the conditioned avoidance response (CAR) was measured (Lande *et al.*, 1973). Using this test, the presence of at least three more stable and behaviorally potent peptides in pituitary extracts was demonstrated and their amino acid compositions were determined. A number of questions arise with regard to these studies: e.g., whether these peptides exist naturally in the pituitary glands or are artifacts, whether such compounds have a physiological role, whether they are released as such from the pituitary gland. The starting material employed in these experiments was a side-product of commercial-scale isolation of porcine ACTH. The extract consists of peptides that fall in a narrow range of physicochemical properties. That some of the peptides were generated during collection and extraction is possible. However, a number of pituitary peptides very susceptible to enzymatic degradation have been isolated from the same extract, suggesting that extensive degradation did not occur. On the other hand, evidence is accumulating that peptides with potent biological activity are derived possibly by enzymatic degradation from big precursor molecules in the pituitary. In fact, the amino acid composition of these peptides suggests that they are generated from lipotropic hormone (β-LPH).

2.2. Posterior Lobectomy on Acquisition and Maintenance of Active Avoidance Behavior—Effect of Pituitary Peptides

In contrast to adenohypophysectomy, the removal of the posterior pituitary, including the intermediate lobe, does not interfere with avoidance learning. However, the extinction of a previously learned avoidance response is markedly facilitated. Posterior-lobectomized rats are unable to maintain the response in the absence of punishment (de Wied, 1965). Escape behavior in the runway is not affected in posterior-lobectomized rats, suggesting that neither motor nor sensory failures are involved (de Wied, 1969). A relatively crude extract of posterior pituitary origin, pitressin, given as a long-acting preparation in amounts that restore water metabolism in these mildly diabetes insipidus rats, also restores the preservation of shuttle-box avoidance behavior. Treatment with purified LVP had the same effect, but so did long-acting preparations of ACTH and α-MSH. It was subsequently found that treatment of posterior-lobectomized rats with pitressin given as a long-acting preparation during acquisition had only the same effect on extinction as treatment during the extinction period, while treatment with ACTH during acquisition resulted in only a small effect on the rate of extinction of the avoidance response. ACTH had a much stronger effect if it was injected during the extinction period (de Wied, 1969). From these experiments, it was inferred that pitressin preserves a CAR irrespective of the time of treatment, while ACTH inhibits extinction only during the period of treatment. Thus, the mechanisms by which these two structurally unrelated principles affect the maintenance of a learned response are basically different. That they are has been amply confirmed in experiments in intact rats, as will be discussed in Sections 3.1 and 4.1.

2.3. Active and Passive Avoidance Behavior in Rats with Hereditary Diabetes Insipidus

The posterior-lobectomized rat, although deficient in posterior lobe principles, still manufactures vasopressin (Moll and de Wied, 1962). The miniature posterior lobe that is formed following posterior lobectomy releases small quantities of vasopressin produced in the remnants of the magnocellular nuclei in the hypothalamus. A superior model to investigate the physiological role of vasopressin in learning and memory is the Brattleboro strain rat, which lacks the ability to synthesize vasopressin due to a mutation in a single pair of autosomal loci (Valtin and Schroeder, 1964; Valtin, 1967). Homozygous hereditary diabetes insipidus rats (HO-DI) were inferior in acquiring a shuttle-box avoidance response as compared with heterozygous littermates (HE-DI) and Wistar strain animals (Bohus *et al.*, 1975a). Extinction in HO-DI rats was fast, significantly less rapid in HE-DI, but slowest in Wistar strain animals. In this respect it is noteworthy that although HE-DI rats are able to synthesize biologically active vasopressin, they have a partial deficit in the process of synthesis and release of this nonapeptide (Moses and Miller, 1970). Like hypophysectomized Wistar rats (Gispen *et al.*, 1970), HO-DI and HE-DI rats are more sensitive to electric shock than intact Wistar strain animals (Bohus *et al.*, 1975a). Habituation in open-field activity was also disturbed in Brattleboro rats, since the rate of ambulation and rearing remained constant during consecutive sessions, in contrast to that of Wistar rats, which declined gradually (Bohus *et al.,* 1975*a*). The rate of acquisition of a one-way active pole-jumping avoidance response appeared to be the same in HO-DI and HE-DI rats, while Wistar strain animals acquired the response significantly faster. Extinction again was very rapid in HO-DI rats, and slowest in Wistar controls. Thus, in the absence of vasopressin, the maintenance of a CAR is severely disturbed. Memory impairment in HO-DI rats is best demonstrated in a one-trial passive avoidance situation. Passive avoidance behavior was absent in these animals when tested 24 hr or later after the learning trial. HE-DI rats showed only a partial defect, and needed more intensive shock to exhibit full retention of the response. However, passive avoidance behavior was exhibited in HO-DI rats when tested immediately after the learning trial, and only partially at 3 hr after the learning trial. Thus, memory rather than learning processes seem to be affected in the absence of vasopressin. These observations suggest that the consolidation of memory is under the control of vasopressin. Administration of AVP or DG-LVP immediately after the learning trial restored passive avoidance behavior in HO-DI rats (de Wied *et al.,* 1975*a*). It should also be kept in mind that HO-DI rats have a hypoactive endocrine system, as indicated by smaller pituitaries, adrenals, and testes, and a lower growth rate, in comparison with those of HE-DI rats. Although such severe endocrine disturbances are not found in posterior-lobectomized rats, the similarity between the behavior of these and Brattleboro rats is striking, and adds further support to the hypothesis that vasopressin plays an important role in memory processes.

3. Behaviorally Active Adrenocorticotropic Hormone Fragments

3.1. Behavioral Effects in Intact Rats

Treatment with ACTH or ACTH fragments generally does not alter the rate of acquisition of shock-motivated active avoidance behavior in intact rats (Kelsey, 1975; Murphy and Miller, 1955). Under some conditions, however, ACTH may improve acquisition. Thus, Beatty *et al.* (1970) reported that ACTH facilitates acquisition of shuttle-box avoidance behavior at a "high" but not at a "moderate" footshock intensity. In contrast, Stratton and Kastin (1974) found that α-MSH improved acquisition of the same behavior at a "low" but not a "high" footshock level. Guth *et al.* (1971) showed that administration of ACTH during acquisition of an appetite-motivated response increased lever-press responding late in the training period. Kastin *et al.* (1975) reported that α-MSH facilitates acquisition of food-rewarded behavior in a multiple T-maze. Improvement of acquisition by $ACTH_{4-10}$ was also found by Isaacson *et al.* (1975) in rats trained to use environmental cues to detect the correct goal box in an elevated X-maze.

Extinction of avoidance behavior seems more senstitive to the treatment with

ACTH. Murphy and Miller (1955) found that ACTH administered during shuttle-box training delayed subsequent extinction of the behavior. A more pronounced effect of ACTH is found, however, when ACTH is given during the extinction period (de Wied, 1967). Although the long-term administration is accompanied by hypercorticism, the influence is independent of the action of ACTH on the adrenal cortex. ACTH is also active on extinction of shuttle-box avoidance behavior in adrenalectomized rats (Miller and Ogawa, 1962). In addition, α-MSH, β-MSH, $ACTH_{1-10}$, and $ACTH_{4-10}$ are as active as natural ACTH in delaying extinction of the avoidance response (de Wied, 1966), while $ACTH_{11-24}$ is ineffective in this respect (de Wied *et al.*, 1968). The same results were obtained in the one-way active pole-jumping avoidance test. These earlier studies revealed that the heptapeptide $ACTH_{4-10}$, which is common to ACTH, α- and β-MSH, and LPH, is the shortest sequence possessing full behavioral activity. These experiments were replicated using saline instead of zinc phosphate as peptide vehicle, thus using short-acting preparations. Exactly the same results were obtained.

ACTH also affects passive avoidance behavior. Lissák *et al.* (1957) were the first to show an effect of ACTH in this respect in dogs. Subsequently, Levine and Jones (1965) found that ACTH affected acquisition of passive avoidance behavior in rats. Using a single step-through, one-trial passive avoidance procedure Ader *et al.* (1972) found that $ACTH_{1-24}$ administered 1 hr prior to the 24-hr retention trial markedly facilitated passive avoidance behavior. The same result was obtained following injection with $ACTH_{1-10}$. A minor effect was found with $ACTH_{11-24}$ or $ACTH_{25-39}$ (Fig. 2). Thus, in this test also, the behaviorally active core of ACTH resides in the N-terminal part of the molecule. Similar effects on passive avoid-

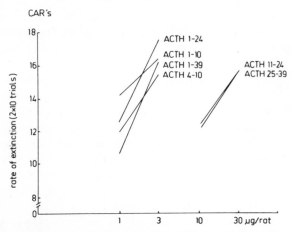

FIG. 2. Potencies of ACTH fragments in delaying extinction of a pole-jumping avoidance response. Rats were trained for 3 consecutive days during 1 daily session of 10 min (10 trials). On the 4th day, an extinction session was run (10 trials; presentation of conditioned stimulus alone, i.e., not followed by the unconditioned stimulus), and all rats that made 8 or more avoidance responses were injected subcutaneously with the respective peptide dissolved in saline. Extinction sessions were run again 2 and 4 hr later, and the total number of positive responses in these 2 extinction sessions were used to determine the potency.

ance behavior were found by Sandman *et al.* (1973) and Dempsey *et al.* (1972) with α-MSH. McGaugh *et al.* (1975) reported effects of ACTH on passive avoidance behavior. The effects depended on the dose used: low doses induced latencies higher than the latency of saline-treated rats, while high doses had the reverse effect.

Approach behavior is also affected by ACTH. ACTH and α-MSH inhibit extinction of an appetite approach response (Sandman *et al.*, 1969; Leonard, 1969; Gray, 1971; Guth *et al.*, 1971). Garrud *et al.* (1974) showed that hungry rats trained to run for food in a runway extinguished significantly slower under the treatment with $ACTH_{1-24}$ or $ACTH_{4-10}$. Similarly, hungry rats trained to press a lever to obtain food on a fixed ratio of reinforcement and injected with synthetic α-MSH were found to have delayed extinction of the task, as compared with control rats (Kastin *et al.*, 1973b, 1974). Although the speed of running may indicate increased motivation, it can also reflect general increased motor activity. Kastin *et al.* (1973a,b), however, found no evidence that it does. Indeed, activity in an open field is not affected by chronic treatment with ACTH fragments (Bohus and de Wied, 1966; Wolthuis and de Wied, 1975).

Albino rats tested in a procedure involving escape from shock in a Y-maze, were trained, after they had achieved the learning criterion, for the other arm of the maze. This reversal learning was facilitated by α-MSH (Sandman *et al.*, 1972). McGaugh *et al.* (1975) found improvement of discrimination learning in rats following ACTH. Many other studies aimed at the elucidation of the extrapigmentary effects of α-MSH have been performed in recent years, in particular by Kastin and associates (for a review, see Kastin *et al.*, 1975). These authors extrapolated the physiological role of MSH on pigment dispersion of the skin of lower vertebrates to an adaptive capacity of this polypeptide in mammals. This might well be an important role of MSH, but also of ACTH and related pituitary principles, which these peptides may have acquired during evolution (de Wied, 1974).

ACTH, $ACTH_{4-10}$, and related peptides attenuate carbon dioxide–induced amnesia for a passive avoidance response when administered prior to the retention test, but not when given prior to acquisition and the induction of amnesia (Rigter *et al.*, 1974). It appeared that $ACTH_{4-10}$ attenuated amnesia when administered within 8 hr prior to the retention test (Rigter *et al.*, 1975). The amnesic effect of $ACTH_{4-10}$ is independent of the nature of the type of amnesia, as it was subsequently demonstrated that reversal of amnesia could also be obtained when electroconvulsive shock was used. Rigter *et al.* (1975) further found that carbon dioxide–induced amnesia for a passive avoidance response remains present over a 2-week period. Irrespective of the duration of the interval between acquisition and retention test, administration of $ACTH_{4-10}$ before the retention test resulted in reversal of amnesia. These observations were interpreted to indicate that ACTH fragments have an effect on retrieval.

Bohus *et al.* (1975b) showed that $ACTH_{4-10}$ delays extinction of a sexually motivated approach response of male rats in a straight runway. Animals treated with $ACTH_{4-10}$ ran faster than controls even in the absence of a receptive female in the goal box. Copulation reward during acquisition, however, appeared to be essential for the effect of $ACTH_{4-10}$, since running activity of males that were prevented from copulation was not affected by $ACTH_{4-10}$. Motivational effects of $ACTH_{4-10}$ on sexual behavior were also found by Meyerson (1975). The urge in the female rat to seek contact with a vigorous male is dependent on ovarian hormones. The cyclic increase in sexual motivation during the estrous state disappears after ovariectomy (Meyerson and Lindström, 1973). $ACTH_{4-10}$ was investigated in ovariectomized rats using a technique to measure how much aversive stimulation (crossing an electric grid) a female was willing to take to reach contact with a sexually active male. Animals given $ACTH_{4-10}$ during the phase of adaptation and training performed at a significantly higher response level than controls when no other treatment (estrogens) was given (Meyerson, 1975).

3.2. Structure–Activity Studies

The 4–10 sequence of ACTH is common to ACTH, MSH, and LPH. Although the sequence has about 10^{-6} times the corticotropic activity of $ACTH_{1-39}$ or $ACTH_{1-24}$ (Schwyzer et al., 1971), this sequence is still considered to be the corticotropically active center of the ACTH molecule. Thus, going from the 4–23 sequence to the 5–24, there is a more than 100-fold decrease in steroidogenic potency (Sayers et al., 1974), and the 5–10 sequence is 30 times less steroidogenic than 4–10 (Schwyzer et al., 1971). The 11–24 sequence has affinity for the receptor in the adrenal membranes, but antagonizes the effect of both $ACTH_{1-39}$ and $ACTH_{1-10}$ (Seelig et al., 1971). $ACTH_{4-10}$ was initially found to be the shortest peptide that delays extinction of the avoidance response. However, further studies on the rate of extinction of the pole-jumping avoidance response revealed that the tetrapeptide $ACTH_{4-7}$ contains the essential elements required for the behavioral effect of ACTH (Greven and de Wied, 1973). Thus, the Trp and Arg residues could be removed without appreciable loss of activity. Interestingly, Trp is essential for MSH activity (Hofmann et al., 1960). These results demonstrate differences in structural requirements for behavioral activity ($ACTH_{4-7}$) and MSH activity ($ACTH_{6-9}$) (Otsuka and Inouye, 1964).

The decapeptide $ACTH_{1-10}$ in which Phe^7 was replaced by its D-enantiomer, appeared to exhibit an effect on extinction of a shuttle-box avoidance response opposite to that of the original all-L fragment (Bohus and de Wied, 1966). This effect was found in intact as well as in hypophysectomized rats. Although [D-Phe^7]$ACTH_{1-18}$ is a strong competitive antagonist of ACTH-stimulated adrenal adenylcyclase activity (Ide et al., 1972), the fact that [D-Phe^7]$ACTH_{1-10}$ was also active in hypophysectomized rats (Bohus and de Wied, 1966) argues against a behavioral antagonism of naturally occurring ACTH analogues. This argument is also supported by the fact that [D-Phe^7]ACTH analogues affect passive avoidance in the same way as [L-Phe^7] fragments, i.e., facilitation of passive avoidance behavior (de Wied, 1974). This finding suggests that [D-Phe^7]ACTH analogues contain an intrinsic new activity on extinction behavior. The heptapeptide $ACTH_{4-10}$ and the tetrapeptide $ACTH_{4-7}$ with a similar [D-Phe^7] substitution appeared to be as potent in facilitating extinction of pole-jumping avoidance behavior as [D-Phe^7]ACTH$_{1-10}$ (de Wied et al., 1975b).

The reversal of the behavioral effect was found only for analogues with the Phe^7 residue in the D-configuration. Successive replacement of the respective amino acid residues in the hexapeptide [Lys^8]ACTH$_{4-9}$ by D-enantiomers failed to facilitate extinction of the pole-jumping avoidance response. The reversal is therefore a privilege of [D-Phe^7]ACTH analogues only. All other D-enantiomer substitutions delayed extinction, as found with the all-L compounds. Such substituted peptides generally caused potentiation of the behavioral effect. This potentiation was most obvious when Lys^8 was replaced by its D-enantiomer. These results again indicate a dissociation between the requirements for behavioral and those for MSH activity. The latter activity increases when the aromatic residues Phe or Trp are replaced by their D-enantiomers, while MSH activity is lost when the basic histidyl or arginyl residues are converted to their D-enantiomers (Koida et al., 1966; Yajima et al., 1966).

Since Phe in position 7 seemed to be crucial for the behavioral effect, replacement of this residue with other amino acids was undertaken in subsequent experiments. Substitution by L-Leu in [Leu^7–Lys^8]ACTH$_{4-9}$ did not impair behavioral activity, while substitution by L-Trp in the same position even increased the inhibitory effect on extinction of the pole-jumping avoidance response. Replacement by L-pentamethylphenylalanine in position 7 further potentiated the behavioral potency. These results suggest that the electron-donor properties of the amino acid residue in position 7 correlate to some extent with behavioral potency. Introduction of the respective D-enantiomers also caused reversal of the behavioral effect, as was found with [D-Phe^7]ACTH analogues. The effect, however, was reduced by substitution with D-Leu or D-Trp, but somewhat augmented following substitution with D-pentamethylphenylalanine.

The substitution of Arg by Lys in position 8, which is accompanied by loss of steroidogenic activity in [Lys^8]ACTH$_{1-24}$ (Tesser et al., 1973) and loss of steroidogenic and MSH activity in [Lys^8]ACTH$_{6-10}$ (Chung and Li, 1967), does not reduce behavioral activity. Substitution of Lys at this position by its D-enantiomer caused a 30-fold potentiation. Replacement of Trp by Phe in position 9, which causes a marked decrease in steroidogenic potency in [Gln^5–Phe^9]ACTH$_{1-20}$–NH$_2$ (Hofmann et al., 1970), did not change the behavioral effect, and in the presence of [D-Lys^8], induced a 100-fold increase. Another change in the molecule that decreases the steroidogenic activity of ACTH and MSH activity is oxydation of the Met^4 residue (Dedman et al., 1955; Lo et al., 1961). This change also gives rise to an increase in behavioral potency. The introduction of the three modifications mentioned above in [Lys^8]ACTH$_{4-9}$ led to a 1000-fold potentiation. The MSH activity of the same molecule appeared to be 1000-fold less, although the steroidogenic activity was not so much reduced as was predicted from the various substitutions. A partial explanation for the effectiveness of the respective substitutions may be found in their protection against enzymatic degradation. Incubation of ^{3}H-labeled ACTH$_{4-9}$ analogues with plasma or brain extracts revealed that the *in vitro* half-life of various substituted analogues of $ACTH_{4-9}$ correlated well with their behavioral potency (Witter *et al.*, 1975).

Earlier observations indicated that [D-Phe^7]$ACTH_{7-10}$ delayed extinction of a shuttle-box avoidance response (de Wied, 1969), but more recent experiments in the pole-jumping test with newly synthesized material demonstrated facilitation of extinction. In both tests, the effect was markedly less than that of [D-Phe^7]$ACTH_{4-10}$. The major breakdown product of the behaviorally potentiated hexapeptide H–Met(O)–Glu–His–Phe–D-Lys–Phe–OH, i.e., Phe–D-Lys–Phe–, still possesses behavioral activity, albeit much less than $ACTH_{4-10}$. That it does suggest that essential features of the behavioral activity are not exclusively restricted to the locus $ACTH_{4-7}$, but are present in the 7–9 area as well. This latter sequence may contain information in a dormant form that needs potentiating modifications (e.g., chain elongation) to become manifest. This possibility is suggested by recent experiments showing that dogfish α-MSH is behaviorally as active as dogfish β-MSH (van Wimersma Greidanus *et al.*, 1975c). These peptides share the sequence His–Phe–Arg–Trp as a common characteristic. This sequence in itself has only minor behavioral effects (Greven and de Wied, 1973). Thus, the essential requirements for the behavioral effects of ACTH may not be restricted to a single locus or active core, but may be present in at least two regions of the molecule that *per se* show only marginal activity. These effects can be potentiated by chain elongation or by introduction of modifications that are believed to make the molecule more resistant to metabolic degradation. Interestingly, Eberle and Schwyzer (1975) reported that α-MSH possesses two message sequences, MSH_{4-10} and MSH_{11-13}, that independently trigger the hormone receptor response for melanin dispersion. In correct covalent combination, the two message sequences act cooperatively to potentiate their activities on the receptor.

3.3. Sites of Action

The sites of the behavioral action of ACTH fragments have been explored in rats bearing lesions in various brain regions, and by implantation of ACTH fragments in the brain. Lesions were made in the thalamic region because this area has been implicated in acquisition and extinction of conditioned avoidance behavior. Bilateral destruction of the nucleus parafascicularis facilitates extinction of shuttle-box avoidance behavior. Rats bearing lesions in this area did not respond to α-MSH with a delay in extinction, as occurs in sham-operated control animals (Bohus and de Wied, 1967a). In subsequent experiments on extinction of a pole-jumping avoidance response, it was found that $ACTH_{4-10}$ also fails to delay extinction of the CAR in rats bearing lesions in the nucleus parafascicularis. The hippocampus is another locus of action of ACTH fragments, since bilateral lesions in the anterodorsal hippocampus prevent the behavioral effects of $ACTH_{4-10}$ (van Wimersma Greidanus and de Wied, 1976). Microinjections of $ACTH_{1-10}$ or $ACTH_{4-10}$, which were not effective systemically, caused inhibition of extinction of the pole-jumping avoidance response when locally applied to the mesencephalic–diencephalic area at the posterior thalamic level (van Wimersma Grei-danus and de Wied, 1971). Implantation in other brain areas such as the ventral and anterior thalamic nuclei, and more rostral and caudal regions, was ineffective. Thus, the parafascicular area seems essential for the behavioral effects of ACTH fragments (van Wimersma Greidanus et al., 1974).

It has been shown that the nonspecific thalamic nuclei, such as the nucleus parafascicularis and the centrum medianum, have an important function in the maintenance of avoidance behavior (Cardo, 1965; Cardo and Valade, 1965). These nuclei may be important in integration of incoming information (Delacour, 1970, 1971). Lesions that destroy these nuclei interfere with avoidance behavior (Bohus and de Wied, 1967b), while electrical stimulation improves avoidance per-formance (Cardo, 1967). Extensive damage to the region of the nucleus posterior thalami produces a severe disturbance in visual discrimination learning, and sev-eral studies suggest that the posterior thalamoventral mesencephalic tract in the rat is functionally important in visually guided behavior (Rich and Thompson, 1965; R. Thompson et al., 1964). Cuts through the posterior thalamic ventromedial midbrain area result in severe retention deficits in visual learning (R. Thompson et al., 1970). In this respect, it is noteworthy that ACTH fragments in animal and man might affect attentive processes, particularly visual attention (Kastin et al., 1973b). Specific uptake of MSH and its metabolites in the reticular nucleus of the thalamus and in the striatum observed by Pelletier et al. (1975) support our find-ings on the relevance of the thalamic area for the behavioral effects of ACTH frag-ments. Since the anterodorsal hippocampus needs to be intact for the behavioral effects of $ACTH_{4-10}$, it is quite possible that the neural substrate of these neuro-peptides in the brain is restricted, not to an anatomical, but to a functional structure. It is likely that midbrain limbic connections need to be intact in order to allow neuropeptides to exert their behavioral effects.

3.4. Electrophysiological Effects

Studies on peripheral nerve and skeletal nuclei also indicate an extraadrenal effect of pituitary polypeptides, an action that may involve neuromuscular func-tions. ACTH increases muscle action potential amplitude and contraction height and delays fatigue in normal, adrenalectomized, and hypophysectomized rats. Adrenalectomy alone that is associated with augmented endogenous levels of ACTH also increases these parameters, while hypophysectomy has the reverse ef-fect (Strand et al., 1973, 1974). The same, in principle, was found with α- and β-MSH and $ACTH_{4-10}$. The heptapeptide [D-Phe^7]$ACTH_{4-10}$ was ineffective. Fa-tigue was more severe in hypophysectomized rats, and the effectiveness of the polypeptides was correspondingly greater (Strand, 1975).

Within the nervous system, ACTH has been shown to activate hypothalamic and midbrain cells (Steiner et al., 1969; Korányi et al., 1971; van Delft and Kitay, 1972), and to increase the excitability of spinal cord cells (Nicolov, 1967). These effects are generally opposite to those evoked by the application of adrenal steroids (Feldman et al., 1961; Pfaff et al., 1971). The interpretation of these studies with ACTH is difficult in view of direct effects and indirect effects mediated by the adrenal cortex. The first evidence for an extra target effect of ACTH on the CNS was provided by studies of Krivoy and Guillemin (Krivoy and Guillemin, 1961; Krivoy *et al.*, 1963), who found that β-MSH augments the evoked potentials in the dorsal root preparation of the cat. This polypeptide also decreased the amplitude of the spontaneous electric discharge of the knife fish (Krivoy *et al.*, 1962). Kastin *et al.* (1968) reported electroencephalographic changes in the human with α-MSH, while Endröczi *et al.* (1970) showed that treatment with ACTH$_{1-24}$, as well as with ACTH$_{1-10}$, of volunteers who had been habituated to sound or visual stimuli led to the disappearance of habituation and induced the same alertness as before habituation. High-voltage, intermediate-frequency activity and disintegration of alpha activity following α-MSH has been reported in rabbits (Dyster-Aas and Krakau, 1965), rats (Sandman *et al.*, 1971), and frogs (Denman *et al.*, 1972). More recent studies in free-moving dogs or rats with recording and stimulating electrodes in the brain indicate that ACTH fragments modify electrical activity in midbrain limbic structures. ACTH fragments have a central activating effect on the hippocampus of dogs (Urban *et al.*, 1974). In the rat, ACTH$_{4-10}$ induced a frequency shift in theta activity from 7.0 to 7.5 Hz in hippocampus and thalamus following stimulation of the reticular formation (Urban and de Wied, 1976). Since similar frequency shifts are produced by increasing the stimulus strength, these findings were interpreted to indicate that ACTH$_{4-10}$ facilitates transmission in midbrain limbic structures. This interpretation suggests that ACTH fragments increase the state of arousal in these structures, which may determine the motivational influence of environmental stimuli and thereby the probability of the generation of stimulus-specific behavioral responses. This view is supported by experiments in which the effects of ACTH$_{4-10}$ on visual evoked potentials in the rat were studied. The latencies of the evoked potential components remained unchanged, as did the amplitudes of the primary potentials. Measured at a wide variety of light intensities, however, the amplitudes of the late components of the visually evoked potential were significantly diminished. This finding indicates that the peptide has an effect on a CNS vigilance-regulating system (Wolthuis and de Wied, 1975). The effects appear to be specific, since spontaneous locomotor behavior is unaffected by ACTH$_{4-10}$. The heptapeptide [D-Phe^7]ACTH$_{4-10}$ had the same effect.

3.5. Effects on Cardiovascular Responses During Emotional Behavior

In order to investigate whether the influence of ACTH fragments is restricted to instrumental behavior only, or is more generalized and involves vegetative responses as well, extinction of a classically conditioned cardiac response was studied under the influence of ACTH$_{4-10}$ (Bohus, 1975). Classic conditioning in the freely moving rat consisted of 3 acquisition and 3 extinction sessions with 12 trials each. The CS was randomly followed by an unavoidable footshock as the unconditioned stimulus. Nonreinforced trials were given during extinction. The

ECG of the rat was recorded through wire leads from transcutaneous electrodes, and the heart rate was measured by off-line measurement of interbeat intervals with the aid of a PDP 8/I computer. Classic conditioning resulted in development of a conditioned bradycardia, which gradually disappeared during extinction. $ACTH_{4-10}$, 1 hr before each extinction session, delayed extinction of the conditioned heart rate response. Thus, $ACTH_{4-10}$ affects not only instrumental but also acquired autonomic responding. If one considers that the conditioned cardiac response is an autonomic correlate of classically conditioned fear, these results raise the question whether the behavioral effects of ACTH fragments are mediated through an influence on classically conditioned fear (see Section 4.6).

Similar studies were performed in a one-trial passive avoidance situation that is facilitated by ACTH fragments (de Wied, 1974). The heart rate of rats during passive avoidance decreased. The rate of the bradycardic response depended on the intensity of the shock presented to the animal during the learning trial (Bohus, 1975). The administration of $ACTH_{4-10}$, which increased avoidance latencies, was associated with tachycardia. According to Bohus (1975), this tachycardic response may be coupled with generalized arousal, while specific fear, as elicited by punishment in the shock compartment, is accompanied by bradycardia.

3.6. Role in Inducing Stretching and Excessive Grooming

In a variety of mammals, a stretching-and-yawning syndrome (SYS) has been observed after intracranial administration of ACTH or peptides derived from the N-terminal part of ACTH (Ferrari *et al.*, 1963; Gessa *et al.*, 1967). The onset of the syndrome in rodents is preceded by a display of increased grooming (Ferrari *et al.*, 1963; Izumi *et al.*, 1973). Recently, the capacity of a variety of peptides to induce excessive grooming in the rat was studied (Gispen *et al.*, 1975a). After injection of the test substance into the third ventricle, the conscious, previously cannulated rat was placed in a glass box. The analysis, which began 15 min after intraventricular administration of the peptide, consisted of a period of 50 min in which the observer determined every 15th sec whether or not the rat displayed an element of its maintenance-behavior repertoire, i.e., vibrating, washing, grooming, scratching, licking paw, licking tail (Gispen *et al.*, 1973). If one of these elements was observed, a positive score was given. Since the predominant element recorded appeared to be grooming, it was preferred to refer to grooming, in keeping with previous reports (Ferrari *et al.*, 1963; Izumi *et al.*, 1973). A rat was considered to have displayed SYS if it had exhibited more than 3 stretchings and yawnings within the observation period. The incidence of SYS reflects only the frequency of relevant episodes, while the intensity and duration may differ from one episode to another. This review presents mainly data on the grooming response. In general, intraventricularly administered $ACTH_{1-24}$ in a dose of $2\mu g$ ($1\mu l$) induces vigorous grooming interrupted by short episodes of SYS. Saline-treated rats invariably fall asleep, and their grooming scores are normally low. The induction of grooming is independent of the adrenals, pituitary, and gonads, and the response can be elicited in both male and female rats (Gispen *et al.*,

1975*a*). Interestingly, in mutant Wistar rats that have lost almost all their fur and have instead a thick "elephantlike" skin, the grooming response can be induced in a similar manner as in intact rats (Gispen and Wiegant, unpublished observations). These data, then, together with the short response delay and the route of administration, were taken to indicate that a central mechanism is responsible for ACTH-induced grooming behavior. It has been reported that excessive grooming and SYS can be elicited by hippocampal stimulation (MacLean, 1957) or implantation of zinc in the ventricles (Izumi *et al.*, 1973). Under both these experimental conditions, however, the display of grooming behavior differs from that induced by ACTH$_{1-24}$ (quantitatively as well as qualitatively).

The biological significance of the observed behavioral response is unclear. It has been suggested that rabbits and rats, despite the induction of SYS, show sexual excitement (Bertolini *et al.*, 1968, 1969; Baldwin *et al.*, 1974). In the present studies, however, penile erections were rarely seen, nor was lordosis behavior, as an indication of sexual excitement.

A variety of peptides in doses equimolar to 3μg ACTH$_{1-24}$ were tested to determine the active sequence of the ACTH molecule for this effect. These studies showed that ACTH$_{1-24}$, ACTH$_{1-16}$–NH$_2$, ACTH$_{1-16}$, α-MSH ([Ac–Ser^2]ACTH$_{1-13}$–NH$_2$),and β-MSH were equipotent. The common denominator, the sequence ACTH$_{4-10}$, was inactive even after application of high doses. These data in part corroborate the findings of Baldwin *et al.* (1974), who concluded that ACTH$_{4-10}$ in rabbits may have some SYS-inducing activity, but certainly acts much weaker than ACTH$_{1-24}$. Although the presence of the ACTH$_{5-10}$ sequence seems of importance, *C*-terminal elongation was necessary to induce excessive grooming. That ACTH$_{1-24}$ is active, while ACTH$_{1-10}$, ACTH$_{11-24}$, and the combined treatment of ACTH$_{1-10}$ and ACTH$_{11-24}$ failed to induce grooming behavior, supports this conclusion, among others. Surprisingly, substitution of the Phe^7 in the *N*-terminal ACTH fragments (ACTH$_{4-10}$ and ACTH$_{1-10}$) by the D-enantiomer resulted in appreciable grooming activity, but not when Arg^8 was replaced by the D-enantiomer. These observations underscore the important role of the Phe residue in the interaction of ACTH analogues with the CNS, and imply that not metabolic stability (Witter *et al.*, 1975) alone, but also intrinsic activity, receptor affinity, and transport properties, determine the ultimate effectiveness of a neuropeptide.

3.7. Interaction with the Opiate Receptor

ACTH and β-MSH can antagonize morphine inhibition of spinal reflex activity (Zimmermann and Krivoy, 1973; Krivoy *et al.*, 1974*a*). Using a Lloyd's preparation to measure spinal reflex activity in the anesthetized cat, these authors found that morphine inhibits the amplitudes of evoked mono- and polysynaptic reflex activity in a dose-dependent manner. Either β-MSH or ACTH$_{1-24}$ injected prior to morphine prevented the depressant action of morphine. Interestingly, β-MSH enhanced reflex activity *per se,* while ACTH$_{1-24}$ only counteracted the spinal reflex inhibition of morphine. These experiments were extended to elegant *in vitro* studies on reflex activity of the isolated spinal cord of the frog (Zimmermann and Krivoy, 1974). Exposure of the *in vitro* preparation to 30 min of morphine reduced the potential amplitude to about 70% of the control value. Incubation with $ACTH_{1-24}$ or $ACTH_{1-24}$ in the presence of morphine failed to alter reflex activity. These *in vitro* data circumvent the problem of *in vivo* treatment with ACTH, and are taken as strong support for an ACTH–morphine interaction at the CNS level.

Terenius and Wahlström (1975) were the first to demonstrate the existence of an endogenous ligand of morphine that was isolated from CSF. The factor appeared to be a peptide. Similar observations were made by others, and recently, Hughes *et al.* (1975) published the structure of two brain peptides with potent opiate activity related to LPH that were designated as *enkephalins* (see Chapter 13).

Terenius showed that $ACTH_{1-28}$ and $ACTH_{4-10}$ have appreciable affinity for the stereospecific opiate-binding site in rat brain synaptosomal plasma membranes (Terenius, 1975). These observations were confirmed and extended. Structure–activity studies pointed to an active site around $ACTH_{4-10}$, with some indication that a second affinity site might be present in a sequence adjacent to 4–10 (Terenius *et al.*, 1975). The initial observation that $ACTH_{11-24}$ possessed affinity could not be replicated, probably due to the quality of the batch that was used in the first experiment (Gispen *et al.*, 1976). Both $[D-Phe^7]ACTH_{4-10}$ and $[D-Phe^7]ACTH_{1-10}$ were active. Interestingly, α-MSH, vasopressin, DG-LVP, Substance P, prolyl–leucyl–glycyl (PLG), and thyrotropin-releasing hormone (TRH) were inactive (Terenius, 1975; Terenius *et al.*, 1975). The calculated dissociation constants were rather high (IC_{50} on the order of 10^{-5}–10^{-6} M), indicating that these ACTH fragments cannot be regarded as powerful endogenous ligands for opiate receptors, if they can be so regarded at all. Nevertheless, the data lend support to findings by Zimmermann and Krivoy (1973) that ACTH and related peptides interfere with morphine in the CNS, and suggest interference by pituitary hormones with the analgesic effect of morphine (Winter and Flataker, 1951; Paroli, 1967; Gispen *et al.*, 1975b). It was found that purified ACTH and $ACTH_{1-24}$ antagonize the analgesic effect of morphine in the rat (Paroli, 1967; Gispen *et al.*, 1975b). The latter authors used the response to inescapable footshock as a test system for analgesic properties, as suggested by Evans (1961). Not only sensoric properties, but also CNS processes affecting the rat's motor responses, are measured under these circumstances (Gispen *et al.*, 1973). Therefore, the classic hot-plate test according to Eddy and Leimbach (1953) was applied in order to further analyze the role of ACTH fragments in counteracting morphine-induced analgesia. Pretreatment with ACTH fragments ($ACTH_{1-24}$, $ACTH_{1-16}$, $ACTH_{4-10}$, and $[D-Phe^7]ACTH_{4-10}$) did not alter the rat's response latency on the hot plate, even if doses of 1 mg/kg were used. Morphine given alone increased the response latency by 13–16 sec. If ACTH fragments were given prior to the morphine treatment, however, response latency was reduced by about 50%, indicating that ACTH fragments counteract the analgesic effect of morphine. Since these fragments, which are devoid of corticotropic activities, have similar effects, the postulated permissive action of cortico-steroids (Gispen *et al.*, 1975b) might not be operating under these circumstances. Interestingly, it was found that $[\text{D-Phe}^7]\text{ACTH}_{4-10}$ was more potent than ACTH_{4-10}, and that ACTH_{11-24} was inactive (Gispen *et al.*, 1976). These data, then, seem to corroborate findings (Terenius, 1975; Terenius *et al.*, 1975) that ACTH fragments have an appreciable affinity for brain opiate receptors *in vitro*.

Interaction between morphine and MSH has been reported at the morphological level as well. In rats, α-MSH elicits an increase in cellular fluorescence in both tuberal and nigral DA neurons within 30 min. These changes are correlated with an increase in single unit activity. α-MSH markedly reduces the increase in fluorescence intensity of morphine. Similar increases are seen following electrical stimulation, cold exposure, and nicotine injection. In mice, α-MSH counteracts the morphine-induced locomotion, and in itself causes sedation (Lichtensteiger and Lienhart, 1975).

As discussed in Section 3.6, excessive grooming and stretching occurs with *N*-terminal ACTH fragments after intraventricular application. The grooming effect of ACTH can be prevented by pretreatment with the specific opiate antagonist naloxone (Blumberg and Dayton, 1973) injected subcutaneously. Naloxone itself does not display excessive grooming. Another selective and powerful opiate antagonist, naltrexone (Blumberg and Dayton, 1973), similarly suppresses ACTH-induced excessive grooming. Interestingly, morphine injected intraventricularly in doses ranging from 0.05 to 0.5 μg also induces a considerable grooming activity, both in intact and hypophysectomized rats. The behavior appeared to be of a similar quality as that produced by ACTH_{1-24}; i.e., the dominant element observed was the head–body grooming movements. In doses higher than 0.5 μg/μl, morphine depressed overall behavior of the rat at least during the first hour after the injection. This finding is in agreement with findings by Ayhan and Randrup (1973), who found that morphine systemically injected in low doses into rats characteristically increased grooming activity. These data support at a behavioral level the notion that opiates and ACTH fragments may have a common denominator in their neurotropic action.

3.8. Biochemical Effects

3.8.1. Adrenocorticotropic Hormone–Brain Cell Surface Interaction; Cyclic Nucleotides

According to the prevailing point of view, ACTH (Sayers *et al.*, 1974) and other polypeptide hormones (Sutherland, 1972) interact with a receptor on the surface of the plasma membrane of the effector cell, resulting in increased production of intracellular cAMP; the cyclic nucleotide (''second messenger''; see Sutherland, 1972) mediates the information of peptide–cell membrane binding, and in this way triggers the biochemical train of events underlying the functional response of the effector cell. The same mechanism of action may be the basis of the CNS effect of ACTH fragments (Gispen and Schotman, 1973; Schotman *et al.*, 1976*a*). In explaining their results with ACTH fragments on neuromuscular function, Strand and Cayer (1975) suggest that these fragments bind to the cell membrane to promote the synthesis of cAMP, which plays an important role in both metabolic and synaptic processes in nervous tissue. This nucleotide has been shown to release calcium from the sarcoplasmic reticulum (Entman *et al.*, 1969), which is the major storehouse for adenylcyclase (Robinson *et al.*, 1971); to restore muscle action potential amplitudes that have been lowered by reserpine or chlorpromazine (Torda, 1974); and to activate both muscle glycogenesis (Schlender *et al.*, 1969) and protein synthesis (Walton *et al.*, 1971). In fact, Bertolini *et al.* (1969) reported that the behavioral effects of intracranially administered ACTH, i.e., induction of stretching and yawning and sexual excitement, could be potentiated by theophylline blockade of brain phosphodiesterase (the enzyme that *in vivo* degrades cAMP to the inert AMP).

Contrary to this hypothesis are the negative results obtained on the action of ACTH on brain adenylcyclase (Burkhard and Gey, 1968) and on cAMP accumulation in brain slices (Forn and Krishna, 1971). Since no detailed information was presented on the properties of the adenylcyclase assay used (e.g., sensitivity), and the effect of ACTH may be detected only in ACTH-sensitive brain regions, it was decided to study the effect of $ACTH_{1-10}$ on cAMP accumulation in slices obtained from the rat posterior hypothalamus containing the nucleus parafascicularis. Slices were preincubated for 30 min at 37°C in 2 ml Krebs–phosphate, pH 7.4, under continuous gassing with pure oxygen. Incubation was started by adding 50 μl medium with or without $ACTH_{1-10}$. After an incubation period of 15 min, the slices were homogenized, and the cAMP was purified in principle according to Mao and Guidotti (1974) and determined by a protein-binding assay according to Gilman (1970). Basal levels of cAMP under these experimental conditions were on the order of 15–30 pmol/mg slice protein. Incubation with 10^{-5} M $ACTH_{1-10}$ resulted in about a 50% elevation of the basal level (Wiegant and Gispen, 1975). These and additional data were taken to indicate that ACTH fragments affect brain metabolism through an increase in the intracellular level of cAMP. Preliminary data in conscious rats on the effect of intraventricular administration of $ACTH_{1-24}$ (1 μg/ μl) on the levels of cAMP in various brain regions 10 and 30 min after injection suggest that there is a marked regional effect of ACTH on brain cAMP content, the highest elevation being found in diencephalon and mesencephalon (Wiegant *et al.*, manuscript in preparation). These results also suggest a peptide–brain plasma membrane interaction. Although specific brain receptor binding for TRH has been established (Burt and Snyder, 1975), specific binding of radioactive-labeled ACTH fragments with brain cell components has not been demonstrated as yet.

3.8.2. Adrenocorticotropic Hormone and Brain Ribonucleic Acid

If the first neurochemical event after ACTH treatment is of the peptide–cyclic nucleotide type, what cellular process is then affected by the second messenger? In a series of experiments, Jakoubek *et al.* (1972) have studied the influence of ACTH on brain macromolecule metabolism. A single high dose of a purified

ACTH preparation resulted in a transient inhibition of uridine incorporation into mouse brain RNA. In intact rats, injection of $ACTH_{1-24}$ led to a small but significant decrease (-12%) in uridine incorporation of brainstem RNA, but not of cerebral and cerebellar RNA (Schotman et al., 1976a). The decrease in RNA labeling was accompanied by a transient decrease in brainstem RNA content 2.5 hr after the injection of $ACTH_{1-24}$. These findings corroborated those of Jakoubek et al. (1972). In adrenalectomized rats (36 hr after surgery), similar treatment stimulated ($+40\%$) instead of inhibited the labeling of brainstem RNA, while brainstem RNA content was not decreased. Changes in uridine incorporation could not be attributed to differences in brain precursor uptake or to gross changes in precursor metabolism in brain tissue (Gispen and Schotman, 1976). $ACTH_{1-10}$ was ineffective both in intact and in adrenalectomized rats.

Chronic treatment of hypophysectomized rats with $ACTH_{1-10}$ also fails to influence the labeling of messengerlike and ribosomal brainstem RNA (Schotman et al., 1972) and the content of brainstem polysomes (Gispen et al., 1971). Neither were Reading and Dewar (1971) able to detect an effect of chronically administered $ACTH_{4-10}$ on the labeling and content of mouse brain RNA. The corticotropic effect of $ACTH_{1-24}$ may be responsible for the differential action of this peptide in intact and adrenalectomized rats. Apparently, in intact rats, increased levels of corticosteroids as generated by $ACTH_{1-24}$ treatment may have inhibited the stimulatory effect of ACTH on uridine incorporation. Depending on the experimental conditions, catabolic effects of glucocorticoids in this respect have been demonstrated (de Kloet and McEwen, 1976). The ineffectiveness of $ACTH_{1-10}$ in stimulating uridine incorporation into brainstem RNA in the various experimental studies (intact, adrenex, hypox) may be explained by assuming that the transport of the peptide to the peptide-sensitive sites in the CNS differs from that of $ACTH_{1-24}$, or that the sensitive sites discriminate between the two sequences. Although the effects of $ACTH_{1-24}$ and $ACTH_{1-10}$ in avoidance behavior are similar, in other situations, differences between these two peptides in interaction with the CNS were noted. $ACTH_{1-24}$ and $ACTH_{1-16}$ accelerate eye-opening in neonatal rats, while $ACTH_{1-10}$ is ineffective (van der Helm-Hylkema, 1973), and intraventricular $ACTH_{1-24}$ induces grooming and stretching and yawning in the rat, while $ACTH_{1-10}$ and $ACTH_{4-10}$ are ineffective even in massive doses (Gispen et al., 1975a). In any case, the data on brain RNA thus far suggest that N-terminal ACTH sequences (4–10 and 1–10) do not affect brain macromolecule metabolism at the level of transcription (uridine incorporation). Experiments were therefore carried out to explore whether these small ACTH fragments might affect brainstem protein metabolism (translation; see the next section).

3.8.3. Adrenocorticotropic Hormone and Brain Protein

Treatment of hypophysectomized rats during a 12-day period with $ACTH_{1-10}$ (zinc phosphate long-acting preparation, 20 μg s.c. every other day) enhances the incorporation of [^{3}H]leucine into protein of the brainstem 5 min after injection of the precursor into the diencephalon, as measured on the day after the last peptide injection (Schotman *et al.*, 1972). Subcellular fractionation of the brainstem homogenate indicated that mainly cytoplasmic proteins were labeled in the 5-min incorporation period (Schotman, 1971). An increase of 28% was found in the labeling of total brainstem protein. This experiment was replicated using another strain of rats, and again a significant increase of overall labeling of brainstem protein was found (Reith, 1975).

The absolute increase in labeling of brain proteins is small. From the literature, however, it can be derived that brain macromolecule metabolism responds only moderately to the effect of various stimuli (see Rees *et al.*, 1974). The hypophysectomized rat shows a decrement of 35% in brainstem protein labeling. Substitution of $ACTH_{1-10}$ nearly completely restores the labeling toward the level found in intact rats (Schotman *et al.*, 1972). In this respect, the effect is as large as one may possibly expect.

It was subsequently decided to explore whether labeling of specific proteins would be influenced by peptide treatment, or whether the peptide effect would be of a more general nature. In view of recent immunoneurochemical data on specific nervous tissue proteins, it was argued that a behaviorally active peptide may be acting on brain protein metabolism by modulating the presence of certain proteins that could be specific for the brain or its function, i.e., behavior (Gispen and Schotman, 1973). The incorporation of $[^3H]$leucine into rat brainstem proteins was thus studied 5 min after the injection of the precursor into the diencephalon of hypophysectomized rats chronically treated with $ACTH_{1-10}$ (in a manner similar to that described above). The homogenate was sequentially extracted with a hypotonic buffer, a nonionic detergent (Triton X100), and an anionic detergent (SDS), resulting in a soluble and two particle-bound protein fractions. Analysis of these protein fractions on polyacrylamide gels revealed that $ACTH_{1-10}$ enhanced the incorporation into all proteins. Superimposed on the general effect, minor differences were found in two low-molecular-weight protein bands (Reith *et al.*, 1975*a*). Thus, chronic treatment with $ACTH_{1-10}$ interferes with overall protein metabolism, rather than with certain protein species in particular.

To substantiate this conclusion, the effect of peptides on leucine incorporation into brainstem protein *in vitro* in slices from the posterior thalamus (see Section 3.8.1) was studied. $ACTH_{1-10}$ in concentrations of 10^{-5} M and 5×10^{-7} M significantly enhanced $[^{14}C]$leucine incorporation (39 and 34%, respectively), while at 10^{-8} M, a tendency to an enhanced incorporation was found (13%) (Reith *et al.*, 1974). Lloyd (1974) reported similar data using $ACTH_{4-10}$ and rat brainstem slices. $ACTH_{4-10}$ in doses of 0.5 and 10 $\mu g/ml$ stimulated the incorporation of $[^{14}C]$leucine into slice proteins by about 90%.

Several authors provided data on the effect of ACTH fragments on brain protein metabolism consistent with the observations discussed above. Chronic treatment of young rats with ACTH induced a complex pattern of neurochemical changes, including a biphasic effect on brain protein content (Palo and Savolainen, 1974). Reading and Dewar (1971) reported that chronic $ACTH_{4-10}$ treatment of intact mice stimulated the incorporation of labeled amino acids into cerebral protein, the incorporation being most pronounced when measured 48 hr after the injection of the precursor. Rudman *et al.* (1974) demonstrated that a single injection of either ACTH or β-MSH increases the rate of accumulation of [^{14}C]valine into mouse brain proteins by 20–100%, 6–24 hr after the intraperitoneal injection of the radioactive precursor. In addition, Dunn and Reed (1975) reported that incorporation of [^{3}H]lysine into mouse brain protein measured 10 min after subcutaneous injection of the precursor was enhanced by 10–20% by a single injection of ACTH$_{1-24}$ or ACTH$_{4-10}$.

Various experiments were performed to determine whether the peptides affected the incorporation process *per se* or other variables that would indirectly account for the observed increased labeling. The possibility that an altered brain uptake of amino acid after peptide treatment might contribute to the observed changes in incorporation into brain proteins was investigated using [^{14}C] α-aminoisobutyric acid as a metabolically inert amino acid analogue. No effects of ACTH$_{1-10}$ on the uptake of this amino acid analogue were found, either *in vivo* after chronic treatment during 12 days or *in vitro* after the addition of the peptide ($\times 10^{-7}$ M) to the slice–incubation mixture. These *in vitro* results were confirmed, using labeled cycloleucine as amino acid analogue (Schotman *et al.*, 1976*b*). In addition, Rudman *et al.* (1974) reported that ACTH and β-MSH did not affect the brain uptake of valine, the concentrations of free amino acids, or the penetration of the metabolically inert amino acid analogue α-aminoisobutyric acid.

The opposite effect on extinction of active avoidance behavior of [L-Phe^7] and [D-Phe^7]ACTH analogues was used to explore whether the changes in protein metabolism would reflect similar opposite effects. Chronic treatment with [D-Phe^7]ACTH$_{1-10}$ indeed decreased the incorporation of leucine into total brainstem protein of hypophysectomized rats (Schotman *et al.*, 1972; Reith, 1975). Reading and Dewar (1971), who reported a stimulation by ACTH$_{4-10}$ of labeling of mouse cerebral proteins, failed, however, to observe an effect of chronically administered [D-Phe^7]ACTH$_{4-10}$ in intact mice. Under *in vitro* conditions using the slice preparation, we were also unable to detect any effect of [D-Phe^7]ACTH$_{1-10}$ on leucine incorporation in doses ranging from 10^{-5} to 10^{-8} M (Reith *et al.*, 1975*b*). Lloyd (1974) had similar experiences with the effect of [D-Phe^7]ACTH$_{4-10}$ on the incorporation of [^{14}C]leucine into slice proteins, although ACTH$_{4-10}$ was active in this respect.

No explanation for the ineffectiveness of the D-enantiomer under *in vitro* conditions can be given, but it may be that the effect of the two isomers on brainstem protein metabolism stems from a different mechanism of action. Since ACTH$_{1-24}$ stimulates protein labeling *in vitro* similarly to ACTH$_{1-10}$ (Reith *et al*, 1975*b*), and the behaviorally inert sequence ACTH$_{11-24}$ is inactive with respect to leucine incorporation into protein under both *in vivo* and *in vitro* conditions (Schotman *et al.*, 1972; Reith *et al.*, 1975*b*), it would appear that the capacity of ACTH fragments to influence active avoidance behavior parallels their effect on brainstem protein synthesis *in vivo*.

3.8.4. Adrenocorticotropic Hormone and Neurotransmission

Torda and Wolff (1952) reported that ACTH increases acetylcholine synthesis in the neuromuscular junction in intact and hypophysectomized rats. In intact rats, ACTH increases noradrenaline (NA) turnover in hypothalamus, cortex, and other parts of the brain (Hökfelt and Fuxe, 1972). Since adrenalectomy increased and hypophysectomy decreased brain NA turnover (for a review, see Versteeg and Wurtman, 1976), it was concluded that ACTH has a direct effect on brain noradrenergic mechanisms. $ACTH_{1-24}$ prevents the decrease in hypothalamic dopamine-β-hydroxylase (DBH) activity following hypophysectomy. ACTH also affects amygdala and brainstem, but not other parts in the brain. Dexamethasone does not restore the low DBH activity of hypophysectomized rats. Low doses of ACTH, which did not induce an increase in plasma corticosterone, were also effective in preventing the hypophysectomy-induced decrease in DBH activity, indicating a direct CNS effect of this polypeptide (Van Loon, 1975). Further studies revealed that $ACTH_{4-10}$ increases NA turnover in the brain (Versteeg, 1973; Leonard, 1974). However, [D-Phe^7]$ACTH_{4-10}$ is not active in this respect (Versteeg, 1973; Leonard, 1974). Although a correlation has been postulated between NA turnover and the rate of extinction of a CAR (Weiss et al., 1970; Hökfelt and Fuxe, 1972), it is unlikely that the behavioral effects of these peptides result from changes in brain NA turnover (Versteeg and Wurtman, 1976). Recently, it was found that $ACTH_{1-24}$ is capable of changing the in vitro phosphorylation of rat brain membrane proteins (Zwiers et al., 1976). Phosphorylation of membrane proteins might alter membrane properties such as ion permeability, and might therefore underlie modulatory influences on neurotransmission, as suggested by Greengard (1975). Finally, one may expect a modulatory influence of ACTH analogues through the observed enhancement of protein synthesis. This may be responsible for changes in transmitter enzyme production, synaptic membrane protein renewal, or sprouting.

4. Behavioral Effects of Vasopressin and Congeners

4.1. Effects on Active and Passive Avoidance Approach, Sexual Behavior, and Amnesia

As in the posterior-lobectomized rat, pitressin affects extinction of the shuttle-box avoidance response in intact rats. Subcutaneous injection of 1 IU long-acting pitressin tannate in oil every 2 days during acquisition or during extinction markedly increased resistance to extinction (de Wied and Bohus, 1966). Rats treated with this preparation maintained a high level of responding during the period of extinction, and during a second extinction period 3 weeks after termination of the first. Similar long-term effects were obtained with LVP in the pole-jumping test. A single injection of 0.6 or 1.8 μg LVP exhibited a long-term dose-dependent effect on extinction of the pole-jumping avoidance response that extended for days beyond the actual presence of the peptide in the organism, indicating again that vasopressin has a long-term effect on the maintenance of avoidance behavior, probably by facilitating memory consolidation (de Wied, 1971b; van Wimersma Greidanus et al., 1973). Other structurally related peptides such as oxytocin or physiologically related peptides such as angiotensin II, insulin, or growth hormone (GH), in comparable amounts, failed to affect extinction of the CAR. Vasopressin congeners such as AVP and DG-LVP had similar effects (de Wied et al., 1972).

In order to determine the critical period for the effect of vasopressin to occur, LVP was injected at 6, 3, and 1 hr before, or immediately, 1, and 6 hr after, the first extinction session. Vasopressin had to be administered within 1 hr before or after the first extinction session to be completely effective (de Wied, 1971b). In fact, vasopressin injection 1 hr before the 1st acquisition session of a pole-jumping avoidance response (King and de Wied, 1974) or immediately after the 3rd acquisition session (de Wied et al., 1974) increased resistance to extinction. The inhibitory effect of vasopressin and vasopressin fragments on extinction of active avoidance behavior could not be explained by a general increase in locomotor activity. No differences in rearing, grooming, or ambulation were observed following administration of LVP as compared with saline treatment in a so-called open field. Moreover, vasopressin affects not only active but also passive avoidance behavior (Ader and de Wied, 1972). LVP injected subcutaneously either immediately after the learning trial or 1 hr prior to the retention test facilitates avoidance learning. In this situation, vasopressin fragments also exhibit a long-term effect. A temporal relationship similar to that in active avoidance behavior was found for retention of passive avoidance behavior, i.e., between 1 hr prior to and 1 hr after the retention session. It was further investigated whether the long-term behavioral effect of LVP is specific for a particular response, or whether generalization occurs to other aversively motivated responses (Bohus et al., 1972). Half the animals were trained in the pole-jumping apparatus in the morning, and subjected to the passive avoidance procedure 6 hr later, in the afternoon. The other half were trained first in the passive avoidance test in the morning, and subjected to the pole-jumping avoidance test in the afternoon. Animals were injected on day 3 of the experiment with either saline or LVP 1 hr prior to the session. Resistance to extinction was seen only in that situation in which the pole-jumping of passive avoidance was performed in the morning, under the influence of the peptide which was given 1 hr prior to the session. No evidence of generalization or transfer of the effect of vasopressin from one behavioral situation to the other was found, indicating that the effect on extinction is restricted to that behavior that is occurring during the time of optimal vasopressin influence.

Garrud et al. (1974) failed to find an effect of LVP and DG-LVP on extinction of a straight runway approach response for food in hungry rats. Although DG-LVP is practically devoid of antidiuretic activities, an interaction between hunger drive and water metabolism could not be excluded. However, the behavior in a continual-punishment situation of food-deprived rats that were trained to hold a lever down for some time to obtain a food reward could be affected by vasopressin. Vasopressin increased the efficiency with which the rats performed (Garrud, 1975). A straight runway may be too simple and therefore inadequate to study the influence of agents on memory processes. The effect of vasopressin was therefore studied in a T-maze. Male rats running for a receptive female chose the correct arm of the maze in a significantly higher percentage following treatment with DG-LVP after each acquisition session. The effect, again, was long-term. Copulation reward appeared to be essential, since nonrewarded rats display no more correct choices than placebo-treated animals (Bohus, 1976). DG-LVP also delays the disappearance of intromission and ejaculatory behavior of male rats following castration. Thus, the effect of vasopressin congeners is certainly not restricted to aversively motivated behavior.

Evidence for an effect on memory processes can also be derived from studies on the protective effect of vasopressin on amnesia. Lande *et al.* (1972) reported that DG-LVP protects against puromycin-induced memory loss in mice, an effect it shares with various vasopressin analogues (Walter *et al.*, 1975). Rigter *et al.* (1974) demonstrated that amnesia for a one-trial passive avoidance response in rats as induced by CO_2 or by electroconvulsive shock is reversed by DG-LVP when injected immediately after the learning trial. These authors suggested that the peptide promotes memory consolidation either by facilitating the consolidation process or by protecting memory consolidation from the adverse effects of the amnesic treatment. The possibility that vasopressin influences retrieval was also considered, since DG-LVP exhibited antiamnesic effects when injected 1 hr prior to the retention test.

4.2. Structure–Activity Studies with Behaviorally Active Vasopressin Fragments

Attempts to determine the active core of the vasopressin molecule that contains the essential requirements for the behavioral effects are met with difficulties due to the variability in purity of available peptides. Activities are therefore expressed as approximated potency (de Wied *et al.*, 1976). The effect of a number of synthetic fragments of vasopressin and oxytocin was measured on extinction of the pole-jumping avoidance response (de Wied, 1971*b*). AVP appeared to be the most potent peptide, followed by LVP (Table 1). Oxypressin also had a considerable activity, amounting to about 30% of that of AVP. The behavioral activity of oxypressin was markedly potentiated by substitution of Leu[8] by Ala. Removal of the C-terminal glycinamide (DG-LVP and DG-AVP) decreased the potency to approximately 50%. Oxytocin and [Arg[8]]vasotocin (AVT) were equally active, but possessed only 20% of the activity of AVP. Elongation of the N-terminal with triglycyl augmented the behavioral effect of oxytocin, but had the opposite effect for AVP. Pressinamide had retained only 10% of the behavioral potency. The C-terminal tripeptide seems to be important, since removal leads to a drastic decrease in potency. Substitution of Leu[8] for Ala in oxypressin potentiates, while substitution of Arg[8] by [D-Arg] markedly decreases, behavioral potency. Various C-terminal tetra-, tri-, and dipeptides do affect extinction of the pole-jumping avoidance response, albeit the activity is only 5% of that of AVP (Walter and de Wied, manuscript in preparation).

Similar structure–activity studies with neurohypophyseal hormones using pro-

TABLE 1. Effect of Various Vasopressin and Oxytocin Analogues on Resistance to Extinction of a Pole-Jumping Avoidance Response

Analogue [a]	Structure	Approximated potency
A	H–Cys–Tyr–Phe–Gln–Asn–Cys–Pro–Arg–Gly–NH$_2$	1
B	H–Cys–Tyr–Phe–Gln–Asn–Cys–Pro–Lys–Gly–NH$_2$	2
C	H–Cys–Tyr–Phe–Gln–Asn–Cys–Pro–Ala–Gly–NH$_2$	3
D	[b] Mpr–Tyr–Phe–Gln–Asn–Cys–Pro–Ala–Gly–NH$_2$	4
E	H–Cys–Tyr–Phe–Gln–Asn–Cys–Pro–Lys–OH	4
F	H–Cys–Tyr–Phe–Gln–Asn–Cys–Pro–Arg–OH	4
G	H–Cys–Tyr–Phe–Gln–Asn–Cys–Pro–Leu–Gly–NH$_2$	5
H	H–Gly–Gly–Gly–Cys–Tyr–Phe–Gln–Asn–Cys–Pro–Arg–Gly–NH$_2$	5
I	[b] Mpr–Tyr–Phe–Gln–Asn–Cys–Pro–Arg–Gly–NH$_2$	5
J	H–Gly–Gly–Gly–Cys–Tyr–Ile–Gln–Asn–Cys–Pro–Leu–Gly–NH$_2$	5
K	H–Cys–Tyr–Ile–Gln–Asn–Cys–Pro–Leu–Gly–NH$_2$	6
L	H–Cys–Tyr–Ile–Gln–Asn–Cys–Pro–Arg–Gly–NH$_2$	6
M	H–Cys–Tyr–Phe–Gln–Asn–Cys–NH$_2$	6
N	H–Cys–Tyr–Phe–Gln–Asn–Cys–Pro–D-Arg–Gly–NH$_2$	7
O	cyclo (Leu–Gly)	7
P	H–Cys–Tyr–Ile–Gln–Asn–Cys–NH$_2$	8
Q	[b] Mpr–Tyr–Phe–Gln–Asn–Cys–Pro–D-Arg–Gly–NH$_2$	9
R	H–Cys–Tyr–Leu–Ser–Asn–Cys–Pro–Arg–Gly–NH$_2$	10
S	H–Leu–Gly–NH$_2$	10
T	H–D-Leu–Gly–NH$_2$	10
U	H–Lys–Pro–Leu–Gly–OH	11
V	[b] Mpr–Tyr–Phe–Leu–Asn–Cys–Pro–D-Arg–Gly–NH$_2$	12

[a] Key: A: AVP; B: LVP; C: [Ala^8]oxypressin; D: [Mpr^1,Ala^8]oxypressin; E: DG-LVP; F: DG-AVP; G: oxypressin; H: [Gly$_3$-Cys^1]AVP; I: [Mpr^1]AVP; J: [Gly$_3$–Cys^1]oxytocin; K: oxytocin; L: [Arg^8]vasotocin; M: pressinamide; N: [D-Arg^8]AVP; P: tocinamide; Q: [Mpr^1,D-Arg^8]AVP (DD-AVP); R: [Leu^3,Ser^4]AVP; V: [Leu^4]DD-AVP.
[b] Mpr: Mercapto propionic acid.

tection against puromycin-induced amnesia in mice (Lande *et al.*, 1972) were performed by Walter *et al.* (1975) at the same time. In these studies, various vasopressin analogues significantly prevented puromycin-induced amnesia. The most active were AVP, LVP, and [Tri–Gly]LVP. [Deamino]LVP, [Leu^4]LVP, and [Asn^{1,6}]LVP were less active. Although there is agreement in both studies that LVP and AVP are the most potent peptides affecting memory, a number of peptides that significantly delay extinction of the pole-jumping avoidance response were inactive in the puromycin-induced amnesia test (oxytocin congeners, pressinamide). Elongation of oxytocin in the amnesia test, however, led to a marked protection against puromycin-induced amnesia. Conversely, a number of peptides active in the amnesia test were markedly less active in the pole-jumping avoidance test (e.g., DD-AVP, [Leu^3,Ser^4]AVP, [Leu^8,Gly^9], [D-Leu^8,Gly^9]). These differences cannot be explained at present, but may be related to the test system used. As mentioned above, pressinamide was inactive in the amnesia test, but had retained 10% of the activity in the pole-jumping test. When this compound was administered via one of the lateral ventricles, only twice as much pressinamide as AVP was needed to induce an equipotent effect on extinction of the CAR (de Wied, unpublished observation). Peptides given by this route may escape metabolic degradation, as is likely to occur when they are administered systemically.

Changes in hydrophobicity, distribution pattern, receptor binding affinity, and metabolic degradation in addition may be responsible for the potency of the memory effect of neurohypophyseal hormones. Receptor studies are therefore needed to determine the active core of the vasopressin molecule in memory processes. It is clear, however, that only part of the molecule is necessary for the behavioral effect. The endocrine and behavioral effects are completely dissociated, since removal of the C-terminal glycinamide almost completely destroys the classic endocrine effects of vasopressin (Lande *et al.*, 1971).

4.3. Cerebrospinal Fluid as a Transport Medium

Morphological and functional evidence has accumulated for the release of hypothalamic hormones from neural tissue into the liquor of the brain ventricular system. Recent studies indicate that neurosecretory axons run to the walls of the brain ventricular system (Sterba, 1974; Rodriguez, 1970). Morphological observations point to a connection between neurosecretory cells and the ependyma of the infundibular recess of the third ventricle. Axons originating in the supraoptic area, filled with neurosecretory substances, have been shown to end in the infundibular recess (Wittkowski, 1968). In addition, vasopressin or vasotocin or both have been shown to be present in rabbit (Heller *et al.*, 1968), in dog and rabbit (Vorherr *et al.*, 1968), and in human CSF (Gupta, 1969; Pavel, 1970; Coculesco and Pavel, 1973; Dogterom *et al.*, manuscript in preparation). With the use of immunocytochemistry techniques, Goldsmith and Zimmerman (Goldsmith and Zimmerman, 1975; Zimmerman *et al.*, 1975) demonstrated that neuronal processes containing granules with neurophysin and vasopressin are present close to portal capillary loops, but also protruding into the third ventricle. How the hormone is transported from the CSF to effector sites is not known. It is possible that different pools of ependymal cells or some circumventricular organs act as mediators in this respect (Sterba, 1974). Various possibilities for the function of vasopressin in the CSF have been suggested, such as in the feedback regulation of its own secretion and in the homeostasis of ionic concentration in the brain. The CSF, however, may be the avenue *par excellence* for vasopressin and other hypothalamic hormones to reach the sites of the behavioral action in limbic midbrain structures. It is also possible, however, that these neuropeptides are transported via a direct hypothalamic limbic pathway, which may be an integral subsystem of the main peptidergic neurosecretory system, as suggested by Sterba (1974). Martin *et al.* (1975) postulated that hypothalamic peptidergic neurons are part of a diffuse neural network that terminates in widespread areas of the brain and that may be important in neurobiological regulation. A number of observations, however, support the notion that the CSF is more important in this respect. Intracerebroventricular administration of vasopressin and fragments affects extinction of a pole-jumping avoidance response at doses 200 times less than those used systemically (de Wied, 1976).

Further support for the assumption that the CSF transports neurohormones to their target tissues in the brain comes from experiments by van Wimersma Greidanus *et al.* (1975*a*). These authors showed that administration of vasopressin antiserum into one of the lateral ventricles immediately after the learning trial induces an almost complete deficit in passive avoidance retention. Oxytocin antiserum, GH antiserum, or normal rabbit serum under the same conditions were ineffective in this respect. Intravenous injection of 100 times as much vasopressin antiserum, which effectively neutralized the systemic effects of vasopressin, as indicated by the virtual absence of vasopressin in the urine and a marked increase in urine production, did not affect passive avoidance behavior. Similar results were obtained on extinction of pole-jumping avoidance behavior when vasopressin antiserum was administered ½ hr prior to each acquisition session. Acquisition of rats treated with vasopressin antiserum tended to be slower than that of animals treated with control rabbit serum. During extinction, treatment was discontinued. Extinction of the avoidance response was significantly faster in rats treated with vasopressin antiserum during acquisition (van Wimersma Greidanus *et al.*, 1975*a*). These findings support the hypothesis that learning can take place in the absence of vasopressin, but that memory consolidation is disturbed, resulting in a disturbed retention.

4.4. Sites of Action

As in the studies with ACTH fragments, the sites of action of the behavioral effects of vasopressin were determined by intracerebral application of the peptide, and by means of lesion experiments. For these studies, the pole-jumping avoidance test was used. Microinjections of 0.1 μg LVP, a dose that was systemically ineffective, were placed in various limbic midbrain structures. Application of LVP in the posterior thalamic area, including the parafascicular nuclei, resulted in increased resistance to extinction of the CAR. Other areas, including ventromedial and anteromedial parts of the thalamus, the posterior hypothalamus, the substantia nigra, the reticular formation, the substantia grisea, the putamen, and the dorsal hippocampal complex, appeared to be ineffective sites (van Wimersma Greidanus *et al.*, 1974, 1975*b*). Lack of effect, however, may not always represent insensitive structures, since a unilateral local microinjection of vasopressin into a restricted brain area may be unable to influence the activity of the function of this area sufficiently to induce behavioral changes. Intraventricular localizations were always effective (van Wimersma Greidanus *et al.*, 1973).

Rats bearing lesions in the parafascicular area were employed for further evaluation of the locus of action of vasopressin. The lesions reduced but did not prevent the behavioral effects of vasopressin. The parafascicular nuclei, therefore, are sensitive to but not essential for, the action of vasopressin. The lesioned rats required only more vasopressin to induce resistance to extinction (van Wimersma Greidanus et al., 1974). Marked behavioral changes have been reported in rats following ablation of the septum (Brady and Nauta, 1953; Fried, 1972) or after dorsal hippocampectomy (Altman et al., 1973). These areas are considered to be involved in learning and memory processes (Altman et al., 1973). Extensive lesions in the rostral septum or in the dorsal hippocampus prevent vasopressin-induced resistance to extinction of the pole-jumping avoidance response (van Wimersma Greidanus et al., 1975b). Smaller lesions of the dorsal hippocampal complex partly inhibited the behavioral effect of vasopressin. These results point to an important role of midbrain limbic structures in vasopressin-induced effects on the consolidation of memory processes. The same was found in rats bearing lesions in the rostral septal area.

4.5. Electrophysiological Effects

There is electrophysiological evidence that vasopressin and oxytocin affect the CNS. Intraventricularly applied vasopressin and oxytocin in low doses affect single- and multiunit activity of hypothalamic neurons and cortical neurons of rabbits in opposite directions (Schulz et al., 1971; Schwarzberg et al., 1973), although Unger and Schwarzberg (1970) found stimulation of hypothalamic neurons following intravenous oxytocin in nonanesthetized rats. Nicoll and Barker (1971) demonstrated inhibition (75%) and excitation (25%) of identified supraoptic neurosecretory cells and excitation of cortical neurons following microiontophoretic application of vasopressin. Cross et al. (1971) found that iontophoretically applied oxytocin elicited an increase in firing rate in 32 out of 37 antidromically identified neurosecretory cells in the nucleus paraventricularis. Vasopressin had no effect in this nucleus or in the supraoptic nucleus, while oxytocin had an inhibitory action. Oxytocin excited the majority of identified paraventricular neurons, but not of neurons that could not be identified antidromically (Moss et al., 1972). Vincent and Arnould (1975) proposed that the neurosecretory product might serve a synaptic function in mediating recurrent inhibition that occurs following osmotic stimulation of the supraoptic neurosecretory neurons. Inhibition was readily elicited by vasopressin administration in spontaneously firing neurosecretory cells, as was the response to osmotic stimulation. Recurrent inhibition, however, still persists in Brattleboro rats, which lack the ability to synthesize vasopressin (Vincent and Arnould, 1975).

Recent electrophysiological studies in free-moving rats have shown that rhythmic slow activity (RSA) of HO-DI rats contains substantially lower hippocampal theta frequencies during paradoxical sleep (PS) than the RSA of HE-DI control rats (Urban and de Wied, 1975). Differences were found in all spectral parameters, and the averaged peak frequency of RSA in HO-DI rats was approximately 1 Hz lower than that in HE-DI animals. Thus, PS fails to produce normal frequency of RSA in the absence of vasopressin, indicating a quantitatively different PS. Administration of DG-AVP enhances the generation of higher frequencies in HO-DI rats, and almost completely restores the distribution of hippocampal theta frequencies. Deprivation of PS has been shown to interfere with the consolidation of learned responses (Leconte and Bloch, 1970; W.C. Stern, 1971; Fishbein, 1971). It might be, therefore, that the impaired memory of HO-DI rats is due to the low quality of PS found in these animals. In addition, Longo and Loizzo (1973) have reported that drugs that facilitate memory functions increase theta frequency in the postlearning period, suggesting that changes in the excitability in hippocampal theta activity may be related to memory consolidation. Landfield et al. (1972) maintain that theta activity in the postlearning period may be a brain state that is optimal for memory storage. Hypophysectomized rats, which also show learning deficits (de Wied, 1964), have shorter PS episodes and lack the normally present PS circadian rhythmicity (Valatx et al., 1975). PS is also markedly disturbed in the chronic pontine cat without hypothalamus or pituitary gland (Jouvet, 1965). These deficits can be restored by treatment with various pituitary principles.

The electrophysiological studies discussed above corroborate the hypothesis that vasopressin and related peptides are involved in the consolidation of memory processes. Interestingly, Chambers et al. (1966) reported that averaged potentials evoked from the midbrain and recorded from the cortex in old rats with permanently implanted electrodes showed a slower multisynaptic response and a higher threshold than those of young rats. These differences could be partly attenuated by chronic treatment with pitressin tannate in oil.

4.6. Effects on Cardiovascular Responses During Emotional Behavior

In a series of experiments meant to dissociate classic and instrumental components underlying the behavioral effects of vasopressin, it appeared that classic conditioning alone is a sufficient behavioral substrate for the long-term effect of vasopressin on avoidance behavior (King and de Wied, 1974), although instrumental conditioning is more important. Vasopressin administration to rats exposed to the conditioned stimulus and the unconditioned stimulus, in the absence of an escape possibility, significantly increased resistance to extinction of subsequent pole-jumping avoidance behavior. In the presence of escape possibilities, however, a single correct response under the influence of vasopressin in the pole-jumping test is more effective in increasing resistance to extinction. The combination of classic and instrumental components therefore seems optimal for the behavioral effect of vasopressin.

As mentioned in Section 3.5, classic conditioning in free-moving rats results in the development of a conditioned bradycardia. Extinction results in a gradual disappearance of the conditioned cardiac response. However, injection of LVP 1 hr before each extinction session delays extinction of the cardiac response. The heart rate accompanying passive avoidance behavior is also affected by vasopressin fragments. DG-LVP facilitates passive avoidance behavior and the bradycardia that occurs at the same time (Bohus, 1974). Specific fear elicited by the punishment in the shock compartment is accompanied by bradycardia, while tachycardia may be coupled with generalized arousal (Bohus, 1974). Thus, vasopressin may affect specific fear responses, while the increase in heart rate that is found following administration of ACTH fragments may under these conditions be mediated by the influences of these neuropeptides on arousal.

4.7. Vasopressin and Morphine Tolerance

It has been suggested that aspects of the development of tolerance to narcotic analgesics might resemble learning processes. At the molecular level, this suggestion received support from observations on similarities in effects of antibiotics on development of tolerance to morphine and retention of CARs. For example, DG-LVP treatment protects against puromycin-induced memory blockade in mice (Lande et al., 1972). Puromycin and other inhibitors of protein synthesis (actinomycin D, anisomycin) induce amnesia (Flood et al., 1975) and prevent memory consolidation (Glassman, 1969; Dunn, 1976). These agents also block the development of tolerance to narcotic analgesics (Cohen et al., 1965; Smith et al., 1966; Cox and Osman, 1970).

Krivoy et al. (1974b) reported that DG-LVP facilitates the development of resistance to morphine analgesia, as measured in mice using a hot-plate assay technique (Krivoy et al., 1974b). The aim of the experiment was to determine whether or not DG-LVP (50 μg s.c.), given under a variety of treatment conditions, affected analgesia produced by morphine (8 mg/kg s.c.). The results obtained were consistent with the interpretation that DG-LVP facilitates development of tolerance to the analgesic action of morphine, if morphine is administered before the peptide treatment. Interestingly, the data suggested that the vasopressin fragment exerts a prolonged action similar to that observed in the avoidance conditioning experiments. DG-LVP did not cause hyperalgesia or alteration of the response to the technique used for evaluating analgesia. Furthermore, Krivoy et al. (1974b) concluded that DG-LVP is not a morphine antagonist, a conclusion that was corroborated by later findings showing that vasopressin congeners do not inhibit morphine binding to the opiate receptor in rat brain synaptosomal plasma membranes (Terenius et al., 1975). Although the dose of DG-LVP used was extremely high, the data indicated that vasopressin fragments may be physiologically involved in the development of tolerance to narcotic analgesics. If this were true, rats with hereditary diabetes insipidus might have difficulty in developing tolerance to the analgesic action of morphine.

To test this hypothesis, the development of resistance to daily injections of morphine was studied in both HO-DI and HE-DI rats, using a hot-plate technique for measuring analgesia (de Wied and Gispen, 1976). HO-DI rats appeared to be slightly more sensitive to the heat stimulus than HE-DI rats. However, when HO-DI rats were tested at a plate temperature of approximately 1.0°C lower than that used for HE-DI rats, the same reaction time on the hot plate was obtained (about 10 sec) as for HE-DI rats. Subsequently, a single injection of morphine (10 mg/kg, i.p.) in both groups delayed the reaction to 25–30 sec, and this delay was noted at least throughout the first 90 min after the injection, but was undetectable 20 hr after the morphine treatment. In this way, the performance on the hot plate after acute morphine was made essentially the same for both groups of rats. The response to chronic treatment consisted of daily injections of morphine, followed every day by a test on the hot plate 30 min after injection. The response of HE-DI rats to this daily treatment diminished in time, and full tolerance had developed after the 5th injection. In HO-DI rats, however, the response to morphine did not diminish in time, and was markedly saved even after 8 consecutive days of morphine treatment.

These data suggest that tolerance to morphine analgesia is impaired in the absence of vasopressin (HO-DI). In subsequent experiments, the possibility was tested that AVP (3 μg s.c.) or DG-LVP (5 μg/100 g s.c.) administered to morphine-treated HO-DI rats immediately after the daily trial on the hot plate would restore tolerance development. Such treatment indeed resulted in the development of tolerance to morphine similar to that observed in HE-DI rats. The vasopressin congeners did not affect the responsiveness to the hot plate *per se,* and their effect could not be due to normalization of water homeostasis of these diabetes insipidus rats, since DG-LVP had the same effect as AVP. Whether this effect is exclusively contained in vasopressin has as yet not been investigated. It is conceivable that other hypothalamic or pituitary principles might share this effect. Interestingly, Ungar *et al.* 1975) reported on the partial characterization of a peptide that induces tolerance to morphine. The exciting aspect of the present studies is that memory molecules seem to operate at more than one level in the organism.

5. General Discussion

5.1. Behavioral Deficiency as a Result of Ablation of the Pituitary Gland; Precursor Molecules; Long- and Short-Term Effects of Neuropeptides

The foregoing presentation reveals the importance of pituitary peptide hormones in acquisition and maintenance of learned behavior. Although removal of the pituitary gland, partially or wholly, leads to a general endocrine and metabolic disturbance, in addition to physical debilitation in case of adenohypophysectomy or total hypophysectomy, the behavioral deficiency that accompanies the absence of the anterior as well as the posterior and intermediate lobe of the pituitary can be readily amended by treatment with pituitary hormones. That fragments of these hormones that in themselves have lost their classic endocrine activities are capable of adequate substitution indicates that the underlying disturbance is first of all caused by the absence of pituitary principles. It may be that peptides such as ACTH, MSH, and vasopressin normally operate in the formation and maintenance of adaptive behavioral responses, but it is also possible that these functions are represented in peptides resembling these entities, or even in structurally unrelated peptides manufactured by the pituitary or generated from big precursor molecules. Such neuropeptides may affect brain structures involved in motivational, learning, and memory processes. The behavioral effects of ACTH, MSH, and vasopressin may then be regarded as an incidental finding that may be followed by the discovery of hitherto unknown neuropeptides. A number of potent peptides, probably derived from LPH that restore avoidance acquisition in hypophysectomized rats have been isolated (Lande *et al.*, 1973). Other evidence for the existence of such new entities has been provided recently by the discovery of two enkephalins in rat brain, one of which resembles the 61–65 sequence of LPH (β-LPH) (Hughes *et al.*, 1975) and the other of which is an opiatelike peptide in the rat pituitary (Teschemacher *et al.*, 1975). Accordingly, studies so far have demonstrated the presence of biologically active peptides in pituitary and brain that affect sleep, learning, memory, motivation, pain, and tolerance to morphine.

As we have seen, the actions of ACTH and vasopressin fragments on avoidance acquisition and extinction, although seemingly similar, are basically different with respect to duration of their effect. This could also be demonstrated in hypophysectomized rats in the shuttle box. Administration of $ACTH_{4-10}$ or LVP restored the impaired avoidance acquisition of these animals (Bohus *et al.*, 1973). Cessation of treatment after 1 week resulted in a progressive decrease of avoidance performance of the $ACTH_{4-10}$-treated group, despite shock reinforcement if no avoidance occurred. In contrast, cessation of LVP treatment did not affect subsequent avoidance behavior. Thus, peptides related to ACTH have a "short-term" and those related to vasopressin a "long-term" effect. This difference has been observed in extinction of avoidance behavior in intact rats as well (Bohus and Lissák, 1968; de Wied and Bohus, 1966; de Wied, 1971*a,b*). The most plausible explanation of the "short-term" effect of ACTH seems to be the temporary restoration of fear motivation of hypophysectomized rats (Weiss *et al.*, 1970; Bohus *et al.*, 1973), that of the "long-term" effect of vasopressin may involve the consolidation of memory processes.

5.2. Significance of the *N*-Terminal Part of Adrenocorticotropic Hormone for Behavior

ACTH analogues affect a variety of behaviors, such as active and passive avoidance behavior, approach behavior, reversal learning, sexually motivated behavior, and amnesia. The essential elements required for the effect of ACTH on active avoidance behavior are contained in the tetrapeptide $ACTH_{4-7}$. However, a second affinity site in this respect seems to be present, since $ACTH_{7-16}$ is as active as $ACTH_{4-7}$ (de Wied *et al.*, 1975*b*). Thus, Phe^7 may be considered as a keyword for the interaction with the receptor. The observed specific reversal of the behavioral effect of ACTH analogues by the introduction of the D-enantiomer of this amino acid residue supports this hypothesis. The finding that Phe^7 can be replaced by Leu or Trp without much loss of activity cannot be explained other than by assuming that these amino acids fulfill similar topochemical requirements necessary for receptor interaction. These considerations should be regarded with care. The structure–activity studies were done *in vivo*. Individual differences in absorption, elimination, and passage in the CNS and interference with endogenous neuropeptides might occur before peptide–receptor interaction takes place.

Of interest is that the structure of TRH is related to tetrapeptide $ACTH_{4-7}$. In fact, TRH delays extinction of the pole-jumping avoidance response, albeit the potency is only ⅓ that of the tetrapeptide. It has been shown that minor changes in the TRH molecule dramatically decrease the intrinsic releasing action (Burgus *et al.*, 1970). In contrast, the behavioral effect is not affected by such measures, since substitution of proline amide by phenylalanine amide or tryptophan amide does not diminish the effect on extinction of the CAR (de Wied *et al.*, 1975*b*). It is possible that some of the recently reported nervous effects of TRH may be shared by those hypothalamic and pituitary peptides that contain related amino acid sequences. Indeed, luteinizing hormone–releasing hormone (LH–RH) appeared to be as potent as $ACTH_{4-7}$ on extinction of the pole-jumping avoidance response (de Wied *et al.*, 1975*b*).

A number of other behaviors are elicited by a direct effect of ACTH analogues in the brain. Intraventricular application of these analogues induces excessive grooming, stretching, and yawning. This effect is again located in the N-terminal part of the ACTH molecule, although $ACTH_{4-10}$ and $ACTH_{1-10}$ are virtually inactive. $ACTH_{5-10}$ is essential, but elongation of the C-terminal part is necessary to express the behavioral effect. Interestingly [D-Phe^7] substitution of N-terminal ACTH analogues ($ACTH_{4-10}$ and $ACTH_{1-10}$) does result in grooming activity. These observations again point to the importance of the Phe^7 residue in the interaction with the CNS.

ACTH, but not $ACTH_{4-10}$, reduces isolation-induced intermale fighting behavior in mice (Brain, 1972). $ACTH_{1-24}$ may also elicit LH release, ovulation, and sexual behavior following intraventricular administration in rabbits (Baldwin *et al.*, 1974). This action is only slightly represented in $ACTH_{4-10}$. These CNS effects apparently need a larger part of the ACTH molecule.

Another interesting effect that suggests the essence of the N-terminal part of ACTH is the interaction of these peptides with the opiate receptor. $ACTH_{4-10}$ appeared to have an appreciable affinity for stereospecific opiate binding sites in rat brain synaptosomal plasma membranes (Terenius *et al.*, 1975). Again, $ACTH_{11-24}$ is not active in this respect. D-Enantiomer ACTH analogues exhibited activities similar to that of their all-L congeners. The significance of these findings at the receptor level was substantiated by the observation that ACTH fragments interfere with the analgesic effect of morphine. In this situation, [D-Phe^7]$ACTH_{4-10}$ is also more potent than $ACTH_{4-10}$, while $ACTH_{11-24}$ is inactive.

It has been suggested that circulating hormones during the first days after birth constitute part of the biochemical climate that determines both development of the CNS and later behavioral flexibility (Schapiro and Vukocivh, 1970). It was found that subcutaneous injection of 3-day-old rats with either natural purified pig ACTH ($ACTH_{1-39}$), synthetic $ACTH_{1-24}$, or the ACTH fragments $ACTH_{1-18}$ and $ACTH_{1-16}$ as long-acting zinc phosphate preparations accelerated the time of eye-opening. This effect is not mediated by the adrenal cortex (van der Helm-Hylkema and de Wied, 1976). It is clear that marked differences exist in receptor requirements among the various central effects of ACTH. Thus, α-MSH, $ACTH_{4-10}$, $ACTH_{4-7}$, and $[H–Met^4(O)–D\text{-}Lys^8–Phe^9]ACTH_{4-9}$, which, like $ACTH_{1-24}$, delay extinction of the pole-jumping avoidance response, do not affect eye-opening. The same peptides, except α-MSH, are inactive in excessive grooming, while $[H–Met^4(O)–D\text{-}Lys^8–Phe^9]ACTH_{4-9}$ has no affinity for the opiate receptor. However, the effects reported thus far reside mainly in the N-terminal part of the ACTH molecule, since $ACTH_{11-24}$ is virtually inactive in the test systems mentioned above.

Other pituitary hormones may have CNS effects. Some preliminary data indicate that GH might have a profound effect in the brain. The offspring of rats treated with GH during pregnancy have larger brains and an increase in cortical neurons (Sara and Lazarus, 1974), and extinguish a shuttle-box avoidance response faster (Block and Essman, 1965). In addition, cats given GH show a selective elevation of PS. Thyrotropin-stimulating hormone (TSH) does not have such an effect (Stern *et al.*, 1975). One may expect that GH and fragments of this pituitary hormone possess important CNS effects. The same may hold for other pituitary peptides. Evidence is accumulating that such neuropeptides are released directly into the CSF. Peptides in the brain can easily reach the CSF, but those generated in the pituitary may be discharged either by retrograde transport along the pituitary stalk or via basilar cysterns that seem to connect the hormone-producing cells of the pituitary gland with the CSF (Allen *et al.*, 1974).

Preliminary studies suggest that ACTH fragments stimulate cAMP levels in posterior thalamic structures. This finding indicates that specific receptors are present in the brain for these neuropeptides. However, specific binding to radioactive-labeled ACTH fragments with brain cell components has not been demonstrated as yet. The small N-terminal ACTH sequences do not affect brain macromolecule metabolism at the level of transcription, but there is ample evidence for a translational effect of these neuropeptides. *In vivo* and *in vitro* studies indicate that ACTH fragments stimulate overall protein synthesis in the brainstem. Thus, ACTH probably elicits conformational changes that induce cAMP production. This stimulates the synthesis of protein metabolism in midbrain limbic structures, which might explain the facilitatory effects on transmission in these structures, as has been demonstrated electrophysiologically (Urban and de Wied, 1976). The information on neurotransmitter activity in the brain is controversial, and certainly not related to the behavioral effects of ACTH fragments. However, the use of more descript areas that has become feasible by the improvement of methods (Versteeg *et al.*, 1975) may reveal the nature of the influence of these neuropeptides on transmission in the brain.

5.3. Significance of Vasopressin in Memory-Consolidation Processes

Vasopressin and fragments of this posterior pituitary principle exert a long-term effect on the maintenance of new behavior patterns. These neuropeptides under certain conditions facilitate acquisition of active avoidance behavior, and increase resistance to extinction of active and passive avoidance behavior and of sexually motivated approach behavior. In addition, vasopressin antagonizes retrograde amnesia in rats as elicited by CO_2 or ECT, and protects against puromycin-induced amnesia in mice. In the absence of vasopressin as found in hereditary diabetes insipidus rats, severe memory disturbances can be demonstrated. Intraventricular administration of vasopressin and fragments facilitates memory consolidation, indicating that the behavioral effect is centrally mediated. In fact, lesions in the rostral septal and dorsal hippocampal area prevent the behavioral effect of vasopressin. Vasopressin antiserum, which is assumed to neutralize the effects of vasopressin in the CNS, prevents memory consolidation. Studies on paradoxical sleep in diabetes insipidus rats that revealed disturbances in hippocampal RSA support the hypothesis that memory processes are under the influence of vasopressin. Facilitation of the bradycardia that accompanies classically conditioned heart rate changes corroborate this view.

Neurophypophyseal hormones have been demonstrated in the CSF and in the portal vessel system. It is conceivable, therefore, that the hypothalamic–neurohypophyseal system serves the general circulation for peripheral effects of posterior pituitary hormones (e.g., kidney, mammae, blood vessels), the portal vessel system for anterior pituitary function (release of various pituitary hormones) and the CSF for CNS activities (memory consolidation). The behavioral effects of vasopressin under physiological conditions may result from an increased release into the CSF in response to emotional stress that accompanies various behaviors, in the same way as its release into the bloodstream is augmented at the retention test of passive avoidance behavior (E.A. Thompson and de Wied, 1973). Although the exact structural requirements are as yet not known, the vasopressin molecule as such is not needed, since removal of C-terminal amino acid residues that destroy the classic endocrine effects does only moderately reduce the behavioral effect. This observation is important, since it makes the use of these molecules in studies on brain function in man much more accessible.

Vasopressin appeared to induce facilitation of the development of resistance to the analgesic action of narcotic analgesics. Conversely, the development of resistance to this action of morphine is severely reduced in the absence of vasopressin. Further studies in this respect are needed before this action is understood. However, that the development of tolerance to narcotic analgesics can be prevented by agents that induce amnesia (Cox and Osman, 1970) suggests that the basic underlying mechanism is the same. Experiments aimed at the elucidation of this effect therefore provide useful information on the molecular events that underlie memory consolidation.

5.4. Behavioral Effects of Releasing Hormones and Other Oligopeptides

Hypothalamic and other brain oligopeptides appear to exhibit various effects in the CNS. For detailed information on the effects of these oligopeptides on brain function, the reader is referred to the excellent review by Prange *et al.* (1976). Thus far, the various releasing hormones have not been connected to specific behaviors except for LRH. The data, however, clearly indicate that these hormones have intrinsic CNS effects. These were demonstrated in a number of pharmacological tests used to screen psychotropic effects of drugs and in neurophysiological, morphological, and brain distribution and neurochemical studies. Some of the CNS effects of a number of these hormones will be discussed briefly. The tripeptide TRH is widely distributed throughout the brain (Stumpf and Sar, 1973; Winokur and Utiker, 1974; Jackson and Reichlin, 1974). This finding suggested that TRH may have functions other than the release of TSH and prolactin. That it may was substantiated by the demonstration that TRH potentiates the behavioral activation induced by pargyline L-dopa (Plotnikoff *et al.*, 1972, 1974; Breese *et al.*, 1974) even in hypophysectomized and thyroidectomized animals. Intraventricular administration of TRH increases locomotion in the rat (Segal and Mandell, 1974) and potentiates L-dopa-induced stereotypes and 5-HTP-induced tremor in mice (Huidobro-Toro *et al.*, 1974, 1975). It also potentiates behavioral changes following increased 5-HT levels in the brain (Green and Grahame-Smith, 1974). It increases the response to a variety of stimuli, aggressiveness, tail-rattling, and "wet-dog shakes," and antagonizes reserpine and oxotremorine hypothermia and prochlorperazine catalepsy and increases the toxicity of pargyline L-dopa (Goujet *et al.*, 1975). In addition, TRH increases strychnine-induced seizure duration (Brown and Vale, 1975). TRH has been found to exert a potent depressant action on the activity of neurons in several areas of the brain (Renaud and Martin, 1975). This action may be explained by a behavioral disinhibition, since the effects of TRH are of an excitatory nature. Such effects were demonstrated in studies showing that the tripeptide antagonized the effects of barbiturates, ethanol, chloral hydrate, and reserpine (Breese *et al.*, 1974), but not the analgesic effect of morphine (Prange *et al.*, 1975).

TRH antagonizes chlorpromazine-induced sedation, muscle relaxation, and hypothermia in mice and rabbits (Kruse, 1975). TRH in itself elicits muscle tremor, excitation, tail-lifting, and piloerection in the rat (Schenkel-Hulliger *et al.*, 1974). Massive doses of TRH induce tremor, piloerection, and eating-like movements of the forelegs in rats (Piva and Steiner, 1972). Barlow *et al.* (1975) reported that TRH decreases food consumption and food-reinforced fixed-ratio barpressing and increases locomotion. These authors did not find effects on self-stimulation behavior or active avoidance behavior. Finally, TRH may act as a discriminative stimulus (Jones *et al.*, 1975). A number of actions of TRH are reminiscent of effects of amphetamine. In this respect, it is of interest that TRH induces head-to-tail rotation in rats pretreated with apomorphine or reserpine, suggesting that TRH activates the nigrostriatal DA system (Cohn, 1975).

The tetradecapeptide somatostatin (SRIF), which is also found in extrahypothalamic brain regions (Hökfelt *et al.*, 1975) and which inhibits the release of GH (Siler *et al.*, 1973; Brazeau *et al.*, 1973) and TSH (Vale *et al.*, 1973), seems to have the opposite effect of TRH. It reduces locomotion (Segal and Mandell, 1974), potentiates pentobarbital sleeping time (Prange *et al.*, 1974) and amobarbital sleeping time (Cohn, 1975), and decreases strychnine-induced seizure duration (Brown and Vale, 1975). It causes sedation and hypothermia (Cohn, 1975). SRIF causes "barrel rolling" behavior in rats following intracerebroventricular administration, in contrast to head-to-tail rotation, which is found following TRH. The "barrel rolling" effect of SRIF can be prevented by atropine (Cohn and Cohn, 1975; Cohn et al., 1975).

The tripeptide prolyl–leucyl–glycine amide (PLG), which inhibits the release of MSH (Kastin et al., 1971; Celis et al., 1971), and which is generated from oxytocin, seems to have effects similar to those of TRH. It is active in the pargyline L-dopa test (Plotnikoff et al., 1971), it antagonizes oxytremorine-induced tremors, and it potentiates L-dopa-induced stereotypes—however, at much lower amounts than TRH (Huidobro-Toro et al., 1974). PLG antagonizes the effect of oxytremorine in intact and hypophysectomized mice (Plotnikoff and Kastin, 1974a). It potentiates L-dopa in antagonizing harmine-induced tremor (Huidobro-Toro et al., 1975). Interestingly, PLG stimulates mounting behavior in rats, as does apomorphine (Plotnikoff and Kastin, 1974b). In cats, PLG decreases locomotor activity and increases stereotyped behavior (sphinxlike position). The same effects are found with L-dopa and amphetamine, and the behavior induced by PLG is indistinguishable from that of the former compounds (North et al., 1973). Attempts to correlate the behavioral effects of TRH, SRIF, and PLG with neurotransmitter changes have so far been not very successful (Spirtes et al., 1975).

A more specific behavioral effect was found with LH–RH. This decapeptide causes the release of gonadotropic hormones (Matsuo et al., 1971). It facilitates lordosis behavior in ovariectomized and hypophysectomized rats (Moss and McCann, 1973; Pfaff, 1973). LH–RH also affects sexual behavior in male rats. It accelerates ejaculation in intact and in castrated rats maintained on testosterone (Moss et al., 1975). Other oligopeptides have been found to exert behavioral effects. Epstein and associates were among the first to report that intracranial injection of the octapeptide angiotensin II elicits drinking in rats (Epstein et al., 1968). These authors showed that intracranial injection of angiotensin II in conscious free-moving rats in normal water balance induces vigorous drinking (for a review, see Fitzsimons, 1972). Substance P is an undecapeptide that is widely distributed in the brain. Its structure was recently characterized by Leeman and associates (Leeman and Mroz, 1975). Otsuka et al. (1975) reviewed the evidence that Substance P is an excitatory transmitter of primary afferent neurons. It abolishes the abstinence syndrome in morphinized mice and tranquilizes aggressive mice (P. Stern and Hadžović, 1973).

A specific class of peptides comprises those that seem to code for memory and that are believed to transfer information to naive animals. A number of these peptides have been isolated from the brain. The original findings by Ungar (1970) on scotophobin, which induced aversion for the dark in naive rats, evoked controversial discussions, but more recent observations by Ungar (1974) and others seem to substantiate the existence of such molecules in the brain. An important objection is based on the claim "one peptide, one behavior." In our hands, a synthetic scotophobinlike peptide appeared to increase resistance to extinction of active and passive avoidance behavior (de Wied *et al.*, 1974). However, a *C*-terminal part of this peptide had similar effects. Moreover, aversive stimulation (footshock) was needed to elicit the behavioral effect of these scotophobinlike principles. A peptide that has also been found in the brain is pGlu–Ala–Gly–Tyr–Ser–Lys–OH, designated as *ameletin,* and that induces sound habituation (Ungar, 1974), has recently been synthesized (Weinstein *et al.*, 1975). Further, peptides have been isolated from the brain, one of which, catabathmophobin, increases stepdown latencies in naive rats, and another of which reverses this effect (Guttman and Cooper, 1975). Future studies may reveal the significance of these peptides in brain function. Finally, evidence has been obtained for the existence of sleep-producing peptides. The CSF of sleep-deprived animals contains hypnogenic material of peptide character (Pappenheimer *et al.*, 1974). Schoenenberger and Monnier (1974) found a hypnogenic peptide in venous blood of rabbits that were brought into a sleeplike state by electrical stimulation of the thalamus. This factor is probably an octapeptide.

5.5. Summary

A great number of laboratory experiments during the last decade disclosed the implication of pituitary and hypothalamic hormones in various brain functions. Such implications became apparent after observations of behavioral disturbances in hypophysectomized rats; in animals with hereditary diabetes insipidus, which lack the ability to synthesize vasopressin; or in rats in which the action of pituitary hormones in the brain had been blocked by the intracerebroventricular administration of, for example, specific vasopressin antiserum. Impaired behavior associated with the pathophysiological conditions mentioned above can be readily amended by treatment with ACTH, MSH, or vasopressin, but also with fragments of these polypeptides, which in themselves are practically devoid of classic endocrine effects. On the basis of these findings, it was postulated that the pituitary manufactures ''neuropeptides,'' which are involved in motivational, learning, and memory processes. It is conceivable that these neuropeptides are generated from precursor molecules. Of more recent date are observations on the CNS effects of various hypothalamic releasing hormones and other brain oligopeptides that suggest that the hypothalamic–pituitary complex releases neuropeptides that carry specific information to brain structures involved in the behavioral repertoire of the organism. Findings so far have demonstrated the existence of neuropeptides that affect learning, memory, and motivation, neuropeptides with opiatelike activity, and peptides affecting sleep and the development of tolerance to morphine, which originate in the pituitary and the brain.

6. References

Ader, R., and de Wied, D., 1972, Effects of vasopressin on active and passive avoidance learning, *Psychon. Sci.* **29:**46–48.

Ader, R., Weijnen, J.A.W.M., and Moleman, P., 1972, Retention of a passive avoidance response as a function of the intensity and duration of electric shock, *Psychon. Sci.* **26:**125–128.

Allen, J.P., Kendall, J.W., McGilvra, R., and Vancura, C., 1974, Immunoreactive ACTH in cerebrospinal fluid, *J. Clin. Endocrinol. Metab.* **38:**586–593.

Altman, J., Brunner, R.L., and Bayer, S.A., 1973, The hippocampus and behavioral maturation, *Behav. Biol.* **8:**557–596.

Applezweig, M.H., and Baudry, F.D., 1955, The pituitary–adrenocortical system in avoidance learning, *Psychol. Rep.* **1:**417–420.

Applezweig, M.H., and Moeller, G., 1959, The pituitary–adrenocortical system and anxiety in avoidance learning, *Acta Psychol.* **15:**602, 603.

Ayhan, I.H., and Randrup, A., 1973, Behavioral and pharmacological studies on morphine-induced excitation of rats. Possible relation to brain catecholamines, *Psychopharmacologia Berlin* **29:**317–328.

Baldwin, D.M., Haun, Ch.K., and Sawyer, Ch.H., 1974, Effects of intraventricular infusions of ACTH$_{1-24}$ and ACTH$_{4-10}$ on LH release, ovulation and behaviour in the rabbit, *Brain Res.* **80:**291–301.

Barlow, T.S., Cooper, B.R., Breese, G.R., Prange, A.J., Jr., and Lipton, M.A., 1975, Effects of thyrotropin releasing hormone (TRH) on behavior: Evidence for an anorexic-like action, *Neurosci. Abstr.* **1:**334.

Barry, H., and Miller, N.R., 1965, Comparison of drug effects on approach, avoidance and escape motivation, *J. Comp. Physiol. Psychol.* **59:**18–24.

Beatty, D.A., Beatty, W.A., Bowman, R.E., and Gilchrist, J.C., 1970, The effects of ACTH, adrenalectomy and dexamethasone on the acquisition of an avoidance response in rats, *Physiol. Behav.* **5:**939–944.

Bélanger, D., 1958, Effets de l'hypophysectomie sur l'apprentissage d'un réaction échappement-évitement, *Can. J. Psychol.* **12:**171–178.

Bertolini, A., Vergoni, W., Gessa, G.L., and Ferrari, W., 1968, Induction of sexual excitement with intraventricular ACTH; permissive role of testosterone in the male rabbit, *Life Sci.* **7**(Part II):1203–1206.

Bertolini, A., Vergoni, W., Gessa, G.L., and Ferrari, W., 1969, Induction of sexual excitement by the action of adrenocorticotrophic hormone in brain, *Nature London* **221:**129–158.

Block, J.B., and Essman, W.B., 1965, Growth hormone administration during pregnancy: A behavioural difference in offspring rats, *Nature London* **205:**1136, 1137.

Blumberg, H., and Dayton, H.B., 1973, Naloxone, naltrexone and related noroxymorphines, in: *Narcotic Antagonists* (M.C. Brande, L.S. Harris, E.L. May, J.P. Smith, and J.E. Villareal, eds.), *Adv. Biochem. Psychopharmacol.* **8:**33–43.

Bohus, B., 1975, Pituitary peptides and autonomic responses, in: *Hormones, Homeostasis, and the Brain* (W.H. Gispen, Tj. B. Van Wimersma Greidanus, B. Bohus, and D. de Wied, eds.), *Progr. Brain Res.* **42:**275–283.

Bohus, B., 1976, Effect of desglycinamide–lysine vasopressin (DG-LVP) on sexually motivated T-maze behavior of the male rat, *Horm. Behav.* in press.

Bohus, B., and de Wied, D., 1966, Inhibitory and facilitatory effect of two related peptides on extinction of avoidance behavior, *Science* **153:**318–320.

Bohus, B., and de Wied, D., 1967*a*, Failure of α-MSH to delay extinction of conditioned avoidance behavior in rats with lesions in the parafascicular nuclei of the thalamus, *Physiol. Behav.* **2:**221–223.

Bohus, B., and de Wied, D., 1967*b*, Avoidance and escape behavior following medial thalamic lesions in rats, *J. Comp. Physiol. Psychol.* **64:**26–29.

Bohus, B., and Lissák, K., 1968, Adrenocortical hormones and avoidance behaviour in rats, *Int. J. Neuropharmacol.* **7:**301–306.

Bohus, B., Ader, R., and de Wied, D., 1972, Effects of vasopressin on active and passive avoidance behavior. *Horm. Behav.* **3:**191–197.

Bohus, B., Gispen, W.H., and de Wied, D., 1973, Effects of lysine vasopressin and ACTH$_{4-10}$ on conditioned avoidance behavior of hypophysectomized rats, *Neuroendocrinology* **11:**137–143.

Bohus, B., van Wimersma Greidanus, Tj.B., and de Wied, D., 1975a, Behavioral and endocrine responses of rats with hereditary hypothalamic diabetes insipidus (Brattleboro strain), *Physiol. Behav.* **14**:609–615.

Bohus, B., Hendrickx, H.H.L., van Kolfschoten, A.A., and Krediet, T.G., 1975b, The effect of ACTH$_{4-10}$ on copulatory and sexually motivated approach behavior in the male rat, in: *Sexual Behavior: Pharmacology and Biochemistry* (M. Sandler and G.L. Gessa, eds.), pp. 269–275, Raven Press, New York.

Brady, J.V., and Nauta, W.J.H., 1953, Subcortical mechanisms in emotional behavior: Affective changes following septal forebrain lesions in the albino rat, *J. Comp. Physiol. Psychol.* **46**:339–346.

Brain, P.F., 1972, Study on the effect of the 4–10 ACTH fraction on isolation induced intermale fighting behavior in the albino mouse, *Neuroendocrinology* **10**:371–376.

Brazeau, P., Vale, W., Burgus, R., Ling, N., Butcher, M., Rivier, J., and Guillemin, R., 1973, Hypothalamic polypeptide that inhibits the secretion of immunoreactive pituitary growth hormone, *Science* **179**:77–79.

Breese, G.R., Cooper, B.R., Prange, A.J., Jr., Cott, J.M., and Lipton, M.A., 1974, Interactions of thyrotropin-releasing hormone with centrally acting drugs, in: *The Thyroid Axis, Drugs and Behavior* (A.J. Prange, Jr., ed.), pp. 115–127, Raven Press, New York.

Brown, M., and Vale, W., 1975, Central nervous system effects of hypothalamic peptides, *Endocrinology* **96**:1333–1336.

Burgus, R., Dunn, T.F., Desiderio, D., Ward, D.N., Vale, W., and Guillemin, R., 1970, Characterization of the hypothalamic hypophysiotropic TSH-releasing factor (TRF) of ovine origin, *Nature London* **226**:321–325.

Burkhard, W.P., and Gey, K.F., 1968, Adenylcyclase in rat brain, *Helv. Physiol. Pharmacol. Acta* **26**:197, 198.

Burt, D.R., and Snyder, S.H., 1975, Thyrotropin releasing hormone (TRH): Apparent receptor binding in rat brain membranes, *Brain Res.* **93**:309–328.

Cardo, B., 1965, Rôle de certain noyaux thalamiques dans l'éboration et la conservation de divers conditionnements, *Psychol. Fr.* **10**:344–351.

Cardo, B., 1967, Effets de la stimulation du noyau parafasciculaire thalamique sur l'acquisition d'un conditionnement d'évitement chez le rat, *Physiol. Behav.* **2**:245–248.

Cardo, B., and Valade, F., 1965, Rôle du noyau thalamique parafasciculaire dans le conservation d'un conditionnement d'évitement chez le rat, *C.R. Acad. Sci. Paris* **261**:1399–1402.

Celis, M.E., Taleisnik, S., and Walter, R., 1971, Regulation of formation and proposed structure of the factor inhibiting the release of melanocyte-stimulating hormone, *Proc. Nat. Acad. Sci. U.S.A.* **68**:1428–1433.

Chambers, W.F., Dunnhue, F.W., Smith, C.G., Blanchard, R.R., Taylor, C.H., and Hill, D.B., 1966, Effect of vasopressin and adrenal steroids on cortical responses evoked at the midbrain level in aged rats, *Gerontologia* **12**:65–73.

Chung, D., and Li, C.H., 1967, Adrenocorticotropins, XXXVII. The synthesis of 8-lysine–ACTH$_{1-17}$–NH$_2$ and its biological properties, *J. Amer. Chem. Soc.* **89**:4208–4213.

Cleghorn, R.A., 1957, Steroid hormones in relation to neuropsychiatric disorders, in: *Hormones, Brain Function and Behavior* (H. Hoogland, ed.), pp. 3–25, Academic Press, New York.

Coculescu, M., and Pavel, S., 1973, Arginine vasotocin-like activity of cerebrospinal fluid in diabetes insipidus, *J. Clin. Endocrinol. Metab.* **36**:1031, 1032.

Cohen, M., Keats, A.S., Krivoy, W.A., and Ungar, G., 1965, Effect of actinomycin D on morphine tolerance, *Proc. Soc. Exp. Biol. Med.* **119**:381–384.

Cohn, M.L., 1975, Acute behavioral changes induced in the rat by the intracerebroventricular administration of thyrotropin releasing factor (TRF) and somatostatin, Society of Toxicology 14th Annual Meeting, Williamsburg, Virginia.

Cohn, M.L., and Cohn, M., 1975, "Barrel rotation" induced by intracerebroventricular injections of somatostatin in the nonlesioned rat, *Fed. Proc. Fed. Amer. Soc. Exp. Biol.* **34**:738.

Cohn, M.L., Cohn, M., and Petro, B., 1975, Neuroendocrine control of rotational behavior in the nonlesioned rat, *Neurosci. Abstr.* **1**:448.

Cox, B.M.,and Osman, O.H., 1970, Inhibition of the development of tolerance to morphine in rats by drugs which inhibit rubonucleic acid or protein synthesis, *Br. J. Pharmacol.* **38**:157–170.

Cross, B.A., Dyball, R.E.J., and Moss, R.L., 1971, Stimulation of paraventricular neurosecretory cells by oxytocin applied iontophoretically, *J. Physiol.* **222**:22, 23P.

Dedman, M.L., Farmer, T.H., and Morris, C.J.O.R., 1955, Oxidation–reduction properties of adreno-corticotrophic hormone, *Biochem. J.* **59**:xii.

De Kloet, R., and McEwen, B.S., 1976, Glucocorticoid interactions with brain and pituitary, in: *Molecular and Functional Neurobiology* (W.H. Gispen, ed.), pp. 258–306, Elsevier, Amsterdam.

Delacour, J., 1970, Specific functions of a medial thalamic structure in avoidance conditioning in the rat, in: *Pituitary, Adrenal and the Brain* (D. de Wied and J.A.W.M. Weijnen, eds.), *Prog. Brain Res.* **32**:158–170, Elsevier, Amsterdam.

Delacour, J., 1971, Effects of medial thalamic lesions in the rat. A review and an interpretation, *Neuropsychologia* **9**:157–174.

Dempsey, G.L., Kastin, A.J., and Schally, A.V., 1972, The effects of MSH on a restricted passive avoidance response, *Horm. Behav.* **3**:333–337.

Denman, P.M., Miller, L.H., Sandman, C.A., Schally, A.V., and Kastin, A.J., 1972, Electrophysiological correlates of melanocyte-stimulating hormone activity in the frog, *J. Comp. Physiol. Psychol.* **80**:59–65.

De Wied, D., 1964, Influence of anterior pituitary on avoidance learning and escape behavior, *Amer. J. Physiol.* **207**:255–259.

De Wied, D., 1965, The influence of the posterior and intermediate lobe of the pituitary and pituitary peptides on the maintenance of a conditioned avoidance response in rats, *Int. J. Neuropharmacol.* **4**:157–167.

De Wied, D., 1966, Inhibitory effect of ACTH and related peptides on extinction of conditioned avoidance behavior in rats, *Proc. Soc. Exp. Biol. Med.* **122**:28–32.

De Wied, D., 1967, Opposite effects of ACTH and glucocorticosteroids on extinction of conditioned avoidance behavior, in: *Excerpta Med. Int. Congr. Ser.*, No. 132, p. 945.

De Wied, D., 1969, Effects of peptide hormones on behavior, in: *Frontiers in Neuroendocrinology,* (W.F. Ganong and L. Martini, eds.), pp. 97–140, Oxford University Press, New York.

De Wied, D., 1971a, Pituitary–adrenal hormones and behavior, in: *Normal and Abnormal Development of Brain and Behavior* (G.B.A. Stoelinga and J.J. van der Weff ten Bosch, eds.), pp. 315–322, Leiden University Press, Leiden, The Netherlands.

De Wied, D., 1971b, Long term effect of vasopressin on the maintenance of a conditioned avoidance response in rats, *Nature London,* **232**:58–60.

De Wied, D., 1974, Pituitary–adrenal system hormones and behavior, in: *The Neurosciences: Third Study Program* (F.O. Schmitt and F.G. Worden, eds.), pp. 653–666, MIT Press, Cambridge, Massachusetts.

De Wied, D., and Bohus, B., 1966, Long term and short term effects on retention of a conditioned avoidance response in rats by treatment with long acting pitressin and α-MSH, *Nature London,* **212**:1484–1486.

De Wied, D., and Gispen, W.H., 1976, Impaired development of tolerance to morphine analgesia in rats with hereditary diabetes insipidus, *Psychopharmacologia Berlin* **46**:27–29.

De Wied, D., Bohus, B., and Greven, H.M., 1968, Influence of pituitary and adrenocortical hormones on conditioned avoidance behavior in rats, in: *Endocrinology and Human Behaviour* (R.P. Michael, ed.), pp. 188–199, Oxford University Press, New York.

De Wied, D., Greven, H.M., Lande, S., and Witter, A., 1972, Dissociation of the behavioral and endocrine effects of lysine vasopressin by tryptic digestion, *Bri. J. Pharmacol.* **45**:118–122.

De Wied, D., van Wimersma Greidanus, Tj.B., and Bohus, B., 1974, The rat supraoptic-neurohypophyseal system and behavior: Role of vasopressin in memory processes, in: *Le Cerveau et les Hormones,* Sér. No. 18, pp. 324–328, Collège de Médecine, Expansion Scientifique Française.

De Wied, D., Bohus, B., and van Wimersma Greidanus, Tj.B., 1975a, Memory deficit in rats with hereditary diabetes insipidus, *Brain Res.* **85**:152–156.

438   D. de Wied and W. H. Gispen

De Wied, D., Witter, A., and Greven, H.M., 1975*b*, Behaviourally active ACTH analogues, *Biochem. Pharmacol.* **24**:1463–1468.

De Wied, D., Bohus, B., Urban, I., van Wimersma Greidanus, Tj.B., and Gispen, W.H., 1976, Pituitary peptides and memory, in: *Peptides: Chemistry, Structure and Biology*, Proceedings of the Fourth American Peptide Symposium (R. Walter and J. Meienhofer, eds.), pp. 635–643, Ann Arbor Science Publishers, Ann Arbor, Michigan.

Dunn, A.H., 1976, The chemistry of learning and the formation of memory, in: *Molecular and Functional Neurobiology* (W.H. Gispen, ed.), pp. 347–387, Elsevier, Amsterdam.

Dunn, A.J., and Rees, H.D., 1975, Amino acid incorporation into brain proteins: The role of hormones in behavior-related changes, *Abstracts of the Fifth International Meeting of the I.S.N.*, Barcelona, Abstract No. 383.

Dyster-Aas, H.K., and Krakau, C.E.T., 1965, General effects of α-melanocyte stimulating hormones in the rabbit, *Acta Endocrinol. Copenhagen* **48**:409–419.

Eberle, A., and Schwyzer, R., 1975, Hormone–receptor interaction. Demonstration of two message sequences (active sites) in α-melanotropin, *Helv. Chim. Acta* **58**:1528–1535.

Eddy, N.B., and Leimbach, D., 1953, Synthetic analgesics. II. Dithienylbutenyl and dithienylbutylamines, *J. Pharmacol. Exp. Ther.* **107**:385–393.

Endröczi, E., Lissák, K., and Fekete, T., 1970, Effects of ACTH on EEG habituation in human subjects, in: *Pituitary, Adrenal and the Brain* (D. de Wied and J.A.W.M. Weijnen, eds.), *Prog. Brain Res.* **32**:254–262, Elsevier, Amsterdam.

Entman, M.L., Levey, G.S., and Epstein, E.S., 1969, Demonstration of adenylcyclase activity in canine cardiac sarcoplasmic reticulum, *Biochem. Biophys. Res. Commun.* **35**:728–733.

Epstein, A.N., Fitzsimons, J.T., and Simons, B.J., 1968, Drinking caused by the intracranial injection of angiotensin into the rat, *J. Physiol.* **196**:98–104.

Evans, W.O., 1961, A new technique for the investigation of some analgesic drugs on a reflexive behavior in the rat, *Psychopharmacologia Berlin* **2**:318–325.

Feldman, S., Todt, J.C., and Porter, R.W., 1961, Effect of adrenocortical hormones on evoked potentials in the brain stem, *Neurology* **11**:109–115.

Ferrari, W., Gessa, G.L., and Vargiu, L., 1963, Behavioral effects induced by intracysternally injected ACTH and MSH, *Ann. N. Y. Acad. Sci.* **104**:330–345.

Fishbein, W., 1971, Disruptive effects of rapid eye movement sleep deprivation on long-term memory, *Physiol. Behav.* **6**:279–282.

Fitzsimons, J.T., 1972, Thirst, *Physiol Rev.* **52**:458–561.

Fitzsimons, J.T., 1975, The renin–angiotensin system and drinking behavior, in: *Hormones, Homeostasis and the Brain* (W.H. Gispen, Tj.B. van Wimersma Greidanus, B. Bohus, and D. de Wied, eds.), *Prog. Brain Res.* **42**:215–233, Elsevier, Amsterdam.

Flood, J.F., Bennett, E.L., and Orme, A.E., 1975, Effects of protein synthesis inhibition on memory for active avoidance training, *Physiol. Behav.* **14**:177–184.

Forn, J., and Krishna, G., 1971, Effect of norepinephrine, histamine and other drugs on cyclic 3′,5′-AMP formation in brain slices of various animal species, *Pharmacology* **5**:193–204.

Fried, P.A., 1972, The septum and behaviour: A review, *Psychol. Bull.* **78**:292–310.

Garrud, P., 1975, Effects of lysine-8-vasopressin on punishment induced suppression of a lever-holding response, in: *Hormones, Homeostasis and the Brain* (W.H. Gispen, Tj.B. van Wimersma Greidanus, B. Bohus, and D. de Wied, eds.), *Prog. Brain Res.* **42**:173–186, Elsevier, Amsterdam.

Garrud, P., Gray, J.A., and de Wied, D., 1974, Pituitary–adrenal hormones and extinction of rewarded behaviour in the rat, *Physiol. Behav.* **12**:109–119.

Gessa, G.L., Pisano, M., Vargiu, L., Crabai, F., and Ferrari, W., 1967, Stretching and yawning movements after intracerebral injections of ACTH, *Rev. Can. Biol.* **26**:229–236.

Gilman, A.G., 1970, A protein binding assay for adenosine 3′,5′-cyclic monophosphate, *Proc. Nat. Acad. Sci. U.S.A.* **67**:305–312.

Gispen, W.H., 1970, Over de relatie tussen het gestoorde voorwaardelijke vluchtgedrag van hypofyseloze ratten en het RNA-metabolisme in de hersenstam, Thesis, University of Utrecht, The Netherlands.

Gispen, W.H., and Schotman, P., 1973, Pituitary–adrenal system, learning and peformance: Some neurochemical aspects, in: *Drug Effects on Neuroendocrine Regulation* (E. Zimmermann, W.H. Gispen, B.H. Marks, and D. de Wied, eds.) *Prog. Brain Res.* **39**:443–459, Elservier, Amsterdam.

Gispen, W.H., and Schotman, P., 1976, ACTH and brain RNA: Changes in content and labeling of RNA in rat brain stem. *Neuroendocrinology,* in press.

Gispen, W.H., van Wimersma Greidanus, Tj.B., and de Wied, D., 1970, Effects of hypophysectomy and $ACTH_{1-10}$ in responsiveness to electric shock in rats, *Physiol. Behav.* **5**:143–146.

Gispen, W.H., de Wied, D., Schotman, P., and Jansz, H.S., 1971, Brainstem polysomes and avoidance performance of hypophysectomized rats subjected to peptide treatment, *Brain Res.* **31**:341–351.

Gispen, W.H., van der Poel, A., and van Wimersma Greidanus, Tj.B., 1973, Pituitary–adrenal influences on behavior. Responses to test situations with or without electric footshock, *Physiol. Behav.* **10**:345–350.

Gispen, W.H., Wiegant, V.M., Greven, H.M., and de Wied, D., 1975a, The induction of excessive grooming in the rat by intraventricular application of peptides derived from ACTH: Structure–activity studies, *Life Sci.* **17**:645–652.

Gispen, W.H., van Wimersma Greidanus, Tj.B., Waters-Ezrin, C., Zimmerman, E., Krivoy, W.A., and de Wied, D., 1975b, Influence of peptides on reduced response of rats to electric footshock after acute administration of morphine, *Eur. J. Pharmacol.* **33**:99–105.

Glassman, E., 1969, The biochemistry of learning: An evaluation of the role of RNA and protein, *Annu. Rev. Biochem.* **38**:605–646.

Goldsmith, P.C., and Zimmerman, E.A., 1975, Ultrastructural localization of neurophysin and vasopressin in the rat median eminence, 57th Meeting of the Endocrine Society, *Endocrinology Suppl.* **96**:A377.

Goujet, M.A., Simon, P., Chermat, R., and Boissier, J.R., 1975, Profil de la TRH en psychopharmacologie expérimentale, *Psychopharmacologia Berlin* **45**:87–92.

Gray, J.A., 1971, Effect of ACTH on extinction of rewarded behaviour is blocked by previous administration of ACTH, *Nature London* **229**:52–54.

Green, A.R., and Grahame-Smith, D.G., 1974, TRH potentiates behavioral changes following increased brain 5-hydroxytryptamine accumulation in rats, *Nature London* **251**:524–526.

Greengard, P., 1975, Cyclic nucleotides, protein phosphorylation and neuronal function, in: *Advances in Cyclic Nucleotide Research,* Vol. 5 (G.I. Drummond, P. Greengard, and G.A. Robinson, eds.), pp. 585–601, Raven Press, New York.

Greven, H.M., and de Wied, D., 1973, The influence of peptides derived from corticotropin (ACTH) on performance. Structure–activity studies, in: *Drug Effects on Neuroendocrine Regulation* (E. Zimmermann, W.H. Gispen, B.H. Marks, and D. de Wied, eds.), *Prog. Brain Res.* **39**:429–442, Elsevier, Amsterdam.

Gupta, K.K., 1969, Antidiuretic hormone in cerebrospinal fluid, *Lancet* **1**:581.

Guth, S., Levine, S., and Seward, J.P., 1971, Appetitive acquisition and extinction effects with exogenous ACTH, *Physiol. Behav.* **7**:195–200.

Guttman, H.N., and Cooper, R.S., 1975, Oligopeptide control of step-down avoidance, *Life Sci.* **16**:915–924.

Harris, R.K., 1973, Acquisition of conditioned avoidance responses by hypophysectomized rats, *J. Comp. Physiol. Psychol.* **82**:254–260.

Heller, H., Hasan, S.H., and Saifi, A.Q., 1968, Antidiuretic activity in the cerebrospinal fluid, *J. Endocrinol.* **41**:273–280.

Hofmann, K., Thompson, T.A., Woolner, M.E., Spühler, G., Yajima, H., Cipera, J.D., and Schwartz, E.T., 1960, Studies on polypeptides. XV. Observations on the relation between structure and melanocyte-expanding activity of synthetic peptides, *J. Amer. Chem. Soc.* **82**:3721–3726.

Hofmann, K. Andreatta, R., Bohn, H., and Moroder, L., 1970, Studies on polypeptides. XLV. Structure–function studies in the β-corticotropin series, *J. Med. Chem.* **13**:339–345.

Hökfelt, T., and Fuxe, K., 1972, On the morphology and neuroendocrine role of the hypothalamic catecholamine neurons, in: *Brain–Endocrine Interaction. Median Eminence: Structure and Function* (K.M. Knigge, D.E. Scott, and A. Weindl, eds.), pp. 181–223, Karger, Basel.

Hökfelt, T., Johansson, O., Fuxe, K., Löfström, A., Goldstein, M., Park, D., Ebstein, R., Fraser, H., Jeffcoate, S., Efendic, S., Luft, R., and Arimura, A., 1975, Mapping and relationship of hypothalamic neurotransmitters and hypothalamic hormones, in: *CNS and Behavioral Pharmacology*, pp. 93–110, Proceedings of the VIth International Congress of Pharmacology, Helsinki.

Hughes, J., Smith, T.W., Kosterlitz, H.W., Fothergill, L.A., Morgan, B.A., and Morris, H.R., 1975, Identification of two related glutapeptides from the brain with potent opiate agonist activity, *Nature London* **258**:557–579.

Huidobro-Toro, J.P., Scotti de Carolis, A., and Longo, V.G., 1974, Action of two hypothalamic factors (TRH, MIF) and of angiotensin II on the behavioral effects of L-DOPA and 5-hydroxytryptophan in mice, *Pharmacol. Biochem. Behav.* **2**:105–109.

Huidobro-Toro, J.P., Scotti de Carolis, A., and Longo, V.G., 1975, Intensification of central catecholaminergic and serotonergic processes by the hypothalamic factors MIF and TRH and by angiotensin II, *Pharmacol. Biochem. Behav.* **3**:235–242.

Ide, M., Tanaka, M., Nakamura, M., and Okabayashi, T., 1972, Stimulation by ACTH analogs of rat adrenal adenylcyclase activity: Correlation with steroidogenic activity, *Arch. Biochem. Biophys.* **149**:189–196.

Isaacson, R.L., Dunn, A.J., Rees, H.D., and Waldock, B., 1975, $ACTH_{4-10}$ and improved use of information in rats, *Physiol. Psychol.* **4**:159–162.

Izumi, K., Donaldson, J., and Barbeau, A., 1973, Yawning and stretching in rats induced by intraventricularly administered zinc, *Life Sci.* **12**:203–210.

Jackson, I.M.D., and Reichlin, S., 1974, Thyrotropin-releasing hormone (TRH): Distribution in hypothalamic and extrahypothalamic brain tissues of mammalian and submammalian chordates, *Endocrinology* **95**:854–862.

Jakoubek, B., Buresova, M., Hajek, I., Etrychova, J., Pavlik, A., and Decicova, A., 1972, Effect of ACTH on the synthesis of rapidly labeled RNA in the nervous system of mice, *Brain Res.* **43**:417–428.

Jones, C.N., Grant, L.D., Prange, A.J., Jr., and Breese, G.R., 1975, Stimulus properties of *d*-amphetamine in the rat: Interaction with thyrotropin-releasing hormone (TRH), *Neurosci. Abstr.* **1**:246.

Jouvet, M., 1965, Étude de la dualité des états de sommeil et des mécanismes de la phase paradoxale, in: *Aspects Anatomo-fonctionelles de la Physiologie du Sommeil*, pp. 397–449, C.N.R.S., Paris.

Kastin, A.J., Kullander, S., Borglin, N.E., Dyster-Aas, K., Dahlberg, B., Ingvar, D., Krakau, C.E.T., Miller, M.C., Bowers, C.Y., and Schally, A.V., 1968, Extrapigmentary effects of MSH in amenorrheic women, *Lancet* **1**:1007–1010.

Kastin, A.J., Schally, A.V., and Viosca, S., 1971, Inhibition of MSH release in frogs by direct application of L-propyl–L-leucyl–glycinamide to the pituitary, *Proc. Soc. Exp. Biol. Med.* **137**:1437–1439.

Kastin, A.J., Miller, M.C., Ferrell, L., and Schally, A.V., 1973*a*, General activity in intact hypophysectomized rats after administration of melanocyte-stimulating hormone (MSH), melatonin and Pro–Leu–Gly–NH$_2$, *Physiol. Behav.* **10**:399–401.

Kastin, A.J., Miller, L.H., Nockton, R., Sandman, C.A., Schally, A.V., and Stratton, L.O., 1973*b*, Behavioral aspects of melanocyte-stimulating hormone (MSH), in: *Drug Effects on Neuroendocrine Regulation* (E. Zimmermann, W.H. Gispen, B.H. Marks, and D. de Wied, eds.), *Prog. Brain Res.* **39**:461–470, Elsevier, Amsterdam.

Kastin, A.J., Dempsey, G.L., Leblanc, B., Dyster-Aas, K., and Schally, A.V., 1974, Extinction of an appetitive operant response after administration of MSH, *Horm. Behav.* **5**:135–139.

Kastin, A.J., Sandman, C.A., Stratton, L.O., Schally, A.V., and Miller, L.H., 1975, Behavioral and electrographic changes in rat and man after MSH, in: *Hormones, Homeostasis and the Brain* (W.H. Gispen, Tj.B. van Wimersma Greidanus, B. Bohus, and D. de Wied, eds.), *Prog. Brain Res.* **42**:143–150, Elsevier, Amsterdam.

Kelsey, J.E., 1975, Role of pituitary–adrenocortical system in mediating avoidance behavior of rats with septal lesions, *J. Comp. Physiol. Psychol.* **88**:271–280.

King, A.R., and de Wied, D., 1974, Localized behavioral effects of vasopressin on maintenance of an active avoidance response in rats, *J. Comp. Physiol. Psychol.* **86**:1008–1018.

Koida, M., Hano, K., and Iso, T., 1966, Evaluation of *in vitro* melanocyte-darkening activities of L-histidyl–L-phenylalanyl–L-arginyl–tryptophyl–glycine and its nine stereoisomers in *Rana nigromaculata, Jpn. J. Pharmacol.* **16**:243–249.

Korányi, I., Beyer, C., and Guzman-Flores, C., 1971. Effect of ACTH and hydrocortisone on multiple unit activity in the forebrain and thalamus in response to reticular stimulation, *Physiol. Behav.* **7**:331–335.

Krivoy, W.A., and Guillemin, R., 1961, On a possible role of β-melanocyte stimulating hormone (β-MSH) in the central nervous system of mammals: An effect of β-MSH in the spinal cord of the cat, *Endocrinology* **69**:170–175.

Krivoy, W.A., Lane, M., Childers, H.A., and Guillemin, R., 1962, On the action of β-melanocyte stimulating hormone (β-MSH) on spontaneous electric discharge of the transparent knife fish, *G. eigenmannia, Experientia (Basel)* **18**:521.

Krivoy, W., Lane, M., and Kroeger, D.C., 1963, The action of certain polypeptides on synaptic transmission, *Ann. N. Y. Acad. Sci.* **104**:312–329.

Krivoy, W.A., Kroeger, D., Taylor, A.N., and Zimmermann, E., 1974a, Antagonism of morphine by β-melanocyte stimulating hormone and by tetracosactin, *Eur. J. Pharmacol.* **27**:339–345.

Krivoy, W.A., Zimmermann, E., and Lande, S., 1974b, Facilitation of development of resistance to morphine analgesia by desglycinamide[9]–lysine vasopressin, *Proc. Nat. Acad. Sci. U.S.A.* **71**:1852–1856.

Kruse, H., 1975, Thyrotropin releasing hormone: Interaction with chlorpromazine in mice, rats and rabbits, *J. Pharmacol. Paris* **6**:249–268.

Lande, S., Witter, A., and de Wied, D., 1971, Pituitary peptides. An octapeptide that stimulates conditioned avoidance acquisition in hypophysectomized rats, *J. Biol. Chem.* **246**:2058–2062.

Lande, S., Flexner, J.B., and Flexner, L.B., 1972, Effect of corticotropin and desglycinamide[9]–lysine vasopressin on suppression of memory by puromycin, *Proc. Nat. Acad. Sci. U.S.A.* **69**:558–560.

Lande, S., de Wied, D., and Witter, A., 1973, Unique pituitary peptides with behavioral-affecting activity, in: *Drug Effects on Neuroendocrine Regulation* (E. Zimmermann, W.H. Gispen, B.H. Marks, and D. de Wied, eds.), *Prog. Brain Res.* **39**:421–427, Elsevier, Amsterdam.

Landfield, P.W., McGaugh, J.L., and Tusa, R.J., 1972, Theta rhythm: A temporal correlate of memory storage processes in the rat, *Science* **175**:87–89.

Leconte, P., and Bloch, 1970, Déficit de la retention d'un conditionnement après privation de sommeil paradoxal chez le rat, *C.R. Acad. Sci. Paris Sér. D* **271**:226–229.

Leeman, S.E., and Mroz, E.A., 1975, Minireview: Substance P, *Life Sci.* **15**:2033–2044.

Leonard, B.E., 1969, The effect of sodium-barbitone, alone and together with ACTH and amphetamine, on the behavior of the rat in the multiple "T" maze, *Int. J. Neuropharmacol.* **8**:427–435.

Leonard, B.E., 1974, The effect of two synthetic ACTH analogues on the metabolism of biogenic amines in the rat brain, *Arch. Int. Pharmacodyn.* **207**:242–253.

Levine S., and Jones, L.E., 1965, Adrenocorticotrophic hormone (ACTH) and passive avoidance learning, *J. Comp. Physiol. Psychol.* **59**:357–360.

Lichtensteiger, W., and Lienhart, R., 1975, Opiates and hypothalamic mechanisms: An ascending influence on diencephalic and mesencephalic dopamine (DA) systems and its interaction with α-MSH, in: *Acute Effects of Narcotic Analgesic Sites and Mechanisms of Action,* Satellite Symposium to the VIth International Congress of Pharmacology, Nokkala, Espoo, Finland, abstract, pp. 26, 27.

Lissák, K., and Bohus, B., 1972, Pituitary hormones and avoidance behavior of the rat, *Int. J. Psychobiol.* **2**:103–115.

Lissák, K., Endröczi, E., and Medgyesi, P., 1957, Somatisches Verhalten und Nebennierenrindentätigkeit, *Pflügers Arch.* **265**:117–124.

Lloyd, G., 1974, The action of neurotropic peptide analogues of the ACTH molecule, *Thesis,* University of Edinburgh, U.K.

Lo, T.-B., Dixon, J.S., and Li, C.H., 1961, Isolation of methionine sulfoxide analogue of α-melanocyte-stimulating hormone from bovine pituitary glands, *Biochim. Biophys. Acta* **53**:584–586.

Longo, V.G., and Loizzo, A., 1973, Effects of drugs on the hippocampal theta rhythm. Possible relationships to learning and memory processes, in: *Pharmacology and the Future of Man*, Fifth International Congress of Pharmacology, Vol. 4, *Brain, Nerves and Synapses* (F.E. Bloom and G.H. Acheson, eds.), pp. 46–54, Karger, Basel.

MacLean, P.D., 1957, Chemical and electrical stimulation of hippocampus in unrestrained animals. II. Behavioral findings, *Amer. Med. Assoc. Arch. Neurol. Psychiatry* **78**:128–142.

Mao, C.C., and Guidotti, A., 1974, Simultaneous isolation of adenosine 3′, 5′-cyclic monophosphate (cAMP) and guanosine 3′, 5′-cyclic monophosphate (cGAP) in small tissue samples, *Anal. Biochem.* **59**:63–68.

Martin, J.B., Renaud, L.P., and Brazeau, P., 1975, Hypothalamic peptides: New evidence for "peptidergic" pathways in the CNS, *Lancet* **II**:393–395.

Matsuo, H., Baba, Y., Nair, R.M.G., Arimura, A., and Schally, A.V., 1971, Structure of the porcine LH- and FSH-releasing hormone. I. The proposed amino-acid sequence, *Biochem. Biophys. Res. Commun.* **43**:1334–1339.

McGaugh, J.L., Gold, P.E., Van Buskirk, R., and Haycock, J., 1975, Modulating influences of hormones and catecholamines on memory storage processes, in: *Hormones, Homeostasis and the Brain* (W.H. Gispen, Tj.B. van Wimersma Greidanus, B. Bohus, and D. de Wied, eds.), *Prog. Brain Res.* **42**:151–162, Elsevier, Amsterdam.

Meyerson, B.J., 1975, The infuence of $ACTH_{4-10}$ on the urge of female rats to seek contact with a sexually active male, *4th Scandinavian Meeting on the Physiology of Behavior*, Oslo, p. 45.

Meyerson, B.J., and Lindström, L., 1973, Sexual behaviour in the female rat. A methodological study applied to the investigation of the effect of estradiol benzoate, *Acta Physiol. Scand. Suppl.* No. 389.

Miller, R.E., and Ogawa, N., 1962, The effect of adrenocorticotropic hormone (ACTH) on avoidance conditioning in the adrenalectomized rat, *J. Comp. Physiol. Psychol.* **55**:211–213.

Moll, J., and de Wied, D., 1962, Observations on the hypothalamo–posthypophyseal system of the posterior lobectomized rat, *Gen. Comp. Endocrinol.* **2**:215–228.

Moses, A.M., and Miller, M., 1970, Accumulation and release of pituitary vasopressin in rats heterozygous for hypothalamic diabetes insipidus, *Endocrinology* **86**:34–41.

Moss, R.L., and McCann, S.M., 1973, Introduction of mating behavior in rats by luteinizing hormone–releasing factor, *Science* **181**:177–179.

Moss, R.L., Dyball, R.E.J., and Cross, B.A., 1972, Excitation of antidromically identified neurosecretory cells of the paraventricular nucleus by oxytocin applied iontophoretically, *Exp. Neurol.* **34**:95–102.

Moss, R.L., McCann, S.M., and Dudley, C.A., 1975, Releasing factors and sexual behavior, in: *Hormones, Homeostasis and the Brain* (W.H. Gispen, Tj.B. van Wimersma Greidanus, B. Bohus, and D. de Wied, eds.), *Prog. Brain Res.* **42**:37–46, Elsevier, Amsterdam.

Murphy, J.V., and Miller, R.E., 1955, The effect of adrenocorticotropic hormone (ACTH) on avoidance conditioning in the rat, *J. Comp. Physiol. Psychol.* **48**:47–49.

Nicoll, R.A., and Barker, J.L., 1971, The pharmacology of recurrent inhibition in the supraoptic neurosecretory system, *Brain Res.* **35**:501–511.

Nicolov, N., 1967, Effect of hydrocortisone and ACTH upon the bioelectric activity of spinal cord, *Folia Med. Plovdiv* **9**:249–255.

North, R.H., Harik, S.I., and Snyder, S.H., 1973, L-Prolyl–L-leucyl–glycinamide (PLG): Influences on locomotor and stereotyped behavior of cats, *Brain Res.* **63**:435–439.

Otsuka, H., and Inouye, K., 1964, Synthesis of peptides related to the *N*-terminal structure of corticotropin. III. The synthesis of L-histidyl–L-phenylalanyl–L-tryptophan, the smallest peptide-exhibiting the melanocyte-stimulating and the lipolytic activities, *Bull. Chem. Soc. Jpn.* **37**:1465–1471.

Otsuka, M., Konishi, S., and Takahashi, T., 1975, Hypothalamic substance P as a candidate for transmitter of primary afferent neurons, *Fed. Proc. Fed. Amer. Soc. Exp. Biol.* **34**:1922–1928.

Palo, J., and Savolainen, H., 1974, The effect of high doses of synthetic ACTH on the rat brain, *Brain Res.* **70**:313–320.

Pappenheimer, J.R., Fencl, V., Karnovsky, M.L., and Koski, G., 1974, Peptides in cerebrospinal fluids and their relation to sleep and activity, in: *Brain Dysfunction in Metabolic Disorders* (F. Plum, ed.), *Res. Publ. Assoc. Nerv. Ment. Dis.* **53**:201–210, Raven Press, New York.

Paroli, E., 1967, Indagini sull'effeto antimorfinico dell'ACTH. I. Relazioni con il corticosurrence ed i livelli ematici degli 11-OH steroidi, *Arch. Ital. Sci. Farmacol.* **13**:234.

Pavel, S., 1970, Tentative identification of arginine vasotocin in human cerebrospinal fluid, *J. Clin. Endocrinol. Metab.* **31**: 369–371.

Pelletier, G., Labrie, F., Kastin, A.J., and Schally, A.V., 1975, Radioautographic localization of radioactivity in rat brain after intracarotid injection of ^{125}I-α-melanocyte-stimulating hormone, *Pharmacol. Biochem. Behav.* **3**:671–674.

Pfaff, D.W., 1973, Luteinizing hormone–releasing factor potentiates lordosis behavior in hypophysectomized ovariectomized female rats, *Science* **182**:1148, 1149.

Pfaff, D.W., Teresa, M., Silva, A., and Weiss, J., 1971, Telemetered recording of hormone effects on hippocampal neurons, *Science* **172**:394, 395.

Piva, F., and Steiner, H., 1972, Bioassay and toxicology of TRH, *Front. Hormone Res.* **1**:11–21.

Plotnikoff, N.P., and Kastin, 1974a, Oxotremorine antagonism by prolyl–leucyl–glycine-amide administered by different routes and several anticholinergics, *Pharmacol. Biochem. Behav.* **2**:417–419.

Plotnikoff, N.P., and Kastin, A.J., 1974b, Pharmacological studies with a tripeptide, prolyl–leucyl–glycine-amide, *Arch. Int. Pharmacodyn. Ther.* **211**(2):211–214.

Plotnikoff, N.P., Kastin, A.J., and Anderson, M.S., 1971, DOPA potentiation by a hypothalamic factor MSH release-inhibiting hormone (MIF), *Life Sci.* **10**:1279–1283.

Plotnikoff, N.P., Prange, A.J., Jr., Breese, G.R., Anderson, M.S., and Wilson, I.C., 1972, Thyrotropin releasing hormone: Enhancement of DOPA activity by a hypothalamic hormone, *Science* **178**:417, 418.

Plotnikoff, N.P., Prange, A.J., Jr., Breese, G.R., and Wilson, I.C., 1974, Thyrotropin releasing hormone: Enhancement of DOPA activity in thyroidectomized rats, *Life Sci.* **14**:1271–1278.

Prange, A.J., Jr., Breese, G.R., Cott, J.M., Martin, B.R., Cooper, B.R., Wilson, I.C., and Plotnikoff, N.P., 1974, Thyrotropin-releasing hormone: Antagonism of pentobarbital in rodents, *Life Sci.* **14**:447–455.

Prange, A.J., Jr., Wilson, I.C., Breese, G.R., and Lipton, M.A., 1975, Behavioral effects of hypothalamic releasing hormones in animals and man, in: *Hormones, Homeostasis and the Brain* (W.H. Gispen, Tj.B. van Wimersma Greidanus, B. Bohus, and D. de Wied, eds.), *Prog. Brain Res.* **42**:1–9, Elsevier, Amsterdam.

Prange, A.J., Jr., Nemeroff, C.B., Lipton, M.A., Breese, G.R., and Wilson, I.C., 1977, Peptides and the central nervous system, in: *Handbook of Psychopharmacology,* (L.L. Iversen, S.D. Iversen, and S.H. Snyder, eds.), Vol. 7, Plenum Press, New York (in press).

Reading, H.W., and Dewar, A.J., 1971, Effects of ACTH$_{4-10}$ on cerebral RNA and protein metabolism in the rat, in: *Third International Meeting of the International Society of Neurochemistry,* Budapest (J. Domonkos, A. Fonyó, I. Huszák, and J. Szentágothai, eds.), abstract, p. 199.

Rees, H.D., Brogan, L.L., Entingh, D.J., Dunn, A.J., Shinkman, P.G., Damstra-Entingh, T., Wilson, J.E., and Glassman, E., 1974, Effect of sensory stimulation on the uptake and incorporation of radioactive lysine into protein of mouse brain and liver, *Brain Res.* **68**:143–156.

Reith, M.E.A., 1975, The effect of behaviorally active ACTH-like peptides on protein biosynthesis in the brainstem of hypophysectomized rats, Thesis, University of Utrecht, The Netherlands.

Reith, M.E.A., Schotman, P., and Gispen, W.H., 1974, Hypophysectomy, ACTH$_{1-10}$ and *in vitro* protein synthesis in rat brain slices, *Brain Res.* **81**:571–575.

Reith, M.E.A., Schotman, P., and Gispen, W.H., 1975a, Incorporation of [^{3}H]leucine into brainstem protein fraction: The effect of a behaviorally active N-terminal fragment of ACTH in hypophysectomized rats, *Neurobiology* **5**:355–368.

Reith, M.E.A., Schotman, P., and Gispen, W.H., 1975b, The neurotropic action of ACTH: Effects of

ACTH-like peptides on the incorporation of leucine into protein of brainstem slices from hypophysectomized rats, *Neurosci. Lett.* **1**:55–59.

Renaud, L.P., and Martin, J.B., 1975, Electrophysiological studies of connections of hypothalamic ventromedial nucleus neurons in the rat: Evidence for a role in neuroendocrine relation, *Brain Res.* **93**:145–151.

Rich, I., and Thompson, R., 1965, The role of hippocampo-septal system, thalamus and hypothalamus in avoidance conditioning, *J. Comp. Physiol. Psychol.* **59**:66–74, 1965.

Rigter, H., van Riezen, H., and de Wied, D., 1974, The effects of ACTH and vasopressin analogues on CO_2-induced retrograde amnesia in rats, *Physiol. Behav.* **13**:381–388.

Rigter, H., Elbertse, R. and van Riezen, H., 1975, Time-dependent anti-amnesic effect of $ACTH_{4-10}$ and desglycinamide–lysine vasopressin, in: *Hormones, Homostasis and the Brain* (W.H. Gispen, Tj.B. van Wimersma Greidanus, B. Bohus, and D. de Wied, eds.), *Prog. Brain Res.* **42**:163–171, Elsevier, Amsterdam.

Robinson, G.A., Butcher, R.W., and Sutherland, E.W., 1971, *Cyclic AMP*, p. 531, Academic Press, New York.

Rodriguez, E.M., 1970, Morphological and functional relationship between the hypothalamo–neurophypophyseal system and cerebrospinal fluid, in: *Aspects in Neuroendocrinology* (W. Bargmann and B. Scharrer, eds.), pp. 265–352, Springer-Verlag, Berlin.

Rudman, D., Scott, J.W., DelRio, A.E., Houser, D.H., and Sheen, S., 1974, Effect of melanotropic peptides on protein synthesis in mouse brain, *Amer. J. Physiol.* **226**:687–692.

Sandman, C.A., Kastin, A.J., and Schally, A.V., 1969, Melanocyte-stimulating hormone and learned appetitive behavior, *Experientia Basel* **25**:1001, 1002.

Sandman, C.A., Kastin, A.J., and Schally, A.V., 1971, Behavioral inhibition as modified by melanocyte-stimulating hormone (MSH) and light–dark conditions, *Physiol. Behav.* **6**:45–48.

Sandman, C.A., Miller, L.H., Kastin, A.J., and Schally, A.V., 1972, Neuroendocrine influence on attention and memory, *J. Comp. Physiol. Psychol.* **80**:54–58.

Sandman, C.A., Alexander, W.D., and Kastin, A.J., 1973, Neuroendocrine influences on visual discrimination and reversal learning in the albino and hooded rat, *Physiol. Behav.* **11**:613–617.

Sara, V.R., and Lazarus, L., 1974, Prenatal action of growth hormone on brain and behaviour, *Nature London* **250**:257, 258.

Sayers, G., Beall, R.J., and Seelig, S., 1974, Modes of action of ACTH, in: *Biochemistry of Hormones* (H.V. Rickenburg, ed.), pp. 25–60, *M.T.P. Int. Rev. Sci. Biochem. Ser.* (H.L. Kornberg and S.C. Philips, eds.), Butterworth University Park Press, London.

Schapiro, S., and Vukocivh, K.R., 1970, Early experience effects upon cortical dentrites: A proposed model for development, *Science* **167**:292–294.

Schenkel-Hulliger, L., Koella, W.P., Hartmann, A., and Maitre, L., 1974, Tremorogenic effect of thyrotropin releasing hormone in rats, *Experientia Basel* **30**:1168–1170.

Schlender, K.K., Wei, S.H., and Villar-Palasi, C., 1969, UDP-glucose: Glycogen α-4-glucosyl-transferase I kinase activity of purified muscle protein kinase. Cyclic nucleotide specificity, *Biochim. Biophys. Acta* **191**:272–278.

Schoenenberger, G.A., and Monnier, M., 1974, Isolation, partial characterization and activity of a humoral "delta-sleep" transmitting factor, in: *Brain and Sleep* (H.M. van Praag and M. Meinardi, eds.), pp. 39–69, Proceedings of the Congress of the Interdisciplinary Society of Biological Psychiatry and the Dutch Branch of the International League Against Epilepsy, Amsterdam, Erven Bohn, Amsterdam.

Schotman, P., 1971, Metabolisme van makromoleculen in de hersenstam van de hypofyseloze rat in relatie tot het voorwaardelijk vluchtgedrag, Thesis, University of Utrecht, The Netherlands.

Schotman, P., Gispen, W.H., Jansz, H.S., and de Wied, D., 1972, Effects of ACTH analogues on macromolecule metabolism in the brainstem of hypophysectomized rats, *Brain Res.* **46**:349–362.

Schotman, P., Reith, M.E.A., van Wimersma Greidanus, Tj.B., Gispen, W.H., and de Wied, D., 1976a, Hypothalamic and pituitary peptide hormones and the central nervous system—with special reference to the neurochemical effects of ACTH, in: *Molecular and Functional Neurobiology* (W.H. Gispen, ed.), pp. 309–344, Elsevier, Amsterdam.

Schotman, P., Reith, M.E.A., and Gispen, W.H., 1976b, The influence of neuropeptides on protein metabolism in the central nervous system, Abstract Federation Meeting of the Dutch Medical Biological Societies, Amsterdam.

Schulz, H., Unger, H., Schwarzberg, H., Pommrich, G., and Stolze, R., 1971, Neuronenaktivität hypothalamischer Kerngebiete von Kaninchen nach intraventrikulärer Applikation von Vasopressin und Oxytocin, *Experientia Basel* **27**:1482, 1483.

Schwarzberg, H., Unger, H., and Schulz, H., 1973, The effect of oxytocin upon the EEG of rabbits, changed by Na-glutamate, *Acta Biol. Med. Ger.* **30**:203–208.

Schwyzer, R., Schiller, P., Seelig, S., and Sayers, G., 1971, Isolated adrenal cells: Log dose response curves for steroidogenesis induced by $ACTH_{1-24}$, $ACTH_{1-10}$, $ACTH_{4-10}$ and $ACTH_{5-10}$. *FEBS Lett.* **19**:229–231.

Seelig, S., Sayers, G., Schwyzer, R., and Schiller, P., 1971, Isolated adrenal cells: $ACTH_{11-24}$, a competitive antagonist of $ACTH_{1-30}$ and $ACTH_{1-10}$, *FEBS Lett.* **19**:233.

Segal, D.S., and Mandell, A.J., 1974, Differential behavioral effects of hypothalamic polypeptides, in: *The Thyroid Axis, Drugs and Behavior* (A.J. Prange, Jr., ed.), pp. 129–133, Raven Press, New York.

Selye, H., 1950, *The Physiology and Pathology of Exposure to Stress*, p. 6, Acta Inc., Montreal.

Siler, T.M., Vandenberg, G., and Yen, S.S.C., 1973, Inhibition of growth hormone release in humans by somatostatin, *J. Clin. Endocrinol. Metab.* **37**:632.

Smith, A.A., Karmin, M., and Gavitt, J., 1966, Blocking effect of puromycin, ethanol, and chloroform on the development of tolerance to an opiate, *Biochem. Pharmacol.* **15**:1877–1879.

Spirtes, M.A., Plotnikoff, N.P., and Kastin, A.J., 1975, Effects of hypothalamic peptides on the brain, in: *CNS and Behavioural Pharmacology*, Vol. 3, pp. 121–129, Proceedings of the Sixth International Congress of Pharmacology, Helsinki, Finland, 1975.

Steiner, F.A., Ruf, K., and Akert, K., 1969, Steroid sensitive neurons in rat brain: Anatomical localization and responses to neurohumours and ACTH, *Brain Res.* **12**:74–85.

Sterba, G., 1974, Das oxytocinerge neurosekretorische System der Wirbeltiere. Beitrag zu einem erweiterten Konzept, *Zool. Jahrb. Abt. Allg. Zool. Physiol. Tiere* **78**:409–423.

Stern, P., and Hadžović, S., 1973, Pharmacological analysis of central actions of synthetic substance P, *Arch. Int. Pharmacodyn.* **202**:259–262.

Stern, W.C., 1971, Acquisition impairments following rapid eye movement sleep deprivation in rats, *Physiol. Behav.* **7**:345–352.

Stern, W.C., Jalowiec, J.E., Shabshelowitz, H., and Morgane, P.J., 1975, Effects of growth hormone on sleep–waking patterns in cats, *Horm. Behav.* **6**:189–196.

Strand, F.L., 1975, The influence of hormones on the nervous system (with special emphasis on polypeptide hormones), *BioScience* **25**:568–577.

Strand, F.L., and Cayer, A., 1975, A modulatory effect of pituitary polypeptides on peripheral nerve and muscle, in: *Hormones, Homeostasis and the Brain* (W.H. Gispen, Tj.B. van Wimersma Greidanus, B. Bohus, and D. de Wied, eds.) *Prog. Brain Res.* **42**:187–194, Elsevier, Amsterdam.

Strand, F.L., Stoboy, H., and Cayer, A., 1973–1974, A possible direct action of ACTH on nerve and muscle, *Neuroendocrinology* **13**:1–20.

Stratton, L.O.,and Kastin, A.J., 1974, Avoidance learning at two levels of motivation in rats receiving MSH, *Horm. Behav.* **5**:149–155.

Stumpf, W.E., and Sar, M., 1973, [³H]TRH and [³H]proline radioactivity localization in pituitary and hypothalamus, *Fed. Proc. Fed. Amer. Soc. Exp. Biol.* **32**, Abstract No. 3.

Sutherland, E.W., 1972, Studies on the mechanism of hormone action, *Science* **177**:401–408.

Terenius, L., 1975, Effect of peptides and aminoacids on dihydromorphine binding to the opiate receptor, *J. Pharm. Pharmacol.* **27**:450–452.

Terenius, L., and Wahlström, A., 1975, Morphine-like ligand for opiate receptors in human CSF, *Life Sci.* **16**:1759–1764.

Terenius, L., Gispen, W.H., and de Wied, D., 1975, ACTH like peptides and opiate receptors in the rat brain: Structure–activity studies, *Eur. J. Pharmacol.* **33**:395–399.

Teschemacher, H., Opheim, K.E., Cox, B.M., and Goldstein, A., 1975, A peptide-like substance from pituitary that acts like morphine, *Life Sci.* **16**:1771–1776.

Tesser, G.I., Maier, R., Schenkel-Hulliger, L., Barthe, P.L., Kamber, B., and Rittel, W., 1973, Biological activity of corticotrophin peptides with homoarginine, lysine or ornithine substituted for arginine in position 8, *Acta Endocrinol. Copenhagen* **74**:56–66.

Thompson, E.A., and de Wied, D., 1973, The relationship between the antidiuretic activity of rat eye plexus blood and passive avoidance behaviour, *Physiol. Behav.* **11**:377–380.

Thompson, R., Rich, I., and Langer, S.K., 1964, Lesion studies on the functional significance of the posterior thalamo-mesencephalic tract, *J. Comp. Neurol.* **123**:29–44.

Thompson, R., Truax, T., and Thorne, M., 1970, Retention of visual learning in the white rat following knife-cuts through the posterior thalamus ventromedial midbrain, *Brain Behav. Evol.* **3**:261–284.

Torda, C., 1974, A potential mechanism for reserpine and chlorpromazine generation of myasthenia gravis–like easy fatigability and parkinsonism involving acetylcholine, dopamine and cyclic AMP, *IRCS Libr. Compend.* (Research on: Clinical Pharmacology and Therapeutics; Neurology and Neurosurgery, Psychiatry and Clinical Psychology) **2**:1111.

Torda, C., and Wolff, H.G., 1952, Effect of pituitary hormones, cortisone and adrenalectomy on some aspects of neuromuscular systems and acetylcholine synthesis, *Amer. J. Physiol.* **169**:140–149.

Ungar, G., 1970, Chemical transfer of learned behavior, *Agents Actions* **1**(4):155–163.

Ungar, G., 1974, Molecular coding of memory, *Life Sci.* **14**:595–604.

Ungar, G., Burzynski, S.R., and Tate, D.F., 1975, Learning-induced brain peptides, in: *Peptides, Chemistry, Structure and Biology* (R. Walter and J. Mienehofer, eds.), pp. 673–677, Proceedings of the Fourth American Peptide Symposium, Ann Arbor Science, Ann Arbor, Michigan.

Unger, H., and Schwarzberg, H., 1970, Untersuchungen über Vorkommen und Bedeutung von Vasopressin and Oxytocin im Liquor cerebrospinalis und Blut für nervöse Funktionen, *Acta Biol. Med. Ger.* **25**:267–280.

Urban, I., and de Wied, D., 1975, Inferior quality of RSA during paradoxical sleep in rats with hereditary diabetes insipidus, *Brain Res.* **97**:362–366.

Urban, I., and de Wied, D., 1976, Changes in excitability of the theta activity generating substrate by $ACTH_{4-10}$ in the rat, *Exp. Brain Res.* **24**:325–334.

Urban, I., Lopes da Silva, F.H., Storm van Leeuwen, W., and de Wied, D., 1974, A frequency shift in the hippocampal theta activity: An electrical correlate of central action of ACTH analogues in the dog, *Brain Res.* **69**:361–365.

Valatx, J.L., Chouvet, G., and Jouvet, M., 1975, Sleep–waking cycle of the hypophysectomized rat, in: *Hormones, Homeostasis and the Brain*, (W. H. Gispen, Tj.B. van Wimersma Greidanus, B. Bohus, and D. de Wied, eds.), *Prog. Brain Res.* **42**:115–120, Elsevier, Amsterdam.

Vale, W., Blackwell, R., Grant, G., and Guillemin, R., 1973, TRF and thyroid hormones on prolactin secretion by rat anterior pituitary cells *in vitro*, *Endocrinology* **93**:26–33.

Valtin, H., 1967, Hereditary hypothalamic diabetes insipidus in rats (Brattleboro strain). A useful experimental model, *Amer. J. Med.* **42**:814–827.

Valtin, H., and Schroeder, H.A., 1964, Familial hypothalamic diabetes insipidus in rats (Brattleboro strain), *Amer. J. Physiol.* **206**:425–430.

Van Delft, A.M.L., and Kitay, J.I., 1972, Effect of ACTH on single unit activity in the diencephalon of intact and hypophysectomized rats, *Neuroendocrinology* **9**:188–196.

Van der Helm-Hylkema, H., 1973, Effecten van vroeg neonataal toegediend ACTH en aan ACTH verwante peptiden op de somatische en gedragsontwikkeling van de rat, Thesis, University of Utrecht, The Netherlands.

Van der Helm-Hylkema, H., and de Wied, D., 1976, Effect of neonatally injected ACTH and ACTH analogues on eye-opening of the rat, *Life Sci.* **18**:1099–1104.

Van Loon, G.R., 1975, Brain catecholamines and ACTH secretion: Studies on brain dopamine beta hydroxylase, in: *CNS and Behavioral Pharmacology* (J. Tuomisto and M.K. Paasonen, eds.), Vol. 3, pp. 111–119. Proceedings of the Fifth International Congress of Pharmacology, Helsinki, Finland.

Van Wimersma Greidanus, Tj. B., and de Wied, D., 1971, Effects of systemic and intracerebral administration of two opposite acting ACTH-related peptides on extinction of conditioned avoidance behavior, *Neuroendocrinology* **7**:291–301.

Van Wimersma Greidanus, Tj. B., and de Wied, D., 1976, Dorsal hippocampus: A site of action of neuropeptides on avoidance behavior? *Pharmacol. Biochem. Behav.* **5**(Suppl.): in press.

Van Wimersma Greidanus, Tj.B., Bohus, B., and de Wied, D., 1973, Effects of peptide hormones on behavior, in: *Progress in Endocrinology, Excerpta Med. Int. Congr. Ser.* No. 273, pp. 197–201, Proceedings of the Fourth International Congress of Endocrinology, Washington D.C., June 18–24, 1972.

Van Wimersma Greidanus, Tj.B., Bohus, B., and de Wied, D., 1974, Differential localization of the influence of lysine vasopressin and of ACTH$_{4-10}$ on avoidance behavior: A study in rats bearing lesions in the parafascicular nuclei, *Neuroendocrinology* **14**:280–288.

Van Wimersma Greidanus, Tj.B., Dogterom, J., and de Wied, D., 1975*a,* Intraventricular administration of anti-vasopressin serum inhibits memory consolidation in rats, *Life Sci.* **16**:637–644.

Van Wimersma Greidanus, Tj.B., Bohus, B., and de Wied, D., 1975*b,* The role of vasopressin in memory processes, in: *Hormones, Homeostasis and the Brain* (W.H. Gispen, Tj.B. van Wimersma Greidanus, B. Bohus, and D. de Wied, eds.), *Prog. Brain Res.* **42**:135–141, Elsevier, Amsterdam.

Van Wimersma Greidanus, Tj.B., Lowry, P.J., Scott, A.P., Rees, L.H., and de Wied, D., 1975*c,* The effects of dogfish MSH's and of corticotrophin-like intermediate lobe peptides (CLIP's) on avoidance behavior in rats, *Horm. Behav.* **6**:319–327.

Versteeg, D.H.G., 1973, Effect of two ACTH-analogs on noradrenaline metabolism in rat brain, *Brain Res.* **49**:483–485.

Versteeg, D.H.G., and Wurtman, R.J., 1976, Synthesis and release of monoamine neurotransmitters: Regulatory mechanisms, in: *Molecular and Functional Neurobiology* (W.H. Gispen, ed.), pp. 201–234, Elsevier, Amsterdam.

Versteeg, D.H.G., van der Gugten, J., and van Ree, J.M., 1975, Regional turnover and synthesis of catecholamines in rat hypothalamus, *Nature London* **256**:502, 503.

Vincent, J.D., and Arnould, E., 1975, Vasopressin as a neurotransmitter in the central nervous system: Some evidence from the supraoptic neurosecretory system, in: *Hormones, Homeostasis and the Brain* (W.H. Gispen, Tj.B. van Wimersma Greidanus, B. Bohus, and D. de Wied, eds.), *Prog. Brain Res.* **42**:57–66, Elsevier, Amsterdam.

Von Zerssen, D., 1976, Mood and behavioral changes under corticosteroid therapy, in: *Psychotropic Action of Hormones,* Proceedings of the First World Congress of Biological Psychiatry, Buenos Aires, Argentina, September 24–28, 1974, Spectrum Publications, Holliswood, New York.

Vorherr, H., Bradbury, M.W.B., Hoghoughi, M., and Kleeman, C.R., 1968, Antidiuretic hormone in cerebrospinal fluid during endogenous and exogenous changes in its blood level, *Endocrinology* **83**:246–250.

Walter, R., Hoffman, P.L., Flexner, J.B., and Flexner, L.B., 1975, Neurohypophyseal hormones, analogs, and fragments: Their effect on puromycin-induced amnesia, *Proc. Nat. Acad. Sci. U.S.A.* **72**:4180–4184.

Walton, G., Gill, G., Abrass, I., and Garren, L., 1971, Phosphorylation of ribosome-associated protein by an adenosine 3′,5′-cyclic monophosphate-dependent protein kinase: Location of the microsomal receptor and protein kinase, *Proc. Nat. Acad. Sci. U.S.A.* **68**:880–884.

Weinstein, B., Bartschot, R.M., Cook, R.M., Tam, P.S., and Guttnab, H.N., 1975, The synthesis of a peptide having the structure attributed to a sound habituating material, *Experientia Basel* **31**:754–756.

Weiss, J.M., McEwen, B.S., Silva, M., and Kalkut, M., 1970, Pituitary–adrenal alterations and fear responding, *Amer. J. Physiol.* **218**:864–868.

Wiegant, V.M., and Gispen, W.H., 1975, Behaviorally active ACTH analogues and brain cyclic AMP, *Exp. Brain Res. Suppl.* **23**:219.

Winokur, A., and Utiker, R.D., 1974, Thyrotropin-releasing hormone: Regional distribution in rat brain, *Science* **185**:265, 266.

Winter, C.A., and Flataker, L., 1951, The effect of cortisone desoxycorticosterone, and adrenocorticotrophic hormone upon the responses of animals to analgesic drugs, *J. Pharmacol. Exp. Ther.* **101**:93.

Witter, A., Greven, H.M., and de Wied, D., 1975, Correlation between structure, behavioral activity and rate of biotransformation of some ACTH$_{4-9}$ analogs, *J. Pharmacol. Exp. Ther.* **193**:853–860.

Wittkowski, W., 1968, Elektronenmikroskopische Studien zur intraventrikulären Neurosekretion in den Recessus infundibularis der Maus, *Z. Zellforsch.* **92**:207–216.

Wolthuis, O.L., and de Wied, D., 1975, ACTH-analogues on motor behavior and visual evoked responses in rats, Abstracts of the Vth Annual Meeting of the American Society of Neuroscience, Abstract No. 776, p. 504.

Yajima, H., Kubo, K., Kinomura, Y., and Lande, S., 1966, Studies on peptides, XI. The effect on melanotropic activity of altering the arginyl residue in L-histidyl–L-phenylalanyl–L-arginyl–L-tryptophylglycine, *Biochim. Biophys. Acta* **127**:545–549.

Zimmermann, E., and Krivoy, W.A., 1973, Antagonism between morphine and the polypeptides ACTH, ACTH$_{1-24}$, and β-MSH in the nervous system, in: *Drug Effects on Neuroendocrine Regulation* (E. Zimmermann, W.H. Gispen, B.H. Marks, and D. de Wied, eds.), *Prog. Brain Res.* **39**:383–394, Elsevier, Amsterdam.

Zimmermann, E., and Krivoy, W.A., 1974, Depression of frog isolated spinal cord by morphine and antagonism by tetracosactin, *Proc. Soc. Exp. Biol. Med.* **146**:575–579.

Zimmerman, E.A., Kozlowski, G.P., and Scott, D.E., 1975, Axonal and ependymal pathways for the secretion of biologically active peptides into hypophysial portal blood, in: *Brain–Endocrine Interaction II* (K.M. Knigge, D.E. Scott, H. Kobayashi, and S. Ishii, eds.), pp. 123–134, Karger, Basel.

Zwiers, H., Veldhuis, H. D., Schotman, P., and Gispen, W. H., 1976, ACTH, cyclic nucleotides and brain protein phosphorylation *in vitro*, *Neurochem. Res.*, in press.

Index